MECHANISMS AND DYNAMICS OF MACHINERY

FOURTH EDITION

Hamilton H. Mabie

Charles F. Reinholtz

Virginia Polytechnic Institute
and State University

JOHN WILEY & SONS
New York   Chichester   Brisbane   Toronto   Singapore

DEDICATED to the late FRED W. OCVIRK
whose contributions to the First and
Second Editions motivated later editions,
and to SALLIE MABIE and JERI REINHOLTZ
whose assistance and forbearance have
made this edition possible.

Library of Congress Cataloging in Publication Data:

Mabie, Hamilton H. (Hamilton Horth), 1914-
  Mechanisms and dynamics of machinery.
  Includes index.
  1. Mechanical movements.   2. Machinery, Kinematics
of.   3. Machinery, Dynamics of.   I. Reinholtz, Charles F.
II. Title.
TJ175.M123   1986   621.8   86-11115
ISBN 0-471-80237-9

Printed in the United States of America

20 19 18 17

About the Authors

HAMILTON H. MABIE, Professor of Mechanical Engineering at Virginia Polytechnic Institute and State University since 1964, received his B.S. degree from the University of Rochester, his M.S. degree from Cornell University, and his Ph.D. degree from Pennsylvania State University.

From 1941 to 1960, Dr. Mabie was on the faculty of the Sibley School of Mechanical Engineering at Cornell University. From 1960 to 1964, he worked at Sandia Laboratory in Albuquerque, New Mexico, where he was engaged in research and development related to nuclear weapons.

In addition to his work in kinematics, Dr. Mabie is engaged in research on gears, torque characteristics of instrument ball bearings, environmental effects on the fatigue life of aluminum, and fretting corrosion of rolling element bearings. He has authored and coauthored many technical papers in these fields. He is a licensed professional engineer and a Life Fellow of The American Society of Mechanical Engineers.

The first edition of *Mechanisms and Dynamics of Machinery* was published by John Wiley & Sons in 1957 and the second in 1963, both with the late F. W. Ocvirk as coauthor. The third edition was published in 1975 and an SI Version in 1978. This fourth edition has Charles F. Reinholtz as coauthor.

CHARLES F. REINHOLTZ is currently Assistant Professor of Mechanical Engineering at Virginia Polytechnic Institute and State University in Blacksburg, Virginia, a position he has held since 1983. He holds B.S., M.S., and Ph.D. degrees from the University of Florida. He has also worked for Burroughs Corporation as a design engineer in the Peripheral Products Group. Professor Rein-holtz has been active in the area of kinematics and mechanism design since 1976. He is a member of The American Society of Mechanical Engineers, The American Society for Engineering Education, and Sigma Xi. He is also a member of Tau Beta Pi and Pi Tau Sigma Honor Societies.

Preface

This textbook has been completely revised and updated. Its contents have been reorganized to better match the sequence of topics typically covered and to reflect the many changes brought about by the use of computers in the classroom. These changes include the use of iterative methods for linkage position analysis and matrix methods for force analysis. BASIC language computer programs, developed on a personal computer, have been added throughout the text to demonstrate the simplicity and power of computer methods. All BASIC programs listed in the text have also been coded in FORTRAN 77 and listed in Appendix Three. The text can now be used with either U.S. units, SI units, or a combination of both. Where the units of an equation must be specified, both a U.S. and an SI form are given. An effort has been made to maintain a balance between analytical and graphical methods.

This edition has been expanded to include a number of new topics. In keeping with the additional emphasis on computer methods, kinematic and dynamic analysis of linkages has been demonstrated using the commercially available Integrated Mechanisms Program (IMP). Analytical cam design material has been expanded to include equations for determining the various disk cam contours. Both the U.S. and metric systems of gearing are covered, and a complete problem set is given for each system of units. A new section that covers nonstandard spur gears cut with a pinion cutter has been added to the chapter on nonstandard gearing. Two new topics have been included in the chapter on gear trains: harmonic drives and power flow through planetary gear trains.

Complex number methods and loop-closure equations have been used more extensively in the velocity and acceleration analysis of linkages. The chapter on force analysis has been thoroughly revised. In addition to the superposition method, it now includes the matrix method, which is a powerful tool when used in conjunction with a computer. In the chapter on balancing, a method for balancing four-bar linkages has been added. The chapter on kinematic synthesis has been revised and expanded to include many new topics, including a general discussion of function generation, path generation, and body guidance, and the problems of branch defect, order defect, and Grashof defect. The final chapter, on spatial mechanisms and robotics, is completely new. The material contained in this chapter is becoming increasingly important in the design of complex automatic production machinery.

We appreciate the many suggestions and helpful comments made by our reviewers: Richard Alexander, Marvin Dixon, and William H. Park. We are indebted to the following instructors at Virginia Polytechnic Institute and State University for their helpful suggestions: Craig A. Rogers, Richard E. Cobb, Edgar G. Munday, Joseph W. David, and Peter J. Leavesly. Finally, we would like to acknowledge the help and encouragement provided by our editors at Wiley, Charity Robey and Bill Stenquist.

HAMILTON H. MABIE

CHARLES F. REINHOLTZ

Contents

Chapter 3
Cams 71

Chapter 4
Spur Gears 128

Chapter One

Introduction

1.1 INTRODUCTION TO THE STUDY OF MECHANISMS

The study of mechanisms is very important. With the continuing advances made in the design of instruments, automatic controls, and automated equipment, the study of mechanisms takes on new significance. *Mechanisms* may be defined as that division of machine design which is concerned with the kinematic design of linkages, cams, gears, and gear trains. *Kinematic design* is design on the basis of motion requirements in contrast to design on the basis of strength requirements. An example of each of the mechanisms listed above will be given in order to present a comprehensive picture of the components to be studied.

A sketch of a linkage is shown in Fig. 1.1. This particular arrangement is known as the slider-crank mechanism. Link 1 is the frame and is stationary, link 2 is the crank, link 3 is the connecting rod, and link 4 is the slider. A common application of this linkage is in the internal-combustion engine where link 4

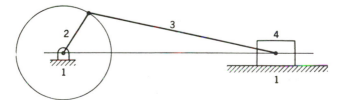

FIGURE 1.1 Slider-crank mechanism.

becomes the piston (Fig. 1.2*a*). This figure also demonstrates how difficult it may be to discern the basic kinematic device when looking at a photograph or a drawing of a complete machine. Figure 1.2*b* shows the *kinematic diagram* of the slider-crank mechanism corresponding to the left-side crankshaft-connecting-rod-piston in the photograph of Fig. 1.2*a*. Such a kinematic diagram is much easier to work with and allows the designer to separate the kinematic considerations from the larger problem of machine design.

FIGURE 1.2*a* Chevrolet V-8 engine showing slider-crank mechanism. (General Motors Corporation).

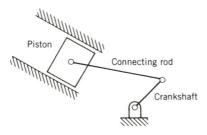

FIGURE 1.2*b* Kinematic diagram of engine mechanism.

Figure 1.3 shows the sketch of a cam and follower. The cam rotates at a constant angular velocity, and the follower moves up and down. On the upward motion the follower is driven by the cam, and on the return motion it is driven by the action of gravity or of a spring. Cams are used in many machines, but one of the most common is the automotive engine where two cams are used per cylinder to operate the intake and exhaust valves, also shown in Fig. 1.2a. A three-dimensional cam is shown in Fig. 1.4. In this cam, the motion of the follower depends not only upon the rotation of the cam but also upon the axial motion of the cam.

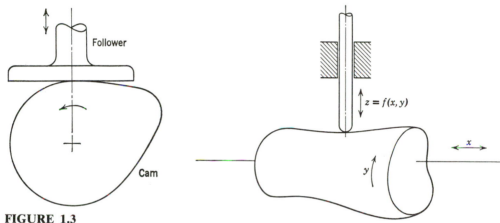

FIGURE 1.3
Two-dimensional cam.

FIGURE 1.4 Three-dimensional cam.

Gears are used in many applications to transmit motion from one shaft to another with a constant angular velocity ratio. Figure 1.5 shows several commonly used gears.

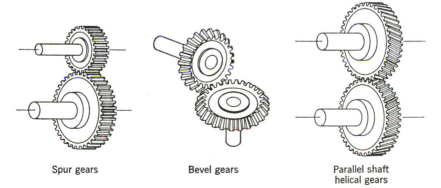

Spur gears   Bevel gears   Parallel shaft
                                     helical gears

FIGURE 1.5 (*continued next page*)

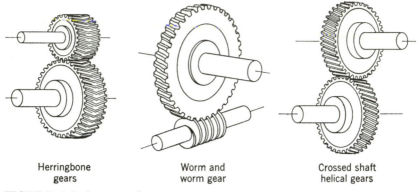

| Herringbone gears | Worm and worm gear | Crossed shaft helical gears |

FIGURE 1.5 (*continued*)

In some cases, the desired reduction in angular velocity is too great to achieve using only two gears. When this occurs, several gears must be connected together to give what is known as a *gear train*. Figure 1.6 shows a gear train where the speed is stepped down in going from gear 1 to gear 2 and again in going from gear 3 to gear 4. Gear 1 is the driver, and gears 2 and 3 are mounted on the same shaft. In many gear trains, it is necessary to be able to shift gears in and out of mesh so as to obtain different combinations of speeds. A good example of this is the automobile transmission where three speeds forward and one in reverse are obtained by shifting two gears.

In devices such as instruments and automatic controls, obtaining the correct motion is all-important. The power transmitted by the elements may be so slight as to be negligible, which allows the components to be proportioned primarily on the basis of motion, strength being of secondary importance.

There are other machines, however, where the kinematic analysis is only one step in the design. After it has been determined how the various machine components will act to accomplish the desired motion, the forces acting upon

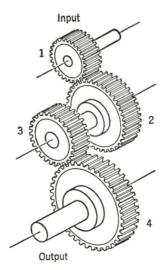

FIGURE 1.6 Gear train.

these parts must be analyzed. From this, the physical size of the parts may be determined. A machine tool is a good example; its strength and rigidity are more difficult to attain than the desired motions.

It is important at this time to define the terms used in the study of mechanisms. This is done in the following section.

1.2 MECHANISM, MACHINE

In the study of mechanisms the terms mechanism and machine will be used repeatedly. These are defined as follows:

A *mechanism* is a combination of rigid or resistant bodies so formed and connected that they move upon each other with definite relative motion. An example is the crank, connecting rod, and piston of an internal-combustion engine as shown diagrammatically in Fig. 1.2*b*.

A *machine* is a mechanism or collection of mechanisms which transmit force from the source of power to the resistance to be overcome. An example is the internal-combustion engine.

1.3 MOTION

In dealing with the study of mechanisms, it is necessary to define the various types of motion produced by these mechanisms.

Plane Motion

Translation

When a rigid body so moves that the position of each straight line of the body is parallel to all of its other positions, the body has motion of translation.

1. **Rectilinear translation.** All points of the body move in parallel straight line paths. When the body moves back and forth in this manner, it is said to reciprocate. This is illustrated in Fig. 1.7, where the slider 4 reciprocates between the limits B' and B''.

2. **Curvilinear translation.** The paths of the points are identical curves parallel to a fixed plane. Figure 1.8 shows the mechanism that was used in connecting

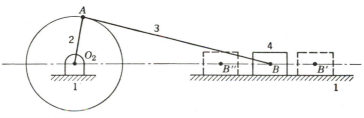

FIGURE 1.7

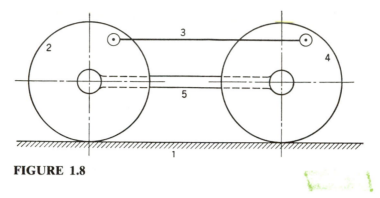

FIGURE 1.8

the drive wheels of the steam locomotive. In this mechanism, link 3 has curvilinear translation, and all points in the body trace out identical cycloids as wheels 2 and 4 roll along track 1. Link 5 moves with rectilinear translation.

Rotation

If each point of a rigid body having plane motion remains at a constant distance from a fixed axis that is perpendicular to the plane of motion, the body has motion of rotation. If the body rotates back and forth through a given angle, it is said to oscillate. This is shown in Fig. 1.9, where link 2 rotates and link 4 oscillates between the positions B' and B''.

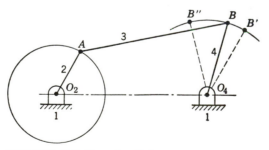

FIGURE 1.9 Four-bar linkage.

Rotation and Translation

Many bodies have motion which is a combination of rotation and translation. Link 3 in Fig. 1.7, links 2 and 4 in Fig. 1.8, and link 3 in Fig. 1.9 are examples of this type of motion.

Helical Motion

When a rigid body moves so that each point of the body has motion of rotation about a fixed axis and at the same time has translation parallel to the axis, the body has helical motion. An example of helical motion is the motion of a nut as the nut is screwed onto a bolt.

Spherical Motion
When a rigid body moves so that each point of the body has motion about a fixed point while remaining at a constant distance from it, the body has spherical motion.

Spatial Motion
A body moving with rotation about three nonparallel axes and translation in three independent directions is said to be undergoing general spatial motion.

1.4 CYCLE, PERIOD, AND PHASE OF MOTION

When the parts of a mechanism have passed through all the possible positions they can assume after starting from some simultaneous set of relative positions and have returned to their original relative positions, they have completed a *cycle* of motion. The time required for a cycle of motion is the *period*. The simultaneous relative positions of a mechanism at a given instant during a cycle are a *phase*.

1.5 PAIRING ELEMENTS

The geometrical forms by which two members of a mechanism are joined together so that the relative motion between these two members is consistent are known as *pairing elements*. If the joint by which two members are connected has surface contact such as a pin joint, the connection is known as a *lower pair*. If the connection takes place at a point or along a line such as in a ball bearing or between two gear teeth in contact, it is known as a *higher pair*. A pair that permits only relative rotation is a *revolute*, or *turning*, *pair*, and one that allows only sliding is a *sliding pair*. A turning pair can be either a lower or a higher pair depending upon whether a pin and bushing or a ball bearing is used for the connection. A sliding pair will be a lower pair as between a piston and cylinder wall.

1.6 LINK, CHAIN

A *link* is a rigid body having two or more pairing elements by means of which it may be connected to other bodies for purposes of transmitting force or motion. Generally, a link is a rigid member with provision at each end for connection to two other links. This may be extended, however, to include three, four, or even more connections. Figures 1.10a, b, and c show these arrangements. Perhaps the extreme case of a multiply connected link is the master rod in a nine-cylinder radial aircraft engine as seen in Fig. 1.10d.

A well-known example of a link with three connections is the bell crank, which can be arranged as shown in Fig. 1.11a or Fig. 1.11b. This link is generally

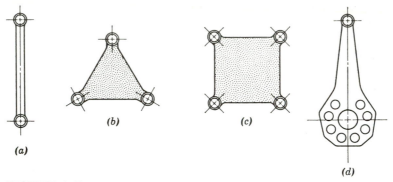

FIGURE 1.10

used for motion reduction and can be proportioned for a given ratio with a minimum of distortion of the required motion.

When a number of links is connected by means of pairs, the resulting system is a kinematic *chain*. If these links are connected in such a way that no motion is possible, a locked chain (structure) results. A constrained chain is obtained when the links are so connected that, no matter how many motion cycles are passed through, the relative motion will always be the same between the links.

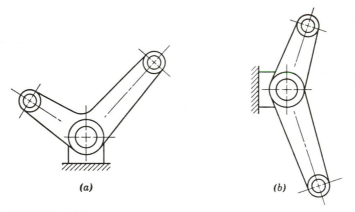

FIGURE 1.11

It is also possible to connect links so that an unconstrained chain results, which means that the motion pattern will vary from time to time depending on the amount of friction present in the joints. If one of the links of a constrained chain is made a fixed link, the result is a mechanism.

1.7 INVERSION

If in a mechanism, the link which was originally fixed is allowed to move and another link becomes fixed, the mechanism is said to be inverted. The inversion of a mechanism does not change the motion of its links relative to each other but does change their absolute motions (relative to the ground).

1.8 TRANSMISSION OF MOTION

In the study of mechanisms, it is necessary to investigate the method in which motion may be transmitted from one member to another. Motion may be transmitted in three ways: (*a*) direct contact between two members such as between a cam and follower or between two gears, (*b*) through an intermediate link or connecting rod, and (*c*) by a flexible connector such as a belt or chain.

The angular velocity ratio is determined for the case of two members in direct contact. Figure 1.12 shows cam 2 and follower 3 in contact at point P. The cam has clockwise rotation, and the velocity of point P as a point on body 2 is represented by the vector $\mathbf{PM_2}$. The line NN' is normal to the two surfaces at point P and is known as the *common normal*, the *line of transmission*, or the *line of action*. The common tangent is represented by TT'. The vector $\mathbf{PM_2}$ is broken into two components, $\mathbf{Pn}$ along the common normal and $\mathbf{Pt_2}$ along the common tangent. Because of the fact that the cam and the follower are rigid members and must remain in contact, the normal component of the velocity of P as a point on body 3 must be equal to the normal component of P as a point on body 2. Therefore, knowing the direction of the velocity vector of P as a point on body 3 to be perpendicular to the radius O_3P and its normal component, it is possible to find the velocity $\mathbf{PM_3}$ as shown in the sketch. From this vector, the angular velocity of the follower may be determined from the relation $V = R\omega$, where V equals the linear velocity of a point moving along a path of radius R and ω equals the angular velocity of the radius R.

In direct-contact mechanisms, it is often necessary to determine the velocity of sliding. From the sketch this can be seen to be the vector difference between the tangential components of the velocities of the points of contact. This difference

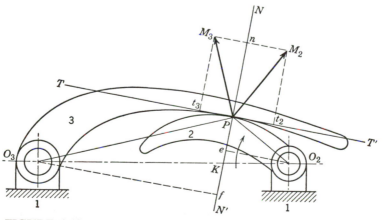

FIGURE 1.12

is given by the distance $\overline{t_2 t_3}$ because the component $\mathbf{Pt}_3$ is opposite in direction to that of $\mathbf{Pt}_2$. If t_2 and t_3 fall on the same side of P, then the distance will subtract. If the contact point P should fall on the line of centers $\overline{O_2 O_3}$, then the vectors $\mathbf{PM}_2$ and $\mathbf{PM}_3$ will be equal and in the same direction. The tangential components must also be equal and in the same direction so that the velocity of sliding will be zero. The two members will then have pure rolling motion. Thus, it may be said that *the condition for pure rolling is that the point of contact shall lie on the line of centers*.

For the mechanism of Fig. 1.12, the motion between the cam and the follower will be a combination of rolling and sliding. Pure rolling can only take place where the point of contact P falls on the line of centers. However, contact at this point may not be possible because of the proportions of the mechanism. Pure sliding cannot occur between cam 2 and follower 3. For this to happen, a point on one link, within the limits of its travel, has to come in contact with all the successive points on the active surface of the other link.

It is possible to determine a relation so that the angular velocity ratio of two members in direct contact can be determined without going through the geometrical construction outlined above. From O_2 and O_3 drop perpendiculars upon the common normal striking it at e and f, respectively. The following relations will be seen to hold:

$$\omega_2 = \frac{PM_2}{O_2 P} \quad \text{and} \quad \omega_3 = \frac{PM_3}{O_3 P}$$

$$\frac{\omega_3}{\omega_2} = \frac{PM_3}{O_3 P} \times \frac{O_2 P}{PM_2}$$

From the fact that triangles $PM_2 n$ and $O_2 Pe$ are similar,

$$\frac{PM_2}{O_2 P} = \frac{Pn}{O_2 e}$$

Also, $PM_3 n$ and $O_3 Pf$ are similar triangles; therefore,

$$\frac{PM_3}{O_3 P} = \frac{Pn}{O_3 f}$$

Therefore,

$$\frac{\omega_3}{\omega_2} = \frac{Pn}{O_3 f} \times \frac{O_2 e}{Pn} = \frac{O_2 e}{O_3 f}$$

With the common normal intersecting the line of centers at K, triangles $O_2 Ke$

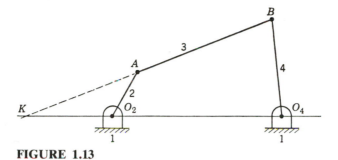

FIGURE 1.13

and O_3Kf are also similar; therefore,

$$\frac{\omega_3}{\omega_2} = \frac{O_2e}{O_3f} = \frac{O_2K}{O_3K} \tag{1.1}$$

Therefore, for a pair of curved surfaces in direct contact, the angular velocities are inversely proportional to the segments into which the line of centers is cut by the common normal. From this it can be seen that *for constant angular velocity ratio the common normal must intersect the line of centers in a fixed point.*

   It is also possible to derive the above relations for the transmission of motion through an intermediate link or connecting rod and for the transmission of motion through a flexible connector. Figures 1.13 and 1.14 show these two cases, respectively, where the angular velocity ratio is given by

$$\frac{\omega_4}{\omega_2} = \frac{O_2K}{O_4K} \tag{1.2}$$

   In Fig. 1.14, the ratio ω_4/ω_2 is independent of the center distance O_2O_4.

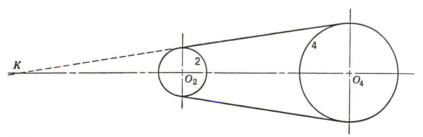

FIGURE 1.14

1.9   MOBILITY, OR NUMBER OF DEGREES OF FREEDOM

Mobility is one of the most fundamental concepts to the study of kinematics. By definition, the *mobility* of a mechanism is the number of degrees of freedom it possesses. An equivalent definition of mobility is the minimum number of independent parameters required to specify the location of every link within a mechanism.

A single link constrained to move with planar motion, such as the one shown in Fig. 1.15a, possesses three degrees of freedom. The x- and y-coordinates of the point P along with the angle θ form an independent set of three parameters describing its location. Two unconnected planar links are shown in Fig. 1.15b. Since each link possesses three degrees of freedom, these two links possess a total of six degrees of freedom. If the two links are pinned together at a point by means of a revolute joint, as shown in Fig. 1.15c, the two-link system will possess only four degrees of freedom. Four independent parameters describing the location of the two links could, for example, be the x- and y-coordinates of the point P_1, the angle θ_1, and the angle θ_2. There are many other parameters that could be used to specify the location of these links, but only four of these can be independent. Once the values of the independent parameters are specified, the position of every point in both links is determined.

In the simple example described above, connecting two planar links with a revolute joint had the effect of removing two degrees of freedom from the system. Stated in another way, a revolute joint permits a single degree of freedom (pure rotation) between the links it connects. Using this type of logic, it is possible to

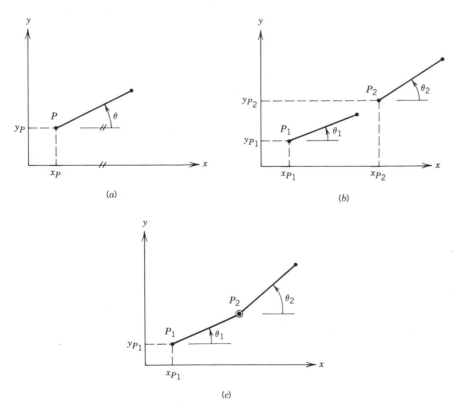

FIGURE 1.15

develop a general equation which will help predict the mobility of any planar mechanism.

For example, a planar mechanism having n links is to be designed. Before any connections are made, the system of n links will have a total of $3n$ degrees of freedom. Recognizing that one link of every mechanism will always be considered to be fixed to the ground removes three degrees of freedom. This leaves the system with a total of $3n - 3$, or $3(n - 1)$, degrees of freedom. Each one-degree-of-freedom joint removes two degrees of freedom from the system. Similarly, each two-degree-of-freedom joint removes one degree of freedom from the system. The total mobility of the system is given by Grubler's equation

$$M = 3(n - 1) - 2f_1 - f_2 \qquad (1.3)$$

where

    M = the mobility, or number of degrees of freedom

    n = the total number of links, including the ground

    f_1 = the number of one-degree-of-freedom joints

    f_2 = the number of two-degree-of-freedom joints

Care must be used when applying this equation because there are a number of special mechanism geometries for which it will not work. Although no all-inclusive rule exists for predicting when the mobility equation may give an incorrect result, special cases often occur when several links of a mechanism are parallel. For example, applying Grubler's equation to the mechanism of Fig. 1.16 gives

$$M = 3(5 - 1) - 2(6) = 0$$

Nevertheless, this device can actually move as a result of its special geometry and is a mechanism with one degree of freedom. It must also be noted that a joint connecting k links at a single point must be counted as $k - 1$ joints. For example, a revolute joint connecting three links at a single point is counted as two joints. Only four types of joints are commonly found in planar mechanisms.

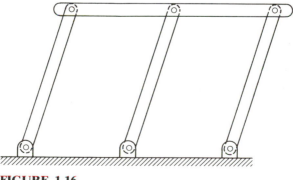

FIGURE 1.16

These are the revolute, the prismatic, and the rolling contact joints (each having one degree of freedom), and the cam or gear joint (having two degrees of freedom). These joints are depicted in Fig. 1.17. The following definitions apply to the actual mobility of a device:

$M \geq$ 1: the device is a mechanism with M degrees of freedom

$M =$ 0: the device is a statically determinate structure

$M \leq -1$: the device is a statically indeterminate structure

Joint Type (Symbol)	Physical Form	Schematic Representation	Degrees of Freedom
Revolute (R)	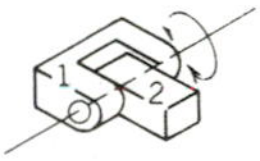	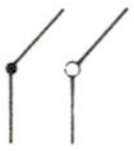	1 (Pure rotation)
Prismatic (P)	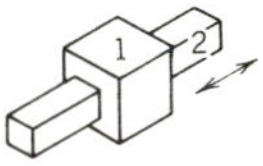		1 (Pure sliding)
Cam or gear	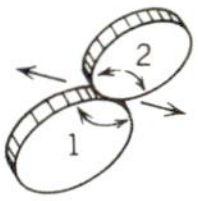	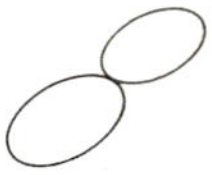	2 (Rolling and sliding)
Rolling contact	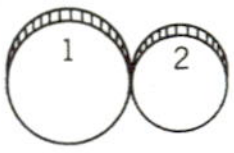	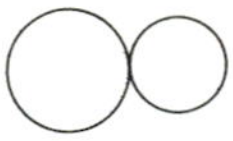	1 (Rolling without sliding)

FIGURE 1.17 Common types of joints found in planar mechanisms.

Example 1.1. Determine the mobility of the four-bar linkage of Fig. 1.18.
    There are four links and four revolute joints, each having one degree of freedom. The mobility is given by

$$M = 3(4 - 1) - 2(4)$$

$$M = 1$$

So this is a one-degree-of-freedom mechanism.

Example 1.2. Determine the mobility of the device of Fig. 1.19.
    There are four links connected by five single-degree-of-freedom joints (the joint connecting three links at a point counts twice). The mobility is given by

$$M = 3(4 - 1) - 2(5)$$

$$M = -1$$

This is a statically indeterminate structure.

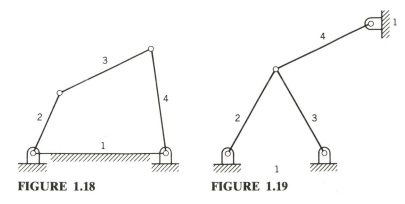

FIGURE 1.18   FIGURE 1.19

Example 1.3. Determine the mobility of the device of Fig. 1.20.

There are three links, two one-degree-of-freedom revolute joints and one two-degree-of-freedom higher-pair joint. In the higher-pair joint, the two contacting bodies may translate along the common tangent to the two surfaces or rotate about the contact point, thus giving two degrees of freedom. The mobility is given by

$$M = 3(3 - 1) - 2(2) - 1(1)$$
$$M = 1$$

This is a one-degree-of-freedom mechanism.

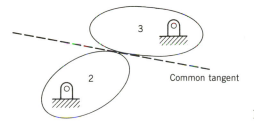

FIGURE 1.20

Problems

1.1. (*a*) If ω_2 = 20 rad/min, calculate the angular velocity of link 3 for the two cases shown in Fig. 1.21. (*b*) Calculate the maximum angle and the minimum angle of the follower with the horizontal.

1.2. Lay out the mechanisms for Problem 1.1 to full scale and graphically determine the velocity of sliding between links 2 and 3. Use a velocity scale of 1 in. = 10 in./min.

1.3. If ω_2 = 20 rad/min for the mechanism shown in Fig. 1.21, using graphical construction, determine the angular velocities of link 3 for one revolution of the cam in 60° increments starting from the position where ω_3 = 0. Plot ω_3 versus cam angle θ letting the scale of ω_3 be 1 in. = 2.0 rad/min and the scale of θ be $\frac{1}{2}$ in. = 60.

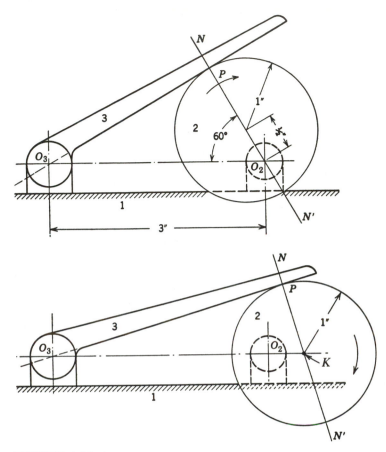

FIGURE 1.21

1.4. (*a*) If ω_2 = 1800 rad/s, calculate the angular velocity of link 3 for the mechanism shown in Fig. 1.22. (*b*) Calculate the maximum angle and the minimum angle of the follower relative to the horizontal.

1.5. For the linkage shown in Fig. 1.23, determine ω_4 and V_B.

1.6. For the linkage shown in Fig. 1.24, determine V_B and ω_4.

1.7. Prove for the linkage shown in Fig. 1.13 that the angular velocities of the driven and driver links are inversely proportional to the segments into which the line of centers is cut by the line of transmission.

1.8. Prove for the belt and pulleys shown in Fig. 1.14 that the angular velocities of the pulleys are inversely proportional to the segments into which the line of centers is cut by the line of transmission.

1.9. In a linkage as shown in Fig. 1.13, the crank 2 is 19 mm long and rotates at a constant angular velocity of 15 rad/s. Link 3 is 38 mm long and link 4 is 25 mm long. The distance between centers O_2 and O_4 is 51 mm. Graphically determine the angular velocity of link

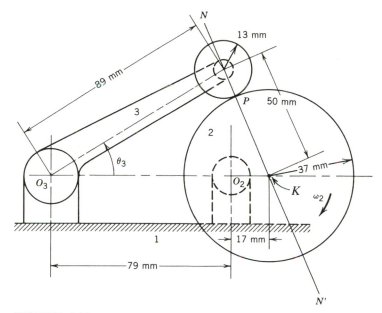

FIGURE 1.22

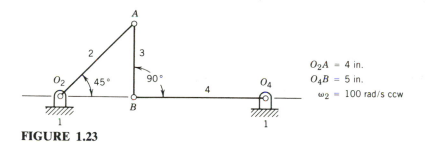

FIGURE 1.23

O_2A = 4 in.
O_4B = 5 in.
ω_2 = 100 rad/s ccw

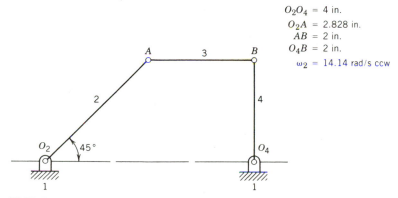

O_2O_4 = 4 in.
O_2A = 2.828 in.
AB = 2 in.
O_4B = 2 in.
ω_2 = 14.14 rad/s ccw

FIGURE 1.24

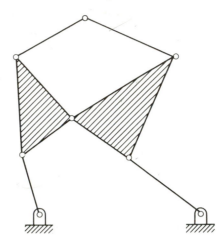

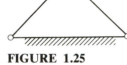

FIGURE 1.25

FIGURE 1.26

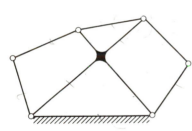

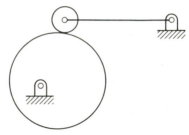

FIGURE 1.27

FIGURE 1.28

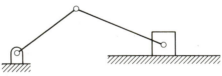

FIGURE 1.29

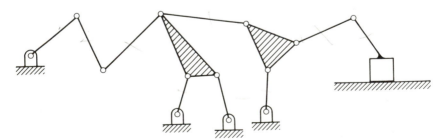

FIGURE 1.30

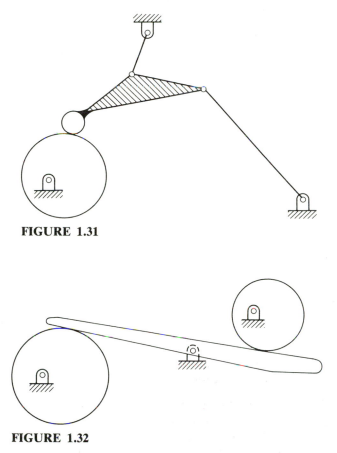

FIGURE 1.31

FIGURE 1.32

4 when link 2 is counterclockwise 45° from the horizontal. State whether or not ω_4 is constant.

1.10. A pulley of 100 mm diam drives one of 200 mm diam by means of a belt. If the angular velocity of the drive pulley is 65 rad/s and the center distance between pulleys is 400 mm, graphically determine the speed of the 200 mm pulley. Will its speed be constant?

1.11. Determine the mobility (number of degrees of freedom) of the devices shown in Figs. 1.25 through 1.32.

Chapter Two

Linkages and Mechanisms

2.1 POSITION ANALYSIS OF THE FOUR-BAR LINKAGE

One of the simplest and most useful mechanisms is the four-bar linkage. A sketch of this linkage is shown in Fig. 2.1. Link 1 is the frame, or ground, and is generally stationary. Link 2 is the driver, which may rotate completely or may oscillate. If link 2 rotates completely, then the mechanism is transforming rotary motion into oscillatory motion. If the crank oscillates, then the mechanism multiplies oscillatory motion.

When link 2 is rotating completely, there is no danger of the linkage locking. However, if link 2 oscillates, care must be taken in proportioning the links to avoid dead points so that the mechanism will not stall in its extreme positions.

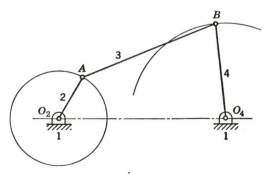

FIGURE 2.1 Four-bar linkage.

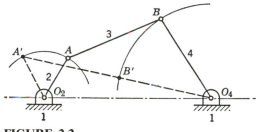

FIGURE 2.2

These dead points will occur when the line of action of the driving force is directed along link 4. This condition is shown by the dashed lines in Fig. 2.2.

If the four-bar mechanism is designed so that link 2 can rotate completely but link 4 is made the driver, dead points will occur, and it is necessary to provide a flywheel to pass through these dead points.

In addition to possible dead points in a four-bar linkage, it is necessary to consider the *transmission angle,* which is the angle between the connecting or coupler link 3 and the output link 4. This is shown in Fig. 2.3*a* as angle γ.

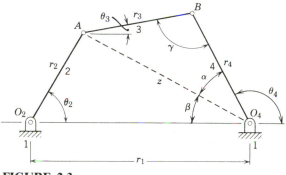

FIGURE 2.3*a*

An equation for the transmission angle can be derived by applying the law of cosines to triangles AO_2O_4 and ABO_4 as follows:

$$z^2 = r_1^2 + r_2^2 - 2r_1r_2 \cos \theta_2 \qquad (2.1)$$

also

$$z^2 = r_3^2 + r_4^2 - 2r_3r_4 \cos \gamma \qquad (2.2)$$

Therefore,

$$r_1^2 + r_2^2 - 2r_1r_2 \cos \theta_2 = r_3^2 + r_4^2 - 2r_3r_4 \cos \gamma$$

and

$$\gamma = \cos^{-1}\left[\frac{r_1^2 + r_2^2 - r_3^2 - r_4^2 - 2r_1r_2\cos\theta_2}{-2r_3r_4}\right]$$

or

$$\gamma = \cos^{-1}\left[\frac{z^2 - r_3^2 - r_4^2}{-2r_3r_4}\right] \tag{2.3}$$

where the value of z is calculated from the first of the two cosine law equations (Eq. 2.1). With the dimensions of the linkage given (i.e., r_1, r_2, r_3, and r_4), γ is a function of only the input angle θ_2. Note that there will be two values of γ corresponding to any one value of θ_2, because the arccosine is a double-valued function. Physically, the second value of γ corresponds to the second mode of assembly, *branch*, or *closure*, of the four-bar linkage, as shown in Fig. 2.3b. For any one value of the input angle θ_2, the four-bar linkage can be assembled in two different ways.

In general, for best transmission of force within the mechanism, links 3 and 4 should be nearly perpendicular throughout the motion cycle. If the transmission angle deviates from $+90°$ or $-90°$ by more than about 45° or 50°, the linkage tends to bind because of friction in the joints; also links 3 and 4 tend to align and may lock. It is especially important to check transmission angles when linkages are designed to operate close to dead points. An illustration of the minimum and the maximum transmission angles for a four-bar linkage is shown in Fig. 2.3c by γ' and γ'', respectively. In this mechanism, link 2 rotates completely and link 4 oscillates.

The output angle of the four-bar linkage (angle θ_4 in Fig. 2.3a) can also be found in closed form as a function of θ_2. By referring to Fig. 2.3a, the law of cosines may be used to express angles α and β as follows:

$$\alpha = \cos^{-1}\left(\frac{z^2 + r_4^2 - r_3^2}{2zr_4}\right) \tag{2.4}$$

$$\beta = \cos^{-1}\left(\frac{z^2 + r_1^2 - r_2^2}{2zr_1}\right) \tag{2.5}$$

The angle θ_4 in Fig. 2.3a is given by

$$\theta_4 = 180° - (\alpha + \beta) \tag{2.6}$$

Great care must be exercised in using this result since both α and β may be either positive or negative angles, depending on which solution is taken for the arccosine function. For the second closure of the linkage (Fig. 2.3b), β must be taken

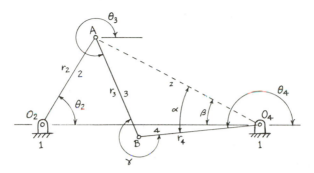

FIGURE 2.3*b*

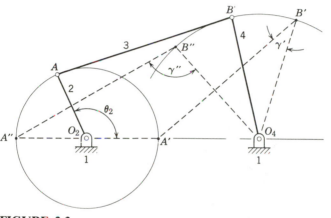

FIGURE 2.3*c*

positive and α must be taken negative in order to use Eq. 2.5. In general, for $0° \leq \theta_2 < 180°$, β should be selected such that $0° \leq \beta < 180°$; and similarly for $180° \leq \theta_2 < 360°$, β should be selected such that $180° \leq \beta < 360°$. With β selected in this way, the two values of α will yield values of θ_4 corresponding to the two distinct closures of the linkage.

The process of finding the variable output angles of a mechanism as functions of the input angle is known as *position analysis*. The method of position analysis just presented is but one of several possible approaches. Another method based on the use of vectors and complex numbers is explored in Appendix 1. All of the methods presented, however, require insights and manipulations to obtain the desired output angle as a function of the input angle. The position analysis problem for linkages containing more than four links can become extremely complicated.

Example 2.1. For the four-bar linkage shown in Fig. 2.4 with $r_1 = 7$ in., $r_2 = 3$ in., $r_3 = 8$ in., $r_4 = 6$ in., and $\theta_2 = 60°$, find the transmission angle, γ, and the output angle, θ_4.

Substituting the known values into the first cosine law equation (Eq. 2.1) gives

$$z^2 = (7)^2 + (3)^2 - 2(7)(3) \cos 60° = 37$$

$$z = 6.083$$

Substituting this value into Eqs. 2.3, 2.4, and 2.5 along with the link dimensions gives

$$\gamma = \arccos \frac{37 - (8)^2 - (6)^2}{-2(8)(6)}$$

$$\gamma = \pm 48.986°$$

$$\alpha = \arccos \frac{37 + (6)^2 - (8)^2}{2(37)^{1/2}(6)}$$

$$\alpha = \pm 82.917°$$

$$\beta = \arccos \frac{37 + (7)^2 - (3)^2}{2(37)^{1/2}(7)}$$

$$\beta = \pm 25.285°$$

Since θ_2 is between 0° and 180°, β must be taken as positive. The values of θ_4 are therefore given by

$$\theta_4 = 180° - (\pm 82.917° + 25.285°)$$

$$\theta_4 = 71.798°, 237.632°$$

Clearly, the first value of θ_4 is correct for the closure shown in Fig. 2.4.

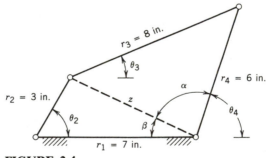

FIGURE 2.4

2.2 FOUR-BAR LINKAGE MOTION AND GRASHOFF'S LAW

The four-bar linkage may take other forms as shown in Fig. 2.5. In Fig. 2.5*a*, the mechanism has been crossed and will give the same type of motion as in Fig. 2.1. In Fig. 2.5*b*, opposite links are all the same length and, therefore, always

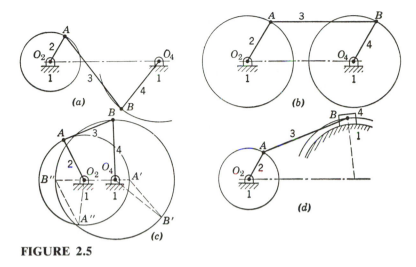

FIGURE 2.5

remain parallel; both links 2 and 4 rotate completely. This type of motion was characteristic of a steam locomotive drive. Figure 2.5c shows another arrangement whereby both the driver and follower rotate continuously. This form of the four-bar linkage is the basis for the drag-link mechanism which will be discussed under the subject of quick-return mechanisms. For rotation of crank 2 at a constant angular velocity, link 4 will rotate at a nonuniform rate. In order to prevent locking of the mechanism, certain relations must be maintained between the links:

$$O_2A \quad \text{and} \quad O_4B > O_2O_4$$

$$(O_2A - O_2O_4) + AB > O_4B$$

$$(O_4B - O_2O_4) + O_2A > AB$$

The second and third relation can be derived from the triangles $O_4A'B'$ and $O_2A''B''$, respectively, and the fact that the sum of two sides of a triangle must be greater than the third side.

Figure 2.5d shows an arrangement whereby link 4 of Fig. 2.1 has been replaced by a sliding block. The motion of the two linkages is identical.

The motion of the four-bar linkage is often characterized by the term *crank rocker* to indicate that crank 2 rotates completely and link 4 oscillates as in Fig. 2.5a. In a similar manner, the term *double crank* means that both link 2 and link 4 rotate completely as in Figs. 2.5b and c. The term *double rocker* indicates that both link 2 and link 4 oscillate as shown in Fig. 2.2.

As a means of determining whether a four-bar linkage will operate as a crank rocker, a double crank, or a double rocker, Grashoff's law can be applied. This law states that if the sum of the lengths of the longest link and the shortest link is *less than* the sum of the lengths of the other two, there will be formed

1. two different crank rockers when the shortest link is the crank and either of the adjacent links is the fixed link
2. a double crank when the shortest link is the fixed link
3. a double rocker when the link opposite the shortest is the fixed link

Also, if the sum of the lengths of the longest and the shortest links is *greater than* the sum of the lengths of the other two, only double-rocker mechanisms will result. Also, if the sum of the longest and shortest links is *equal to* the sum of the other two, the four possible mechanisms are similar to those of 1, 2, and 3 above. However, in this last case the center lines of the links can become collinear so that the driven link can change direction of rotation unless some means is provided to avoid it. Such a linkage is shown in Fig. 2.5*b*, where the links become collinear along the line of centers O_2O_4. At this position, the direction of rotation of the driven link 4 could change unless inertia carried link 4 through this point.

2.3   POSITION ANALYSIS OF LINKAGES USING LOOP CLOSURE EQUATIONS AND ITERATIVE METHODS

It is possible to analyze the majority of mechanisms by using methods such as the one described in section 2.1. These are known as closed-form methods; that is, a finite number of calculations are required to find a theoretically exact solution. Unfortunately, however, it is difficult to develop a computer-aided analysis package using this approach, because each different type of mechanism generally requires a separate analysis method and a separate computer program. Because of this, several of the commercially available mechanism analysis programs have been developed based on iterative methods. Iterative methods attempt to converge on a solution by repetitive calculations. For this reason, it is not known beforehand how many calculations will be required, or even if a solution can be found. The basic concepts of iterative mechanism analysis will now be illustrated by way of a four-bar linkage example.

Consider the four-bar linkage of Fig. 2.6 with the *x*-coordinate axis along link 1 which is fixed. Since the links of this mechanism form a closed loop, the *x*- and *y*-components of the links must sum to zero. This may be expressed as follows:

x-Components:
$$r_1 + r_4 \cos \theta_4 - r_2 \cos \theta_2 - r_3 \cos \theta_3 = 0 \qquad (2.7)$$

y-Components:
$$r_4 \sin \theta_4 - r_2 \sin \theta_2 - r_3 \sin \theta_3 = 0 \qquad (2.8)$$

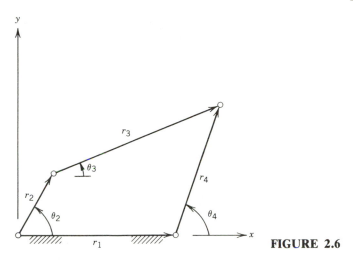

FIGURE 2.6

In position analysis, the link lengths r_1, r_2, r_3, and r_4 are known, and the problem is to find the angles θ_3 and θ_4 for a given value of θ_2. Thus, Eqs. 2.7 and 2.8 appear to be a simple set of two equations in the two unknowns θ_3 and θ_4. The complication is that these equations are transcendental, and a simple linear solution is not possible.

Note that the above equations will be satisfied only for those particular values of θ_3 and θ_4 that close the mechanism loop. These values are often called the roots of the equation. For any values of θ_3 and θ_4 other than the roots, these equalities will not be satisfied, so that in general

$$r_1 + r_4 \cos \theta_4 - r_2 \cos \theta_2 - r_3 \cos \theta_3 = f_1(\theta_3, \theta_4) = f_1(\boldsymbol{\theta}) \qquad \textbf{(2.9)}$$

$$r_4 \sin \theta_4 - r_2 \sin \theta_2 - r_3 \sin \theta_3 = f_2(\theta_3, \theta_4) = f_2(\boldsymbol{\theta}) \qquad \textbf{(2.10)}$$

where the shorthand notation $\boldsymbol{\theta} = \theta_3, \theta_4$ has been used.

Finding the roots of these equations is now equivalent to finding the values of θ_3 and θ_4 for which $f_1(\boldsymbol{\theta})$ and $f_2(\boldsymbol{\theta})$ are simultaneously equal to zero. At this point, a trial solution could be used to find the roots. A far more efficient procedure, however, is to use a linear approximation to the functions in seeking successively improved solutions.

Suppose, for example, that values of the angles θ_3 and θ_4 have been selected arbitrarily. In general, these will not be roots of the equations. There exist, however, some values $\Delta\theta_3$ and $\Delta\theta_4$ which, when added to θ_3 and θ_4, will give the roots. This can be expressed as follows:

$$f_i(\theta_3 + \Delta\theta_3, \theta_4 + \Delta\theta_4) = f_i(\boldsymbol{\theta} + \boldsymbol{\Delta}) = 0 \qquad i = 1, 2 \qquad \textbf{(2.11)}$$

A linear approximation to this function is obtained by taking the first two terms of its Taylor series expansion about the point θ_3, θ_4:

$$f_i(\theta + \Delta) = f_i(\theta) + \left(\frac{\partial f_i(\theta)}{\partial \theta_3}\right)\Delta\theta_3 + \left(\frac{\partial f_i(\theta)}{\partial \theta_4}\right)\Delta\theta_4 \quad i = 1, 2 \quad (2.12)$$

It is now possible to solve for the values of $\Delta\theta_3$ and $\Delta\theta_4$ that will drive this linear function to zero. If the linear function is a reasonable approximation to the original function, these values should also cause the original function to be approximately equal to zero. Setting the linear function equal to zero gives

$$f_i(\theta) + \left(\frac{\partial f_i(\theta)}{\partial \theta_3}\right)\Delta\theta_3 + \left(\frac{\partial f_i(\theta)}{\partial \theta_4}\right)\Delta\theta_4 = 0 \quad i = 1, 2 \quad (2.13)$$

By substituting $i = 1$ and $i = 2$ in Eq. 2.13 and by rearranging terms, the following equations result:

$$\left(\frac{\partial f_1(\theta)}{\partial \theta_3}\right)\Delta\theta_3 + \left(\frac{\partial f_1(\theta)}{\partial \theta_4}\right)\Delta\theta_4 = -f_1(\theta) \quad (2.14)$$

$$\left(\frac{\partial f_2(\theta)}{\partial \theta_3}\right)\Delta\theta_3 + \left(\frac{\partial f_2(\theta)}{\partial \theta_4}\right)\Delta\theta_4 = -f_2(\theta) \quad (2.15)$$

Once an initial estimate has been made for the values of θ_3 and θ_4, the values of $f_1(\theta)$ and $f_2(\theta)$ can be calculated from Eqs. 2.9 and 2.10. The partial derivatives needed in Eqs. 2.14 and 2.15 are found to be

$$\frac{\partial f_1(\theta)}{\partial \theta_3} = r_3 \sin \theta_3$$

$$\frac{\partial f_1(\theta)}{\partial \theta_4} = -r_4 \sin \theta_4$$

$$\frac{\partial f_2(\theta)}{\partial \theta_3} = -r_3 \cos \theta_3$$

$$\frac{\partial f_2(\theta)}{\partial \theta_4} = r_4 \cos \theta_4$$

$$(2.16)$$

Since the partial derivatives found in Eq. 2.16 are evaluated at the estimated values of θ_3 and θ_4, Eqs. 2.14 and 2.15 are actually two linear equations in the two unknowns $\Delta\theta_3$ and $\Delta\theta_4$. Solving these equations simultaneously yields the values of $\Delta\theta_3$ and $\Delta\theta_4$ which, when added to the estimated values θ_3 and θ_4, will make the approximate linear function equal to zero. Although, in general, these will not be the same as the roots of the original function, they will be an improved estimate. By using this improved estimate, a second linear approximation is made to the function and a new set of values for θ_3 and θ_4 is calculated. This process is repeated until the roots of the approximate function produce values of the original function which are nearly equal to zero. The method just described is one of the best known and most often used numerical root-finding techniques. It is known as the Newton–Raphson method. The following numerical example will help to show the details of this method and its application to mechanism analysis.

Example 2.2. For the four-bar linkage shown in Fig. 2.6, solve the position analysis problem using the Newton–Raphson root-finding method. Use $\theta_2 = 60°$, and use link dimensions $r_1 = 7$ in., $r_2 = 3$ in., $r_3 = 8$ in., and $r_4 = 6$ in.

     Before analysis can proceed, initial estimates of θ_3 and θ_4 on which to iterate must be obtained. Normally, position analysis begins at some known starting position of the mechanism and proceeds by incrementing the input angle by some small amount. The values of θ_3 and θ_4 at the previous position are usually a good estimate of the corresponding values at the present position. Another approach is to estimate these values graphically. To demonstrate the rapid convergence of the Newton–Raphson method, values of θ_3 and θ_4 will be selected which are known to be far from the true roots. The value of θ_3 will be estimated at $0°$, and the value of θ_4 will be estimated at $100°$. Substituting these estimated values along with known dimensions of the linkage in the loop closure equations (Eqs. 2.9 and 2.10) and also in the expressions for the partial derivatives (Eq. 2.16) gives

$$f_1(\boldsymbol{\theta}) = 7 + 6 \cos 100° - 3 \cos 60° - 8 \cos 0° = -3.542$$

$$f_2(\boldsymbol{\theta}) = 6 \sin 100° - 3 \sin 60° - 8 \sin 0° = 3.311$$

$$\frac{\partial f_1(\boldsymbol{\theta})}{\partial \theta_3} = 8 \sin 0° = 0$$

$$\frac{\partial f_1(\boldsymbol{\theta})}{\partial \theta_4} = -6 \sin 100° = -5.909$$

$$\frac{\partial f_2(\boldsymbol{\theta})}{\partial \theta_3} = -8 \cos 0° = -8.000$$

$$\frac{\partial f_2(\boldsymbol{\theta})}{\partial \theta_4} = 6 \cos 100° = -1.042$$

Substituting these values into Eqs. 2.14 and 2.15 yields the following linear equations in the unknowns $\Delta\theta_3$ and $\Delta\theta_4$:

$$(0) \, \Delta\theta_3 + (-5.909) \, \Delta\theta_4 = 3.542$$

$$(-8.000) \, \Delta\theta_3 + (-1.042) \, \Delta\theta_4 = -3.311$$

Solving for $\Delta\theta_3$ and $\Delta\theta_4$ gives

$$\Delta\theta_3 = 0.492 \text{ rad} = 28.185° \qquad \Delta\theta_4 = -0.599 \text{ rad} = -34.344°$$

Upon adding these to the estimated values of θ_3 and θ_4, the following improved estimates are obtained:

$$\theta_3 = 0° + 28.185° = 28.185°$$

$$\theta_4 = 100° - 34.344° = 65.656°$$

Values of the functions and the partial derivatives are recalculated using these new values, and a second set of approximate θ values is obtained. This process is repeated until the values of the $f_1(\theta)$ and $f_2(\theta)$ are equal to zero, or until no further improvement can be obtained. A flowchart of this iterative process is shown in Fig. 2.7. The corresponding computer program, written in BASIC, is shown in Fig. 2.8. The results of this program at each iteration are given in Table 2.1. It is evident from this table that the Newton–Raphson method converges rapidly for this example.

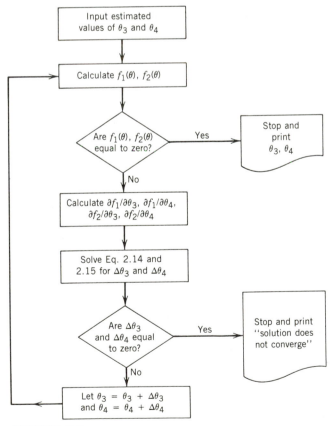

FIGURE 2.7

```
10 '*****************************************************************
20 '* MECHANISM DESIGN - DISPLACEMENT ANALYSIS (5/27/85)
30 '* -   Uses Newton-Raphson root finding method to determine unknown
40 '*     angles of links 3 & 4 of a four bar linkage.
50 '* -   Mabie and Reinholtz, 4th Ed.
60 '* -   Program revised by - Steve Wampler (6/ 5/85)
70 '*****************************************************************
80 CLS ' clear the screen then ask user to discribe mechanism
90 INPUT "Enter angular displacement of link 2 (degrees)";THETA2
100 INPUT "Guess angular displacement for link 3 (degrees)";THETA3
110 INPUT "Guess angular displacement for link 4 (degrees)";THETA4
120 INPUT "Enter link lengths r1,r2,r3,r4";R1,R2,R3,R4
130 PRINT:PRINT" THETA3    THETA4    FUNC.1    FUNC.2    DELT3    DELT4"
140 DEG2RAD=3.14159/180 'constant to convert from degrees to radians
150 THETA2=THETA2*DEG2RAD:THETA3=THETA3*DEG2RAD:THETA4=THETA4*DEG2RAD
160 FUNC.1=1 'force next WHILE statement to be true
170 WHILE ABS(FUNC.1)>.001 OR ABS(FUNC.2)>.001 'loop until roots found
180     ' Evaluate loop equations
190     FUNC.1=R1+(R4*COS(THETA4))-(R2*COS(THETA2))-(R3*COS(THETA3))
200     FUNC.2=(R4*SIN(THETA4))-(R2*SIN(THETA2))-(R3*SIN(THETA3))
210     ' Evaluate partial derivatives
220     DF1DT3=R3*SIN(THETA3)   'Partial of func. 1 w/respect to theta3
230     DF1DT4=-R4*SIN(THETA4)  'Partial of func. 1 w/respect to theta4
240     DF2DT3=-R3*COS(THETA3)  'Partial of func. 2 w/respect to theta3
250     DF2DT4=R4*COS(THETA4)   'Partial of func. 2 w/respect to theta4
260     ' Now solve 2 eq.s in 2 unknowns with Cramer's Rule.
270     DEL=DF1DT3*DF2DT4-DF1DT4*DF2DT3   'calc. del function
280     DELTA.THETA4=(DF2DT3*FUNC.1-DF1DT3*FUNC.2)/DEL
290     DELTA.THETA3=-(DF2DT4*FUNC.1-DF1DT4*FUNC.2)/DEL
300     ' Output the results
310     PRINT USING "####.### ";THETA3/DEG2RAD,THETA4/DEG2RAD;
320     PRINT USING "####.### ";FUNC.1,FUNC.2;
330     PRINT USING "####.### ";DELTA.THETA3/DEG2RAD,DELTA.THETA4/DEG2RAD
340     ' make new guess for both theta 3 and theta 4
350     THETA3=THETA3+DELTA.THETA3:THETA4=THETA4+DELTA.THETA4
360 WEND ' do loop again if roots have not been found
370 PRINT:LINE INPUT "Press RETURN to rerun program ...";A$:RUN
```

FIGURE 2.8

TABLE 2.1 Results of the Iterative Analysis Program

θ_3	θ_4	$f_1(\theta)$	$f_2(\theta)$	$\partial f_1/\partial\theta_3$
0.000	100.000	−3.542	3.311	0.000
28.185	65.656	0.922	−0.910	3.778
22.897	71.663	0.018	−0.015	3.113
22.812	71.798	0.000	−0.000	3.102

$\partial f_1/\partial\theta_4$	$\partial f_2/\partial\theta_3$	$\partial f_2/\partial\theta_4$	$\Delta\theta_3$	$\Delta\theta_4$
−5.909	−8.000	−1.042	28.185	−34.344
−5.467	−7.051	2.473	−5.287	6.008
−5.695	−7.370	1.888	−0.085	0.134
−5.700	−7.374	1.874	−0.000	−0.000

2.4   LINKAGE ANALYSIS USING THE INTEGRATED MECHANISMS PROGRAM (IMP)

As mentioned in the previous section, several commercially available mechanism analysis programs have been developed based on iterative methods of solving the loop closure equations. One of the most widely used of these programs is the Integrated Mechanisms Program, known as IMP. This program was developed by Sheth and Uicker[1] and is currently distributed by Structural Dynamics Research Corporation, a subsidiary of General Electric CAE International.[2] Professor Uicker and his associates have developed another IMP program which is being distributed by JML Research Inc.[3] The IMP system is capable of analyzing displacements, velocities, accelerations, and forces in a wide variety of two- and three-dimensional rigid link mechanisms. It must be emphasized that the use of this program is no substitute for a solid understanding of basic kinematic principles. It can, however, relieve the designer of many routine calculations and provide analysis capabilities far beyond those attainable using hand calculation or user-written programs. For these reasons, programs such as IMP are rapidly becoming indispensable tools for industrial designers.

The following example will illustrate how this program can be set up to analyze the angular displacements of the input and output links of a four-bar linkage.

Example 2.3. In the four-bar linkage shown in Fig. 2.9*a*, link 2 is the driver and rotates completely, and link 4 oscillates. Use the IMP program to determine the angles which correspond to the extreme positions of link 4.

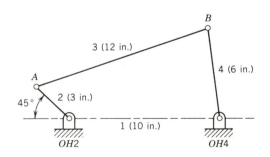

FIGURE 2.9*a*

[1] P. N. Sheth, and J. J. Uicker, "IMP (Integrated Mechanisms Program), A Computer-Aided Design Analysis System for Mechanisms and Linkages," *Journal of Engineering for Industry, Trans. ASME,* Vol. 94, May 1972, pp. 454–464.

[2] "IMP, Integrated Mechanisms Program," Structural Dynamics Research Corporation, Milford, OH, February, 1979.

[3] "THE INTEGRATED MECHANISMS PROGRAM (IMP): A Problem Oriented Language for the Computer-Aided Design and Analysis of Mechanical Systems," JML Research Inc., 1984.

The designations *OH*2, *A*, *B*, and *OH*4 in Fig. 2.9*a* represent turning pairs or revolutes (joints) and permit only relative rotation. These positions would contain the bearings in an actual mechanism. The ends of each link terminate in a point which is the center of the revolute. In Fig. 2.9*b*, link 2 is defined by points *OO*2 and *AA*2, link 3 by points *AA*3 and *BB*3, and link 4 by *BB*4 and *OO*4. (This use of a single letter, for example, *A*, to designate a revolute and of a double letter, *AA*, to designate a point is chosen for convenience to avoid confusion in specifying the model for the mechanism.) As illustrated in Chapter 1, point *AA* is a point on both links 2 and 3, and point *BB* is common to both links 3 and 4. In a similar manner, point *OO*2 is common to links 1 and 2, and point *OO*4 is common to links 1 and 4. The additional labeling of the mechanism to specify these points is shown in Fig. 2.9*b*. It is very important that the revolutes and points be clearly distinguished.

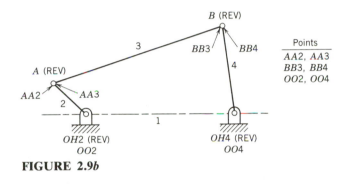

FIGURE 2.9*b*

The statements for the revolutes to be listed in the input to the IMP program follow:

```
GROUND = FRAME
REVOLUTE ( FRAME , LNK2 ) = OH2
REVOLUTE ( LNK2 , LNK3 ) = A
REVOLUTE ( LNK3 , LNK4 ) = B
REVOLUTE ( LNK4 , FRAME ) = OH4
```

It should be noted in the above listing that the pattern is link 2 relative to frame, link 3 relative to link 2, link 4 relative to link 3, and frame relative to link 4. By starting with frame and ending with frame, the requirement that the loop must close is satisfied.

The next step is to determine the coordinates of the revolutes and label them as shown in Fig. 2.9*c*. Also the orientation of two local coordinate systems attached to the links on each side of the revolutes must be chosen. All data for revolute joints must be given relative to the global reference frame.

The data for each revolute must now be listed in the input to the IMP program with a data:revolute statement. This will contain (*a*) the coordinates of the revolute, which is also the origin of the two local coordinate systems; (*b*) a point on the common local positive *z*-axes of these systems; and (*c*) and (*d*) points on the positive *x*-axes of the local coordinate systems for the first and second links named in the corresponding revolute statement. The easiest way to give the directions in item (*c*) and (*d*) is to go from *OH*2 to *A* for the *x*-direction along link 2 for revolute *OH*2 and for revolute *A* to go back from

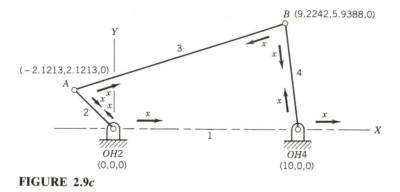

FIGURE 2.9c

A to $OH2$. The directions of the x-coordinates for the other revolutes are handled in a similar manner. The data for the revolutes are as follows:

```
DATA:REVOLUTE(OH2)=0,0,0/0,0,1/1,0,0/-2.1213,2.1213,0
DATA:REVOLUTE(A)=-2.1213,2.1213,0/-2.1213,2.1213,1/0,0,0/$
               9.2242,5.9388,0
DATA:REVOLUTE(B)=9.2242,5.9388,0/9.2242,5.9388,1/$
               -2.1213,2.1213,0/10,0,0
DATA:REVOLUTE(OH4)=10,0,0/10,0,1/9.2242,5.9388,0/12,0,0
```

The data for the points given in the local coordinate systems of the associated joints are listed next.

```
POINT(LNK2)=OO2,AA2
DATA:POINT(OO2,OH2)=0,0,0
DATA:POINT(AA2,A)=0,0,0
POINT(LNK3)=AA3,BB3
DATA:POINT(AA3,A)=0,0,0
DATA:POINT(BB3,B)=0,0,0
POINT(LNK4)=BB4,OO4
DATA:POINT(BB4,B)=0,0,0
DATA:POINT(OO4,OH4)=0,0,0
ZOOM(7)=5,1.5,0
RETURN
```

The IMP program was run on a VAX 11/780 computer using the above input listing. The minimum displacement angle for the output joint $OH4$ is shown in Fig. 2.9d. The value of the angle for joint $OH2$ is 15.68° (positive because it is taken as link 2 relative to frame, ccw). IMP defines counterclockwise angles as positive. The value of the angle for joint $OH4$ is $-42.55°$ (negative because it is taken as frame relative to link 4, cw). The maximum displacement angle for the joint $OH4$ is shown in Fig. 2.9e. The value of the angle for joint $OH2$ is 216.25°, and the angle for joint $OH4$ is $-117.55°$.

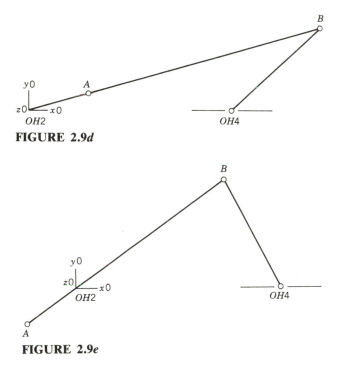

FIGURE 2.9d

FIGURE 2.9e

2.5 SLIDER-CRANK MECHANISM

This mechanism is widely used and finds its greatest application in the internal-combustion engine. Figure 2.10a shows a sketch in which link 1 is the frame (considered fixed), link 2 is the crank, link 3 is the connecting rod, and link 4 is the slider. With the internal-combustion engine, link 4 is the piston upon which gas pressure is exerted. This force is transmitted through the connecting rod to the crank. It can be seen that there will be two dead points during the cycle, one at each extreme position of piston travel. In order to overcome these, it is necessary to attach a flywheel to the crank so that the dead points can be passed. This mechanism is also used in air compressors where an electric motor drives the crank which in turn drives the piston that compresses the air.

In considering the slider crank, it is often necessary to calculate the displacement of the slider and its corresponding velocity and acceleration. Equations for displacement, velocity, and acceleration are derived using Fig. 2.10b:

$$
\begin{aligned}
x &= R + L - R \cos \theta - L \cos \phi \\
&= R(1 - \cos \theta) + L(1 - \cos \phi) \\
&= R(1 - \cos \theta) + L\left[1 - \sqrt{1 - \left(\frac{R}{L}\right)^2 \sin^2 \theta}\right]
\end{aligned}
\qquad \textbf{(2.17)}
$$

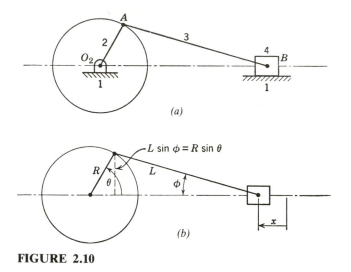

FIGURE 2.10

In order to simplify the above expression, the radical can be approximated by replacing it with the binomial series

$$(1 \pm B^2)^{1/2} = 1 \pm \frac{1}{2} B^2 - \frac{B^4}{2 \cdot 4} \pm \frac{1 \cdot 3B^6}{2 \cdot 4 \cdot 6} - \frac{1 \cdot 3 \cdot 5B^8}{2 \cdot 4 \cdot 6 \cdot 8} \pm \cdots$$

where $B = (R/L) \sin \theta$.

In general, it is sufficiently accurate to use only the first two terms of the series.

Therefore,

$$\sqrt{1 - \left(\frac{R}{L}\right)^2 \sin^2 \theta} = 1 - \frac{1}{2} \left(\frac{R}{L}\right)^2 \sin^2 \theta \qquad \text{(approximately)}$$

and

$$x = R(1 - \cos \theta) + \frac{R^2}{2L} \sin^2 \theta$$

where $\theta = \omega t$ because ω is constant; and

$$V = \frac{dx}{dt} = R\omega \left[\sin \theta + \frac{R}{2L} \sin 2\theta \right] \qquad \textbf{(2.18)}$$

$$A = \frac{d^2x}{dt^2} = R\omega^2 \left[\cos \theta + \frac{R}{L} \cos 2\theta \right] \qquad \textbf{(2.19)}$$

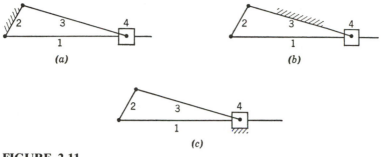

FIGURE 2.11

   It is possible to fix some link other than 1 on the slider crank and thus obtain three inversions, which are shown in Fig. 2.11. In Fig. 2.11a, the crank is held fixed and all the other links are allowed to move. This gives a mechanism that was used in early aircraft engines. They were known as rotary engines because the crank was stationary and the cylinders rotated about the crank. A more modern application of this inversion is in the Whitworth mechanism, which will be discussed under quick-return mechanisms. Figure 2.11b shows an inversion in which the connecting rod is held fixed. This inversion in modified form is the basis for the crank-shaper mechanism to be discussed later. The third inversion where the slider is held fixed, Fig. 2.11c, is sometimes used in the hand farm pump.
   A variation of the slider-crank mechanism can be affected by increasing the size of the crank pin until it is larger than the shaft to which it is attached and at the same time offsetting the center of the crank pin from that of the shaft. This enlarged crank pin is called an *eccentric* and can be used to replace the crank in the original mechanism. Figure 2.12 shows a sketch where point A is the center of the eccentric and point O the center of the shaft. The motion of this mechanism with the equivalent crank length OA is identical with that of the slider crank. One serious disadvantage of this mechanism, however, is the problem of proper lubrication between the eccentric and the rod. This limits the amount of power that can be transmitted.

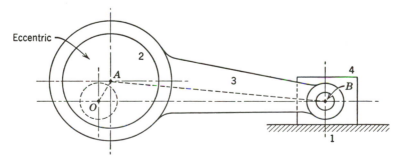

FIGURE 2.12

2.6 SCOTCH YOKE

This mechanism is one which will give simple harmonic motion. Its early application was on steam pumps, but it is now used as a mechanism on a test machine to produce vibrations. It is also used as a sine-cosine generator for computing elements. Figure 2.13*a* shows a sketch of this mechanism. Figure 2.13*b* shows the manner in which simple harmonic motion is generated. The radius *r* rotates at a constant angular velocity ω_r, and the projection of the point *P* upon the *x*-axis (or *y*-axis) moves with simple harmonic motion. The displacement from where the circle cuts the *x*-axis and increasing to the left is

$$x = r - r\cos\theta_r \qquad \text{where } \theta_r = \omega_r t \qquad (2.20)$$

Therefore,

$$x = r(1 - \cos\omega_r t)$$

$$V = \frac{dx}{dt} = r\omega_r \sin\omega_r t = r\omega_r \sin\theta_r \qquad (2.21)$$

$$A = \frac{d^2x}{dt^2} = r\omega_r^2 \cos\omega_r t = r\omega_r^2 \cos\theta_r \qquad (2.22)$$

Another mechanism which will give simple harmonic motion is a circular cam (eccentric) with a flat-faced radial follower. This is discussed in the following chapter.

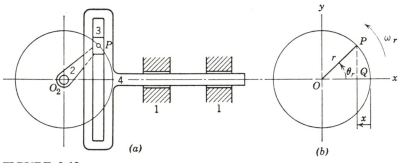

(a)   *(b)*

FIGURE 2.13

2.7 QUICK-RETURN MECHANISMS

These mechanisms are used on machine tools to give a slow cutting stroke and a quick return stroke for a constant angular velocity of the driving crank and are combinations of simple linkages such as the four-bar linkage and the slider-crank mechanism. An inversion of the slider crank in combination with the conventional slider crank is also used. In the design of quick-return mechanisms, the ratio of

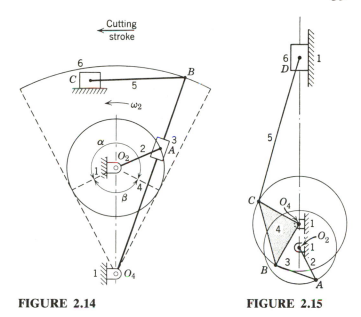

FIGURE 2.14   **FIGURE 2.15**

the crank angle for the cutting stroke to that for the return stroke is of prime importance and is known as the *time ratio*. To produce a quick return of the cutting tool, this ratio must obviously be greater than unity and as large as possible. As an example, the crank angle for the cutting stroke for the mechanism shown in Fig. 2.14 is labeled α, and that for the return stroke is labeled β. With the assumption that the crank operates at a constant speed, the time ratio is, therefore, α/β, which is much greater than unity.

There are several types of quick-return mechanisms which are described as follows:

Drag Link

This is developed from the four-bar linkage and is shown in Fig. 2.15. For a constant angular velocity of link 2, link 4 will rotate at a nonuniform velocity. Ram 6 will move with nearly constant velocity over most of the upward stroke to give a slow upward stroke and a quick downward stroke when driving link 2 rotates clockwise.

Whitworth

This is a variation of the first inversion of the slider crank in which the crank is held fixed. Figure 2.16 shows a sketch of the mechanism, and both links 2 and 4 make complete revolutions.

Crank Shaper

This mechanism is a variation of the second inversion of the slider crank in which the connecting rod is held fixed. Figure 2.14 shows the arrangement in which

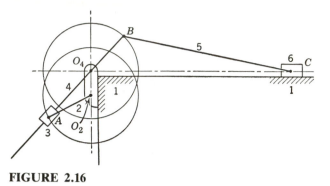

FIGURE 2.16

link 2 rotates completely and link 4 oscillates. If the distance O_2O_4 is shortened until it is less than the crank, the mechanism will revert to the Whitworth.

Offset Slider Crank

The slider crank can be offset as shown in Fig. 2.17, which will give a quick return motion. However, the amount of quick return is very slight, and the mechanism would only be used where space was limited and the mechanism had to be simple.

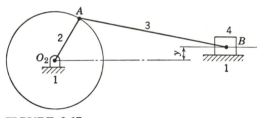

FIGURE 2.17

2.8 TOGGLE MECHANISM

This mechanism has many applications where it is necessary to overcome a large resistance with a small driving force. Figure 2.18 shows the mechanism; links 4 and 5 are of the same length. As the angles α decrease and links 4 and 5 approach being collinear, the force F required to overcome a given resistance P decreases as shown by the following relation:

$$\frac{F}{P} = 2 \tan \alpha \qquad (2.23)$$

It can be seen that for a given F as α approaches zero, P approaches infinity. A stone crusher utilizes this mechanism to overcome a large resistance with a small force. This mechanism can be used statically as well as dynamically, as is seen in numerous toggle clamping devices for holding work pieces.

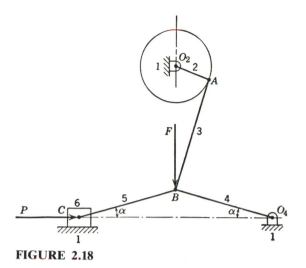

FIGURE 2.18

2.9 OLDHAM COUPLING

This mechanism provides a means for connecting two parallel shafts which are out of line a small amount so that a constant angular velocity ratio can be transmitted from the drive shaft to the driven shaft. A sketch is shown in Fig. 2.19. This mechanism is an inversion of the Scotch yoke.

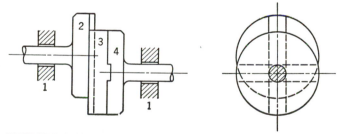

FIGURE 2.19

2.10 STRAIGHT-LINE MECHANISMS

As the name suggests, these mechanisms are designed so that a point on one of the links will move in a straight line. This straight line will be either an approximate or a theoretically correct straight line, depending on the mechanism.

An example of an approximate straight-line mechanism is the Watt, which is shown in Fig. 2.20. Point P is so located that the segments AP and BP are inversely proportional to the lengths O_2A and O_4B. Therefore, if links 2 and 4 are equal in length, point P must be the midpoint of link 3. Point P will trace

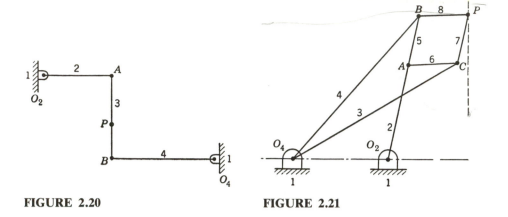

FIGURE 2.20   **FIGURE 2.21**

out a path in the form of a figure 8. Part of this path will very nearly approach a straight line.

The Peaucellier mechanism is one which will generate an exact straight line. Figure 2.21 shows a sketch where links 3 and 4 are equal. Links 5, 6, 7, and 8 are also equal, and link 2 equals the distance O_2O_4. Point P will trace out an exact straight-line path.

Straight-line mechanisms have many applications; notable among these are the mechanisms for engine indicators and for electrical switch gear equipment.

2.11   PANTOGRAPH

This mechanism is used as a copying device. When one point is made to follow a certain path, another point on the mechanism will trace out an identical path that is enlarged or reduced. Figure 2.22 shows a sketch. Links 2, 3, 4, and 5 form a parallelogram, and point P is on an extension of link 4. Point Q is on link 5 at the intersection of a line drawn from O to P. As point P traces out a path, point Q will trace out a similar path to a reduced scale.

This mechanism finds many applications in copying devices, particularly in engraving or profiling machines. One use of the profiling machine is in making dies or molds. Point P serves as a finger and traces out the contour of a template while a rotating endmill is placed at Q to machine the die to a smaller scale.

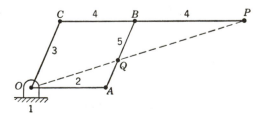

FIGURE 2.22

2.12 CHAMBER WHEELS

This mechanism takes several forms, which fall into two classifications. The first type consists of two lobed wheels operating within a casing. The Roots blower, as shown in Fig. 2.23, is an example of this type. The rotors are cycloids and are driven by a pair of meshing gears of equal size in back of the case. In the modern application, the Roots blower has three lobes on each rotor and is used for a low-pressure supercharger on Diesel engines.

The other class of chamber wheels has only one rotor placed eccentrically within the casing and is generally a variation of the slider-crank mechanism. Figure 2.24 shows a sketch of this type. The mechanism shown was originally designed for a steam engine, but its modern application is in the form of a pump.

Another example of the second type of chamber wheel is shown in Fig. 2.25, which illustrates the principle of the *Wankel engine*. In this mechanism, the expanding gases act upon the three-lobed rotor, which revolves directly on the eccentric and transmits torque to the output shaft through the eccentric which is integral with the shaft. The phase relation between the rotor and the rotation of

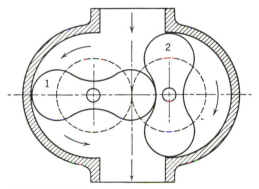

FIGURE 2.23

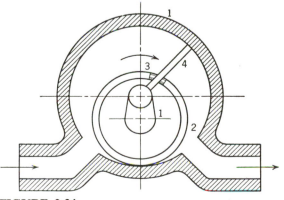

FIGURE 2.24

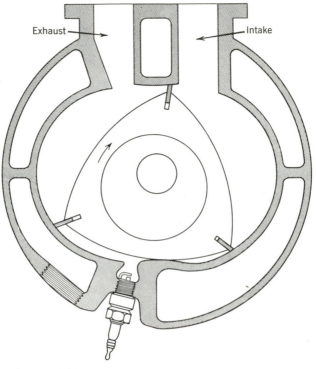

FIGURE 2.25

the eccentric shaft is maintained by a pair of internal and external gears (not shown) so that the orbital motion of the rotor is properly controlled.

2.13 HOOKE'S COUPLING

This coupling is used to connect two intersecting shafts. It is also known as a universal joint and has its widest use in the automotive field. A sketch of the coupling is shown in Fig. 2.26, and a commercial model is illustrated in Fig. 2.27. In Fig. 2.26, link 2 is the driver and link 4 the follower. Link 3 is a cross piece that connects the two yokes. It can be shown that, although both shafts must complete a revolution in the same length of time, the angular velocity ratio of the two shafts is not constant during the revolution but varies as a function of the angle β between the shafts and of the angle of rotation θ of the driver. The relation is given as

$$\frac{\omega_4}{\omega_2} = \frac{\cos \beta}{1 - \sin^2 \beta \sin^2 \theta} \tag{2.24}$$

A plot of this equation in polar coordinates for a quarter revolution of the

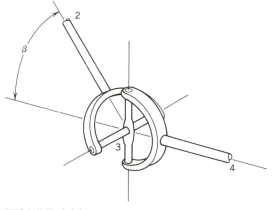

FIGURE 2.26

driving shaft is shown in Fig. 2.28, which clearly indicates the effect of a large angle β between the shafts.

It is possible to connect two shafts by two Hooke's couplings and an intermediate shaft such that the uneven velocity ratio of the first coupling will be canceled out by the second. Figure 2.29 shows this application when the two shafts 2 and 4, which are to be connected, do not lie in the same plane. The connection must be made so that the driver and driven shafts 2 and 4 make equal angles β with the intermediate shaft 3. Also, the yokes on shaft 3 must be connected in such a way that when one yoke lies in the plane of shafts 2 and 3, the other yoke lies in the plane of shafts 3 and 4. If the two shafts to be connected lie in the same plane, then the yokes on the intermediate shaft will be parallel.

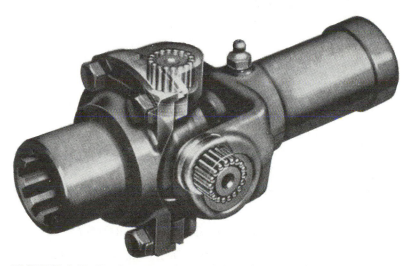

FIGURE 2.27 Hooke-type universal joint. (Courtesy of Mechanics Universal Joint Division, Borg-Warner Corp.)

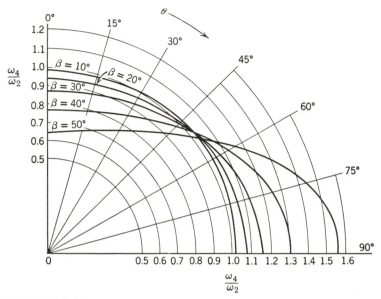

FIGURE 2.28

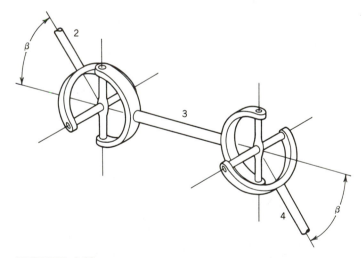

FIGURE 2.29

An application of two universal joints connecting shafts that lie in the same plane is the Hotchkiss automotive drive, which is used on most rear-wheel drive cars.

2.14 CONSTANT-VELOCITY UNIVERSAL JOINTS

Engineers have been considering for many years the development of a single universal joint capable of transmitting a constant-velocity ratio. Several joints that were variations of the Hooke principle were proposed, one as early as 1870, with the intermediate shaft reduced to zero length. As far as is known, however, joints of this design have never been used to any extent commercially.

With the development of the front-wheel drive for automotive vehicles, the need for a universal joint which was capable of transmitting a constant angular-velocity ratio was increased. It was true that two Hooke's couplings and an intermediate shaft could be used, but this was not entirely satisfactory. With a drive such as is required on a front wheel of an automobile, where the angle β is sometimes quite large, the changing conditions made it almost impossible to obtain constant angular-velocity ratio. The need for a constant-velocity universal joint was met by the introduction of the Weiss and the Rzeppa joints in this country and by the Tracta joint in France. The Weiss joint was first patented in 1925, the Rzeppa in 1928, and the Tracta in 1933. Operation of these joints is not based on the same principle as the Hooke coupling.

A *Bendix–Weiss joint* is shown in Fig. 2.30. As shown in the figure, grooves that are symmetrical with respect to each other about the center lines of the shafts are formed in the surfaces of the prongs of the yokes, and four steel balls are located between these prongs at a point where the axes of the grooves in one yoke intersect the axes of the grooves in the other yoke. Power is transmitted from the driver to the follower through these balls. A fifth ball with a slot provides for locking of the parts in assembly as well as for taking end thrust. In operation, the balls will automatically shift their positions as the angular displacement of the two shafts is varied, so that the plane containing the centers of the balls will always bisect the angle between the two shafts. A constant angular-velocity ratio will therefore result from this condition. A photograph of a Bendix–Weiss joint is shown in Fig. 2.33.

A bell-type *Rzeppa joint* (pronounced "sheppa") is shown in Fig. 2.31. The joint consists of a spherical housing and an inner race with corresponding grooves in each part. Six steel balls inserted in these grooves transmit torque from driver to follower. The balls are located in curved grooves in the races and are positioned by a cage between the races. Centers of curvature for the grooved races are offset

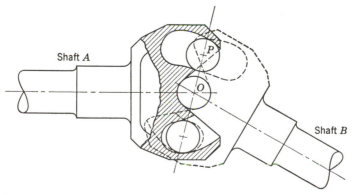

FIGURE 2.30

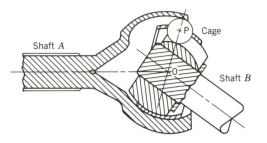

FIGURE 2.31

in opposite directions from the joint center along the shaft axes. The offsets control the positions of the balls so that their centers always lie in a plane which bisects the angles between the shafts. With the centers of the balls in this plane, the joint will transmit a constant angular-velocity ratio. A photograph of a Rzeppa joint is shown in Fig. 2.33.

A *Tracta joint* is shown in Fig. 2.32. It consists of four parts: two shafts with forked ends and two hemispherical parts, one of which has a tongue and the other a groove to receive the tongue. In addition, each of the hemispherical bodies is provided with a groove that permits the connection of a fork. The forks subtend an angle greater than 180° so as to be self-locking when assembled. The tongue and the tongue groove are at right angles to the grooves which admit the forks. By means of the union of the tongue and groove when the joint is assem-

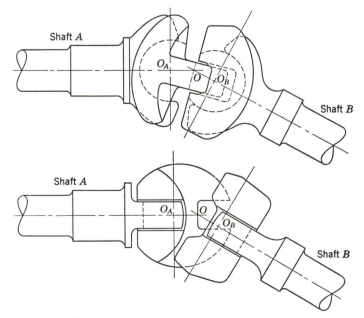

FIGURE 2.32

bled, the axes of the hemispherical parts must always remain in the same plane. When the joint is assembled, the forks are free to rotate about the axes of the hemispherical bodies, which lie in the plane of the tongue and groove.

The joint is held in proper alignment for industrial application by two spherical housings not shown. When assembled, these provide a ball joint type of housing that support the shafts so that their axes will intersect at all times at a point equidistant from the centers of the hemispherical members. With this alignment, the Tracta joint will transmit motion with a constant-velocity ratio. A photograph of a Tracta joint is shown in Fig. 2.33.

In addition to the constant-velocity joints discussed above, another type of joint has been developed known as the *tri-pot joint*. The tri-pot joint has a cylindrical housing with three partially cylindrical, equally spaced axial bores. The axial bores house a spider which has three trunions, with a ball mounted on each trunion. The contact points between the balls and housing bores always lie in a plane which bisects the angle between the two shafts. Constant angular

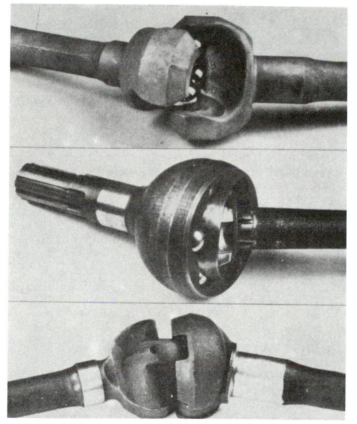

FIGURE 2.33 Constant-velocity universal joints: Bendix–Weiss, Rzeppa, and Tracta.

Rzeppa

Tripot

FIGURE 2.34 Front-wheel drive unit. (Courtesy of Saginaw Steering Gear Division, General Motors Corporation.)

velocity will therefore be transmitted between the input and the output shaft at any angle between the shafts. The spider is usually splined to one shaft, and the housing is bolted to the other shaft.[4]

A tri-pot joint in combination with a Rzeppa joint is used extensively in automotive front-wheel drives where the Rzeppa is used for the outboard and the tri-pot for the inboard joint. A front-wheel drive unit of a Rzeppa and a tri-pot joint is shown in Fig. 2.34. In front-wheel drive vehicles, if the front wheels are designed to have independent suspension, it is necessary to use two constant-velocity joints per axle to allow for the suspension motion and the steering angle of the wheel.

2.15 INTERMITTENT-MOTION MECHANISMS

There are many instances where it is necessary to convert continuous motion into intermittent motion. One of the foremost examples is the indexing of a work table on a machine tool so as to bring a new work piece before the cutters with

---

[4]*Machine Design,* June 28, 1984 (1984 Mechanical Drives Reference Issue), "Universal Joints," pp. 72–75.

each index of the table. There are several ways of accomplishing this type of motion.

Geneva Wheel

This mechanism is very useful in producing intermittent motion because the shock of engagement is minimized. Figure 2.35 shows a sketch where plate 1, which rotates continuously, contains a driving pin P that engages in a slot in the driven member 2. In the sketch, 2 is turned one-quarter revolution for each revolution of plate 1. The slot in member 2 must be tangential to the path of the pin upon engagement in order to reduce shock. This means that angle O_1PO_2 will be a right angle. It can also be seen that angle β is one half of the angle turned through by member 2 during the indexing period. For the case shown, β is 45°.

It is necessary to provide a locking device so that when member 2 is not being indexed, it will not tend to rotate. One of the simplest ways of accomplishing this is to mount a locking plate upon plate 1 whose convex surface will mate with the concave surface of member 2 except during the indexing period. It is necessary to cut the locking plate back to provide clearance for member 2 as it swings through the indexing angle. The clearance arc in the locking plate will be equal to twice the angle α.

If one of the slots in member 2 is closed, then plate 1 can make only a limited number of revolutions before the pin P strikes the closed slot and motion ceases. This modification is known as the Geneva stop and is used in watches and similar devices to prevent overwinding.

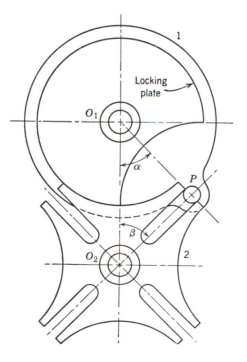

FIGURE 2.35

Ratchet Mechanism

This mechanism is used to produce intermittent circular motion from an oscillating or reciprocating member. Figure 2.36 shows the details. Wheel 4 is given intermittent circular motion by means of arm 2 and driving pawl 3. A second pawl 5 prevents 4 from turning backward when 2 is rotated clockwise in preparation for another stroke. The line of action PN of the driving pawl and tooth must pass between centers O and A as shown in order to have the pawl 3 remain in contact with the tooth. The line of action (not shown) for the locking pawl and tooth must pass between centers O and B. This mechanism has many applications, particularly in counting devices.

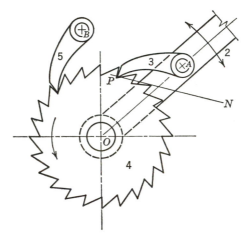

FIGURE 2.36

Intermittent Gearing

This mechanism finds application where the loads are light and shock is of secondary importance. The driving wheel will carry one tooth, and the driven member will carry a number of tooth spaces to produce the required indexing angle. Figure 2.37 shows this arrangement. A locking device must be employed to

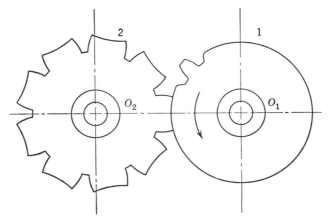

FIGURE 2.37

prevent wheel 2 from rotating when not indexing. One method is shown in the figure; the convex surface of wheel 1 mates with the concave surface between the tooth spaces on member 2.

Escapements

This type of mechanism is one in which a toothed wheel, to which torque is applied, is allowed to rotate in discrete steps by the action of a pendulum. Because of this action, the mechanism can be used as a timing device and as such finds its widest application in clocks and watches. A second application is its use as a governor to control displacement, torque, or velocity.

There are many types of escapements, but the one that is used in watches and clocks because of its high accuracy is the *balance wheel escapement* shown in Fig. 2.38.

The balance wheel and hairspring constitute a torsional pendulum with a fixed period (time of oscillation through one cycle). The escape wheel is driven by a mainspring and gear train (not shown) and has intermittent clockwise rotation as governed by the lever. For every complete oscillation of the balance wheel, the lever allows the escape wheel to advance one tooth. The escape wheel therefore counts the number of times the balance wheel oscillates and also supplies

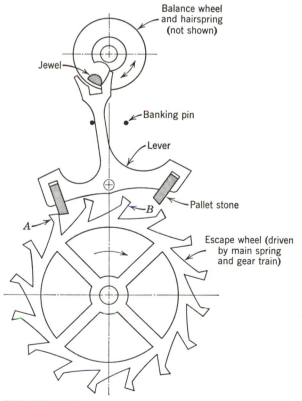

FIGURE 2.38

energy through the lever to the balance wheel to make up for friction and windage losses.

To study the motion of this mechanism through one cycle, consider the lever held against the left banking pin by the escape wheel tooth A acting on the left pallet stone. The balance wheel rotates counterclockwise so that its jewel strikes the lever driving it clockwise. The motion of the lever causes the left pallet stone to slip past and unlock the escape wheel tooth A. The wheel now rotates clockwise, with the top of tooth A giving an impulse to the bottom of the left pallet stone as it slides under it. From this impulse the lever now begins to drive the jewel, thereby giving energy to the balance wheel to maintain its motion.

After the escape wheel rotates a short distance, it comes to rest again as tooth B strikes the right pallet stone, which has been lowered due to rotation of the lever. The lever strikes the right banking pin and stops, but the balance wheel continues to rotate until its energy is overcome by tension in the hairspring, by pivot friction, and by air resistance.

The force of escape wheel tooth B on the right pallet stone keeps the lever locked against the right banking pin. The balance wheel completes its swing, reverses direction, and returns with clockwise motion. The jewel now strikes the left side of the lever notch and drives the lever counterclockwise. This action unlocks tooth B, which gives an impulse to the lever through the right pallet stone. After a short rotation of the escape wheel, it comes to rest again as tooth A strikes the left pallet stone.

The balance wheel escapement is also known as the detached lever escapement because the balance wheel is free and out of contact with the lever during most of its oscillation. Because of this relative freedom of the balance wheel, the escapement has an accuracy of $\pm1\%$.

For more information on escapements and their applications, consult one of the many references on the subject.

2.16 COMPUTING ELEMENTS

At one time, mechanical computing elements found wide application in analog computers for the solution of complicated equations. They were used in the control of guided missiles, weapon fire control, bombsights, and many other systems, both commercial and military. While electronic computing systems have largely replaced mechanical systems, there are many instances where mechanical devices are preferable because they do not require electrical power. With this advantage, mechanical units are particularly suitable for use in hazardous environments and in remote locations such as oil and gas pipeline systems.

Mechnical computing elements, in addition to their ability to generate particular mathematical functions, are also used to produce various types of motions in production machinery. Notable examples of these are integrators, contour cams, noncircular or contour gears, and differentials. These mechanisms have high reliability and long life.

2.17 INTEGRATORS

A mechanism for integration is shown in Fig. 2.39. Disk 2 rotates driving the balls which are positioned by the ball carriage 3. The balls, in turn, drive roller 4. Pure rolling action is maintained between the disk and the balls and between the roller and the balls. The input variables are the rate of rotation of disk 2 and the axial displacement r of the balls. The output of roller 4 is the result. The action of the mechanism therefore gives the relation

$$R\,d\theta_4 = r\,d\theta_2$$

because the linear distance traveled by the top ball on disk 2 must be equal to that traveled by the bottom ball on roller 4. Integrating the preceding equation gives

$$\theta_4 = \frac{1}{R}\int r\,d\theta_2 \tag{2.25}$$

where r is a function of θ_2. The value $1/R$ is the *integrator constant* and is very important in the design of an integrator system. The unit can also be used as a multiplier by taking r as a constant during each operation. The unit will then generate $\theta_4 = (r/R)\theta_2$.

Equation 2.25 can also be expressed in terms of x, y, and z. Let the rotation θ_2 be represented by x, the ball carriage position r by y, which equals $f(x)$, and the output θ_4 by z. Substituting these quantities into Eq. 2.25 gives

$$z = \frac{1}{R}\int y\,dx \tag{2.26}$$

These quantities are shown schematically in Fig. 2.40.

In the integrator, input x and output z are shaft rotations, whereas input y is a linear distance from the ball carriage to the center of the disk. To provide the axial motion necessary for y, a lead screw is often used. By so doing, the

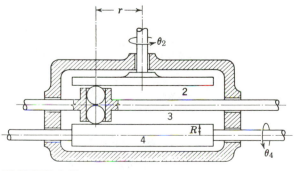

FIGURE 2.39

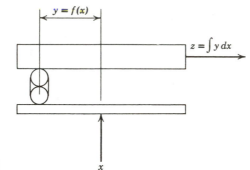

$$z = \int y \, dx$$

FIGURE 2.40

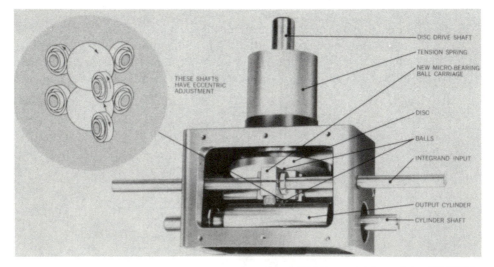

FIGURE 2.41 An integrator. (Courtesy of LIBRASCOPE, a division of the SINGER Company.)

rotation of the screw, which is proportional to the carriage position, can be used to represent *y*. Therefore, the input and the output will all be shaft rotations. A commercial integrator is shown in Fig. 2.41. The theory of contour cams is presented in Chapter 3.

2.18 SYNTHESIS

In the linkages studied in this chapter, the proportions have been given and the problem has been to analyze the motion produced by the linkage. It is quite a different matter, however, to start with a required motion and to try to proportion a mechanism to give this motion. This procedure is known as the *synthesis of mechanisms*.

The following section describes a typical mechanism design problem and illustrates how synthesis can be applied to find a solution. Chapter 11, Introduction To Synthesis, describes the synthesis problem in more general terms and outlines a variety of graphical and analytical methods of solution.

2.19 CASE STUDY IN MECHANISM DESIGN: THE HYDROMINER[5]

The design of mechanisms and their implementation into machines is often a complex iterative process where many related factors must be considered at each stage of design. It would therefore be impossible to condense the details of a real design into a single section of a textbook. It is instructive however to examine a few of the detailed kinematic considerations that go into the design of a machine. The example presented in this section shows that, although the kinematics of a device may be simple, the design of a complete machine is usually quite involved.

The value of coal as a long-term source of energy is well documented. Equally well documented are the dirty and hazardous conditions under which coal is typically mined. In an effort to overcome these problems, the U.S. Department of the Interior commissioned the University of Missouri–Rolla to develop a coal-mining device that used jets of high-pressure water rather than mechanical saw blades to cut the coal. The principle of operation of this mining device, known as the *hydrominer*, is shown in Fig. 2.42. Water jets undercut the coal. A moving wedge is forced into the opening made by the water jets, and large cantilever beam sections of the coal are mechanically broken off and carried away on a conveyor. The water jets are superior to sawing because dust is eliminated and the risk of gas explosions caused by sparks is reduced.

The design of the hydrominer calls for three high-pressure water jets to oscillate vertically along the edge of the plow, as shown in Fig. 2.43. Each spray arm pivots about a high-pressure swivel joint and is required to swing through an arc of 29° and to oscillate at a frequency of up to 200 cycles/s. The nozzles operate at a pressure of 10,000 psi with an exit diameter of 0.004 in., producing

[5]C. R. Barker, "Hydrominer Spray Arm Drive System Design," *Proceedings of the Fifth OSU Applied Mechanisms Conference,* 1977.

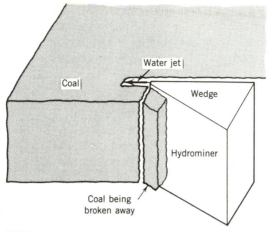

FIGURE 2.42

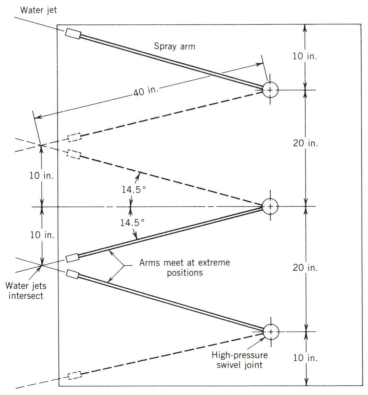

FIGURE 2.43

an exit velocity of over 1000 ft/s for the water jet. This design is able to cut a slot in the coal that is 2 in. wide, 20 in. long, and from 24 to 30 in. ahead of the plow.

In the initial design of the hydrominer, the oscillation of the spray arms was controlled by hydraulic cylinders driven by a combined electric and hydraulic control system. It soon became apparent, however, that this system would be too slow and would result in jerky arm motions that would damage the coherence of the jet. It was also felt that this system would be unreliable when operating in a wet and dirty environment. Consequently, an alternate solution using a four-bar linkage drive system was investigated.

The four-bar linkage is a logical choice because it is simple and reliable. Using a crank rocker-type linkage provides the required oscillatory motion of the output arm while the input link is driven at constant angular velocity. This eliminates the need for a control system to regulate the arm motion. Several restrictions were imposed on the design of the linkage, however. It was required that the motion of the output arms be smooth and continuous so that the coherence of the water jets would not be interrupted. Obviously, the linkage was required to fit within the physical dimensions of the hydrominer and not to interfere with structural members of the device. Finally, to enhance the cutting action at the extreme ends of travel of the spray arms, the middle arm was required to meet the outer arms at its two extreme positions. In other words, the middle arm had to be 180° out of phase with the outside arms.

In designing a driving linkage, several important characteristics of the motion must be recognized. First, the velocity of the spray arm must be zero at the extreme ends of the oscillation cycle. This is easily seen by noting that the angular position of the arm reaches a maximum value at one end and reaches a minimum value at the other end. At these extremes, the velocity (the time derivative of position) must equal zero. Also, it is desirable to make the forward and return strokes of the spray arm occur in approximately the same length of time. Since the input link will rotate at constant angular velocity, this requirement can be satisfied by making the position of the input link at the two extreme positions differ by 180°. A final assumption is that the spray arm itself will function as one link of the mechanism.

A kinematic diagram of the proposed concept (a four-bar linkage) appears in Fig. 2.44. This mechanism is similar to the linkage analyzed in Section 2.3 but with the ground link inclined at an angle θ_1. Writing the loop closure equations gives

x-Components:

$$r_2 \cos \theta_2 + r_3 \cos \theta_3 - r_4 \cos \theta_4 - r_1 \cos \theta_1 = 0 \qquad \textbf{(2.27)}$$

y-Components:

$$r_2 \sin \theta_2 + r_3 \sin \theta_3 - r_4 \sin \theta_4 - r_1 \sin \theta_1 = 0 \qquad \textbf{(2.28)}$$

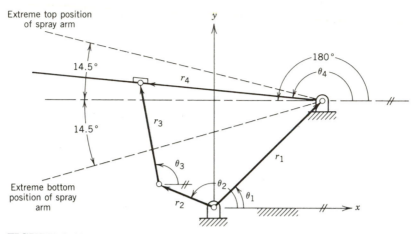

FIGURE 2.44

Taking the time derivatives of these two equations gives the velocity loop closure equations

$$-\omega_2 r_2 \sin \theta_2 - \omega_3 r_3 \sin \theta_3 + \omega_4 r_4 \sin \theta_4 = 0 \qquad (2.29)$$

$$\omega_2 r_2 \cos \theta_2 + \omega_3 r_3 \cos \theta_3 - \omega_4 r_4 \cos \theta_4 = 0 \qquad (2.30)$$

noting that θ_1 is constant.

The following parameters are substituted into these equations at each of the two positions:

Position 1 (extreme top position):

$$\theta_2 = \theta_{21} \qquad \theta_3 = \theta_{31} \qquad \omega_3 = \omega_{31} \qquad \theta_4 = 165.5° \qquad \omega_4 = 0$$

Position 2 (extreme bottom position):

$$\theta_2 = \theta_{21} + 180° \qquad \theta_3 = \theta_{32} \qquad \omega_3 = \omega_{32} \qquad \theta_4 = 194.5° \qquad \omega_4 = 0$$

Substituting these values into the velocity equations (Eqs. 2.29 and 2.30) gives a total of four equations in seven unknowns (θ_{21}, θ_{31}, ω_{31}, θ_{32}, ω_{32}, r_2, and r_3). Dividing all four equations by $r_2\omega_2$ and considering r_3/r_2 and ω_3/ω_2 as one variable each (i.e., using r_2 and ω_2 as scale factors) results in four equations in five unknowns. Solving for θ_{31} and θ_{32} in terms of θ_{21} gives

$$\theta_{31} = \theta_{21} \qquad \text{and} \qquad \theta_{32} = \theta_{21}$$

which means links 2 and 3 are in line in the two positions of interest. This result is important because it shows that the two extreme positions of a crank rocker-type four-bar linkage occur when the input link and the coupler link are collinear.

Using this result and substituting the known values at the two positions into Eqs. 2.27 and 2.28 gives four equations in six unknowns (θ_{21}, θ_1, r_1, r_2, r_3, and r_4). These may be written out as follows:

$$r_2 \cos \theta_{21} + r_3 \cos \theta_{21} - r_4 \cos 165.5° - r_1 \cos \theta_1 = 0 \qquad (2.31)$$

$$r_2 \sin \theta_{21} + r_3 \sin \theta_{21} - r_4 \sin 165.5° - r_1 \sin \theta_1 = 0 \qquad (2.32)$$

$$r_2 \cos (\theta_{21} + 180°) + r_3 \cos \theta_{21} - r_4 \cos 194.5° - r_1 \cos \theta_1 = 0 \qquad (2.33)$$

$$r_2 \sin (\theta_{21} + 180°) + r_3 \sin \theta_{21} - r_4 \sin 194.5° - r_1 \sin \theta_1 = 0 \qquad (2.34)$$

Subtracting Eq. 2.33 from Eq. 2.31 and noting that $\cos (\theta_{21} + 180°) = -\cos \theta_{21}$ gives

$$2r_2 \cos \theta_{21} = 0$$

which means that either $\theta_{21} = 90°$ or $\theta_{21} = -90°$ if $r \neq 0$. Substituting the positive root for θ_{21} into Eqs. 2.32 and 2.34 and adding the two together gives

$$r_1 \sin \theta_1 = r_3 \qquad (2.35)$$

Similarly, subtracting Eq. 2.34 from Eq. 2.32 gives

$$4r_2 = r_4 \qquad (2.36)$$

Finally, substituting this result into Eq. 2.31 gives

$$r_1 \cos \theta_1 = 3.873 r_2 \qquad (2.37)$$

Equations 2.35, 2.36, and 2.37 show that there are two free choices available to the designer. For example, r_3 and r_4 can be selected by the designer to meet other requirements of the system, and r_1, θ_1, and r_2 are calculated from these equations. Figure 2.45 shows a family of six possible solutions to the design equations, all having $r_3 = 10$ in. Figure 2.46 shows the final design of the drive system and the method for achieving the proper phase relation between the middle arm and the outer arms.

Problems

2.1. In the four-bar linkage shown in Fig. 2.1, let $O_2O_4 = 2$ in., $O_2A = 2\frac{1}{2}$ in., $AB = 1\frac{1}{2}$ in., and $O_4B = 1\frac{3}{4}$, $2\frac{3}{4}$, and $3\frac{1}{4}$ in. Sketch the mechanism full size for the three sets of dimensions, and determine for each case whether links 2 and 4 rotate or oscillate. In the case of oscillation, determine the limiting positions.

2.2. In the four-bar linkage shown in Fig. 2.1, link 2 is to rotate completely and link 4 oscillate through an angle of 75°. Link 4 is to be 114 mm long; and when it is at one

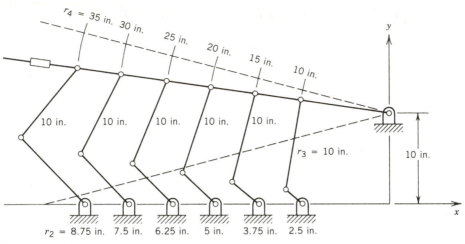

FIGURE 2.45

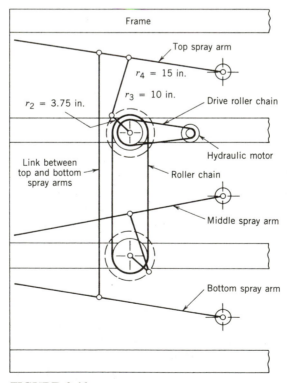

FIGURE 2.46

extreme position, the distance O_2B is to be 102 mm and at the other extreme position 229 mm. Determine the length of links 2 and 3, and draw the mechanism to scale as a check. Determine the maximum and the minimum transmission angles.

2.3. If for the drag-link mechanism shown in Fig. 2.5c, O_2A = 76.2 mm, AB = 102 mm, and O_4B = 127 mm, what can be the maximum length of O_2O_4 for proper operation of the linkage?

2.4. In the four-bar mechanism shown in Fig. 2.47, the guide is part of the fixed link and its centerline is a circular arc of radius R. Draw the mechanism full size and, using graphical construction, determine the magnitude of the angular velocity ω_4 of the slider when the mechanism is in the phase shown and ω_2 is 1 rad/s. Give the sense of ω_4.

2.5. Considering the slider-crank mechanism shown in Fig. 2.10b, derive equations for the displacement, velocity, and acceleration of the slider as a function of R, L, θ, ω, and ϕ. Do not make approximations. Let ω be constant.

2.6. The approximate equation for the displacement of the slider in the slider-crank mechanism is $x = R(1 - \cos\theta) + (R^2/2L)\sin^2\theta$, and $\theta = \omega t$ because ω is constant. Derive the equations for the velocity and acceleration of the slider if ω is not constant.

2.7. Write a computer program to calculate the slider displacement, velocity, and acceleration of the slider crank shown in Fig. 2.10. Use both the exact equations and the approximate equations. Let R = 2 in., L = 8 in., n_2 = 2400 rpm. Calculate displacement, velocity, and acceleration at 10° intervals of θ from 0° to 360°.

2.8. A slider-crank mechanism has a crank length R of 50 mm and operates at 250 rad/s. Calculate the maximum values of velocity and acceleration and determine at what crank angles these maximums occur for connecting rod lengths of 200, 230, and 250 mm. Use approximate equations, and assume ω constant.

2.9. Write a computer program to compare simple harmonic motion of the Scotch yoke (Fig. 2.13) with the motion of the slider crank. Let n = 1800 rpm, R = 2 in., L = 8 in., for the slider crank and r = 2 in. for the Scotch yoke. Vary the angle θ from 0° to 360° (ccw) and calculate displacement, velocity, and acceleration at each value of θ. Use approximate equations for the slider crank, and assume ω constant.

2.10. In the mechanism shown in Fig. 2.48, neglect the connecting-rod effect (assume

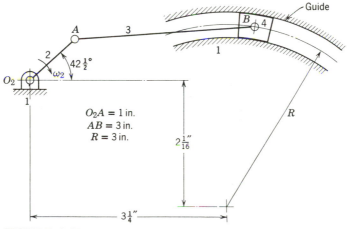

$O_2A = 1$ in.
$AB = 3$ in.
$R = 3$ in.

FIGURE 2.47

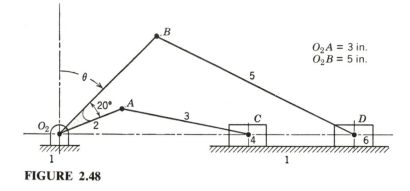

$O_2A = 3$ in.
$O_2B = 5$ in.

FIGURE 2.48

connecting rod infinitely long) and determine an expression for the relative motion of the two sliders. This relation should be a function of time and reduced to a single trigonometric term, with ω constant.

2.11. If link 2 in the Scotch yoke shown in Fig. 2.13a rotates at 100 rpm, determine the maximum velocity and maximum acceleration of link 4 if its stroke is 100 mm.

2.12. Shown in Fig. 2.49 is a modified Scotch yoke mechanism in which the guide of the yoke is a circular arc of radius r and R is the crank radius. Derive an expression for the displacement x of the yoke (link 4) in terms of θ, R, and r. Indicate the displacement x on the sketch.

2.13. Considering the drag-link quick-return mechanism shown in Fig. 2.15, determine the velocity (ft/min) of the slider 6 for a complete revolution of the crank 2 in 45° increments. The crank rotates clockwise at 100 rpm. Use a scale of 4 in. = 12 in., and let $O_2O_4 = 3$ in., $O_2A = 4\frac{1}{2}$ in., $AB = 5\frac{1}{2}$ in., $BC = 8\frac{1}{2}$ in., $O_4B = 6$ in., $O_4C = 6$ in., and $CD = 18\frac{1}{2}$ in. Determine ω_4 graphically using the principle of the transmission of motion, and then calculate the velocity of the slider using the slider-crank equation.

2.14. Using the proportions of the drag-link quick-return mechanism given in Problem

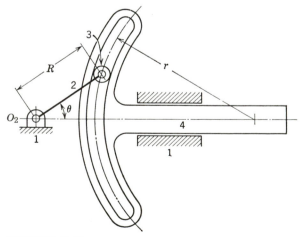

FIGURE 2.49

2.13, graphically determine the length of stroke of slider 6, and the time ratio of advance to return. Use a scale of 3 in. = 12 in.

2.15. For the Whitworth quick-return mechanism shown in Fig. 2.16, graphically determine the length of stroke of slider 6 and the time ratio of advance to return. Use a scale of 3 in. = 12 in., and let $O_2O_4 = 2\frac{1}{2}$ in., $O_2A = 5$ in., $O_4B = 5$ in., and $BC = 18$ in.

2.16. For the crank-shaper mechanism shown in Fig. 2.14, graphically determine the length of stroke and the time ratio of advance to return. Use a scale of 3 in. = 12 in., and let $O_2O_4 = 16$ in., $O_2A = 6$ in., $O_4B = 26$ in., $BC = 12$ in., and the distance from O_4 to the path of $C = 25$ in.

2.17. Design a Whitworth quick-return mechanism to have a length of stroke of 12 in. and a time ratio of 11/7. Use a scale of 3 in. = 12 in.

2.18. Design a crank-shaper mechanism to have a length of stroke of 12 in. and a time ratio of 11/7. Use a scale of 3 in. = 12 in.

2.19. For the quick-return mechanism shown in Fig. 2.50, derive an expression for the displacement x of the slider (link 5) as a function only of θ of the driving link (link 2) and the constant distances shown.

2.20. Shown in Fig. 2.51 is a quick-return mechanism in which link 2 is the driver. Link 5 moves to the right during a working stroke and to the left during a quick-return stroke. Draw the mechanism to full scale and, using graphical construction, determine (*a*) the

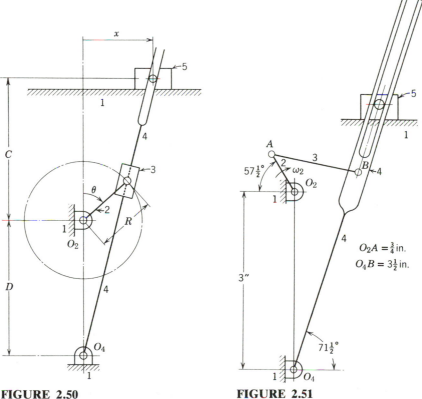

FIGURE 2.50   **FIGURE 2.51**

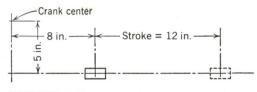

FIGURE 2.52

angular velocity ratio ω_4/ω_2 when the mechanism is in the phase shown and (b) the time ratio of the mechanism.

2.21. Derive the equations for displacement, velocity, and acceleration for the offset slider-crank mechanism shown in Fig. 2.17. They should be in form similar to Eqs. 2.17, 2.18, and 2.19.

2.22. Calculate the length of the crank and of the connecting rod for an offset slider-crank mechanism to satisfy the conditions shown in Fig. 2.52.

2.23. For the offset slider-crank mechanism shown in Fig. 2.53, calculate (a) the length of stroke of slider 4, (b) the distance O_2B when the slider is in its extreme left position, and (c) the time ratio of working stroke to return stroke.

2.24. Referring to Fig. 2.18 and considering only links 4, 5, and 6 of the toggle mechanism shown, write a computer program to illustrate the force development of this mechanism. Let F be a constant value of 10 lb. *Suggestion:* Use Eq. 2.23 and vary α from 10° to near 0°.

2.25. Plot the path of point P in the Watt straight-line mechanism shown in Fig. 2.20 if $O_2A = 2$ in., $O_4B = 3$ in., $AP = 1\frac{1}{2}$ in., $BP = 1$ in., and links 2 and 4 are perpendicular to link 3.

2.26. Referring to Fig. 2.20, graphically determine the proportions of the Watt straight-line mechanism that will give an approximate straight-line motion of point P over a length of 5 in.

2.27. Prove that point P in the Peaucellier straight-line mechanism shown in Fig. 2.21 will trace true straight-line motion.

2.28. Prove that points P and Q in the pantograph shown in Fig. 2.22 will trace similar paths.

2.29. In the pantograph shown in Fig. 2.54, point Q is to trace a 76.0 mm path while P

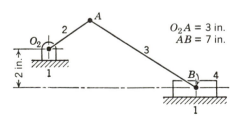

$O_2A = 3$ in.
$AB = 7$ in.

FIGURE 2.53

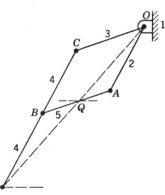

FIGURE 2.54

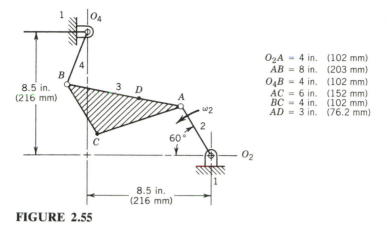

$O_2A = 4$ in. (102 mm)
$AB = 8$ in. (203 mm)
$O_4B = 4$ in. (102 mm)
$AC = 6$ in. (152 mm)
$BC = 4$ in. (102 mm)
$AD = 3$ in. (76.2 mm)

FIGURE 2.55

traces a 203 mm path. If OP is to have a maximum working distance of 394 mm, design a pantograph to give the required motion using a scale of 10 mm = 30 mm. Draw the mechanism in its two extreme positions, and dimension the links.

2.30. For the mechanism shown in Fig. 2.55, determine the angular positions of the output link (link 4) when the input link (link 2) is at an angle of 60°.

2.31. For the linkage of Fig. 2.56, construct a table showing the angles θ_3 and θ_4 as a function of θ_2 for 10° increments of θ_2 from 0° to 360°. Clearly indicate those values of θ_2 for which the mechanism will not assemble.

2.32. Find the range of angular positions for the input link (link 2) and for the output link (link 4) for the four-bar linkage shown in Fig. 2.57.

2.33. For the linkage shown in Fig. 2.58, find the angular positions of link 2 when link 4 is in the position shown. Be sure to consider both closures of the mechanism.

2.34. Determine the angular velocity of the crank of the slider-crank mechanism shown in Fig. 2.59.

2.35. For the linkage in Fig. 2.60, determine the values of θ_4 and γ when (a) $\theta_2 = -30°$; (b) $\theta_2 = 0°$; (c) $\theta_2 = 30°$.

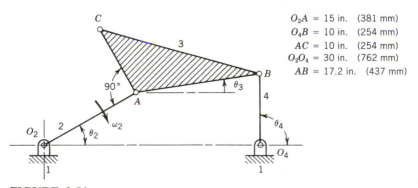

$O_2A = 15$ in. (381 mm)
$O_4B = 10$ in. (254 mm)
$AC = 10$ in. (254 mm)
$O_2O_4 = 30$ in. (762 mm)
$AB = 17.2$ in. (437 mm)

FIGURE 2.56

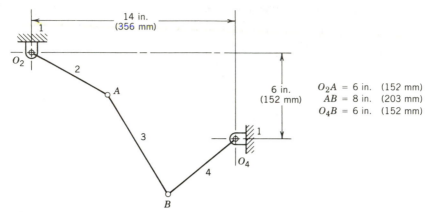

$$O_2A = 6 \text{ in.} \quad (152 \text{ mm})$$
$$AB = 8 \text{ in.} \quad (203 \text{ mm})$$
$$O_4B = 6 \text{ in.} \quad (152 \text{ mm})$$

FIGURE 2.57

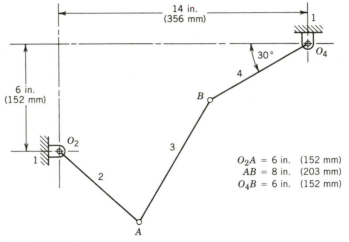

$$O_2A = 6 \text{ in.} \quad (152 \text{ mm})$$
$$AB = 8 \text{ in.} \quad (203 \text{ mm})$$
$$O_4B = 6 \text{ in.} \quad (152 \text{ mm})$$

FIGURE 2.58

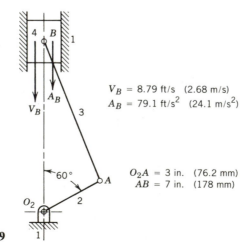

$$V_B = 8.79 \text{ ft/s} \quad (2.68 \text{ m/s})$$
$$A_B = 79.1 \text{ ft/s}^2 \quad (24.1 \text{ m/s}^2)$$

$$O_2A = 3 \text{ in.} \quad (76.2 \text{ mm})$$
$$AB = 7 \text{ in.} \quad (178 \text{ mm})$$

FIGURE 2.59

68

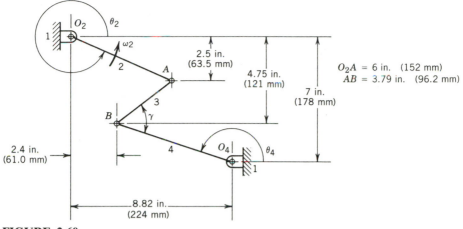

O_2A = 6 in. (152 mm)
AB = 3.79 in. (96.2 mm)

FIGURE 2.60

2.36. For the combined linkage mechanism in Fig. 2.61, determine the maximum and minimum angular positions for the output link (link 6) during full rotation of the crank (link 2). Also determine the angular positions of the crank when the output link is at its extreme positions.

2.37. A Hooke's coupling connects two shafts at an angle of 135° (β = 45°), as shown in Fig. 2.26. If the angular velocity of the drive shaft is constant at 100 rpm, calculate the maximum and minimum velocity of the driven shaft.

2.38. Derive equations that describe the angular displacement and angular velocity of the driven member of a Geneva mechanism (Fig. 2.35) from the point where the driving pin engages the driven wheel to the point of disengagement. Find $\beta = f(\alpha)$ and $d\beta/d\alpha = f(\alpha)$, and use $(d\beta/d\alpha)(d\alpha/d\tau) = d\beta/d\tau$ to determine an equation for the angular velocity of the driven member.

2.39. Using the equations derived in Problem 2.38, write a computer program and calculate the values of β and ω_2 for α varying from 60° to 0° in increments of 10°. Let α at the point of first contact = 60°, O_1P = $1\frac{3}{4}$ in., O_1O_2 = $3\frac{1}{2}$ in., n_1 = 1000 rpm (constant).

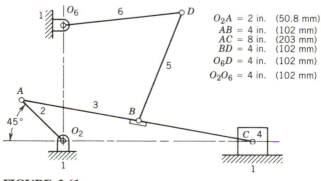

O_2A = 2 in. (50.8 mm)
AB = 4 in. (102 mm)
AC = 8 in. (203 mm)
BD = 4 in. (102 mm)
O_6D = 4 in. (102 mm)
O_2O_6 = 4 in. (102 mm)

FIGURE 2.61

2.40. Lay out a Geneva wheel mechanism to satisfy the following conditions: the driver is to rotate continuously while the driven member rotates intermittently, making a quarter revolution for every revolution of the driver. The distance between the centers of the driving and driven shafts is to be $3\frac{1}{2}$ in. Let the diameter of the driving pin be $\frac{3}{8}$ in. The diameters of the driving and driven shafts are to be $\frac{5}{8}$ in. and 1 in., with keyway for a $\frac{3}{16} \times \frac{3}{16}$ in. and $\frac{1}{4} \times \frac{1}{4}$ in. key, respectively. Show a hub on each member with the hub on the driver shown in back of the plate. Let the diameters of the hubs be $1\frac{3}{4}$ to 2 times the diameters of the bores. Dimension the angles α and β.

Chapter Three

Cams

Cams play a very important part in modern machinery and are extensively used in internal-combustion engines, machine tools, mechanical computers, instruments, and many other applications. A cam may be designed in two ways: (*a*) to assume the required motion for the follower and to design the cam to give this motion, or (*b*) to assume the shape of the cam and to determine what characteristics of displacement, velocity, and acceleration this contour will give.

The first method is a good example of synthesis. In fact, designing a cam mechanism from the desired motion is an application of synthesis that can be solved every time. However, after the cam is designed, it may be difficult to manufacture. This difficulty of manufacture is eliminated in the second method by making the cam symmetrical and by using shapes for the cam contours that can be generated. This is the type of cam that is used in automotive applications where the cam must be produced accurately and cheaply.

Only the design of cams with specified motion will be treated. For the automotive-type cam where the contour is specified, the reader is referred to the reference below.[1] Cams with specified motion can be designed graphically and in certain cases analytically. Graphical procedures will be considered first.

3.1 CAM CLASSIFICATION AND NOMENCLATURE

Grubler's mobility equation (Eq. 1.3) can be used to create a countless variety of mechanisms containing cam pairs. In practice, however, the majority of cam pairs are found in simple cam-and-follower mechanisms which contain only three

---

[1]H. A. Rothbart, *Cams,* John Wiley & Sons, New York, 1956.

links, the two links of the cam pair and a ground link. The material in this chapter deals only with cam-and-follower systems having three links; these are often referred to simply as cam mechanisms. Cam mechanisms may be classified by the type of cam or by the shape, motion, or location of the follower. The simplest and most often-used cam mechanism is a rotating-disk cam with either a reciprocating or oscillating follower. Several other types of cams are in common use, and these are discussed later in this chapter. Figure 3.1 shows a disk cam with six different follower arrangements.

Figure 3.1a shows a disk cam with an in-line knife edge follower. The follower is considered to be *in-line* (or *radial*) when its centerline passes through the center of cam rotation. This type of follower is of theoretical interest but is not of great practical importance because it generally produces high contact stresses. Figure 3.1b shows a disk cam with an in-line roller follower. Figure 3.1c shows a disk cam with an offset roller follower. In each of the cam-and-follower mechanisms of Figs. 3.1a, b, and c, the cam rotates while the follower reciprocates. Figure 3.1d shows a disk cam with an oscillating roller follower. Figure 3.1e shows a disk cam with a reciprocating flat-faced follower. In this case, there is no need to distinguish between in-line and offset followers because they are kinematically equivalent; any follower shaft parallel to the one shown will produce the same output motion. However, it may be necessary to change the length of the follower face when the follower is offset. Figure 3.1f shows a disk cam with

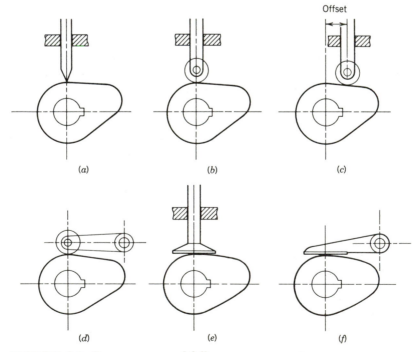

FIGURE 3.1 Common cam-and-follower arrangements.

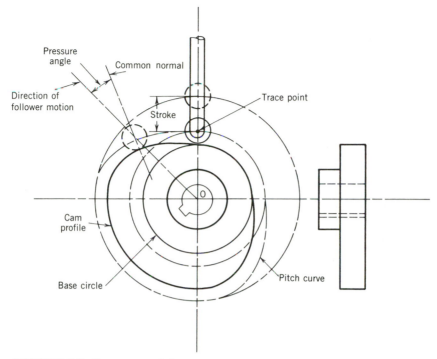

FIGURE 3.2 Cam nomenclature.

an oscillating flat-faced follower. Many other possible cam-and-follower arrangements are described by Chen.[2]

Figure 3.2 shows the nomenclature used to describe a typical cam mechanism. The *trace point* is a point on the follower that corresponds to the contact point of a fictitious knife edge follower. The trace point of a roller follower is the center of the roller. The *pitch curve* is the path of the trace point relative to the cam. The *base circle* is the smallest circle tangent to the cam surface about the center of cam rotation. The *pressure angle* is the angle between the direction of motion of the trace point and the common normal (the *line of action*) to the contacting surfaces. The pressure angle is a measure of the instantaneous force transmission properties of the mechanism. The *throw*, or *stroke*, is the distance between the two extreme positions of the follower.

3.2 DISK CAM WITH RADIAL FOLLOWER (GRAPHICAL DESIGN)

Figure 3.3 shows a disk cam with a radial flat-faced follower. As the cam rotates at a constant angular velocity in the direction shown, the follower moves upward a distance of 1 in. with the displacements shown in half a revolution of the cam.

---

[2]F. Y. Chen, *Mechanics and Design of Cam Mechanisms*, Pergamon Press, New York, 1982.

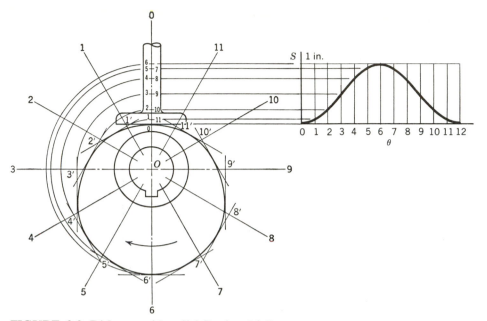

FIGURE 3.3 Disk cam with radial flat-faced follower.

The return motion is to be the same. To determine the cam contour graphically, it will be necessary to invert the mechanism and to hold the cam stationary while the follower moves around it. This will not affect the relative motion between the cam and the follower, and the procedure is as follows:

1. Rotate the follower about the center of the cam in a direction opposite to that of the proposed cam rotation.
2. Move the follower radially outward the correct amount for each division of rotation.
3. Draw the cam outline tangent to the polygon that is formed by the various positions of the follower face.

Unfortunately, in the last step, there is no graphical way of determining the contact point between the cam and the follower, and it must be determined by eye with the use of a French curve. The length of the follower face must also be determined by trial. Occasionally, a combination of displacement scale and minimum radius of cam is selected that gives a cam profile with a sharp corner or cusp. This cusp can be eliminated by modifying the displacement scale or by increasing the minimum radius of the cam.

Figure 3.4a shows the same type of cam with a roller follower. With this type of follower, the center of the roller will move with the prescribed motion. The principles of construction are the same as for the flat-faced follower with the exception that the cam is drawn tangent to the various positions of the roller follower. From Fig. 3.4a, it can also be seen that the line of action from the cam

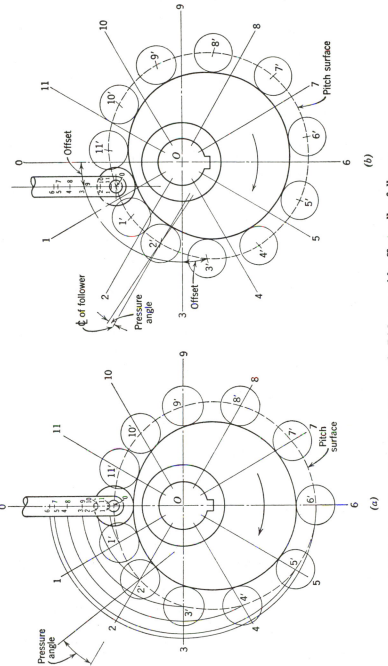

FIGURE 3.4 (*a*) Disk cam with radial roller follower. (*b*) Disk cam with offset roller follower.

to the follower cannot be along the axis of the follower except when the follower is dwelling (no motion up or down). This produces a side thrust on the follower and may result in deflection and jamming of the follower stem. The maximum value of the pressure angle, the angle between the line of action and the center line of the follower, must be kept as small as possible, especially in light mechanisms. In most cases, the pressure angle should not exceed approximately 30° for acceptable operation. Although it is possible to measure the maximum pressure angle from the graphical construction of a cam, it is often difficult to determine this maximum analytically. For this reason, a nomogram for finding maximum pressure angles is given in a later section on analytical cam design. The pressure angle for any radial flat-faced follower is a constant. For the follower shown in Fig. 3.3, where the follower face is at right angles to the stem, the pressure angle is zero so that the side thrust on the follower is negligible compared to that on a roller follower. Pressure angles may be reduced by increasing the minimum radius of the cam so that the follower travels a greater linear distance on the cam for a given rise. This is analogous to increasing the length of an inclined plane for a given rise in order to decrease the angle of ascent. Also, in a cam with a roller follower, the radius of curvature of the pitch surface must be larger than the radius of the roller; otherwise, the cam profile will become pointed.

In both the flat-faced and the roller follower, the follower stems are sometimes offset instead of being radial as shown in Figs. 3.3 and 3.4a. This may be done for structural reasons or, in the case of the roller follower, for the purpose of reducing the pressure angle on the upward stroke. It should be noted, however, that although the pressure angle on the upward stroke is reduced, the pressure angle on the downward stroke is increased. Figure 3.4b shows a cam designed with the follower offset and with the same displacement scale and minimum cam radius as in Fig. 3.4a.

3.3 DISK CAM WITH OSCILLATING FOLLOWER (GRAPHICAL DESIGN)

Figure 3.5 shows a disk cam with an oscillating flat-faced follower. Using the same principle of construction as in the disk cam with a radial follower, the follower is rotated about the cam. At the same time the follower must be rotated about its own center through the required displacement angle for each position. There are several ways of rotating the follower about its own center. The method shown in Fig. 3.5 is to use the intersection of two radii (for example, point 3′) to determine a point on the rotated position of the follower face. The first of these two radii (cam center to position 3 on displacement scale) is swung from the cam center. The second radius (follower center to displacement scale) is swung from the follower center which has been rotated into position 3. The intersection of these two radii gives point 3′. Because an infinite number of lines can be drawn through point 3′, it is necessary to have additional information to locate the correct position of the follower face through point 3′. As shown in the figure, this was supplied by a circle tangent to the follower face which has been

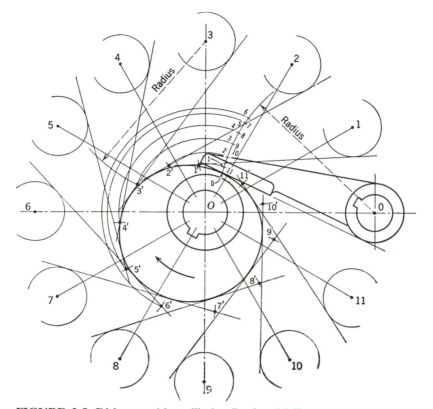

FIGURE 3.5 Disk cam with oscillating flat-faced follower.

extended in the zero position. For the follower design shown, this circle happens to coincide with the outside of the follower hub. The radius of this circle is then swung from each of the rotated positions of the follower center. For position 3, the follower face is drawn through point 3' tangent to the rotated circle of the follower hub. By repeating this process, the polygon of the various positions of the follower face is obtained from which the cam is drawn.

Figure 3.6 shows a disk cam with oscillating roller follower. The procedure for determining the points labeled with primes (for example, point 3') is similar to that of Fig. 3.5. In this case, however, the primed points are the centers of the rotated roller follower. After these roller circles are drawn, the cam can now be drawn tangent to them. It should be noted that in an actual design, smaller cam divisions would be used so that the circles would intersect each other to minimize cam contour error. It should also be mentioned that the same procedure can be used in designing a cam with an oscillating roller follower as was used for a cam with an offset translating follower.

Although most of the cams in use are of the types mentioned, there are many others, some of which find wide application. Three of these are discussed in the following sections.

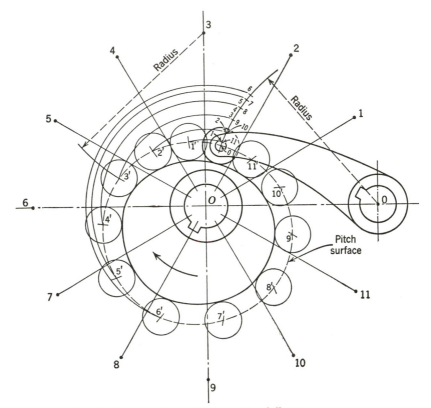

FIGURE 3.6 Disk cam with oscillating roller follower.

3.4 POSITIVE-RETURN CAM (GRAPHICAL DESIGN)

With a disk cam and radial follower, it is often necessary to return the follower in a positive manner and not by the action of gravity or of a spring. Figure 3.7 shows a cam of this type where the cam positively controls the motion of the follower not only during the outward motion but also on the return stroke. Of necessity, the return motion must be the same as the outward motion, but in the opposite direction. This cam is also spoken of as a constant-breadth cam.

This type of cam may also be designed using two roller followers instead of the flat-faced followers. If it is necessary to have the return motion independent of the outward motion, two disks must be used, one for the outward motion and one for the return motion. These double-disk cams can be used with either roller or flat-faced followers.

3.5 CYLINDER CAM (GRAPHICAL DESIGN)

This type of cam finds many applications, particularly on machine tools. Perhaps the most common example, however, is on level-winding fishing reels. Figure

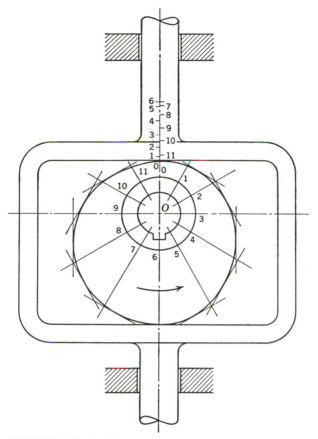

FIGURE 3.7 Positive-return cam.

3.8 shows a sketch where the cylinder rotates completely about its axis imparting motion to a follower which is guided by a groove in the cylinder.

3.6 INVERSE CAM (GRAPHICAL DESIGN)

Occasionally, it is advantageous to reverse the role of the cam and the follower and to let the follower drive the cam. This inversion finds application in sewing machines and other mechanisms of similar nature. Figure 3.9 shows a sketch of a plate cam where the arm oscillates, causing reciprocation of the block by action of a roller in the cam groove.

3.7 CAM DISPLACEMENT CURVES

Before a cam contour can be determined, it is necessary to select the motion with which the follower will move in accordance with the requirements of the system. If operation is to be at slow speed, the motion may be any one of several

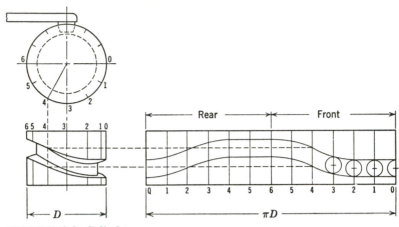

FIGURE 3.8 Cylinder cam.

common motions, for example, parabolic (constant acceleration and deceleration), parabolic with constant velocity, simple harmonic, or cycloidal.

Parabolic motion has the lowest theoretical acceleration for a given rise and cam speed for the motions listed, and for this reason it has been used for many cam profiles. However, in slow-speed work, this has little significance. Parabolic motion may or may not have equal intervals of acceleration and deceleration, depending on requirements. Parabolic motion may also be modified to include an interval of constant velocity between the acceleration and deceleration; this is often spoken of as *modified constant velocity*.

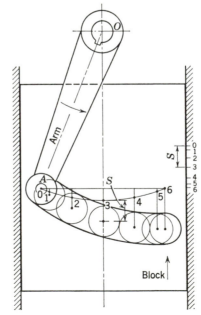

FIGURE 3.9 Inverse cam.

Simple harmonic motion has the advantage that, with a radial roller follower, the maximum pressure angle will be smaller than with parabolic motion with equal time intervals or with cycloidal motion. This will allow the follower to be less rigidly supported and more overhung in its construction. Less power will also be needed to operate the cam.

After the follower motion has been selected, it is necessary to determine the displacement scale and to mark it off on the follower axis as shown in Fig. 3.3. The scale increments may be calculated, but they are more easily determined graphically by plotting a displacement–time graph.

In plotting the displacement–time graph, it is necessary to first determine the inflection point if the motion is parabolic or a modification thereof. For simple harmonic and cycloidal motion, the inflection point is automatically determined by the method of generation of the curve. The inflection point for parabolic motion will be at the midpoint of the displacement scale and of the time scale if the intervals are equal. To find the inflection points where the parabolic motion has been modified is a little more complicated, as shown below.

Consider a point moving with modified constant velocity where it starts from rest with constantly accelerated motion, next has constant velocity, and finally comes to rest with constantly decelerated motion. The inflection points may be found by specifying the time intervals or the displacement intervals corresponding to each type of motion. Figure 3.10 shows a graphical means for finding the inflection points A and B when the time intervals are given. Figure 3.11 shows the construction for displacement intervals. From the relations $S = \frac{1}{2}At^2$, $V = At$, and $S = Vt$, it is possible to prove the validity of the construction shown in Figs. 3.10 and 3.11.

After the inflection points have been determined as, for example, in Fig. 3.11, the constantly accelerated portion $0A$ of the displacement curve is constructed as shown in Fig. 3.12, where the displacement L (corresponding to S_1 of Fig. 3.11) is divided into the same number of parts as is the time scale, in this case four parts. The deceleration portion BC of the curve in Fig. 3.11 is constructed in a similar manner for its particular displacement S_3 and corresponding time interval.

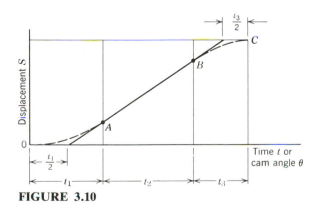

FIGURE 3.10

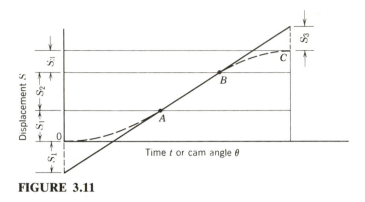

FIGURE 3.11

Figure 3.13 shows simple harmonic motion $[S = r(1 - \cos \omega_r t)]$ for a displacement L in six time intervals. In Fig. 3.13, it should be noted that if the cam rotates through half a revolution while the follower moves through the displacement L, the angular velocity ω_r of the rotational radius r equals the angular velocity ω of the cam, and the equation for follower displacement can be written as $S = r(1 - \cos \omega t) = r(1 - \cos \theta)$. If the cam rotates only a quarter revolution for the displacement L, $\omega_r = 2\omega$ and $S = r(1 - \cos 2\theta)$. Therefore, it can be seen that the relation between ω_r and ω can be expressed as

$$\frac{\omega_r}{\omega} = \frac{180°}{\text{degrees of cam rotation for follower rise } L}$$

A circular cam (eccentric) will impart simple harmonic motion to a radial flat-faced follower because the contact point between the cam and follower is always over the geometrical center of the cam.

Figure 3.14 shows the construction for cycloidal motion

$$S = L\left(\frac{\theta}{\beta} - \frac{1}{2\pi} \sin 2\pi \frac{\theta}{\beta}\right)$$

for a displacement L in six time intervals. The radius of the construction circle

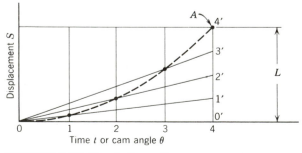

FIGURE 3.12 Parabolic motion.

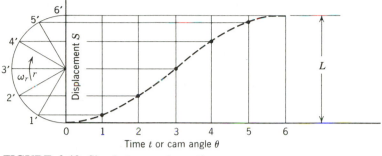

FIGURE 3.13 Simple harmonic motion.

is $L/2\pi$. The circumference of this circle is divided into the same number of parts as is the time scale, in this case six parts. The six marks on the circumference are projected horizontally onto the vertical diameter of the circle. The marks on the vertical diameter are then projected parallel to $0A$ to the corresponding line on the time axis.

For cams operating at higher speeds, the selection of the motion of the cam follower must be based not only on the displacement but also on the forces acting on the system as a result of the motion selected. For many years, cam design was concerned only with moving a follower through a given distance in a certain length of time. Speeds were low so that accelerating forces were unimportant. With the trend toward higher machine speeds, however, it has become necessary to consider the dynamic characteristics of the system and to select a cam contour that will minimize the dynamic loading and prevent separation of cam and follower; this topic is presented in Section 9.20.

As an example of the importance of dynamic loading, consider parabolic motion. On the basis of inertia forces, this motion would seem to be very desirable because of its low acceleration. However, the acceleration increases from zero

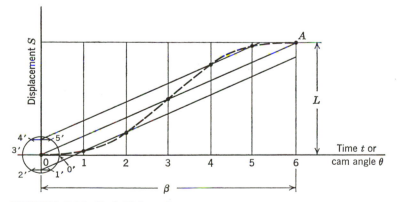

FIGURE 3.14 Cycloidal motion.

to its constant value almost instantaneously, which results in a high rate of application of load. The rate of change of acceleration is determined by the third derivative of the displacement and has been given the name "jerk." Therefore, jerk is an indication of the impact characteristic of the loading; it may be said that impact has jerk equal to infinity. Lack of rigidity and backlash in the system also tend to increase the effect of impact loading. In parabolic motion where jerk is infinite, this impact occurs twice during the cycle and has the effect of a sharp blow on the system, which may set up undesirable vibrations as well as causing structural damage.

As a means of avoiding infinite jerk and its deleterious effect on the cam train, a system of cam design has been developed by Kloomok and Muffley that utilizes three analytic functions: (*a*) cycloid (and half-cycloid), (*b*) harmonic (and half-harmonic), and (*c*) eighth-power polynomial. Plots of the displacement, velocity, and acceleration curves of these functions are given in Figs. 3.15, 3.16, and 3.17. The curves have continuous derivatives at all intermediate points. Therefore, acceleration changes gradually, and jerk is finite. Infinite jerk is avoided at the ends by matching accelerations. It should be noted that the velocities will also match because no discontinuities can appear in the displacement time curve. As an example, when a rise follows a dwell, the zero acceleration at the end of the dwell is matched by selecting a curve having zero acceleration at the start of the rise. The acceleration required at the end of the rise is determined by the succeeding condition. If a fall follows immediately, the rise can end in a fairly high value of deceleration because this can be matched precisely by a curve having the same deceleration for the start of the fall.

The selection of profiles to suit particular requirements is made according to the following criteria:

1. The cycloid provides zero acceleration at both ends of the action. Therefore, it can be coupled to a dwell at each end. Because the pressure angle is relatively high and the acceleration returns to zero unnecessarily, two cycloids should not be coupled together.

2. The harmonic provides the lowest peak acceleration and pressure angle of the three curves. Therefore, it is preferred when the acceleration at both start and finish can be matched to the end acceleration of the adjacent profiles. Because acceleration at the midpoint is zero, the half-harmonic can often be used where a constant-velocity rise follows an acceleration. However, a dwell cannot be inserted in the motion between H-5 and H-6 of Fig. 3.16 because jerk becomes infinite. The half-harmonic can also be coupled to a half-cycloid or to a half-polynomial.

3. The eighth-power polynomial has a nonsymmetrical acceleration curve and provides a peak acceleration and pressure angle intermediate between the harmonic and the cycloid.

In Figs. 3.15, 3.16, and 3.17, the units of velocity and acceleration are given in *inches per degree* and in *inches per degree squared*. The unit of degrees instead

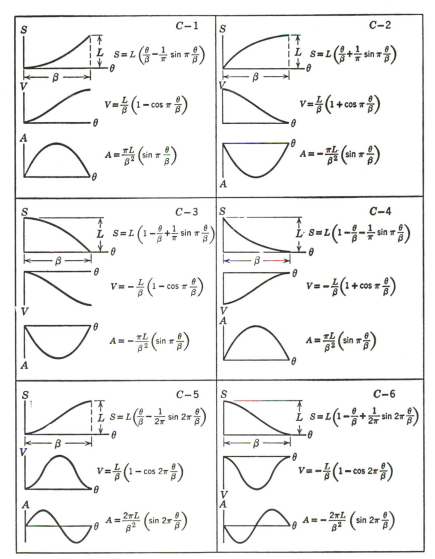

FIGURE 3.15 Cycloidal motion characteristics: S = displacement, inches; V = velocity, inches per degree; A = acceleration, inches per degree squared. **(M. Kloomok and R. V. Muffley, "Plate Cam Design—with Emphasis on Dynamic Effects," *Prod. Eng.*, February 1955.)** *N. B.* **For SI units,** S = displacement, millimeters; V = velocity, millimeters per degree; A = acceleration, millimeters per degree squared.

of *seconds* was selected to make it unnecessary to consider the angular velocity of the cam until the follower motions are selected. To obtain velocity and acceleration in terms of time, the velocity from the curves V (in./deg) can easily be converted to in./s by the relation V (in./s) = $(180/\pi)\,\omega V$ (in./deg), where ω is cam velocity (rad/s). In a similar manner, A (in./s^2) = $(180/\pi)^2\omega^2 A$ (in./deg^2).

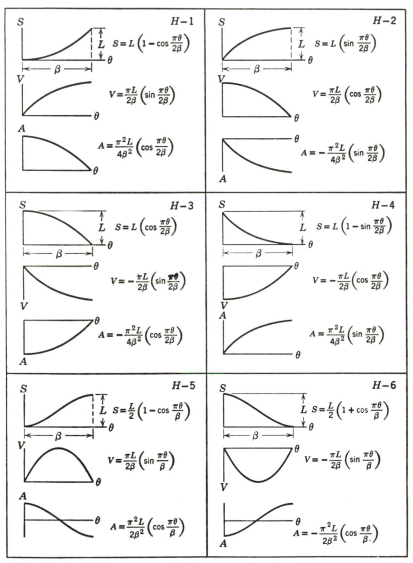

FIGURE 3.16 Harmonic motion characteristics: S = displacement, inches; V = velocity, inches per degree; A = acceleration, inches per degree squared. **(M. Kloomok and R. V. Muffley, "Plate Cam Design—with Emphasis on Dynamic Effects,"** *Prod. Eng.,* **February 1955.)** *N. B.* **For SI units,** S = displacement, millimeters; V = velocity, millimeters per degree; A = acceleration, millimeters per degree squared.

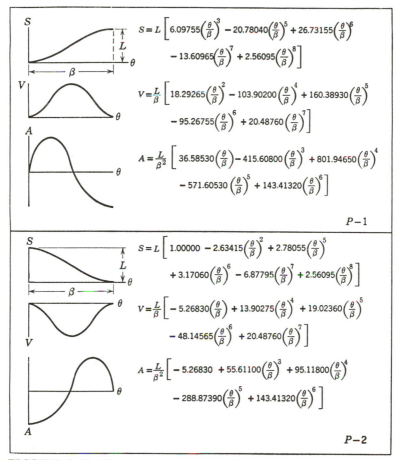

$$S=L\left[6.09755\left(\frac{\theta}{\beta}\right)^3 - 20.78040\left(\frac{\theta}{\beta}\right)^5 + 26.73155\left(\frac{\theta}{\beta}\right)^6\right.$$
$$\left. - 13.60965\left(\frac{\theta}{\beta}\right)^7 + 2.56095\left(\frac{\theta}{\beta}\right)^8\right]$$

$$V=\frac{L}{\beta}\left[18.29265\left(\frac{\theta}{\beta}\right)^2 - 103.90200\left(\frac{\theta}{\beta}\right)^4 + 160.38930\left(\frac{\theta}{\beta}\right)^5\right.$$
$$\left. - 95.26755\left(\frac{\theta}{\beta}\right)^6 + 20.48760\left(\frac{\theta}{\beta}\right)^7\right]$$

$$A=\frac{L}{\beta^2}\left[36.58530\left(\frac{\theta}{\beta}\right) - 415.60800\left(\frac{\theta}{\beta}\right)^3 + 801.94650\left(\frac{\theta}{\beta}\right)^4\right.$$
$$\left. - 571.60530\left(\frac{\theta}{\beta}\right)^5 + 143.41320\left(\frac{\theta}{\beta}\right)^6\right]$$

P–1

$$S=L\left[1.00000 - 2.63415\left(\frac{\theta}{\beta}\right)^2 + 2.78055\left(\frac{\theta}{\beta}\right)^5\right.$$
$$\left. + 3.17060\left(\frac{\theta}{\beta}\right)^6 - 6.87795\left(\frac{\theta}{\beta}\right)^7 + 2.56095\left(\frac{\theta}{\beta}\right)^8\right]$$

$$V=\frac{L}{\beta}\left[-5.26830\left(\frac{\theta}{\beta}\right) + 13.90275\left(\frac{\theta}{\beta}\right)^4 + 19.02360\left(\frac{\theta}{\beta}\right)^5\right.$$
$$\left. - 48.14565\left(\frac{\theta}{\beta}\right)^6 + 20.48760\left(\frac{\theta}{\beta}\right)^7\right]$$

$$A=\frac{L}{\beta^2}\left[-5.26830 + 55.61100\left(\frac{\theta}{\beta}\right)^3 + 95.11800\left(\frac{\theta}{\beta}\right)^4\right.$$
$$\left. - 288.87390\left(\frac{\theta}{\beta}\right)^5 + 143.41320\left(\frac{\theta}{\beta}\right)^6\right]$$

P–2

FIGURE 3.17 Eighth-power polynomial motion characteristics: S = displacement, inches; V = velocity, inches per degree; A = acceleration, inches per degree squared. (M. Kloomok and R. V. Muffley, "Plate Cam Design—with Emphasis on Dynamic Effects," *Prod. Eng.*, February 1955.) *N. B.* For SI units, S = displacements, millimeters; V = velocity, millimeters per degree; A = acceleration, millimeters per degree squared.

Example 3.1. A roller follower is to move through a total displacement and return with no dwells in the cycle. Because of the operation performed by the mechanism, a portion of the outward motion must be at constant velocity. Determine the motion curves to be used. Refer to Fig. 3.18a.

AB: Use half-cycloid C-1 to provide zero acceleration at start of motion A and at B where connection is made to constant velocity portion of curve.

BC: Constant velocity.

CD: Use half-harmonic H-2, which will couple at C to the constant velocity section with zero acceleration and provide minimum pressure angle over the rest of the curve.

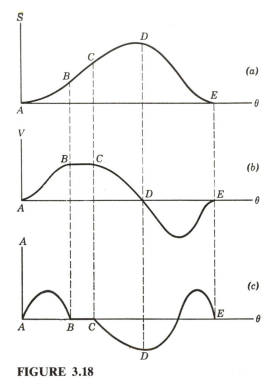

FIGURE 3.18

DE: Use polynomial *P*-2 to match the deceleration of the harmonic at *D* and to provide a zero acceleration juncture at the end of the cycle at *E*.

The velocities and accelerations are matched, and their curves are shown in Figs. 3.18*b* and 3.18*c*. From Fig. 3.18*c*, it can be seen that jerk is finite throughout the cycle.

3.8 CAM DISPLACEMENT CURVES —ADVANCED METHODS

The creation of smooth cam profiles without discontinuities in velocity, acceleration, and higher derivatives is critical to the satisfactory operation of all cams. In the previous section it was shown how segments of simple curves such as cycloidal, harmonic, and polynomial can be pieced together to give continuous acceleration curves. In addition, the importance of minimizing peak values of acceleration, and hence minimizing peak dynamic loads, was discussed. In some high-speed applications, the methods described in the previous section may not be sufficient, however. A continuous acceleration curve may have a discontinuous jerk curve, as shown in Fig. 3.19. Such discontinuities tend to induce vibrations, which may result in noise, wear, and reduced precision of operation. Discontinuities in higher derivatives may also produce undesirable effects.

A wide variety of methods have been proposed to address the above-men-

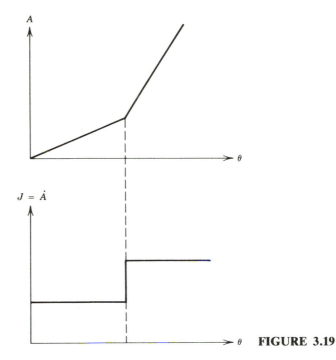

FIGURE 3.19

tioned problems. A logical extension of the methods examined in the previous section is the use of higher-power algebraic polynomials. These polynomial curves are quite versatile, and, in most cases, determination of coefficients is straight-forward. The major disadvantage of the polynomial curves is that they do not directly allow local control of the motion. Another logical extension of the material in the previous section is to create other types of composite curves. Perhaps the simplest of these is the so-called trapezoidal acceleration curve shown in Fig. 3.20. This curve produces smooth velocity and displacement curves and results in finite jerk. Trapezoidal segments can be used in composite with other types of curves to produce smooth cam profiles with good acceleration properties.[3] As a further extension of this approach, it is possible to construct a jerk curve composed of straight-line segments. This produces continuous jerk and smooth curves for acceleration, velocity, and displacement.

With the advent of the programmable calculator and the digital computer, a number of numerical methods have been developed for creating and modifying cam motion curves. In most cases, this is actually a process of smoothing out unacceptable portions of acceleration curves created by other methods. A finite number of points on the original acceleration curve are selected, and an approximate curve is fitted through these points. This new curve is then modified

---

[3]D. Tesar and G. K. Matthew, *The Dynamic Synthesis, Analysis, and Design of Modeled Cam Systems,* Lexington Books, Lexington, Mass., 1976.

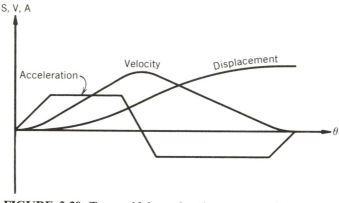

FIGURE 3.20 Trapezoidal acceleration curve and computer-generated plots of the corresponding velocity and displacement curves.

by moving points or adding points through which it must pass. Popular methods of this type include finite differences,[4] Johnson's method,[4] the finite-integration method,[4] and the B-spline technique.[5]

3.9 DISK CAM WITH RADIAL FLAT-FACED FOLLOWER (ANALYTICAL DESIGN)

The treatment of the flat-faced follower allows the actual cam outline to be determined analytically. In the graphical method, the points of contact between the cam and the follower are unknown, and it is difficult to determine their exact location as the cam outline is drawn in. Also, the minimum radius of the cam to prevent cusps can only be determined by trial. In the analytical method, which was developed by Carver and Quinn, these disadvantages are overcome, and three valuable characteristics of the cam may be determined: (*a*) parametric equations of the cam contour; (*b*) minimum radius of the cam to avoid cusps; and (*c*) location of the point of contact which gives the length of the follower face. Of these, the first has little practical application, but the other two give information from which the cam can be produced. The development of these characteristics follows.

Figure 3.21 shows a cam with a radial flat-faced follower. The cam rotates with a constant angular velocity. The contact point between the cam and follower is at x, y, which is a distance l from the radial center line of the follower. The displacement of the follower from the origin is given by the following equation:

$$R = C + f(\theta) \tag{3.1}$$

[4] F. Y. Chen, *Mechanics and Design of Cam Mechanisms*, Pergamon Press, New York, 1982.
[5] M. N. Sanchez and J. Garcia de Jalon, "Application of B-Spline Functions to the Motion Specification of Cams," *ASME Paper* 80-DET-28, 1980.

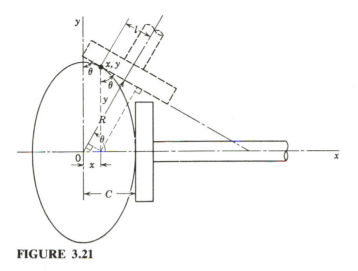

FIGURE 3.21

where the minimum radius of the cam is represented by C and $f(\theta)$ represents the desired motion of the follower as a function of the angular displacement of the cam.

The equation for the length of contact l can be easily determined from the geometry of Fig. 3.21. From the triangles shown,

$$R = y \sin \theta + x \cos \theta \tag{3.2}$$

and

$$l = y \cos \theta - x \sin \theta \tag{3.3}$$

The right side of Eq. 3.3 is the derivative with respect to θ of the right side of Eq. 3.2. Therefore,

$$l = \frac{dR}{d\theta} = \frac{d}{d\theta}[C + f(\theta)]$$

and

$$l = f'(\theta) \tag{3.4}$$

If the displacement diagram is given by a mathematical equation $S = f(\theta)$, then R and l are easily determined from Eqs. 3.1 and 3.4. From Eq. 3.4, it can be seen that the minimum length of the follower face is independent of the minimum radius of the cam. Also, the point of contact is at its greatest distance from the centerline of the follower when the velocity of the follower is a maximum. When the follower moves away from the cam center with positive velocity, l is positive and contact occurs above the axis of the follower in Fig. 3.21. When the follower moves toward the cam center, the velocity is negative and the resulting negative value of l indicates that contact is below the axis of the follower.

To determine the equations for x and y for the cam contour, it is only necessary to solve Eqs. 3.2 and 3.3 simultaneously, which gives

$$x = R \cos \theta - l \sin \theta$$

and

$$y = R \sin \theta + l \cos \theta$$

By substituting the values of R and l from Eqs. 3.1 and 3.4, respectively,

$$x = [C + f(\theta)] \cos \theta - f'(\theta) \sin \theta \tag{3.5}$$

$$y = [C + f(\theta)] \sin \theta + f'(\theta) \cos \theta \tag{3.6}$$

The minimum radius C to avoid a cusp or point on the cam surface can be easily determined analytically. A cusp occurs when both $dx/d\theta$ and $dy/d\theta = 0$. When this happens, a point is formed on the cam as shown at x, y in Fig. 3.22. To demonstrate this, consider that the centerline of the follower has rotated through angle θ and that contact between the follower face and the cam occurs at point x, y. When the follower is further rotated through a small angle $d\theta$, the point of contact (x, y) does not change because of the cusp and is still at x, y. Thus, it can be seen that $dx/d\theta = dy/d\theta = 0$.

By differentiating Eqs. 3.5 and 3.6,

$$\frac{dx}{d\theta} = -[C + f(\theta) + f''(\theta)] \sin \theta \tag{3.7}$$

$$\frac{dy}{d\theta} = [C + f(\theta) + f''(\theta)] \cos \theta \tag{3.8}$$

Equations 3.7 and 3.8 can become zero simultaneously only when

$$C + f(\theta) + f''(\theta) = 0$$

Therefore, to avoid cusps,

$$C + f(\theta) + f''(\theta) > 0 \tag{3.9}$$

The sum $f(\theta) + f''(\theta)$ must be inspected for all values of θ to determine its minimum algebraic value. It is necessary to use the minimum value so that C will be sufficiently large to ensure that Eq. 3.9 does not become zero for any

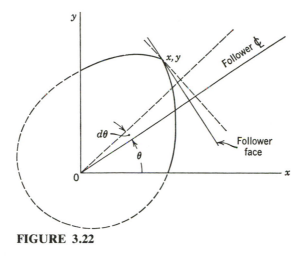

FIGURE 3.22

value of θ. The sum may be positive or negative. If positive, C will be negative and have no practical significance. In this case, the minimum radius will be determined by the hub of the cam rather than by the function $f(\theta)$.

Points on the cam profile may be determined from Eqs. 3.5 and 3.6, which give the Cartesian coordinates, or by calculating R and l for various values of θ. In general, the second method is easier, but in either case the points have to be connected by use of a French curve to obtain the cam outline. In actual practice, however, it is seldom necessary to draw the cam profile to scale. After the minimum radius C has been determined and the displacements R of the follower calculated, the cam can be generated. For the generating process, the length of the milling cutter must exceed twice the maximum value of l. During cutting, the axis of the milling cutter is parallel to the plane of the cam.

Example 3.2. To illustrate the method of writing the displacement equations, consider the following conditions: a flat-faced follower is driven through a total displacement of $1\frac{1}{2}$ in. At the start of the cycle (zero displacement), the follower dwells for $\pi/2$ rad. It then moves $1\frac{1}{2}$ in. with cycloidal motion (Kloomok and Muffley curve C-5) in $\pi/2$ rad. The follower dwells for $\pi/2$ rad and returns $1\frac{1}{2}$ in. with cycloidal motion (C-6) in $\pi/2$ rad. A sketch of the displacement diagram is shown in Fig. 3.23.

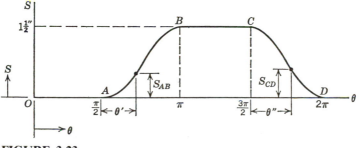

FIGURE 3.23

For the cycloid C-5, the Kloomok and Muffley curves give

$$S = L\left(\frac{\theta}{\beta} - \frac{1}{2\pi} \sin \frac{2\pi\theta}{\beta}\right)$$

It should be mentioned that in writing the relation $S = f(\theta)$, the value of S must always be measured from the abscissa and the value of θ from the ordinate. In the preceding equation, however, θ is measured from point A in Fig. 3.23 and not point O. Therefore, rewrite the equation using θ' as shown in Fig. 3.23:

$$S_{AB} = L\left(\frac{\theta'}{\beta} - \frac{1}{2\pi} \sin \frac{2\pi\theta'}{\beta}\right)$$

It is possible to transfer the origin from point A to point O by substituting the relation

$$\theta' = \theta - \frac{\pi}{2}$$

Therefore,

$$S_{AB} = L\left[\frac{(\theta - \pi/2)}{\beta} - \frac{1}{2\pi} \sin \frac{2\pi(\theta - \pi/2)}{\beta}\right]$$

By substituting $L = 1\frac{1}{2}$ in. and $\beta = \pi/2$ rad,

$$S_{AB} = \frac{3}{\pi}\left(\theta - \frac{\pi}{2}\right) - \frac{3}{4\pi} \sin(4\theta - 2\pi)$$

For the cycloidal curve C-6,

$$S_{CD} = L\left(1 - \frac{\theta''}{\beta} + \frac{1}{2\pi} \sin 2\pi \frac{\theta''}{\beta}\right)$$

where

$$\theta'' = \theta - \frac{3\pi}{2}$$

$$L = 1\frac{1}{2} \text{ in.}$$

$$\beta = \frac{\pi}{2}$$

Therefore,

$$S_{CD} = 6 - \frac{3\theta}{\pi} + \frac{3}{4\pi} \sin(4\theta - 6\pi)$$

It should be observed that with the combinations of dwell and cycloidal motion used, velocities and accelerations are matched and jerk is finite throughout the cycle.

Example 3.3. As an example of how the minimum radius C and the length of the follower face are determined, consider a flat-faced radial follower which moves out and back 50.8 mm with simple harmonic motion for half a revolution of the cam. Two motion cycles of the follower occur for one revolution of the cam.

Only one displacement equation (H-5) is necessary to specify the follower motion:

$$S = \frac{L}{2}\left(1 - \cos \pi \frac{\theta}{\beta}\right)$$

where

$$L = 50.8 \text{ mm}$$

and

$$\beta = \frac{\pi}{2}$$

Therefore,

$$S = f(\theta) = 25.4(1 - \cos 2\theta)$$
$$f'(\theta) = 50.8 \sin 2\theta$$

and

$$f''(\theta) = 101.6 \cos 2\theta$$

To find the minimum radius, the sum $C + f(\theta) + f''(\theta)$ must be greater than zero. By substituting for $f(\theta)$ and $f''(\theta)$ and by simplifying,

$$C + 25.4 + 76.2 \cos 2\theta > 0$$

The sum of $25.4 + 76.2 \cos 2\theta$ will be a minimum at $\theta = \pi/2$, which gives

$$C + 25.4 - 76.2 > 0$$

or

$$C > 50.8 \text{ mm}$$

The length of the follower face is determined from

$$l = f'(\theta) = 50.8 \sin 2\theta$$
$$l_{max} = 50.8 \text{ mm}$$

Because the motion is symmetrical, the theoretical length of the follower face is 50.8 mm on each side of the centerline. An additional amount must be added to each side of the follower to prevent contact from occurring at the very end of the face.

3.10 DISK CAM WITH RADIAL ROLLER FOLLOWER (ANALYTICAL DESIGN)

The analytical determination of the pitch surface of a disk cam with a radial roller follower presents no difficulties. In Fig. 3.24, the displacement of the center of the follower from the center of the cam is given by the following equation:

$$R = R_0 + f(\theta) \tag{3.10}$$

where R_0 is the minimum radius of the pitch surface of the cam and $f(\theta)$ is the radial motion of the follower as a function of cam angle. Once the value of R_0 is known, it is an easy matter to determine the polar coordinates of the centers of the roller follower from which the cam may be generated.

A method for checking this type of cam for pointing has been developed by Kloomok and Muffley, which considers the radius of curvature ρ of the pitch surface and the radius of the roller R_r. These values are shown in Fig. 3.25 together with the radius of curvature ρ_c of the cam surface. If in Fig. 3.25 ρ is held constant and R_r is increased, ρ_c will decrease. If this is continued until R_r equals ρ, then ρ_c will be zero and the cam becomes pointed as shown in Fig. 3.26a. As R_r is further increased, the cam becomes undercut as shown in Fig. 3.26b and the motion of the follower will not be as prescribed. Therefore, to prevent a point or an undercut from occurring on the cam profile, R_r must be less than ρ_{min}, where ρ_{min} is the minimum value of ρ over the particular segment of profile being considered. If there are several types of motion through which the follower passes, each case must be checked separately. Because it is impossible

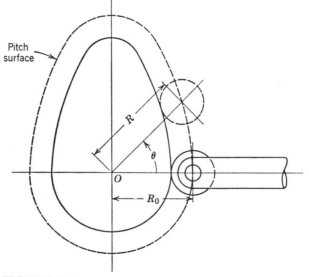

FIGURE 3.24

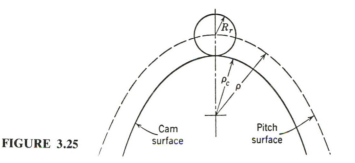

FIGURE 3.25

to undercut a concave portion of a cam, only the convex portions need to be investigated.

The radius of curvature at a point on a curve expressed in polar coordinates can be given by

$$\rho = \frac{[R^2 + (dR/d\phi)^2]^{3/2}}{R^2 + 2(dR/d\phi)^2 - R(d^2R/d\phi^2)}$$

where $R = f(\phi)$ and the first two derivatives are continuous. This equation can be used for finding the radius of curvature of the pitch surface of the cam. For this case, $f(\theta) = f(\phi)$. From Eq. 3.10,

$$R = R_0 + f(\theta)$$

$$\frac{dR}{d\theta} = f'(\theta)$$

$$\frac{d^2R}{d\theta^2} = f''(\theta)$$

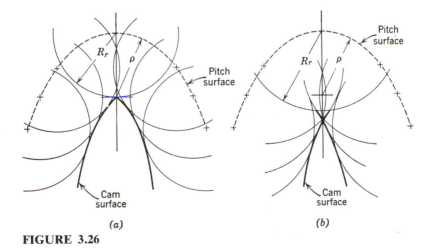

(a)

(b)

FIGURE 3.26

Therefore,

$$\rho = \frac{\{R^2 + [f'(\theta)]^2\}^{3/2}}{R^2 + 2[f'(\theta)]^2 - R[f''(\theta)]} \tag{3.11}$$

Equation 3.11 may be evaluated to find the expression for ρ for a particular type of motion. However, to prevent points or undercuts on the cam profile, ρ_{min} must be determined. Differentiation of Eq. 3.11 with its various functions to obtain minima gives very complex transcendental equations. For this reason, three sets of curves are given that show the plot of ρ_{min}/R_0 versus β for various values of L/R_0. In these curves, β is the total angular rotation of the cam for a complete event, and L is the lift. Figure 3.27 shows the graph for cycloidal motion, Fig. 3.28 for simple harmonic motion, and Fig. 3.29 for eighth-power polynomial motion. By means of these curves, it is possible to determine whether or not ρ_{min} is greater than R_r.

Example 3.4. A radial roller follower is to move through a total displacement of $L = 0.60$ in. with cycloidal motion while the cam rotates $\beta = 30°$. The follower dwells for 45° and then returns with cycloidal motion in 70°. Check the cam for pointing or undercutting if the radius R_r of the roller is 0.25 in. and the minimum radius R_0 of the pitch surface is 1.50 in.

$$\frac{L}{R_0} = \frac{0.60}{1.50} = 0.40$$

The outward motion will govern because of its smaller β. Therefore, from Fig. 3.27 for $L/R_0 = 0.40$ and $\beta = 30°$,

$$\frac{\rho_{min}}{R_0} = 0.22$$

and

$$\rho_{min} = 0.22 \times 1.50 = 0.33 \text{ in.}$$

The cam will not be pointed or undercut because $\rho_{min} > R_r$.

As mentioned previously, pressure angle is an important consideration when designing cams with roller followers. It is necessary to keep the maximum pressure angle as small as possible, and this maximum has been set at 30°. However, higher values are occasionally used when conditions permit. Although it is possible to make a layout of the cam and measure the maximum pressure angle, analytical methods are to be preferred. Several methods are available, one of which has been developed by Kloomok and Muffley, whereby the pressure angle can be determined analytically for either a radial roller follower or an oscillating roller follower. Only the radial roller follower will be treated here.

For the disk cam and radial roller follower shown in Fig. 3.30, the pressure

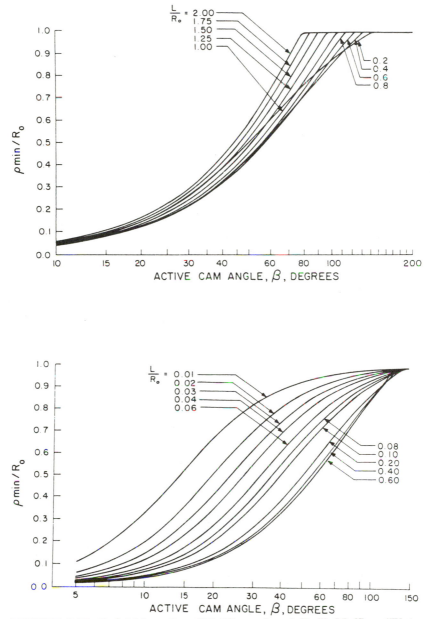

FIGURE 3.27 Cycloidal motion. [M. Kloomok and R. V. Muffley, "Plate Cam Design—Radius of Curvature," *Prod. Eng.*, September 1955, as revised by M. A. Ganter and J. J. Uicker, Jr., "Design Charts for Disk Cams with Reciprocating Radial Roller Followers," *ASME Trans., Journal of Mechanical Design*, 101 (3), July 1979.]

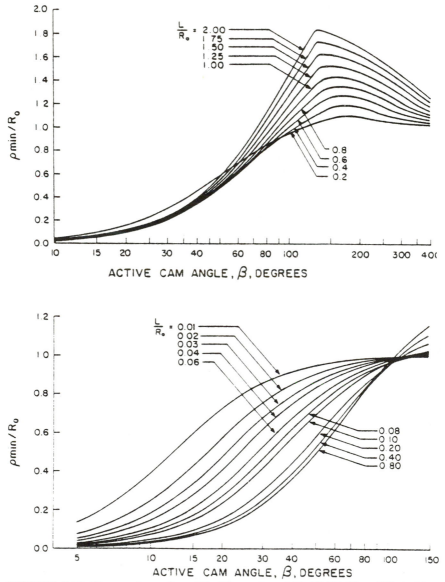

FIGURE 3.28 Harmonic motion. [M. Kloomok and R. V. Muffley, "Plate Cam Design—Radius of Curvature," *Prod. Eng.*, September 1955, as revised by M. A. Ganter and J. J. Uicker, Jr., "Design Charts for Disk Cams with Reciprocating Radial Roller Followers," *ASME Trans., Journal of Mechanical Design*, 101 (3), July 1979.]

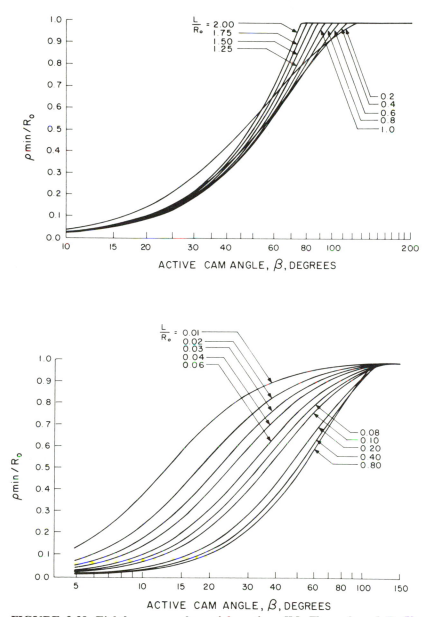

FIGURE 3.29 Eighth-power polynomial motion. [M. Floomok and R. V. Muffley, "Plate Cam Design—Radius of Curvature," *Prod. Eng.*, September 1955, as revised by M. A. Ganter and J. J. Uicker, Jr., "Design Charts for Disk Cams with Reciprocating Radial Roller Followers," *ASME Trans., Journal of Mechanical Design*, 101 (3), July 1979.]

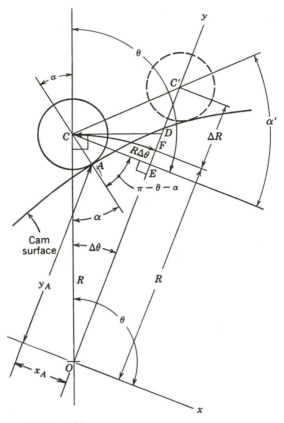

FIGURE 3.30

angle OCA is denoted by α and the center of the cam by O. The cam is assumed stationary, and the follower center rotates clockwise from position C to C' through a small angle $\Delta\theta$. From the sketch,

$$\alpha' = \tan^{-1}\frac{C'E}{CE}$$

As $\Delta\theta$ approaches zero, angles OCE and ACC' approach 90°. At the same time, CD approaches CF, which equals $R\,\Delta\theta$, and both approach CE. Therefore,

$$\lim_{\Delta\theta\to0}\alpha' = \tan^{-1}\frac{1}{R}\frac{dR}{d\theta}$$

Because the sides of α and α' become mutually perpendicular when $\Delta\theta$ approaches zero, α' becomes equal to α. Therefore,

$$\alpha = \tan^{-1}\frac{1}{R}\frac{dR}{d\theta} = \tan^{-1}\left[\frac{f'(\theta)}{R_0 + f(\theta)}\right] \qquad \textbf{(3.12)}$$

An expression for α may be determined from Eq. 3.12 for any type of motion. To solve for the maximum pressure angle is often very difficult, however, because of the resulting complex transcendental equation. For this reason, Kloomok and Muffley use a nomogram developed by E. C. Varnum, which is given in Fig. 3.31; β and L/R_0 are parameters as previously defined. From this chart, the maximum value of pressure angle may be determined for the three types of motion.

Points on the surface of the cam may also be determined by using Fig. 3.30. The coordinates of the point C are given by

$$x_C = R \cos \theta$$
$$y_C = R \sin \theta$$

(3.13)

The coordinates of the point of contact (point A) are obtained from the x and y projections of line segment CA and from the distances x_C and y_C as follows:

$$x_A = x_C + R_r \cos (\pi - \theta - \alpha)$$
$$y_A = y_C - R_r \sin (\pi - \theta - \alpha)$$

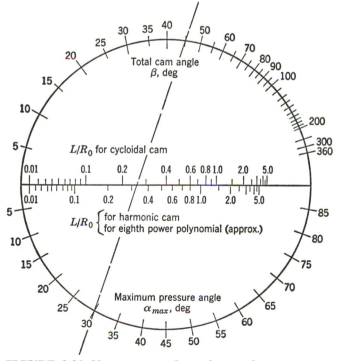

FIGURE 3.31 Nomogram to determine maximum pressure angle in a disk cam with a radial roller follower. (Courtesy of E. C. Varnum, Barber-Colman Company.)

where R_r is the radius of the roller. Simplifying these expressions using trigonometric identities leads to

$$x_A = x_C - R_r \cos (\theta + \alpha)$$
$$y_A = y_C - R_r \sin (\theta + \alpha)$$

(3.14)

Example 3.5. A radial roller follower is to move through a total displacement of 0.75 in. with cycloidal motion while the cam rotates 45°. The follower dwells for 30° and then returns with cycloidal motion in 60°. Find the value of R_0 to limit α_{max} to 30°. The outward motion will govern because of its smaller β.

For $\beta = 45°$ and $\alpha_{max} = 30°$,

$$\frac{L}{R_0} = 0.26 \qquad \text{(from Fig. 3.31)}$$

Therefore,

$$R_0 = \frac{0.75}{0.26} = 2.88 \text{ in.}$$

If space does not permit such a value of R_0, β can be increased and the cam run faster to maintain the lift time at a constant value.

The analytical calculation of cam-and-follower parameters using a hand calculator becomes tedious when more than a few cam angles must be considered. Fortunately, such repetitive calculations are conveniently carried out on a digital computer. The example below shows a computer program developed for the case of a disk cam and translating roller follower. This program was written in BASIC on an IBM personal computer. Although this program is specifically written for cycloidal rise, dwell, and cycloidal return, it would be an easy matter to generalize this to include other types of motion specifications or other follower configurations.

Example 3.6. A radial roller follower is to rise through a total displacement of 50.0 mm with C-5 cycloidal motion while the cam rotates 180°. The follower dwells for the next 90° and then returns 50.0 mm with C-6 cycloidal motion in 90° of cam rotation. The minimum radius R_0 of the pitch surface is 25.0 mm. Write a computer program to calculate the displacement S, velocity V, and acceleration A of the follower at each 10° of cam rotation. The program should also calculate the radius of the pitch surface R, the radius of curvature (ρ), and pressure angle (α) at each 10° of cam rotation.

Solution. The BASIC program shown in Fig. 3.32 was developed to solve this problem. Running this program on an IBM personal computer produced the output of Table 3.1. Note from the output that the maximum pressure angle during rise will be 35.6° at a cam angle of $\theta = 70°$ and that the maximum pressure angle during return will be 54.2° at $\theta = 320°$. These values are too high for most applications. The designer may wish to improve

```
10  '*************************************************************
20  '     BASIC program for cam design (3/27/85)
30  '     -  Disk cam with radial roller follower
40  '     -  Cycloidal rise - Dwell - Cycloidal return
50  '     -  Mabie and Reinholtz, 4th Ed.
60  '     -  Program revised by Steve Wampler (5/28/85)
70  '*************************************************************
80  INPUT "Minimum pitch radius";R0
90  INPUT "Total follower displacement";L
100 INPUT "Rise angle (in degrees)";DEG.RISE
110 INPUT "End of dwell angle (in degrees)";DEG.DWELL
120 INPUT "Angle increment (in degrees)";DEG.INC
130 PI=3.1415926#:TWO.PI=2*PI:PRINT
140 PRINT "  INPUT ANG      DISPL       VELOCITY      ACCEL";
150 PRINT "        RADIUS      CURVATURE     PRESS ANG"
160 PRINT "     (THETA)        (S)          (V)         (A)";
170 PRINT "          (R)          (RHO)        (ALPHA)"
180 '**** C-5 rise ****
190 BETA1=DEG.RISE-0
200 FOR THETA=0 TO DEG.RISE STEP DEG.INC
210 S=L*((THETA/BETA1)-(1/TWO.PI)*SIN(TWO.PI*THETA/BETA1))
220 V=(L/BETA1)*(1-COS(TWO.PI*THETA/BETA1))
230 A=((TWO.PI*L)/(BETA1^2))*SIN(TWO.PI*THETA/BETA1)
240 GOSUB 420 'Calculate R, RHO, ALPHA and print results
250 NEXT THETA
260 '**** dwell ****
270 FOR THETA=DEG.RISE TO DEG.DWELL STEP DEG.INC
280 S=L:V=0:A=0
290 GOSUB 420 'Calculate R, RHO, ALPHA and print results
300 NEXT THETA
310 '**** C-6 return ****
320 BETA2=360-DEG.DWELL
330 FOR THETA=DEG.DWELL TO 360 STEP DEG.INC
340 T=THETA-DEG.DWELL
350 S=L*((1-(T/BETA2))+(1/TWO.PI)*SIN(TWO.PI*T/BETA2))
360 V=-(L/BETA2)*(1-COS(TWO.PI*T/BETA2))
370 A=-(TWO.PI*L/BETA2^2)*(SIN(TWO.PI*T/BETA2))
380 GOSUB 420 'Calculate R, RHO, ALPHA and print results
390 NEXT THETA
400 END
410 '-----------------------------------------------------------
420 'Suboutine to calculate R, RHO, ALPHA and print results
430 R=R0+S
440 VR=V*180/PI:AR=A*(180/PI)^2  'Convert degrees to radians
450 RHO=(((R^2)+(VR^2))^(3/2))/((R^2)+(2*(VR^2))-R*AR)
460 ALPHA=(180/PI)*ATN(VR/R)
470 PRINT USING "  ##.##^^^^";THETA,S,V,A,R,RHO,ALPHA
480 RETURN
```

these pressure angles by increasing the base circle radius. Once the program is written, such changes require a minimum of effort. Running the program using $R_0 = 50.0$ mm resulted in maximum pressure angles of 24.0° during rise and 41.6° during return. Also, note that the pitch surface curvature ρ has a minimum tabulated value of 14.1 mm at 350° of cam rotation. Since this is a concave portion of the cam surface, undercutting will not occur here, and so the limiting value of ρ is actually 25.0 mm. In practice, a table like Table 3.1 should be developed using input angle steps of 1° or 2° rather than 10°. This will give more accurate extreme values of pressure angle and curvature.

Another advantage of using a computer is the ability to quickly generate plots of such things as displacements, velocities, accelerations, and the cam pitch surface. For this example problem, a computer-generated plot of the cam pitch surface is shown in Fig. 3.33. Such graphical output is valuable because it provides quick, easy-to-interpret visual feedback.

TABLE 3.1 Output Generated by the BASIC Program of Fig. 3.32

INPUT ANG (THETA)	DISPL (S)	VELOCITY (V)	ACCEL (A)	RADIUS (R)	CURVATURE (RHO)	PRESS ANG (ALPHA)
0.00E+00	0.00E+00	0.00E+00	0.00E+00	2.50E+01	2.50E+01	0.00E+00
1.00E+01	5.61E-02	1.68E-02	3.32E-03	2.51E+01	4.42E+01	2.19E+00
2.00E+01	4.40E-01	6.50E-02	6.23E-03	2.54E+01	1.10E+02	8.33E+00
3.00E+01	1.44E+00	1.39E-01	8.40E-03	2.64E+01	2.17E+02	1.68E+01
4.00E+01	3.27E+00	2.30E-01	9.55E-03	2.83E+01	1.17E+02	2.50E+01
5.00E+01	6.05E+00	3.26E-01	9.55E-03	3.11E+01	6.91E+01	3.10E+01
6.00E+01	9.78E+00	4.17E-01	8.40E-03	3.48E+01	5.40E+01	3.45E+01
7.00E+01	1.43E+01	4.91E-01	6.23E-03	3.93E+01	4.86E+01	3.56E+01
8.00E+01	1.95E+01	5.39E-01	3.32E-03	4.45E+01	4.67E+01	3.48E+01
9.00E+01	2.50E+01	5.56E-01	1.46E-09	5.00E+01	4.60E+01	3.25E+01
1.00E+02	3.05E+01	5.39E-01	-3.32E-03	5.55E+01	4.58E+01	2.91E+01
1.10E+02	3.57E+01	4.91E-01	-6.23E-03	6.07E+01	4.60E+01	2.49E+01
1.20E+02	4.02E+01	4.17E-01	-8.40E-03	6.52E+01	4.66E+01	2.01E+01
1.30E+02	4.39E+01	3.26E-01	-9.55E-03	6.89E+01	4.79E+01	1.52E+01
1.40E+02	4.67E+01	2.30E-01	-9.55E-03	7.17E+01	5.01E+01	1.04E+01
1.50E+02	4.86E+01	1.39E-01	-8.40E-03	7.36E+01	5.35E+01	6.18E+00
1.60E+02	4.96E+01	6.50E-02	-6.23E-03	7.46E+01	5.85E+01	2.86E+00
1.70E+02	4.99E+01	1.68E-02	-3.32E-03	7.49E+01	6.54E+01	7.34E-01
1.80E+02	5.00E+01	0.00E+00	-2.93E-09	7.50E+01	7.50E+01	0.00E+00
1.90E+02	5.00E+01	0.00E+00	0.00E+00	7.50E+01	7.50E+01	0.00E+00
2.00E+02	5.00E+01	0.00E+00	0.00E+00	7.50E+01	7.50E+01	0.00E+00
2.10E+02	5.00E+01	0.00E+00	0.00E+00	7.50E+01	7.50E+01	0.00E+00
2.20E+02	5.00E+01	0.00E+00	0.00E+00	7.50E+01	7.50E+01	0.00E+00
2.30E+02	5.00E+01	0.00E+00	0.00E+00	7.50E+01	7.50E+01	0.00E+00
2.40E+02	5.00E+01	0.00E+00	0.00E+00	7.50E+01	7.50E+01	0.00E+00
2.50E+02	5.00E+01	0.00E+00	0.00E+00	7.50E+01	7.50E+01	0.00E+00
2.60E+02	5.00E+01	0.00E+00	0.00E+00	7.50E+01	7.50E+01	0.00E+00
2.70E+02	5.00E+01	0.00E+00	0.00E+00	7.50E+01	7.50E+01	0.00E+00
2.80E+02	4.96E+01	-1.30E-01	-2.49E-02	7.46E+01	3.57E+01	-5.71E+00
2.90E+02	4.67E+01	-4.59E-01	-3.82E-02	7.17E+01	2.87E+01	-2.01E+01
3.00E+02	4.02E+01	-8.33E-01	-3.36E-02	6.52E+01	3.30E+01	-3.62E+01
3.10E+02	3.05E+01	-1.08E+00	-1.33E-02	5.55E+01	4.36E+01	-4.81E+01
3.20E+02	1.95E+01	-1.08E+00	1.33E-02	4.45E+01	5.75E+01	-5.42E+01
3.30E+02	9.78E+00	-8.33E-01	3.36E-02	3.48E+01	1.07E+02	-5.39E+01
3.40E+02	3.27E+00	-4.59E-01	3.82E-02	2.83E+01	-4.23E+01	-4.29E+01
3.50E+02	4.40E-01	-1.30E-01	2.49E-02	2.54E+01	-1.41E+01	-1.63E+01
3.60E+02	-2.40E-06	0.00E+00	1.17E-08	2.50E+01	2.50E+01	0.00E+00

3.11 DISK CAM WITH OSCILLATING ROLLER FOLLOWER (ANALYTICAL DESIGN)

In Fig. 3.34 is seen the start of a layout of a disk cam with an oscillating roller follower. The displacement angle ψ is a function of the cam angle θ. Although the cam rotates through the angle θ for the displacement angle ψ, the radius R rotates through the angle ϕ. By specifying values of R and ϕ, it is possible to generate the cam.

From Fig. 3.34, it can be seen that

$$\phi = \theta - \lambda \tag{3.15}$$

where

$$\lambda = \beta - \Gamma \tag{3.16}$$

Angle β is a constant for the system, and its equation can be derived from triangle

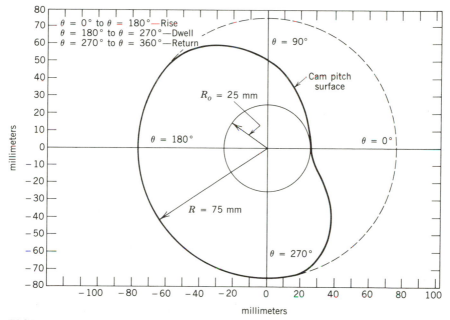

FIGURE 3.33 Cam pitch surface generated using an IBM Personal Computer and an IBM Instruments XY/749 Digital Plotter.

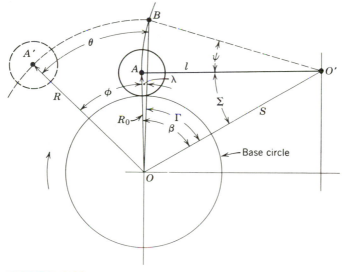

FIGURE 3.34

OAO' as

$$\cos \beta = \frac{S^2 + R_0^2 - l^2}{2SR_0} \tag{3.17}$$

where S, R_0, and l are fixed dimensions.

Angle Γ is a function of R; its equation can be derived from triangle OBO' as

$$\cos \Gamma = \frac{S^2 + R^2 - l^2}{2SR} \tag{3.18}$$

An equation for R can also be written from triangle OBO' as follows:

$$R^2 = l^2 + S^2 - 2lS \cos (\psi + \Sigma) \tag{3.19}$$

Angle Σ is a constant determined from triangle OAO' as

$$\cos \Sigma = \frac{l^2 + S^2 - R_0^2}{2lS} \tag{3.20}$$

and angle ψ is the displacement angle for a particular cam angle θ. Therefore, from the preceding equations, the values of R and ϕ can be calculated for given values of cam angle θ and their corresponding angles of displacement ψ.

In designing this type of cam, it is necessary to check for undercutting and for the maximum pressure angle. Equations for the radius of curvature and the pressure angle can best be developed by using complex variables. Figure 3.35 shows the sketch of a disk cam and oscillating roller follower with the radius of curvature of the pitch surface designated as ρ and the pressure angle as α. Point O is the center of the cam, point D the center of curvature, and point O' the center of oscillation of the follower. The angular displacement of the follower from the horizontal is σ, which is given by the equation

$$\sigma = \sigma_0 + f(\theta) \tag{3.21}$$

where $f(\theta)$ is the desired angular displacement of the follower from the reference angle σ_0 (not shown). From Fig. 3.35, the pressure angle α is given by

$$\alpha = \sigma - \frac{\pi}{2} - \gamma$$

Substituting Eq. 3.21 for σ,

$$\alpha = [\sigma_0 + f(\theta)] - \frac{\pi}{2} - \gamma \tag{3.22}$$

To obtain an expression for angle γ, two independent position equations are

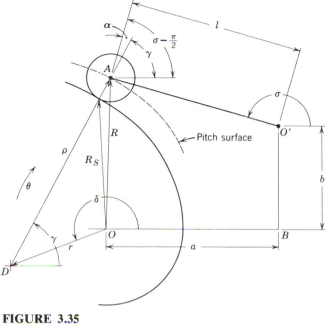

FIGURE 3.35

written for the center of the roller A. One is written by following the path O to D to A, and the other by going from O to B to O' to A. The equation for the first path (O–D–A) is given by

$$\mathbf{R} = re^{i\delta} + \rho e^{i\gamma}$$

$$= r(\cos \delta + i \sin \delta) + \rho(\cos \gamma + i \sin \gamma) \qquad (3.23)$$

The equation for the second path (O–B–O'–A) is given by

$$\mathbf{R} = a + bi + le^{i\sigma}$$

$$= a + bi + l(\cos \sigma + i \sin \sigma) \qquad (3.24)$$

By separating real and imaginary parts of Eqs. 3.23 and 3.24, it follows that

$$r \cos \delta + \rho \cos \gamma = a + l \cos \sigma \qquad (3.25)$$

$$r \sin \delta + \rho \sin \gamma = b + l \sin \sigma \qquad (3.26)$$

By differentiating Eqs. 3.25 and 3.26 with respect to θ,

$$-r \sin \delta \frac{d\delta}{d\theta} - \rho \sin \gamma \frac{d\gamma}{d\theta} = -l \sin \sigma \frac{d\sigma}{d\theta}$$

$$r \cos \delta \frac{d\delta}{d\theta} + \rho \cos \gamma \frac{d\gamma}{d\theta} = l \cos \sigma \frac{d\sigma}{d\theta}$$

For an infinitesimal rotation of the cam, ρ may be considered to remain constant. Thus, point D, the center of curvature of the cam at the point of contact, and r may be regarded as fixed to the cam for an incremental rotation $d\theta$. Therefore, the magnitude of $d\delta$ is equal to $d\theta$; and since δ decreases as θ increases, it follows that $d\delta/d\theta = -1$. Also, $d\sigma/d\theta = f'(\theta)$. Therefore,

$$r \sin \delta - \rho \sin \gamma \frac{d\gamma}{d\theta} = -lf'(\theta) \sin \sigma \qquad (3.27)$$

$$-r \cos \delta + \rho \cos \gamma \frac{d\gamma}{d\theta} = lf'(\theta) \cos \sigma \qquad (3.28)$$

By eliminating $d\gamma/d\theta$ from Eqs. 3.27 and 3.28,

$$\tan \gamma = \frac{r \sin \delta + lf'(\theta) \sin \sigma}{r \cos \delta + lf'(\theta) \cos \sigma}$$

The terms $r \cos \delta$ and $r \sin \delta$ can be evaluated from Eqs. 3.25 and 3.26 to give

$$\tan \gamma = \frac{b + l \sin \sigma[1 + f'(\theta)]}{a + l \cos \sigma[1 + f'(\theta)]} \qquad (3.29)$$

which, when substituted in Eq. 3.22, will give the pressure angle α. To find α_{max}, it will be necessary to work out design charts similar to those given by Kloomok and Muffley.

To find the radius of curvature ρ, it is necessary to first differentiate Eq. 3.29 with respect to θ. Substituting $d\gamma/d\theta$ from Eq. 3.28 and with the aid of Eqs. 3.21, 3.25, and 3.29, the following equation for ρ is obtained:

$$\rho = \frac{[C^2 + D^2]^{3/2}}{(C^2 + D^2)[1 + f'(\theta)] - (aC + bC)f'(\theta) + (a \sin \sigma - b \cos \sigma)lf''(\theta)} \qquad (3.30)$$

where

$$C = a + l \cos \sigma[1 + f'(\theta)]$$

$$D = b + l \sin \sigma[1 + f'(\theta)]$$

To avoid undercutting, ρ must be greater than the radius of the roller. Therefore, it must be possible to determine ρ_{min} for each portion of the cam profile. In order to do this, it will be necessary to work out design charts similar to those given by Kloomok and Muffley.

Once the radius of curvature has been found, points on the surface of the cam are easily determined from Fig. 3.35:

$$\mathbf{R}_s = re^{i\delta} + (\rho - R_r)e^{i\gamma} \qquad (3.31)$$

where $\mathbf{R}_s$ is the vector locating the contact point and R_r is the radius of the roller follower.

3.12 CONTOUR CAMS

The application of this type of cam is primarily in the design of computer and mechanical control systems. A sketch is shown in Fig. 3.36. With this type of cam, the members roll upon each other without sliding; this facilitates the design for two reasons: (*a*) the contact point *P* will always lie on the line of centers, and (*b*) both surfaces will roll on each other through the same distance. By making use of these factors, equations for the distance from the cam centers to the contact point can easily be derived.

In Fig. 3.36, R_2 and R_3 are the instantaneous distances from the cam centers to the point of contact and *C* is the fixed distance between centers. If cam 2 rotates through a small angle $d\theta_2$ and cam 3 through $d\theta_3$, the point of contact on cam 2 will move through $R_2\, d\theta_2$, and that on cam 3 will move through $R_3\, d\theta_3$. For pure rolling,

$$R_2\, d\theta_2 = R_3\, d\theta_3$$

Also,

$$R_2 + R_3 = C$$

Therefore,

$$R_2 = \frac{C}{1 + (d\theta_2/d\theta_3)} \tag{3.32}$$

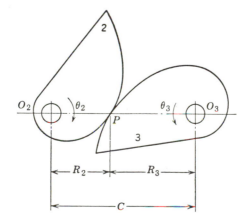

FIGURE 3.36

and

$$R_3 = \frac{C}{1 + (d\theta_3/d\theta_2)} \tag{3.33}$$

These cams can be used to generate several types of functions, three of which are described below.

1. *Square Function.* To generate the square function,

$$\theta_3 = k\theta_2^2$$

$$\frac{d\theta_3}{d\theta_2} = 2k\theta_2$$

and

$$\frac{d\theta_2}{d\theta_3} = \frac{1}{2k\theta_2}$$

Therefore,

$$R_2 = \frac{2kC\theta_2}{1 + 2k\theta_2}$$

and

$$R_3 = \frac{C}{1 + 2k\theta_2}$$

From the equations for R_2 and R_3, the cam contours may be determined that will generate the given square function. If the cams are operated in reverse, square roots are obtained.

2. *Logarithmic Function.* To generate the logarithm,

$$\theta_3 = \log_{10} \theta_2$$

$$\theta_3 = \frac{1}{2.303} \ln \theta_2$$

$$\frac{d\theta_3}{d\theta_2} = \frac{1}{2.303\theta_2}$$

and

$$\frac{d\theta_2}{d\theta_3} = 2.303\theta_2$$

Therefore,

$$R_2 = \frac{C}{1 + 2.303\theta_2}$$

and

$$R_3 = \frac{2.303C\theta_2}{1 + 2.303\theta_2}$$

From these equations, the cam contours may be determined that will generate the given logarithm. Operation in reverse will give antilogs.

3. *Trigonometric Function.* To illustrate the generation of a trigonometric function, consider

$$\theta_3 = \tan \theta_2$$

$$\frac{d\theta_3}{d\theta_2} = \sec^2 \theta_2$$

and

$$\frac{d\theta_2}{d\theta_3} = \frac{1}{\sec^2 \theta_2} = \cos^2 \theta_2$$

Therefore,

$$R_2 = \frac{C}{1 + \cos^2 \theta_2}$$

and

$$R_3 = \frac{C \cos^2 \theta_2}{1 + \cos^2 \theta_2}$$

If one refers to the equations for R_2 and R_3 developed for the three functions, it is evident that in (1), $R_2 = 0$ when $\theta_2 = 0$ and in (2), $R_3 = 0$ when $\theta_2 = 0$. In (3), $R_3 = 0$ when $\theta_2 = 90°$. When one of the radii goes to zero, an impractical design results. With the functions illustrated, the fact that the scale of θ_2 cannot start at zero in the first two cases nor extend to 90° in the third case will probably not limit the generation of these functions. There are cases, however, where such limitations would prove a disadvantage and a means must be found for eliminating this problem when necessary. Another problem that sometimes arises when designing contour cams is that with certain functions the value of $d\theta_3/d\theta_2$ may become equal to -1, which makes the radii R_2 and R_3 infinite. Either of these problems must be avoided if it occurs in the working range of the function. This can be accomplished by offsetting the function by a constant which can later be subtracted by a differential. As an example, consider the function

$$\theta_3 = \sin^2 \theta_2$$

and

$$\frac{d\theta_3}{d\theta_2} = 2 \sin \theta_2 \cos \theta_2$$

Therefore,

$$R_2 = \frac{C(2 \sin \theta_2 \cos \theta_2)}{2 \sin \theta_2 \cos \theta_2 + 1}$$

and

$$R_3 = \frac{C}{1 + 2 \sin \theta_2 \cos \theta_2}$$

When θ_2 equals zero, $R_2 = 0$; when θ_2 equals 135°, $d\theta_3/d\theta_2 = -1$. To avoid these conditions, the function may be offset by a constant $k\theta_2$ such that

$$\theta_3' = \sin^2 \theta_2 + k\theta_2$$

and

$$\frac{d\theta_3'}{d\theta_2} = 2 \sin \theta_2 \cos \theta_2 + k$$

After generation of the new function, $k\theta_2$ would be subtracted to give the original function $\theta_3 = \sin^2 \theta_2$.

If large torques are to be transmitted, the cams can be replaced by gears having pitch surfaces identical to the cam contours. This substitution is possible because of the pure rolling action of the cams. Such gears are known as *contour gears*, or *noncircular gears*. A photograph of a pair of noncircular gears is shown in Fig. 3.37.

FIGURE 3.37 Noncircular gears. (Sample gears courtesy of Cunningham Corporation.)

3.13 THREE-DIMENSIONAL CAMS

A sketch of a three-dimensional cam is shown in Fig. 3.38, where the displacement z of the follower is a function of both the rotation y and the translation x of the cam.

A three-dimensional cam can easily be designed to solve the equation $Q = 0.05a\sqrt{h}$, which expresses the flow through an orifice (ft³/s) in terms of orifice area a (in.²) and pressure head h (ft).

Table 3.2 has been developed to give a range of values for the parameters from which the cam can be designed.

Figure 3.39 shows the orientation of the values of the parameter a around the circumference of the cam with the value of $a = 1.0$ in.² taken at the top of the cam. Figure 3.40 shows a vertical axial section through the cam which extends through $a = 1.0$ in.² to $a = 1.30$ in.². Figures 3.41 and 3.42 show transverse sections through the cam at $h = 25$ ft and $h = 49$ ft, respectively. The axial section shows Q as a function of h and the transverse sections show Q as a

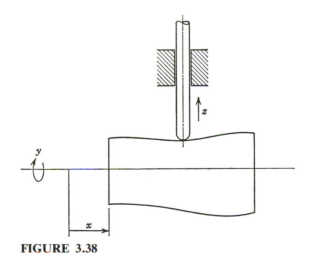

FIGURE 3.38

TABLE 3.2 Flow Q (ft^3/s)

h, ft	a, in.^{2}										
	1.00	1.05	1.10	1.15	1.20	1.25	1.30	1.35	1.40	1.45	1.50
1	0.05	0.0525	0.055	0.0575	0.060	0.0625	0.065	0.0675	0.070	0.0725	0.075
4	0.10	0.105	0.110	0.115	0.120	0.125	0.130	0.135	0.140	0.145	0.150
9	0.15	0.158	0.165	0.173	0.180	0.188	0.195	0.203	0.210	0.218	0.225
16	0.20	0.210	0.220	0.230	0.240	0.250	0.260	0.270	0.280	0.290	0.300
25	0.25	0.263	0.275	0.288	0.300	0.313	0.325	0.338	0.350	0.363	0.375
36	0.30	0.315	0.330	0.345	0.360	0.375	0.390	0.405	0.420	0.435	0.450
49	0.35	0.368	0.385	0.403	0.420	0.438	0.455	0.473	0.490	0.508	0.525

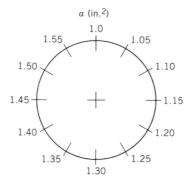

FIGURE 3.39

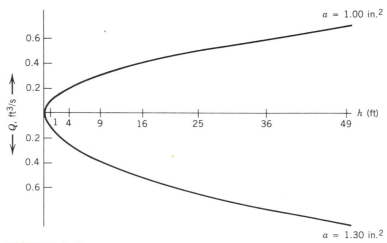

FIGURE 3.40

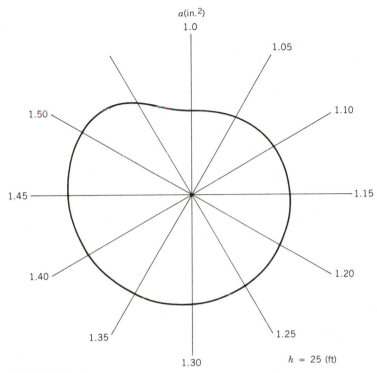

FIGURE 3.41

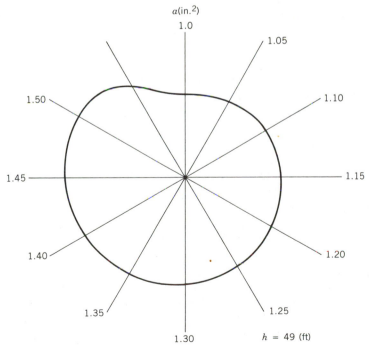

FIGURE 3.42

function of *a*. The design of the cam can be completed by generating additional axial and transverse sections.

The production of a three-dimensional cam is very difficult because of the accuracy and hand finishing required. After the displacements of the follower have been specified for the desired increments of rotation and translation of the cam, a casting is made approximating the desired shape. Using a cutting tool the same size and shape as the follower, the cam blank is set up in a cam miller, and a cut is made at each data point. By proper rotation and translation of the cam and by bringing the cutting tool down the correct displacement for each data point, the cutter will simulate the follower in its relation with the cam. In this manner, an accurate point will be spotted on the cam contour. According to Rothbart, as many as 15,000 points are sometimes required with accuracy of ±0.0004 in. After the data points have been spotted, the cam is next finished by hand filing, followed by polishing with emery paper.

3.14   CAM PRODUCTION METHODS

The graphical method of cam design is limited to slow-speed applications. The production of this type of cam depends upon the accuracy of the design layout and upon the method used in following this layout as a template. At one extreme, the layout of the cam is scribed on a steel plate and the cam cut out with a band saw. At the other extreme is production by a milling cutter whose motion is guided by a tracer moving over the cam outline on a copy of the layout drawing. The drawing over which this tracer moves may be made several times the actual size to improve the accuracy of copying. In either of these cases, the cam profile must be finished by hand.

Graphical design and the resulting copying method of production are not sufficiently accurate for high-speed cams. For this reason, attention has turned to analytical cam design and the method it offers for cam generation. If it is possible to calculate the follower displacements for small increments of cam rotation, the cam profile can be generated on a milling machine or on a jig borer with the cutter assuming the role of the follower. If the follower is to be a roller, the axis of the cutter will be perpendicular to the plane of the cam with the cutter the same size as the roller. If a flat-faced follower is to be used, the axis of the cutter will be parallel to the plane of the cam. In either case, the cutter can be given the correct position corresponding to the cam rotation angle. Naturally, the smaller the increments of the cam angle, the better the cam finish will be. Generally, increments of 1° are used, which leave tiny scallops or flats on the cam that must be removed by hand. Automatic numerically controlled cam milling machines have been developed which index the cam a fraction of a degree, with the cutter advancing by tenths of thousandths of an inch. Although the machine operates in discrete steps, the steps are so minute as to give the appearance of continual operation. It is hoped that the cam surface finish produced by a machine of this kind will be of such quality as to allow elimination of the hand-finishing operation. This type of machine will also produce a cam much more rapidly than a jig borer when both machines are using the same increments of cam angle.

In the preceding discussions, it was assumed that the cam being generated was the cam to be used in the final application. Where several machines of the same model are being produced and many copies of a cam are needed, it is generally more practical to generate what is known as a master cam and to use this master with a cam duplicating machine. The master cam is often made several times the actual size.

Problems

3.1. A disk cam rotating clockwise drives a radial flat-faced follower through a total displacement of $1\frac{1}{2}$ in. with the following lift figures:

CAM ANGLE, deg	LIFT, in.
0	0.00
30	0.10
60	0.37
90	0.75
120	1.13
150	1.40
180	1.50
210	1.40
240	1.13
270	0.75
300	0.37
330	0.10
360	0.00

Lay out the cam using a minimum radius of 1 in. Determine the length of the follower face (symmetrical). After the length has been found by trial, add $\frac{1}{8}$ in. to each end to positively ensure proper contact.

3.2. A disk cam rotating counterclockwise drives a radial roller follower through a total displacement of $1\frac{1}{2}$ in. Lay out the cam using the lift figures from Problem 3.1 and a minimum radius of 1 in. The roller diameter is to be $\frac{7}{8}$ in. By trial determine the magnitude and position of the maximum pressure angle.

3.3. A disk cam rotating clockwise drives an offset flat-faced follower through a total displacement of $1\frac{1}{2}$ in. Lay out the cam using the lift figures from Problem 3.1. The centerline of the follower is offset $\frac{1}{2}$ in. to the left of and parallel to the vertical centerline of the cam. The minimum radius of the cam is to be 1 in. Determine the length of the follower face (symmetrical). After the length has been found by trial, add $\frac{1}{8}$ in. to each end to positively ensure proper contact.

3.4. A disk cam rotating counterclockwise drives an offset roller follower through a total displacement of $1\frac{1}{2}$ in. Lay out the cam using the lift figures in Problem 3.1. The centerline of the follower is offset $\frac{1}{2}$ in. to the right of and parallel to the vertical centerline of the cam. The minimum radius of the cam is to be 1 in. and the roller diameter $\frac{7}{8}$ in. By trial determine the maximum pressure angle during the outward motion and during the return motion.

3.5. A disk cam rotating clockwise drives an oscillating flat-faced follower through a total angle of 20° with the following displacement figures:

CAM ANGLE, deg	FOLLOWER ANGLE, deg
0	0.0
30	1.5
60	5.5
90	10.0
120	14.5
150	18.5
180	20.0
210	18.5
240	14.5
270	10.0
300	5.5
330	1.5
360	0.0

Lay out the cam using a minimum radius of 30 mm. The center of the hub of the follower is to be 80 mm to the right of the center and on the horizontal centerline of the cam similar to Fig. 3.5. The distance from the center of the follower hub to the arc of the displacement scale is to be 70 mm. Determine the length of the follower face. After the length has been found by trial, add 3 mm to each end to positively ensure proper contact. Assuming a bore of 16 mm, a hub of 25 mm, and a 5 mm key, draw in the rest of the follower to reasonable proportions.

3.6. A disk cam rotating counterclockwise drives an oscillating roller follower through a total angle of 20°. Lay out the cam using the displacement figures from Problem 3.5 and a minimum radius of 1 in. The center of the hub of the follower is to be 3 in. to the right of the center and on the horizontal centerline of the cam similar to Fig. 3.6. The diameter of the roller is to be $\frac{3}{4}$ in. and the distance from the center of the hub to the center of the roller is to be $2\frac{7}{8}$ in. Using a bore of $\frac{5}{8}$ in., a hub of 1 in., and a $\frac{3}{16} \times \frac{3}{16}$ in. key, draw in the rest of the follower to reasonable proportions.

3.7. A positive return cam rotating clockwise drives a flat-faced yoke as shown in Fig. 3.7. The lift figures for the outward motion are as follows:

CAM ANGLE, deg	LIFT, mm
0	0.00
30	1.27
60	4.32
90	9.65
120	17.0
150	23.4
180	25.4

Lay out the cam using a minimum radius of 25 mm. Using reasonable proportions, complete the sketch of the follower.

3.8. A positive-return cam rotating counterclockwise drives a yoke with roller followers. Lay out the cam using the lift figures from Problem 3.7 for the outward motion and a minimum radius of 25 mm. The diameters of the rollers are to be 19 mm. Using reasonable proportions, complete the yoke that carries the roller followers.

3.9. An oscillating roller follower moving through a total angle of 60° drives an inverse cam as shown in Fig. 3.9, with the following displacement figures:

FOLLOWER ANGLE, deg	CAM DISPLACEMENT, in.
0.0	0.00
4.5	0.06
16.0	0.24
30.0	0.50
44.0	0.76
55.5	0.94
60.0	1.00

Lay out the groove in the cam block if the cam is to move upward and to the right at an angle of 45° as the follower moves counterclockwise. The follower moves symmetrically about the vertical centerline. The distance from the center of the roller follower to the center of oscillation is 3 in., and the diameter of the roller is $\frac{3}{8}$ in. The cam block is 3 × 4 in.

3.10. Prove that the method for finding inflection points when the time intervals are known as shown in Fig. 3.10 is correct.

3.11. Prove that the method for finding inflection points when the displacement intervals are known as shown in Fig. 3.11 is correct.

3.12. Prove that the method of construction for parabolic motion shown in Fig. 3.12 is correct.

3.13. Draw the displacement–time graph for a follower that rises through a total displacement of $1\frac{1}{2}$ in. with constant acceleration for a three-sixteenth revolution, constant velocity for a quarter revolution, and constant deceleration for a quarter revolution of the cam. After dwelling for a sixteenth revolution, the follower returns with simple harmonic motion in a quarter revolution of the cam. Use an abscissa 4 in. long.

3.14. Draw the displacement–time graph for a follower that rises 19 mm with simple harmonic motion in a quarter revolution, dwells for an eighth revolution, and then rises 19 mm with simple harmonic motion in a quarter revolution of the cam. The follower dwells for a sixteenth revolution and then returns 38 mm with parabolic motion in a quarter revolution, followed by a dwell for a sixteenth revolution of the cam. Use an abscissa 160 mm long.

3.15. Draw the displacement–time graph for a follower that rises $1\frac{1}{2}$ in. in a half revolution of the cam such that the first $\frac{3}{8}$ in. is constant acceleration, constant velocity for the next $\frac{3}{4}$ in., and constant deceleration for the remaining $\frac{3}{8}$ in. The return motion is simple harmonic in a half revolution of the cam. Use an abscissa 6 in. long.

3.16. Draw the displacement–time graph for a follower that moves through a total displacement of 32 mm with constant acceleration for 90° and constant deceleration for 45° of cam rotation. The follower returns 16 mm with simple harmonic motion in 90°, dwells for 45°, and returns the remaining 16 mm with simple harmonic motion in 90° of cam rotation. Use an abscissa 160 mm long.

3.17. The radial flat-faced follower shown in Fig. 3.43 is made to reciprocate by the action of a circular disk cam rotating about the axis at O_2. (*a*) Derive expressions for the follower displacement R and for the distance l to the point of contact in terms of the cam angle θ, the radius r of the cam, and the offset distance h. (*b*) Make a sketch of the displacement R versus cam angle θ for one cycle of the cam. Label the lift L of the follower as the

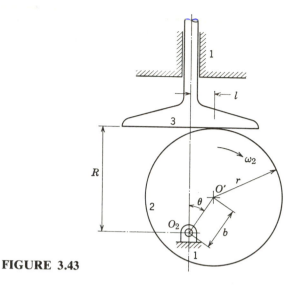

FIGURE 3.43

distance from the follower's lowest position to its highest position. Give the magnitude of L. (c) Name the type of follower motion produced by the cam.

3.18. A radial follower is actuated by a cam rotating at 1 rad/s. The follower starts from rest and moves through a distance of 50 mm with simple harmonic motion while the cam turns 120°. The follower dwells for the next 120° and then returns with simple harmonic motion during the remaining 120° of cam rotation. Using an abscissa of 150 mm and increments of cam angle of 30°, plot the displacement, velocity, acceleration, and jerk curves on the same axis.

3.19. From the relation for simple harmonic motion, derive the equation for displacement S for motion classification H-5 of Fig. 3.16.

3.20. Derive equations which will allow the Kloomok and Muffley equations to be used to determine follower velocities and accelerations when the angular velocity of the cam is not constant.

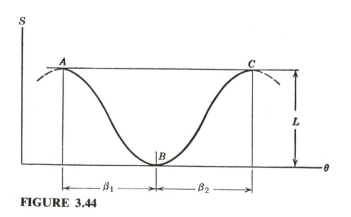

FIGURE 3.44

3.21. A follower is to have cyclical motion according to the displacement curve shown in Fig. 3.44. The displacement and velocity requirements are as follows:

POINT A	POINT B	POINT C
$S = L$	$S = 0$	$S = L$
$V = 0$	$V = 0$	$V = 0$

Recommend the curves to be used for the displacement graph and the relation between β_1 and β_2 to match accelerations at point B and at points A and C.

3.22. A follower dwells and then passes through the motion cycle shown in Fig. 3.45 and dwells again. The motion requirements are as follows:

POINT A	POINT B	POINT C
$S = 0$	$S = L$	$S = 0$
$V = 0$	$V = 0$	$V = 0$
$A = 0$	$A = A_1$	$A = 0$

Recommend the curves to be used for the displacement graph and the relation between β_1 and β_2 to match accelerations at point B.

3.23. A follower dwells, rises with acceleration, rises with constant velocity, rises with deceleration, and then dwells again as shown in Fig. 3.46. The motion requirements are as follows:

POINT A	POINT B	POINT C	POINT D
$S = 0$	$S = L_1$	$S = L_1 + L_2$	$S = L_1 + L_2 + L_3$
$V = 0$	$V = V_1$	$V = V_1$	$V = 0$
$A = 0$	$A = 0$	$A = 0$	$A = 0$

Recommend the curves to be used for the displacement graph and the relation between β_1, β_2, and β_3 to match velocities at points B and C.

3.24. For the motion given in Fig. 3.18a of Example 3.1, let β_1 denote the angle for curve AB, β_2 the angle for BC, β_3 the angle for CD, and β_4 the angle for DE. Also let L_1 be the rise AB, L_2 the rise BC, L_3 the rise CD, and L_4 the fall DE. Determine the relation between β_3 and β_4 to match accelerations at point D.

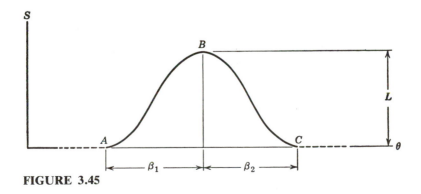

FIGURE 3.45

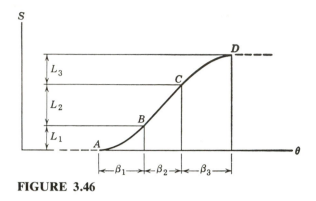

FIGURE 3.46

3.25. Determine (a) the relation between the cam angles β_1, β_2 and the lifts L_1, L_2 to match a cycloidal curve C-1 with a constant velocity curve, and (b) the relation to match a constant velocity curve with curve C-4.

3.26. Set up the equations for relating the lifts L_1, L_2 and cam angles β_1, β_2 for transferring from (a) cycloidal to harmonic motion, (b) cycloidal to constant velocity, (c) harmonic to cycloidal, and (d) harmonic to constant velocity. Transfer motions only when acceleration equals zero.

3.27. Determine (a) the relation between cam angles β_1, β_2 and lifts L_1, L_2 to match a cycloidal curve C-1 with a harmonic curve H-2, and (b) the relation to match curve H-3 with curve C-4.

3.28. Determine (a) the relation between cam angles β_1, β_2 and lifts L_1, L_2 to match a harmonic curve H-1 with a cycloidal curve C-2, and (b) the relation to match curve C-3 with curve H-4.

3.29. Determine (a) the relation between cam angles β_1, β_2 and lifts L_1, L_2 to match a harmonic curve H-1 with a constant-velocity curve, and (b) the relation to match a constant-velocity curve with curve H-4.

3.30. A follower is to have a period of constant-velocity motion during its outward travel and again on its return. Is it possible to use harmonic curves with these constant-velocity curves and not have jerk become infinite? If so, recommend the curves to be used and sketch the displacement graph showing the motions.

3.31. Determine (a) the relation between cam angles β_1, β_2 and lifts L_1, L_2 to match a harmonic curve H-5 with an eighth-power polynomial curve P-2, and (b) the relation to match a harmonic curve H-2 to an eighth-power polynomial curve P-2.

3.32. Select a combination of cycloidal motion, harmonic, and eighth-power polynomial that will not produce infinite jerk.

3.33. Determine (a) the relation between cam angles β_1, β_2 and lifts L_1, L_2 to match an eighth-power polynomial curve P-1 with a harmonic curve H-6, and (b) the relation to match an eighth-power polynomial curve P-1 with a harmonic curve H-3.

3.34. Select a combination of harmonic motion and eighth-power polynomial that will not produce infinite jerk.

3.35. A follower moves with harmonic motion (H-1) a distance of 25 mm in $\pi/4$ rad of cam rotation. The follower then moves 25 mm more with cycloidal motion (C-2) to complete its displacement. The follower dwells and returns 25 mm with cycloidal motion

(C-3) and then moves the remaining 25 mm with harmonic motion (H-4) in $\pi/4$ rad. (a) Find the intervals of cam rotation for the cycloidal motions and the dwell by matching velocities and accelerations. (b) Determine the equation for S as a function of θ for each type of motion. These equations should be written so that the displacement measured from the zero position can be calculated for any cam angle by use of the proper equation.

3.36. In the displacement diagram of Fig. 3.47, it is desired to achieve the full lift of 1.5 in. of a radial flat-faced follower by matching the cycloidal curve C-1 with the harmonic curve H-2. (a) Using the data given in the diagram, determine the angle β_2 for the harmonic event so that both velocity and acceleration of the follower will be matched at B where the two events are joined. (b) Determine the maximum theoretical length of follower face needed for the two events shown.

3.37. A disk cam drives a radial flat-faced follower with simple harmonic motion. The follower moves out and back in one revolution of the cam. If the total displacement is 50 mm and the minimum radius 25 mm, determine the parametric (x and y) equations of the cam contour. Eliminate the parameter to obtain the equation of the curve, which is the cam contour. Determine the theoretical length of the follower face.

3.38. A flat-faced radial follower is driven through a total displacement of 1.6 in. The follower moves upwards 0.40 in. with constant acceleration for 60°, 0.80 in. with constant velocity for 60°, and the remaining 0.40 in. with constant deceleration for another 60° of cam rotation. The follower dwells for 45° and returns with simple harmonic motion as the cam completes its revolution.

For each type of motion, write an equation expressing displacement S as a function of cam angle θ. These equations should be written so that the displacement, measured from the zero position, can be calculated for any cam angle by the use of the proper equation. Calculate the minimum radius C and the maximum length of contact l_{max} for each type of motion. Specify the minimum radius of the cam and the length of the follower face.

3.39. A flat-faced radial follower is driven through a total displacement of 38 mm. The follower moves upwards 25 mm with constant acceleration for 120° and the remaining 13 mm with constant deceleration for 60° of cam rotation. The follower returns with simple harmonic motion in 90° and dwells for the remainder of the revolution of the cam. Complete the solution as outlined in Problem 3.38.

3.40. In the sketch shown in Fig. 3.48, the disk cam is used to position the radial flat-faced follower in a computing mechanism. The cam profile is to be designed to give a

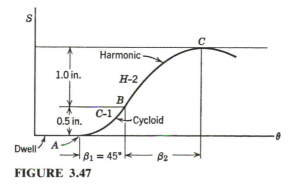

FIGURE 3.47

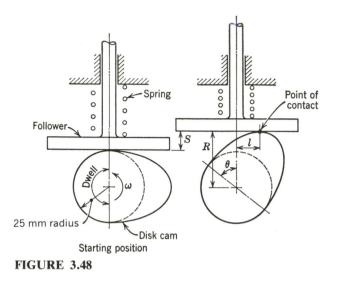

FIGURE 3.48

follower displacement S for a counterclockwise cam rotation θ according to the function $S = k\theta^2$ starting from dwell. For 60° cam rotation from the starting position, the lift of the follower is 10 mm. By analytical methods, determine the distances R and l when the cam has been turned 45° from the starting position. Also calculate whether cusps in the cam profile would occur in the total cam rotation of 60°.

3.41. A radial roller follower is driven through a total displacement of 25 mm with simple harmonic motion in a half revolution of the cam. The return motion is the same in a half revolution of the cam. Using a minimum radius R_0 of the pitch surface of 38 mm and a roller diameter of 19 mm, calculate a set of lift figures for the center of the follower for 15° increments of cam angle and lay out the cam full size. Calculate pressure angles to determine contact points.

3.42. A radial roller follower moves through a total displacement of 50 mm with cycloidal motion in 135° of cam rotation. The follower dwells for the next 90° and then returns 50 mm with cycloidal motion in 135° of cam rotation. Using a minimum radius R_0 of the pitch surface of 25 mm, calculate with a computer the displacement, velocity, acceleration, and pressure angle for the follower for each 10° rotation of the cam.

3.43. A radial roller follower is to move through a total displacement of $L = 0.75$ in. with harmonic motion while the cam rotates $\beta = 30°$. Check the cam for pointing if the radius R_r of the roller is 0.25 in. and the minimum radius R_0 of the pitch surface is 1.875 in.

3.44. A radial roller follower is to move through a total displacement of $L = 6.5$ mm with harmonic motion while the cam rotates $\beta = 45°$. The radius R_r of the roller is 6.5 mm. Determine the limiting R_0 that will give a pointed cam profile during this event.

3.45. A radial roller follower moves through a total displacement of $L = 0.75$ in. with cycloidal motion while the cam rotates $\beta = 30°$. Determine the radius of curvature ρ of the pitch surface when $\theta = 15°$. The radius R_r of the roller is 0.25 in. and R_0 is 1.875 in.

3.46. A radial roller follower is to move through a total displacement of $L = 19$ mm with harmonic motion while the cam rotates $\beta = 30°$. Find the value of R_0 to limit α_{max} to 30°.

3.47. By use of Eq. 3.12 and the appropriate expressions for R and $dR/d\theta$, develop the equation for α for cycloidal motion. Using the data from Example 3.5, calculate the pressure angle α when $\theta = 22.5°$.

3.48. A radial roller follower is to move through a total displacement of $L = 16$ mm with cycloidal motion while the cam rotates $\beta = 30°$. Assuming $R_0 = 38$ mm, determine α_{max}. If α_{max} is too large and if space requirements dictate that R_0 cannot be increased, make other recommendations to limit α_{max} to 30°.

3.49. Using the displacement figures of Problem 3.5, calculate the values of R and ϕ for a disk cam with oscillating roller follower. The cam is to rotate counterclockwise and has a minimum radius of 1 in. The diameter of the roller is to be $\frac{3}{4}$ in., and the distance from the center of the hub of the follower to the center of the roller is $2\frac{7}{8}$ in. The center of the hub is 3 in. to the right of the center of the cam. Let the zero position of the follower fall on the vertical centerline of the cam. Lay out the cam full size from the calculated values of R and ϕ, and check it graphically.

3.50. In the preceding problem (3.49), $\psi = 0.174(1 - \cos \theta)$ rad approximately. Using this relation, calculate the pressure angle at position 3.

3.51. Using the relation for ψ as a function of θ as given in Problem 3.50 and data from Problem 3.49, calculate the pressure angle for position 0 and check graphically.

3.52. Using the relation for ψ as a function of θ from Problem 3.50 and data from Problem 3.49, calculate the radius of curvature for position 2.

Chapter Four

Spur Gears

4.1 INTRODUCTION TO INVOLUTE SPUR GEARS

In considering two curved surfaces in direct contact, it has been shown that the angular-velocity ratio is inversely proportional to the segments into which the line of centers is cut by the line of action or common normal to the two surfaces in contact. If the line of action always intersects the line of centers at a fixed point, then the angular-velocity ratio remains constant. This is the condition that is desired when two gear teeth mesh together: The angular-velocity ratio must be constant. It is possible to assume the form of the tooth on one gear and, by applying the above principle (the common normal intersects the line of centers at a fixed point), to determine the outline of the mating tooth. Such teeth would be considered *conjugate teeth,* and the possibilities are limited only by one's ability to form the teeth. Of the many shapes possible, only the cycloid and the involute have been standardized. The cycloid was used first but has been replaced by the involute for all applications except watches and clocks. The involute has several advantages, the most important of which are its ease of manufacture and the fact that the center distance between two involute gears may vary without changing the velocity ratio. The involute system of gearing is discussed in detail in the following paragraphs. A pair of involute spur gears is shown in Fig. 4.1.

Consider two pulleys connected by a crossed wire as shown in Fig. 4.2. It is evident that the two pulleys will turn in opposite directions and that the angular-velocity ratio will be constant provided the wire does not slip and will depend upon the inverse ratio of diameters. It is also seen that the angular-velocity ratio will not change when the center distance is changed. For convenience, assume that one side of the wire is removed and a piece of cardboard is attached to wheel

FIGURE 4.1 Spur gears. (Courtesy of Philadelphia Gear Works.)

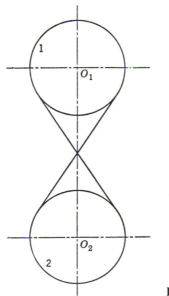

FIGURE 4.2

1 (Fig. 4.3a). Place a pencil at a point Q on the wire and turn wheel 2 counter-clockwise. Relative to the ground, point Q will trace a straight line, whereas relative to wheel 1, Q will trace an involute on the cardboard. The same involute could be generated by cutting the wire at Q and unwrapping the wire from the wheel 1, keeping the wire taut. If a cardboard is now attached to wheel 2 (Fig. 4.3b) and the process is repeated, an involute is generated on the cardboard of wheel 2. If the cardboards are now cut along the involute, one side of a tooth is formed on both wheels 1 and 2. The involute on wheel 1 can now be used to drive the involute on wheel 2. The angular-velocity ratio will be constant because the line of action, which by the method of constructing the involute is normal to the involutes at the point of contact Q, cuts the line of centers at a fixed point. As in the case of the pulleys with the crossed wire, the angular-velocity ratio is inversely proportional to the diameters of the wheels. If the center distance is changed, involute 1 will still drive involute 2, but different portions of the two involutes will now be in contact. As long as the diameters of the wheels are not changed, the velocity ratio will be the same as before.

The circles that were used as a basis for generating the involutes are known as *base circles,* and they are the heart of the involute system of gearing. In Fig. 4.4, the angle that is included by a line perpendicular to the line of action through the center of the base circle and a line from O_1 to Q (or O_2 to Q) is known as the *involute pressure angle* and is a dimension of the point on the involute at which contact is taking place. If in Fig. 4.4 the point of intersection of the line of action and the line of centers is labeled P, the angular-velocity ratio will be

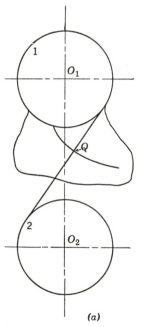

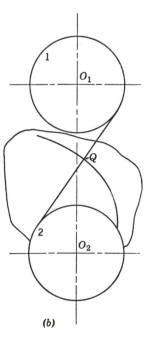

(a)   (b)

FIGURE 4.3

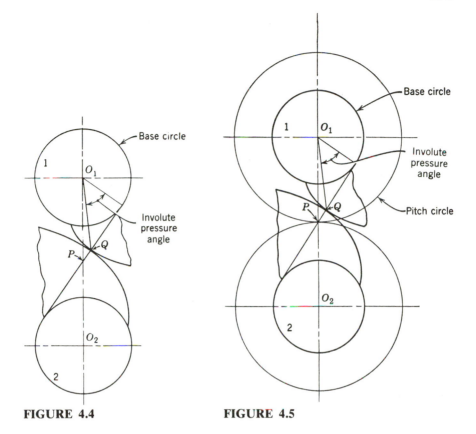

FIGURE 4.4   FIGURE 4.5

inversely proportional to the segments into which this point divides the line of centers.

It is possible to draw circles through point P using first O_1 as a center and then O_2. Figure 4.5 shows this condition. Point P is called the *pitch point*, and the circles which pass through this point are known as *pitch circles*. It can be proved that, as involute 1 drives involute 2, the two pitch circles will move together with pure rolling action. Because the segments into which point P divides the line of centers are now the radii of the pitch circles, the angular-velocity ratio is inversely proportional to the radii of the two pitch circles. If the diameter of pitch circle 1 is D_1 and that of circle 2 is D_2, then $\omega_1/\omega_2 = D_2/D_1$. It will be shown in a later section that the number of teeth on a gear is directly proportional to the pitch diameter. Therefore, $\omega_1/\omega_2 = D_2/D_1 = N_2/N_1$.

4.2 INVOLUTOMETRY

In considering the involute for a tooth form, it is necessary to be able to calculate certain properties of the involute.

Figure 4.6 shows an involute which has been generated from a base circle

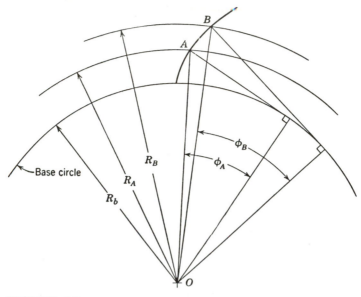

FIGURE 4.6

of radius R_b. The involute contains two points A and B with corresponding radii R_A and R_B and involute pressure angles ϕ_A and ϕ_B. It is an easy matter to work out a relationship for the above factors because the base circle radius remains constant no matter which point is under consideration. Therefore,

$$R_b = R_A \cos \phi_A \qquad\qquad (4.1)$$

or

$$R_b = R_B \cos \phi_B$$

and

$$\cos \phi_B = \frac{R_A}{R_B} \cos \phi_A \qquad\qquad (4.2)$$

From Eq. 4.2, it is possible to determine the involute pressure angle at any point of known radius on the involute.

Figure 4.7 shows the sketch of Fig. 4.6 extended to include the whole gear tooth. From this sketch it will be possible to develop an equation for finding the tooth thickness at any point B, given the thickness at point A.

From the principle of the generation of an involute, arc DG is equal to

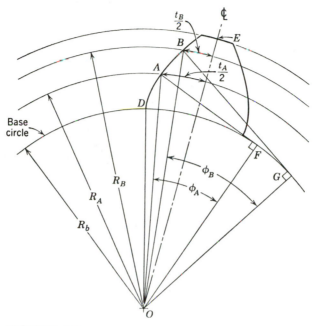

FIGURE 4.7

length BG. Therefore,

$$\angle DOG = \frac{\text{arc } DG}{OG} = \frac{BG}{OG}$$

$$\tan \phi_B = \frac{BG}{OG}$$

Thus,

$$\angle DOG = \tan \phi_B$$

Also,

$$\angle DOB = \angle DOG - \phi_B$$

$$= \tan \phi_B - \phi_B$$

It can also be shown that

$$\angle DOA = \tan \phi_A - \phi_A$$

The expression $\tan \phi - \phi$ is called an *involute function* and is sometimes written inv ϕ. It is easy to calculate the involute function when the angle is known; ϕ is expressed in radians. However, it is very difficult to convert from inv ϕ to ϕ; for this reason, extensive tables of involute functions have been published. See Appendix 2.

By referring again to Fig. 4.7,

$$\angle DOE = \angle DOB + \frac{\frac{1}{2}t_B}{R_B}$$

$$= \text{inv } \phi_B + \frac{t_B}{2R_B}$$

Also

$$\angle DOE = \angle DOA + \frac{\frac{1}{2}t_A}{R_A}$$

$$= \text{inv } \phi_A + \frac{t_A}{2R_A}$$

From the above relations,

$$t_B = 2R_B \left[\frac{t_A}{2R_A} + \text{inv } \phi_A - \text{inv } \phi_B \right] \tag{4.3}$$

By means of Eq. 4.3, it is possible to calculate the tooth thickness at any point on the involute, given the thickness at some other point. An interesting application of this equation is to determine the radius at which the tooth becomes pointed.

4.3   SPUR GEAR DETAILS

To continue the study of involute gearing, it is necessary to define the basic elements of a gear as shown in Figs. 4.8a and 4.8b. It should also be mentioned that the smaller of two gears in mesh is called the *pinion;* the pinion is generally the driver. If the pitch radius R of a gear becomes infinite, a rack results as seen in Fig. 4.8c and Fig. 4.9. The side of the rack tooth is a straight line, which is the form taken by an involute when it is generated on a base circle of infinite radius. From Fig. 4.8a, the *base pitch* p_b is the distance from a point on one tooth to the corresponding point on the next tooth measured on the base circle. The *circular pitch p* is defined in the same way, except that it is measured on the pitch circle. The *addendum a* and *dedendum b* are radial distances measured as shown. The portion of the flank below the base circle is approximately a radial line. The tooth curve is the line of intersection of the tooth surface and the pitch surface.

Although it is impossible to show on the sketches of Fig. 4.8, backlash is

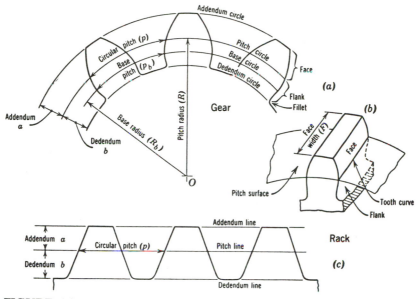

FIGURE 4.8

FIGURE 4.9 Spur pinion and rack. (Models courtesy of Illinois Gear & Machine Company.)

an important consideration in gearing. *Backlash* is the amount by which the width of a tooth space exceeds the thickness of the engaging tooth on the pitch circles. Theoretically, backlash should be zero, but practically some allowance must be made for thermal expansion and tooth error. Unless otherwise stated, zero backlash is to be assumed in this text. In a later section, the method for calculating backlash for a change in center distance will be given.

4.4 CHARACTERISTICS OF INVOLUTE ACTION

From the discussion of the generation of the involute, it was seen that the common normal to the two involute surfaces is tangent to the two base circles. This common normal is also referred to as the *line of action*. The beginning of contact occurs where the line of action intersects the addendum circle of the driven gear, and the end of contact occurs where the line of action intersects the addendum circle of the driver. That this occurs is evident from Fig. 4.10, which shows a pair of teeth just coming into contact and the same pair as they later go out of contact (shown dotted). Point A is the beginning of contact and point B the end of contact. The path of the point of contact is therefore along the straight line APB. Point C is where the tooth profile (gear 1) at the beginning of contact cuts the pitch circle. Point C' is where the profile at the end of contact cuts the pitch circle. Points D and D' are similar points on gear 2. The arcs CC' and DD' are called *arcs of action* and must be equal for pure rolling action of the pitch circles to take place as mentioned earlier. The angles of motion are generally broken into two parts as shown in Fig. 4.10, where α is the *angle of approach* and β the *angle of recess*. The angle of approach does not in general equal the angle of recess. For continuous driving to take place, the arc of action must be equal to or greater than the circular pitch. If this is true, then a new pair of teeth will come into action before the preceding pair goes out of action.

The ratio of the arc of action to the circular pitch is known as the *contact ratio*. The contact ratio for involute gears is also equal to the ratio of the *length of action* (that is, the distance from the beginning to the end of contact measured

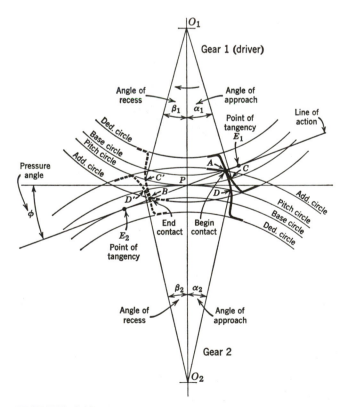

FIGURE 4.10

on the line of action) to the base pitch and is generally calculated in this way as will be shown later. Considered physically, the contact ratio is the average number of teeth in contact. If, for example, the ratio is 1.60, it does not mean that there are 1.60 teeth in contact. It means that there are alternately one pair and two pairs of teeth in contact, and on a time basis the number averages 1.60. The theoretical minimum value of the contact ratio is 1.00. This value, of course, must be increased for actual operating conditions. Although it is difficult to quote specific values because of the many conditions involved, 1.40 has been used as a practical minimum, with 1.20 for extreme cases. It should be noted, however, that the lower the contact ratio, the higher the degree of accuracy needed in machining the profiles to secure quiet running.

Figure 4.10 also shows an angle ϕ, which is formed by the line of action and a line perpendicular to the line of centers at the pitch point P. This angle is known as the *pressure angle of the two gears in mesh* and must be differentiated from the involute pressure angle of a point on an involute. When the two gears are in contact at the pitch point, the pressure angle of the gears in mesh and the involute pressure angles of the two involutes in contact at the pitch point will be equal. These angles can be seen in Fig. 4.11.

An equation for the length of action Z can be derived from Fig. 4.11,

where

$$A = \text{beginning of contact}$$
$$B = \text{end of contact}$$
$$E_1 \text{ and } E_2 = \text{points of tangency of line of action and base circles}$$
$$R_o = \text{outside radius}$$
$$R_b = \text{base radius}$$
$$\phi = \text{pressure angle}$$
$$C = \text{center distance}$$

From the figure,

$$Z = AB = E_1B + E_2A - E_1E_2$$

Therefore,

$$Z = \sqrt{(R_{o_1})^2 - (R_{b_1})^2} + \sqrt{(R_{o_2})^2 - (R_{b_2})^2} - C \sin \phi \qquad \textbf{(4.4)}$$

The base pitch p_b is given by

$$p_b = \frac{2\pi R_b}{N} \qquad \textbf{(4.5)}$$

where

$$R_b = \text{base radius}$$
$$N = \text{number of teeth}$$

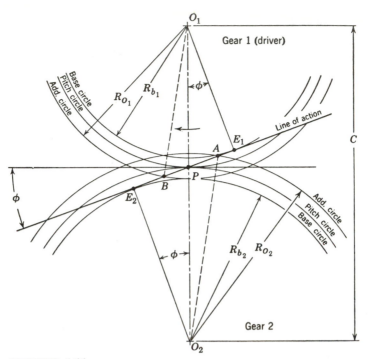

FIGURE 4.11

The contact ratio m_p is then

$$m_p = \frac{Z}{p_b} \tag{4.6}$$

The equation for the length of action for a rack and pinion can be derived in a similar manner as

$$Z = \sqrt{(R_o)^2 - (R_b)^2} - R \sin \phi + \frac{a}{\sin \phi} \tag{4.7}$$

where

R = pitch radius

a = addendum

If it seems odd to calculate a contact ratio by dividing a straight-line measurement by a circular measurement, consider the sketches in Fig. 4.12. In Fig. 4.12a are shown two adjacent teeth on one gear of a mating pair. The base pitch p_b is dimensioned on the base circle in accordance with its definition. A straight-line distance on the line of action is also designated p_b. From the manner in which two adjacent involutes would be generated, it can be seen that the two distances marked p_b must be equal. Therefore, the base pitch can also be considered as

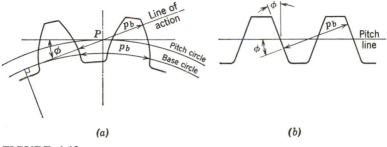

(a) *(b)*

FIGURE 4.12

the normal distance between the corresponding sides of adjacent teeth. Figure 4.12b illustrates how the base pitch is measured on a rack.

Example 4.1. A pinion of 24 teeth drives a gear of 60 teeth at a pressure angle of 20°. The pitch radius of the pinion is 1.5000 in., and the outside radius is 1.6250 in. The pitch radius of the gear is 3.7500 in., and the outside radius is 3.8750 in. Calculate the length of action and contact ratio.

Solution.

$$Z = \sqrt{(R_{o_1})^2 - (R_{b_1})^2} + \sqrt{(R_{o_2})^2 - (R_{b_2})^2} - C \sin \phi$$

$$R_{o_1} = 1.6250 \text{ in.}$$

$$R_{b_1} = R_1 \cos \phi = 1.5000 \cos 20° = 1.4095 \text{ in.}$$

$$R_{o_2} = 3.8750 \text{ in.}$$

$$R_{b_2} = R_2 \cos \phi = 3.75 \cos 20° = 3.5238 \text{ in.}$$

$$C \sin \phi = (1.500 + 3.750) \sin 20° = 1.7956 \text{ in.}$$

$$Z = \sqrt{1.6250^2 - 1.4095^2} + \sqrt{3.8750^2 - 3.5238^2} - 1.7956$$

$$= \sqrt{2.6406 - 1.9867} + \sqrt{15.0156 - 12.4172} - 1.7956$$

$$= 0.8099 + 1.6115 - 1.7956 = 0.6258 \text{ in.}$$

Therefore,

$$Z = AB = 0.6258 \text{ in.}$$

$$m_p = \frac{Z}{p_b} \quad \text{and} \quad p_b = \frac{2\pi R_{b_1}}{N_1} = \frac{2\pi \times 1.4095}{24} = 0.3689 \text{ in.}$$

Therefore,

$$m_p = \frac{0.6258}{0.3689} = 1.6964$$

4.5 INTERFERENCE IN INVOLUTE GEARS

It was mentioned previously that an involute starts at the base circle and is generated outward. It is therefore impossible to have an involute inside the base circle. The line of action is tangent to the two base circles of a pair of gears in mesh, and these two points represent the extreme limits of the length of action. These two points are spoken of as *interference points*. If the teeth are of such proportion that the beginning of contact occurs before the interference point is met, then the involute portion of the driven gear will mate with a noninvolute portion of the driving gear, and *involute interference* is said to occur. This condition is shown in Fig. 4.13; E_1 and E_2 show the interference points that should limit the length of action, A shows the beginning of contact, and B shows the end of contact. It is seen that the beginning of contact occurs before the interference point E_1 is met; therefore, interference is present. The tip of the driven tooth will gouge out or undercut the flank of the driving tooth as shown by the dotted line. There are several ways of eliminating interference, one of which is to limit the addendum of the driven gear so that it passes through the interference point E_1, thus giving a new beginning of contact. If this is done in this case, interference will be eliminated.

Involute interference is undesirable for several reasons. Interference and the resulting undercutting not only weaken the pinion tooth but may also remove a small portion of the involute adjacent to the base circle, which may cause a serious reduction in the length of action.

The conditions for interference of a rack and a pinion will now be discussed.

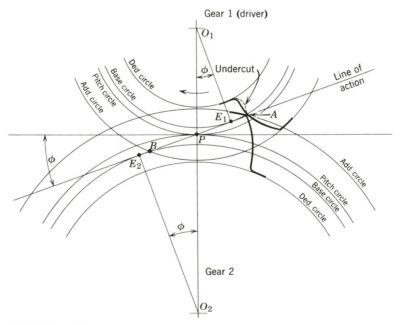

FIGURE 4.13

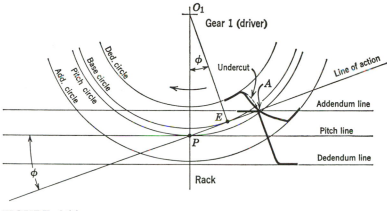

FIGURE 4.14

In Fig. 4.14 is shown a pinion and rack in mesh. The point of tangency of the line of action and the base circle of the pinion is labeled as the interference point E, the same as in the case for the pinion and gear. The interference point therefore fixes the maximum addendum for the rack for the pressure angle shown. With the rack addendum as shown in Fig. 4.14, contact begins at A, and undercutting will occur as shown by the dotted line. If the addendum of the rack extends only to the line that passes through the interference point E, then the interference point becomes the beginning of contact, and interference is eliminated.

It can also be seen from Fig. 4.14 that, if a gear of finite radius with the same addendum as the rack (rack addendum now passing through interference point) were to mesh with the pinion, the beginning of contact would occur on the line of action at some place between the pitch point P and the interference point E. Therefore, there would be no possibility of interference between the pinion and gear. It may be concluded then that, if the number of teeth on the pinion is such that it will mesh with a rack without interference, it will mesh without interference with any other gear having the same or a larger number of teeth.

Although involute interference and its resulting undercutting should be avoided, a small amount might be tolerated if it does not reduce the contact ratio for a pair of mating gears below a suitable value. However, the problem of determining the length of action when undercutting has occurred is difficult, and it cannot be calculated from Eq. 4.4. It can be seen from Fig. 4.11 and Eq. 4.4 that if the value of either radical is greater than $C \sin \phi$, interference will exist.

4.6 GEAR STANDARDIZATION

Thus far, no attempt has been made to bring up the question of standardizing spur gears for facilitating the development of interchangeable gears; the foregoing discussions applied to spur gears in general with no thought to interchangeability.

Tied in with the question of interchangeability is the manner in which the gears are to be cut.

There are several ways of machining spur gears, the oldest of which is to use a form-milling cutter to remove the material between the teeth as the gear blank is indexed through a complete revolution on a milling machine. This method produces a composite involute and cycloidal profile and finds application primarily in the cutting of replacement gears which cannot be economically obtained from standard sources. This method is also used to produce gears with large-size teeth that cannot be cut on conventional gear generators. Modern spur gears are gen-

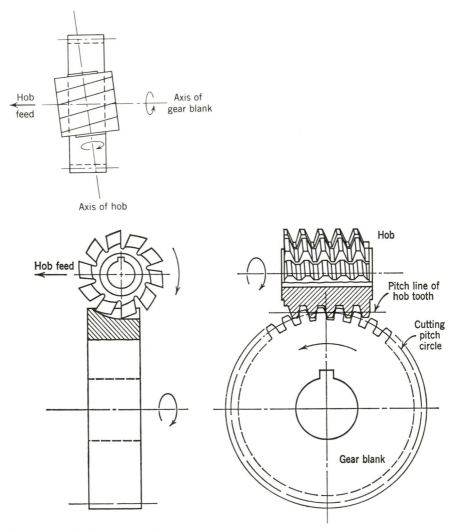

FIGURE 4.15a Generating a spur gear with a hob.

FIGURE 4.15b **Hobbing operation. (Photograph courtesy of Barber-Colman.)**

erated to produce an involute tooth profile. The two most common methods for producing modern spur gears are the method of hobbing and the Fellows method of shaping. The principles of hobbing and of the Fellows method are illustrated for cutting external gears in Figs. 4.15 and 4.16, respectively. To cut small internal gears, it is necessary to use the Fellows method; however, if space is available, large internal gears can be hobbed. The Fellows method is also used for cutting shoulder gears where space is insufficient at one end of the teeth to allow run-out of a hob, as shown in Fig. 4.15a.

As gear technology developed, a means of classifying the cutters and the gears produced was sought. The method that was adopted in the United States was that of specifying the ratio of the number of teeth to the pitch diameter. This ratio was given the name *diametral pitch* and is expressed as

$$P_d = \frac{N}{D} \tag{4.8}$$

where

N = number of teeth

D = pitch diameter, in.

Although the units of diametral pitch are teeth per inch, it is not customary to give the units when specifying numerical values.

In Europe, the method of classification is to specify the ratio of the pitch diameter to the number of teeth, and this ratio is designated as *module*. The module is therefore the reciprocal of the diametral pitch and is expressed as

$$m = \frac{D}{N} \tag{4.9}$$

where

D = pitch diameter, mm

N = number of teeth

m = module

The numerical values of modules are specified in units of millimeters.

It should be noted that diametral pitch and module are defined as ratios and are not physical distances that can be measured on a gear. Circular pitch, on the other hand, has been previously defined as the distance measured along the pitch circle from a point on one tooth to the corresponding point on the next tooth. The relation between circular pitch and diametral pitch or module can be expressed as follows:

$$p = \frac{\pi D}{N} = \frac{\pi}{P_d} \qquad \text{(U.S.)} \tag{4.10}$$

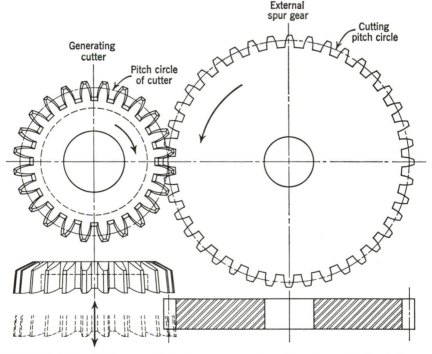

FIGURE 4.16a Fellows method of gear generating. (Courtesy of Fellows Corporation.)

FIGURE 4.16b **Shaping operation. (Photograph courtesy of Barber-Colman.)**

and

$$p = \frac{\pi D}{N} = \pi m \qquad \text{(metric)} \qquad \textbf{(4.11)}$$

where

p = circular pitch

P_d = diametral pitch

m = module

For the purpose of specifying gear cutters, the values of diametral pitch and module were generally taken as whole numbers. The following is a list of commercially available gear hobs in diametral pitches with $14\frac{1}{2}°$ and/or 20° pressure angles:

2, $2\frac{1}{2}$, 3, $3\frac{1}{2}$, 4, 5, 6, 7, 8, 9, 10, 12, 14, 16, 18, 20, 22, 24, 26, 28, 30, 32, 36, 40, 42, 48, 50, 64, 72, 80, 96, 120

Finer pitches may be specified by even increments up to 200. Pitches commonly used for precision gears for instruments are 48, 64, 72, 80, 96, and 120. The

AGMA (American Gear Manufacturers Association) also lists diametral pitches of $\frac{1}{2}$ and 1, although hobs of these sizes are not usually carried as stock items by tool manufacturers. The following is a list of standard hobs in metric modules (20° pressure angle):

> 1, 1.25, 1.50, 1.75, 2, 2.25, 2.50, 2.75, 3, 5, 6, 8, 10, 12, 16, 20

When cutters were standardized, a pressure angle of $14\frac{1}{2}°$ was adopted. This was a carry-over from the process of casting gears which used $14\frac{1}{2}°$ because sin $14\frac{1}{2}°$ approximates $\frac{1}{4}$, which was convenient for pattern layout. Later, a pressure angle of 20° was also adopted. Both $14\frac{1}{2}°$ and 20° have been used for many years, but the tendency in recent years has been toward using 20° in preference to $14\frac{1}{2}°$. It will be shown in a later section that it is possible to cut a pinion with fewer teeth without undercutting when using a 20° pressure angle hob in contrast to one of $14\frac{1}{2}°$. As a result of the trend toward higher pressure angles, the AGMA has adopted 20° and 25° for coarse-pitch gearing (1–19.99 P_d) and 20° for fine-pitch gearing (20–200 P_d). British and German metric standards specify 20° pressure angle. The Society of Automotive Engineers in their Aerospace Standard AS1560 (February 1979) for metric gears recommends a 20° pressure angle for general purposes. Pressure angles of 22.5° and 25° are also included because these high-pressure angles are in use for aerospace gearing.

The tooth proportions of U.S. Standard involute spur gears are given in Table 4.1.

Table 4.2 gives the proportions for the full depth $14\frac{1}{2}°$ and the stub 20° spur gears. Although these gears are seldom specified in new designs, they are essential for replacement gears in older machinery.

Because of the design of gear shaper cutters, such cutters are classified not only according to diametral pitch or module but also according to pitch diameter and number of teeth. Table 4.3 shows a tabulation of standard spur gear shaper cutters classified by diametral pitch, and Table 4.4 shows a listing of metric spur gear shaper cutters.

The British standard metric modules are shown in Table 4.5. Tooth proportions are as follows:

Addendum (a)	$1.000\,m$
Dedendum (b)	$1.250\,m$
Pressure angle (ϕ)	20°

The German standard metric modules are shown in Table 4.6. Tooth proportions are as follows:

Addendum (a)	$1.000\,m$
Dedendum (b)	$1.157\,m$ or $1.167\,m$
Pressure angle (ϕ)	20°

TABLE 4.1 Tooth Proportions—Involute Spur Gears

	Coarse Pitch (1–19.99 P_d) AGMA 201.02 August 1974 20° or 25° Full Depth	Fine Pitch (20–200 P_d) AGMA 207.06 November 1977 20° Full Depth
Addendum (a)	$\dfrac{1.000}{P_d}$	$\dfrac{1.000}{P_d}$
Dedendum (b)	$\dfrac{1.250}{P_d}$	$\dfrac{1.200}{P_d} + 0.002$ (min)
Clearance (c) (dedendum − addendum)	$\dfrac{0.250}{P_d}$	$\dfrac{0.200}{P_d} + 0.002$ (min)[a]
Working depth (h_k) (twice addendum)	$\dfrac{2.000}{P_d}$	$\dfrac{2.000}{P_d}$
Whole depth (h_t) (addendum + dedendum)	$\dfrac{2.250}{P_d}$	$\dfrac{2.200}{P_d} + 0.002$ (min)
Fillet radius of basic rack (r_f)	$\dfrac{0.300}{P_d}$	Not given
Tooth thickness (t)	$\dfrac{1.5708}{P_d}$	$\dfrac{1.5708}{P_d}$

[a]For shaved or ground teeth, $c = 0.350/P_d + 0.002$ (min).

TABLE 4.2 Tooth Proportions for Full Depth $14\frac{1}{2}°$ and Stub 20° Spur Gears

	$14\frac{1}{2}°$ Full Depth	20° Stub
Addendum (a)	$\dfrac{1.000}{P_d}$	$\dfrac{0.800}{P_d}$
Dedendum (b)	$\dfrac{1.157}{P_d}$	$\dfrac{1.000}{P_d}$
Clearance (c)	$\dfrac{0.157}{P_d}$	$\dfrac{0.200}{P_d}$
Fillet radius (r_f)	$\dfrac{0.209}{P_d}$	$\dfrac{0.304}{P_d}$
Tooth thickness (t)	$\dfrac{1.5708}{P_d}$	$\dfrac{1.5708}{P_d}$

TABLE 4.3 Spur Gear Shaper Cutters[a]

Diametral Pitch	Pitch Diameter, in.				Number of Teeth			
(a) 14½° Pressure Angle								
4	4		5		16		20	
5	4				20			
6	4		5		24		30	
8	4		5		32		40	
10	3	4	5		30	40	50	
12	3		4		36		48	
16	3		4		48		64	
20	3		4		60		80	
24	3				72			
32	3				96			
(b) 20° Pressure Angle								
3	4				12			
4	4	5	6		16	20	24	
5	4	5	6		20	25	30	
6	4	5	6		24	30	36	
8	3	4	5	6	24	32	40	48
10	3	4	5		30	40	50	
12	3	4	5		36	48	60	
14	4				56			
16	3		4		48		64	
18	4				72			
20	3		4		60		80	
24	3				72			
32	3				96			

[a]The following cutters are also manufactured to Fellows fine-pitch precision limits: 32, 48, 64, 72, 80, 96, and 120 diametral pitch. Courtesy of Fellows Corporation.

From the fact that gear cutters in both the U.S. and metric systems were taken generally as whole numbers, conversion from diametral pitches to module millimeters does not give whole-number values. See Table 4.7.

The metric symbols used to denote the proportions of spur gears vary considerably from those recommended by AGMA. Table 4.8 shows the comparison between the AGMA and the proposed International Standard ISO 701. Similar tables are given for bevel, helical, and worm gears in Chapter 6.

If spur gears are cut with standard cutters, they will be interchangeable if the following conditions are satisfied:

1. The diametral pitches or modules of the cutters used to produce the gears are equal.

**TABLE 4.4 Metric Spur Gear Shaper Cutters:
20° Pressure Angle—Full Depth**[a]

Module	Diametral Pitch	Pitch Diameter, in.	Number of Teeth
1.0	25.400	2.992	76
1.5	16.933	2.953	50
2.0	12.700	2.992	38
2.5	10.160	2.953	30
3.0	8.466	3.071	26
3.5	7.257	3.031	22
4.0	6.350	4.094	26
4.5	5.644	3.897	22
5.0	5.080	3.937	20
6.0	4.233	4.252	18
8.0	3.175	5.039	16

[a]Courtesy of Illinois Tool Works.

**TABLE 4.5 British Standard Normal
Metric Modules**[a]

Preferred Modules	Second-Choice Modules
1	1.125
1.25	1.375
1.5	1.75
2	2.25
2.5	2.75
3	3.5
4	4.5
5	5.5
6	7
8	9
10	11
12	14
16	18
20	22
25	28
32	36
40	45
50	

[a]The values are in millimeters. Wherever possible, the preferred modules should be applied rather than those of second choice. B.S. 436: Part 2: 1970. From *Machinery Handbook,* 22nd edition, 1984, p. 823.

**TABLE 4.6 German
Standard Metric Modules**[a]

0.3	2.5	8	27
0.4	2.75	9	30
0.5	3	10	33
0.6	3.25	11	36
0.7	3.5	12	39
0.8	3.75	13	42
0.9	4	14	45
1	4.5	15	50
1.25	5	16	55
1.5	5.5	18	60
1.75	6	20	65
2	6.5	22	70
2.25	7	24	75

[a]The values are in millimeters. DIN-867. From *Machinery Handbook,* 22nd edition, 1984, p. 966.

TABLE 4.7 Diametral Pitch and Metric Module

Diametral Pitch	Module Millimeters	Diametral Pitch	Module Millimeters
0.5000	50.8000	11	2.3091
0.7500	33.8667	12	2.1167
1	25.4000	13	1.9538
1.2500	20.3200	14	1.8143
1.5000	16.9333	15	1.6933
1.7500	14.5143	16	1.5875
2	12.7000	17	1.4941
2.2500	11.2889	18	1.4111
2.5000	10.1600	19	1.3368
2.7500	9.2364	20	1.2700
3	8.4667	24	1.0583
3.5000	7.2571	32	0.7938
4	6.3500	40	0.6350
5	5.0800	48	0.5292
6	4.2333	64	0.3969
7	3.6286	72	0.3528
8	3.1750	80	0.3175
9	2.8222	96	0.2646
10	2.5400	120	0.2117

TABLE 4.8 Spur Gear Symbols

	AGMA	ISO 701
Number of teeth	N	z
Pitch radius	R	r
Pitch diameter	D	d'
Outside radius	R_o	r_a
Outside diameter	D_o	d_a
Base radius	R_b	r_b
Face width	F	b
Addendum	a	h_a
Dedendum	b	h_f
Circular pitch	p	p
Base pitch	p_b	p_b
Pressure angle	ϕ	α
Length of action	Z	g_α
Contact ratio (transverse)	m_p	ϵ_a
Center distance	C	a
Working depth	h_k	—
Whole depth	h_t	h
Tooth thickness	t	s
Clearance	c	c
Backlash	B	j_t

2. The pressure angles of the cutters used are equal.
3. The gears have the same addendums and the same dedendums.
4. Tooth thicknesses of the gears are equal to one half the circular pitch.

The term *standard gear* is often used and means that it was cut by one of the standard cutters listed previously and that the tooth thickness is equal to the tooth space which equals one half of the circular pitch. Standard gears are interchangeable. The spur gears that are offered for sale in gear manufacturers' catalogs are standard gears. However, most of the gears used in the automotive and the aircraft industries are nonstandard to give certain advantages over standard gears. Chapter 5 presents the subject of nonstandard spur gears and shows that they can be cut with standard cutters.

4.7 MINIMUM NUMBER OF TEETH TO AVOID INTERFERENCE

The question of interference was considered previously for the meshing of a pinion and gear and for a pinion and rack. From the discussion of Fig. 4.14, it was found that if there were no interference between a pinion and rack, there would be no interference between the pinion and a gear the same size as the pinion or larger. Naturally, this is assuming the same tooth proportions for the two cases. When considering a standard gear where the tooth proportions are those given in the tables, it is possible to calculate the minimum number of teeth in a pinion that will mesh with a rack without involute interference. To solve for this limiting case, the addendum line of the rack is passed through the interference point of the pinion.

In Fig. 4.17, the essential features of a pinion and rack for this case are shown. The pitch point is notated by P and the interference point by E. Therefore,

$$\sin \phi = \frac{\overline{PE}}{R}$$

also

$$\sin \phi = \frac{a}{\overline{PE}} = \frac{k/P_d}{\overline{PE}}$$

where k is a constant that, when divided by the diametral pitch, gives the addendum ($a = k/P_d$). For the full-depth system, $k = 1.00$; and for the stub system, $k = 0.80$. Multiplying the two equations for $\sin \phi$ together gives

$$\sin^2 \phi = \frac{k}{RP_d}$$

But

$$P_d = \frac{N}{2R}$$

where N = number of teeth. Therefore,

$$\sin^2 \phi = \frac{2k}{N}$$

and

$$N = \frac{2k}{\sin^2 \phi} \tag{4.12}$$

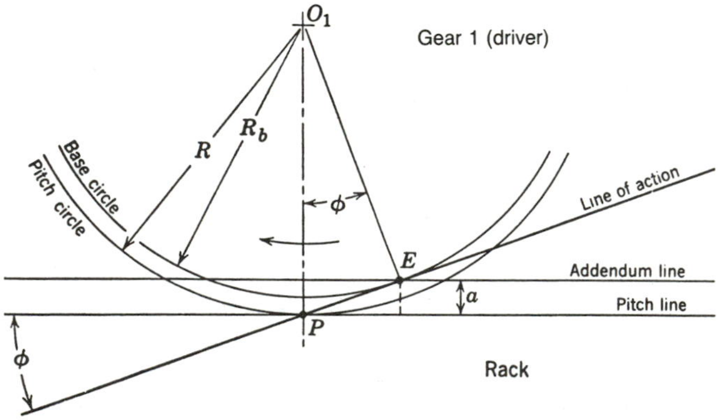

FIGURE 4.17

TABLE 4.9 Minimum Number of Teeth to Mesh With a Rack Without Undercutting

	$14\frac{1}{2}°$ Full Depth	$20°$ Full Depth	$20°$ Stub	$25°$ Full Depth
N	32	18	14	12

From this equation, the smallest number of teeth for a pinion that will mesh with a rack without interference can be calculated for any given standard tooth system. These are shown in Table 4.9 for the common systems. Because the values in the table were calculated for a pinion meshing with a rack, they can also be used as minimums for a pinion meshing with a gear without danger of interference.

Because the tooth action of a hob cutting a spur gear is similar to that of a pinion meshing with a rack, Eq. 4.12 can be used to determine the minimum numbers of teeth that can be cut without undercutting. For this case, the value of k must be increased from 1.000 to allow for cutting the necessary clearance between mating gears. This results in $k = 1.157$ for $14\frac{1}{2}°$ gears and $k = 1.250$ for $20°$ and $25°$ gears. Therefore, the minimum number of teeth than can be hobbed is 37 teeth for $14\frac{1}{2}°$, 21 teeth for $20°$, and 14 teeth for $25°$ pressure angles. Figure 4.18 shows two computer-generated plots of a severely undercut spur gear having 10 teeth cut with a $20°$ full-depth hob.

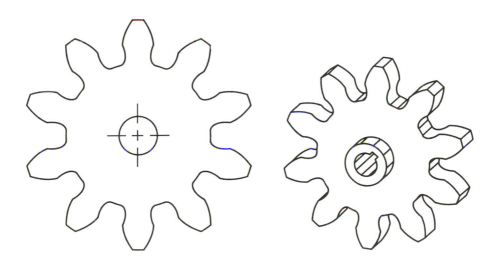

FIGURE 4.18 Undercut standard spur gear with $N = 10$ produced by a hob of $P_d = 1$ and $\phi = 20°$. Figures generated using an IBM 7375 pen plotter. (Courtesy of CAD-CAM Laboratory at Virginia Polytechnic Institute and State University.)

Determining the minimum number of teeth that a pinion cutter can generate on a gear without undercutting is more difficult than determining the minimum number of teeth when the gear is cut by a rack or a hob. An equation for determining the approximate number of teeth can be developed from Fig. 4.19. In this figure, the addendum circle of gear 2 passes through the interference point E of gear 1. The following relation for the outside radius of gear 2 can be written as

$$R_{o_2} = \sqrt{(R_{b_2})^2 + C^2 \sin^2 \phi}$$

By substituting,

$$R_{o_2} = R_2 + a = \frac{N_2 + 2k}{2P_d}$$

$$R_{b_2} = R_2 \cos \phi = \frac{N_2}{2P_d} \cos \phi$$

and

$$C = R_1 + R_2 = \frac{N_1 + N_2}{2P_d}$$

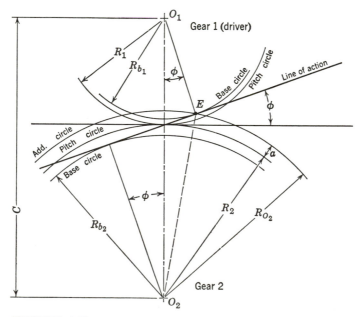

FIGURE 4.19

Therefore,

$$\frac{N_2 + 2k}{2P_d} = \sqrt{\left(\frac{N_2}{2P_d}\right)^2 \cos^2 \phi + \left(\frac{N_1 + N_2}{2P_d}\right)^2 \sin^2 \phi}$$

and

$$(N_2 + 2k)^2 = (N_2)^2 \cos^2 \phi + (N_1 + N_2)^2 \sin^2 \phi$$

Expanding and using the relation $\sin^2 \phi + \cos^2 \phi = 1$ gives the following equation from which the largest spur gear (N_2) that can be meshed with a given gear (N_1) without causing involute interference on gear 1 can be determined:

$$N_2 = \frac{4k^2 - (N_1)^2 \sin^2 \phi}{2N_1 \sin^2 \phi - 4k} \tag{4.13}$$

Equation 4.13 can be expanded to give

$$2N_2N_1 \sin^2 \phi - 4kN_2 = 4k^2 - (N_1)^2 \sin^2 \phi$$

and

$$(N_1)^2 \sin^2 \phi + 2N_2N_1 \sin^2 \phi - 4k(N_2 + k) = 0$$

Thus,

$$(N_1)^2 + 2N_2N_1 - \frac{4k}{\sin^2 \phi}(N_2 + k) = 0 \tag{4.14}$$

Equation 4.14 can be simplified as follows: If $\phi = 14\frac{1}{2}°$, then

$$(N_1)^2 + 2N_2N_1 - 63.8k(N_2 + k) = 0 \tag{4.15}$$

If $\phi = 20°$, then

$$(N_1)^2 + 2N_2N_1 - 34.2k(N_2 + k) = 0 \tag{4.16}$$

A curve has been plotted in Fig. 4.20 from Eq. 4.16, showing the relation of N_1 as a function of N_2 for 20° full-depth teeth ($k = 1$). This curve can also be used to approximate the minimum numbers of teeth that can be cut by a pinion cutter by considering N_1 as the number of teeth being cut in the gear and N_2 as the number of teeth in the pinion cutter. However, the values of N_1 will only be approximate because the outside radius of gear 2 used to develop Eq. 4.13 was taken as $R_{o_2} = R_2 + a$. If gear 2 is considered a pinion cutter, its outside radius

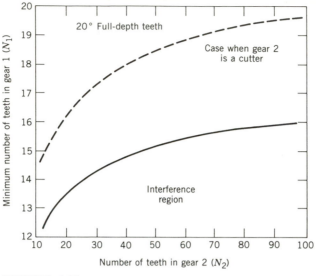

FIGURE 4.20

must be increased to cut clearance on gear 1. In other words, the addendum of the cutter must be equal to the dedendum of the gear being cut. In addition, as can be seen in Table 4.1, the equation for clearance is not identical for coarse-pitch and for fine-pitch gears.

A dotted curve has been added to Fig. 4.20 to show the relation of N_1, the number of teeth generated without undercutting, to the number of teeth N_2 assumed in the cutter when clearance has been added. For this case, clearance for course-pitch gears was used, and calculations were made using Eq. 4.16 with $k = 1.250$.

4.8 DETERMINATION OF BACKLASH

In Fig. 4.21a is shown the outline of two standard gears meshing at the standard center distance

$$C = \frac{N_1 + N_2}{2P_d} \qquad \text{(U.S.)}$$

$$C = \frac{(N_1 + N_2)}{2} m \qquad \text{(metric)}$$

with zero backlash. The pitch circles at which these two gears operate are the pitch circles at which they were cut, and their radii are given by $R = N/2P_d$. The cutting pitch circles are also known as *standard pitch circles*. The pressure

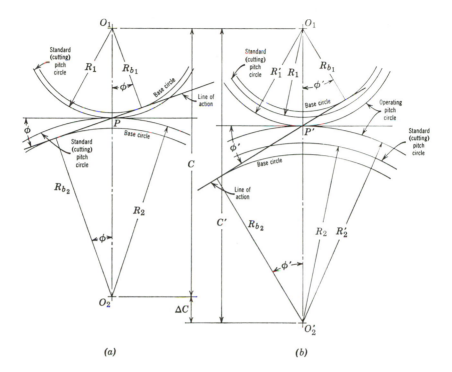

(a)   (b)

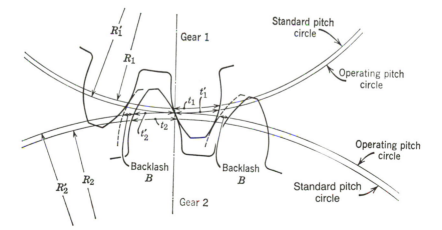

(c)

FIGURE 4.21

angle ϕ at which the gears will operate is the pressure angle at which they were cut, that is, $14\frac{1}{2}°$, 20°, or 25°. In other words, the cutting and operating pitch circles are identical as well as the cutting and operating pressure angles.

Figure 4.21*b* shows the condition where the two gears have been pulled apart a distance ΔC to give a new center distance C'. The line of action now crosses the line of centers at a new pitch point P'. It can be observed that the standard or cutting pitch circles (radii R_1 and R_2) are no longer tangent to each other. Also, the pitch point P' divides the center distance C' into segments which are inversely proportional to the angular velocity ratio. These segments become the radii R_1' and R_2' of new pitch circles that are tangent to each other at point P'. These circles are known as *operating pitch circles,* and equations for their radii can be determined from

$$\frac{\omega_1}{\omega_2} = \frac{N_2}{N_1} = \frac{R_2'}{R_1'}$$

and

$$R_1' + R_2' = C'$$

to give

$$R_1' = \left(\frac{N_1}{N_1 + N_2}\right) C'$$

and

$$R_2' = \left(\frac{N_2}{N_1 + N_2}\right) C'$$

In addition to the change in pitch circles, the pressure angle also increases. Angle ϕ' is known as the *operating pressure angle* and is larger than the cutting pressure angle ϕ. An equation for the operating pressure angle ϕ' can easily be derived from Fig. 4.21*b* as follows:

$$C' = \frac{R_{b_1} + R_{b_2}}{\cos \phi'} = (R_1 + R_2) \frac{\cos \phi}{\cos \phi'} = C \frac{\cos \phi}{\cos \phi'}$$

or

$$\cos \phi' = \frac{C}{C'} \cos \phi \qquad\qquad (4.17)$$

Also,

$$\Delta C = C' - C$$

$$= C\frac{\cos\phi}{\cos\phi'} - C$$

$$= C\left[\frac{\cos\phi}{\cos\phi'} - 1\right] \tag{4.18}$$

When the gears are operated under the condition of Fig. 4.21*b*, backlash will be present as shown in Fig. 4.21*c*. The angular-velocity ratio will not be affected as long as the gears remain in mesh. If the direction of rotation is reversed, however, lost motion will be encountered. An equation for backlash may be derived from the fact that the sum of the tooth thicknesses plus backlash must be equal to the circular pitch all measured on the operating pitch circle. From Fig. 4.21*c*, the following equation may be written:

$$t_1' + t_2' + B = \frac{2\pi R_1'}{N_1} = \frac{2\pi R_2'}{N_2} \tag{4.19}$$

where

t' = tooth thickness on operating pitch circle

B = backlash

R' = radius of operating pitch circle

N = number of teeth

From Eq. 4.3 developed in the section on involutometry,

$$t_1' = 2R_1'\left[\frac{t_1}{2R_1} + \text{inv }\phi - \text{inv }\phi'\right]$$

$$= \frac{R_1'}{R_1}t_1 - 2R_1'(\text{inv }\phi' - \text{inv }\phi) \tag{4.20}$$

$$t_2' = 2R_2'\left[\frac{t_2}{2R_2} + \text{inv }\phi - \text{inv }\phi'\right]$$

$$= \frac{R_2'}{R_2}t_2 - 2R_2'(\text{inv }\phi' - \text{inv }\phi) \tag{4.21}$$

where

t = tooth thickness on standard or cutting pitch circle ($t = p/2 = \pi/2P_d$)

R = radius of standard or cutting pitch circle ($R = N/2P_d$)

φ = cutting pressure angle ($14\frac{1}{2}°$, 20°, 25°)

φ' = operating pressure angle

Also,

$$\frac{R_1}{R_1'} = \frac{R_2}{R_2'} = \frac{C}{C'} \qquad (4.22)$$

and

$$C' = R_1' + R_2' \qquad (4.23)$$

Substituting Eqs. 4.20, 4.21, 4.22, and 4.23 into Eq. 4.19 and remembering that

$$\frac{2\pi R}{N} = p = \frac{\pi}{P_d}$$

$$B = \frac{C'}{C}\left[\frac{\pi}{P_d} - (t_1 + t_2) + 2C(\text{inv } \varphi' - \text{inv } \varphi)\right] \qquad \text{(U.S.)} \qquad (4.24)$$

$$B = \frac{C'}{C}\left[\pi m - (t_1 + t_2) + 2C(\text{inv } \varphi' - \text{inv } \varphi)\right] \qquad \text{(metric)}$$

For standard gears,

$$t_1 = t_2 = \frac{p}{2} = \frac{\pi}{2P_d} \qquad \text{(U.S.)}$$

$$t_1 = t_2 = \frac{p}{2} = \frac{\pi m}{2} \qquad \text{(metric)}$$

and Eq. 4.24 simplifies to

$$B = 2C'(\text{inv } \varphi' - \text{inv } \varphi) \qquad (4.25)$$

Equation 4.24 should be used if the gears are not standard, that is, if $t_1 \neq t_2$. Nonstandard gears will be presented in Chapter 5. Recommended values for backlash can be found in the AGMA *Gear Handbook,* Volume 1, 390.03 (March 1980).

Example 4.2. (*a*) A 3-module, 20° pinion of 24 teeth drives a gear of 60 teeth. Calculate the length of action and contact ratio if the gears mesh with zero backlash.

$$R_1 = \frac{N_1 m}{2} = \frac{24 \times 3}{2} = 36.000 \text{ mm}$$

$$R_2 = \frac{N_2 m}{2} = \frac{60 \times 3}{2} = 90.000 \text{ mm}$$

$$R_{b_1} = R_1 \cos \phi = 36.000 \cos 20° = 33.829 \text{ mm}$$

$$R_{b_2} = R_2 \cos \phi = 90.000 \cos 20° = 84.572 \text{ mm}$$

$$a_1 = a_2 = m = 3.0000 \text{ mm}$$

$$R_{o_1} = R_1 + a_1 = 39.000 \text{ mm}$$

$$R_{o_2} = R_2 + a_2 = 93.000 \text{ mm}$$

$$C = \left(\frac{N_1 + N_2}{2}\right)m = \left(\frac{24 + 60}{2}\right)3 = 126.00 \text{ mm}$$

$$Z = \sqrt{(R_{o_1})^2 - (R_{b_1})^2} + \sqrt{(R_{o_2})^2 - (R_{b_2})^2} - C \sin \phi$$

$$= \sqrt{39.000^2 - 33.829^2} + \sqrt{93.000^2 - 84.572^2} - 126.00 \sin 20°$$

$$= \sqrt{1521.0 - 1144.4} + \sqrt{8649.0 - 7152.4} - 43.095$$

$$= 19.406 + 38.686 - 43.095 = 14.997 \text{ mm}$$

$$m_p = \frac{Z}{p_b} \text{ and } p_b = \frac{2\pi R_{b_1}}{N_1} = \frac{2\pi \times 33.829}{24} = 8.8564 \text{ mm}$$

Therefore,

$$m_p = \frac{14.997}{8.8564} = 1.6934$$

(*b*) If the center distance is increased 0.5000 mm, calculate the radii of the operating pitch circles, the operating pressure angle, and the backlash produced.

$$C' = C + \Delta C = 126.00 + 0.5000 = 126.50 \text{ mm}$$

$$R_1' = \left(\frac{N_1}{N_1 + N_2}\right)C' = \left(\frac{24}{24 + 60}\right) \times 126.50 = 36.143 \text{ mm}$$

$$R_2' = C' - R_1' = 126.50 - 36.143 = 90.357 \text{ mm}$$

$$\cos \phi' = \frac{C \cos \phi}{C'} = \frac{126.00 \cos 20°}{126.50}$$

$$\phi' = 20.61°$$

$$B = 2C'(\text{inv } \phi' - \text{inv } \phi)$$

$$= 2 \times 126.50(\text{inv } 20.61° - \text{inv } 20°)$$

$$= 2 \times 126.50(0.016362 - 0.014904)$$

$$= 0.3689 \text{ mm}$$

4.9 INTERNAL (ANNULAR) GEARS

In many applications, an internal involute gear is meshed with a pinion instead of using two external gears to derive certain advantages. Perhaps the most important advantage is that of compactness of the drive. Also for the same tooth proportions, internal gears will have greater length of contact, greater tooth strength, and lower relative sliding between meshing teeth than external gears.

In an internal gear, the tooth profiles are concave instead of convex as in an external gear. Because of this shape, a type of interference can occur that is not possible in an external gear or a rack. This interference is known as *fouling* and occurs between inactive profiles as the teeth go in and out of mesh. Fouling will occur when there is not sufficient difference between the numbers of teeth on the internal gear and on the pinion. Figure 4.22 shows a pinion meshing with an internal gear. They are so nearly the same size that fouling occurs at points a, b, c, d, and e. When an internal gear is cut, a Fellows cutter is used with two fewer teeth than the gear being cut. This automatically relieves the tips of the internal gear teeth to prevent fouling at points a, b, c, d, and e. Involute interference can also occur between active profiles the same as in external gears. This will be discussed in the following paragraph.

Figure 4.23 shows two teeth in contact from Fig. 4.22 with the line of action tangent to the base circle of the gear at point f and tangent to the base circle of the pinion at point g. An involute profile for the gear can begin at point f, but the involute for the pinion cannot begin until point g. Point g is therefore the first possible point of contact without involute interference and determines the maximum addendum of the gear. Point h, the intersection of the addendum circle of the pinion and the line of action, is the end of contact, and the length of action is gPh. It should be mentioned that the relation $P_d = N/D$ holds for an internal gear as well as for an external gear. Figure 4.24 shows a photograph of an internal gear being cut by hobbing.

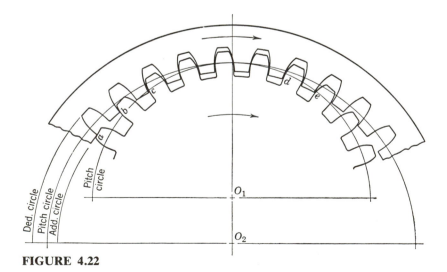

FIGURE 4.22

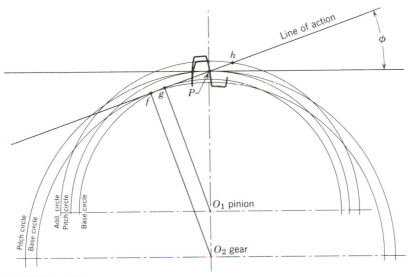

FIGURE 4.23

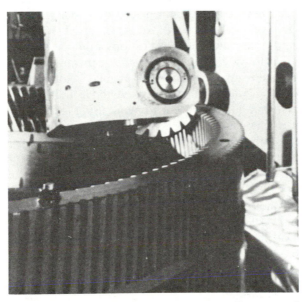

FIGURE 4.24 Hobbing internal gear. (Courtesy of Cincinnati Gear Company.)

4.10 CYCLOIDAL GEARS

Even though the cycloidal gear has been largely replaced by the involute, the cycloidal profile possesses certain advantages worthy of note. These are discussed briefly below.

Cycloidal gears do not have interference, and a cycloidal tooth is in general

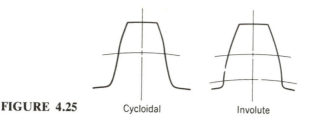

FIGURE 4.25   Cycloidal   Involute

stronger than an involute tooth because it has spreading flanks in contrast to the radial flanks of an involute tooth. Also, cycloidal teeth have less sliding and therefore less wear. Figure 4.25 shows a cycloidal gear tooth and, for comparison, an involute tooth. However, an important disadvantage of cycloidal gearing is the fact that for a pair of cycloidal gears there is only one theoretically correct center distance for which they will transmit motion at a constant angular-velocity ratio. Another disadvantage is that, although it is possible to hob a cycloidal gear, the hob is not as easily made as an involute hob because cycloidal rack teeth are not straight-sided as are involute rack teeth. For this reason, involute gears can be produced more accurately and more cheaply than cycloidal gears.

Involute gears have completely replaced cycloidal gears for power transmission. However, cycloidal gears are extensively used in watches, clocks, and certain instruments in cases where the question of interference and strength is a prime consideration. In watches and clocks, the gear train from the power source to the escapement steps up or increases its angular velocity ratio with the gear driving the pinion. In a watch, this step-up may be as high as 5000:1. The gears will therefore be so small that in order to avoid using excessively small teeth, it is necessary to use pinions (the driven gears in this case) having as few as six or seven teeth. The tooth profile of these pinions must also be capable of action over 60° of rotation. For this purpose, cycloidal gears are used in preference to involute gears. The problem of center distance and angular-velocity ratio is not important in this case because the whole train as governed by the escapement comes to rest and then starts again several times per second. The operation of the train thus involves such large changes of momentum that the effect of tooth form on change of momentum is negligible. The effect of tooth form on consistency of velocity ratio is thus unimportant in itself.

Problems—U.S. Standard

4.1. An involute is generated on a base circle having a radius R_b of 4.000 in. As the involute is generated, the angle which corresponds to inv ϕ varies from 0° to 15°. For increments of 3° for this angle, calculate the corresponding pressure angle ϕ and radius R for points on the involute. Plot the series of points in polar coordinates and connect with a smooth curve to give the involute.

4.2. Write a computer program for Problem 4.1 letting R_b = 3.000, 4.000, and 5.000 in. Determine the corresponding values of pressure angle ϕ and radius R for each value of R_b.

4.3. The thickness of an involute gear tooth is 0.3140 in. at a radius of 3.500 in. and a pressure angle of $14\frac{1}{2}°$. Calculate the tooth thickness and radius at a point on the involute which has a pressure angle of 25°.

4.4. If the involutes that form the outline of a gear tooth are extended, they will intersect and the tooth become pointed. Determine the radius at which this occurs for a tooth which has a thickness of 0.2620 in. at a radius of 4.000 in. and a pressure angle of 20°.

4.5. The thickness of an involute gear tooth is 0.1960 in. at a radius of 2.000 in. and a pressure angle of 20°. Calculate the tooth thickness on the base circle.

4.6. The pitch radii of two spur gears in mesh are 2.000 and 2.500 in., and the outside radii are 2.250 and 2.750 in., respectively. The pressure angle is 20°. Make a full-size layout of these gears as shown in Fig. 4.10, and label the beginning and end of contact. The pinion is the driver and rotates clockwise. Determine and label the angles of approach and recess for both gears.

4.7. A pinion of 2.000 in. pitch radius rotates clockwise and drives a rack. The pressure angle is 20°, and the addendum of the pinion and of the rack is 0.2000 in. Make a full-size layout of these gears, and label the beginning and end of contact. Determine and label the angle of approach and recess for the pinion.

4.8. Two equal spur gears of 48 teeth mesh together with pitch radii of 4.000 in. and addendums of 0.1670 in. If the pressure angle is $14\frac{1}{2}°$, calculate the length of action Z and the contact ratio m_p.

4.9. Contact ratio is defined either as the arc of action divided by the circular pitch or as the ratio of the length of action to the base pitch. Prove that

$$\frac{\text{Arc of action}}{\text{Circular pitch}} = \frac{\text{Length of action}}{\text{Base pitch}}$$

4.10. Verify Eq. 4.7 for the length of action Z for a pinion driving a rack in terms of the pitch radius R, the base radius R_b, the addendum a, and the pressure angle ϕ.

4.11. A pinion with a pitch radius of 1.500 in. drives a rack. The pressure angle is $14\frac{1}{2}°$. Calculate the maximum addendum possible for the rack without having involute interference on the pinion.

4.12. A 24-tooth pinion cut with a 12-pitch, 20° full-depth hob drives a 40-tooth gear. Calculate the pitch radii, base radii, addendum, dedendum, and tooth thickness on the pitch circle.

4.13. An 18-tooth pinion cut with an 8-pitch, 25° hob drives a 45-tooth gear. Calculate the pitch radii, base radii, addendum, dedendum, and the tooth thickness on the pitch circle.

4.14. A 42-tooth pinion cut with a 120-pitch, 20° full-depth hob drives a 90-tooth gear. Calculate the contact ratio.

4.15. If the radii of a pinion and gear are increased so that each becomes a rack, the length of action theoretically becomes a maximum. Determine the equation for the length of action under these conditions and calculate the maximum contact ratio for $14\frac{1}{2}°$, 20°, and 25° full-depth systems.

4.16. A 20-tooth pinion cut with a 4-pitch, 20° stub hob drives a rack. Calculate the pitch radius, base radius, working depth, whole depth, and the tooth thickness of the rack on the pitch line.

4.17. A 20° full-depth rack has an addendum of 0.2500 in. Calculate the base pitch, and show it as a dimension on a full-size sketch of a portion of the rack.

4.18. Determine the number of teeth in a $14\frac{1}{2}°$ involute spur gear so that the base circle diameter will equal the dedendum circle diameter.

4.19. Determine the following for a pair of standard spur gears in mesh: (a) an equation for the center distance C as a function of the numbers of teeth and diametral pitch; (b) the various combinations of 20° full-depth gears that can be used to operate at a center distance of 5.000 in. with an angular velocity ratio of 3:1. The diametral pitch is not to exceed 12 and the gears are not to be undercut. The gears are to be hobbed.

4.20. A 30-tooth pinion cut with a 6-pitch, 25° hob drives a rack. Calculate the length of action and the contact ratio.

4.21. A 24-tooth pinion cut with a 2-pitch, 20° full-depth hob drives a rack. If the pinion rotates counterclockwise at 360 rpm, determine graphically the velocity of sliding between the pinion tooth and the rack tooth at the beginning of contact, pitch point, and at the end of contact. Use a scale of 1 in. = 10 ft/s.

4.22. Two shafts whose axes are 8.500 in. apart are to be coupled together by standard spur gears with an angular velocity ratio of 1.5:1. Using a diametral pitch of 6, select two pairs of gears to best fit the above requirements. What change in the given data would have to be allowed if each set were to be used?

4.23. An 8-pitch, $14\frac{1}{2}°$ hob is used to cut a spur gear. The hob is right-handed with a lead angle of 2°40′, a length of 3.000 in., and an outside diameter of 3.000 in. Make a full-size sketch of the hob cutting a 48-tooth spur gear. The gear blank is $1\frac{1}{2}$ in. wide. Show the pitch cylinder of the hob on top of the gear blank with the pitch helix of the hob in correct relation to the pitch element of the gear tooth. Show three tooth elements on the gear and $1\frac{1}{2}$ turns of the thread on the hob; position these elements by means of the normal circular pitch. Label the axis of the hob and gear blank, the lead angle of the hob, and the direction of rotation of the hob and gear blank.

4.24. For a pressure angle of 22.5° in the full-depth system, calculate the minimum number of teeth in a pinion to mesh with a rack without involute interference. Also, calculate the number of teeth in a pinion to mesh with a gear of equal size without involute interference.

4.25. A 24-tooth pinion cut with an 8-pitch, 20° full-depth hob drives a 56-tooth gear. Determine the outside radii so that the addendum circle of each gear passes through the interference point of the other. Calculate the value of k for each gear.

4.26. Two equal gears cut with a 5-pitch, 20° full-depth hob mesh together such that the addendum circle of each gear passes through the interference point of the other. If the contact ratio is 1.622, calculate the number of teeth and the outside radius on each gear.

4.27. Two equal gears cut with a 20° full-depth hob mesh at the standard center distance. The addendum circle of each gear passes through the interference point of the other. Derive an equation for k as a function of N, where N is the number of teeth and k is a constant which, when divided by the diametral pitch, gives the addendum.

4.28. In the sketch of a standard gear shown in Fig. 4.26, the teeth are 20° full depth. If the pitch diameter is 4.800 in. and the diametral pitch is 5, calculate the radius of the pin which contacts the profile at the pitch point. Compute the diameter D_M measured over two opposite pins.

4.29. A 40-tooth pinion cut with a 10-pitch, $14\frac{1}{2}°$ hob meshes with a rack with no backlash. If the rack is pulled out 0.0700 in., calculate the backlash produced.

4.30. An 18-tooth pinion cut with a 12-pitch, 20° full-depth hob drives a gear of 54 teeth. If the center distance at which the gears operate is 3.050 in., calculate the operating pressure angle.

4.31. A 36-tooth pinion cut with a 10-pitch, $14\frac{1}{2}°$ hob drives a 60-tooth gear. If the center

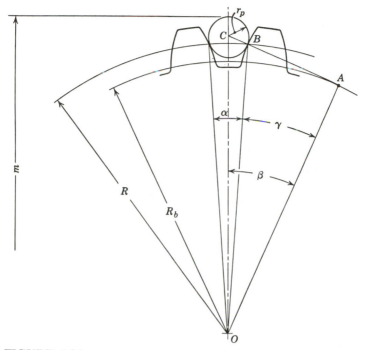

FIGURE 4.26

distance is increased by 0.0250 in., calculate (*a*) the radii of the operating pitch circles, (*b*) the operating pressure angle, and (*c*) the backlash produced.

4.32. A 24-tooth pinion cut with a 4-pitch, 20° stub hob drives a 40-tooth gear. Calculate (*a*) the maximum theoretical distance that these gears can be drawn apart and still mesh together with continuous driving, and (*b*) the backlash on the new pitch circles when the gears are drawn apart the amount calculated in part *a*.

4.33. A pinion with 24 teeth has a tooth thickness of 0.2550 in. at a cutting pitch radius of 1.500 in. and a pressure angle of 20°. A gear having 40 teeth has a tooth thickness of 0.2300 in. at a cutting pitch radius of 2.500 in. and a pressure angle of 20°. Calculate the pressure angle and center distance if these gears are meshed together without backlash.

4.34. A 15-tooth pinion cut with a 10-pitch, 25° hob drives a 45-tooth gear. Using a computer, calculate the backlash produced when the center distance is increased from 3.000 in. to 3.030 in. by increments of 0.0010 in.

4.35. A 96-pitch pinion of 34 teeth drives a gear with 60 teeth. If the center distance is increased by 0.0050 in., compare the backlash produced with pressure angles of $14\frac{1}{2}°$, 20°, and 25°.

Problems—Metric

4.1m. An involute is generated on a base circle having a radius R_b of 102 mm. As the involute is generated, the angle that corresponds to inv ϕ varies from 0 to 15°. For increments of 3° for this angle, calculate the corresponding pressure angle ϕ and radius

R for points on the involute. Plot this series of points in polar coordinates and connect with a smooth curve to give the involute.

4.2m. Write a computer program for Problem 4.1m, letting R_b = 76.2, 102, and 127 mm. Determine the corresponding values of pressure angle ϕ and radius *R* for each value of R_b.

4.3m. The thickness of an involute gear tooth is 7.98 mm at a radius of 88.9 mm and a pressure angle of $14\frac{1}{2}°$. Calculate the tooth thickness and radius at a point on the involute which has a pressure angle of 25°.

4.4m. If the involutes that form the outline of a gear tooth are extended, they will intersect and the tooth becomes pointed. Determine the radius at which this occurs for a tooth which has a thickness of 6.65 mm at a radius of 102 mm and a pressure angle of 20°.

4.5m. The thickness of an involute gear tooth is 4.98 mm at a radius of 50.8 mm and a pressure angle of 20°. Calculate the tooth thickness on the base circle.

4.6m. The pitch radii of two spur gears in mesh are 51.2 mm and 63.9 mm, and the outside radii are 57.2 mm and 69.9 mm, respectively. The pressure angle is 20°. Make a full-size layout of these gears as shown in Fig. 4.10, and label the beginning and end of contact. The pinion is the driver and rotates clockwise. Determine and label the angles of approach and recess for both gears.

4.7m. A pinion of 50.0 mm pitch radius rotates clockwise and drives a rack. The pressure angle is 20°, and the addendum of the pinion and of the rack is 5.00 mm. Make a full-size layout of these gears, and label the beginning and end of contact. Determine and label the angle of approach and recess for the pinion.

4.8m. Two equal spur gears of 48 teeth mesh together with pitch radii of 96.0 mm and addendums of 4.00 mm. If the pressure angle is 20°, calculate the length of action *Z* and the contact ratio m_p.

4.9m. Contact ratio is defined either as the arc of action divided by the circular pitch or as the ratio of the length of action to the base pitch. Prove that

$$\frac{\text{Arc of action}}{\text{Circular pitch}} = \frac{\text{Length of action}}{\text{Base pitch}}$$

4.10m. Verify Eq. 4.7 for the length of action *Z* for a pinion driving a rack in terms of the pitch radius *R*, the base radius R_b, the addendum *a*, and the pressure angle ϕ.

4.11m. A pinion with a pitch radius of 38.0 mm drives a rack. The pressure angle is 20°. Calculate the maximum addendum possible for the rack without having involute interference on the pinion.

4.12m. A 2-module, 20° pinion of 24 teeth drives a 40-tooth gear. Calculate the pitch radii, base radii, addendum, dedendum, and tooth thickness on the pitch circle.

4.13m. A 3-module, 20° pinion of 18 teeth drives a 45-tooth gear. Calculate the pitch radii, base radii, addendum, dedendum, tooth thickness on the pitch circle, and the contact ratio.

4.14m. A 0.2-module, 20° pinion of 42 teeth drives a gear of 90 teeth. Calculate the contact ratio.

4.15m. If the radii of a pinion and gear are increased so that each becomes a rack, the length of action theoretically becomes a maximum. Determine the equation for the length of action under these conditions and calculate the maximum contact ratio for $14\frac{1}{2}°$, 20°, and 25° full-depth systems.

4.16m. A 6-module, 20° pinion of 20 teeth drives a rack. Calculate the pitch radius, base radius, working depth, whole depth, and tooth thickness of the rack on the pitch line.

4.17m. A 20° rack has an addendum of 6.00 mm. Calculate the base pitch, and show it as a dimension on a full-size sketch of a portion of the rack.

4.18m. Determine the approximate number of teeth in a 20° involute spur gear so that the base circle diameter will equal the dedendum circle diameter.

4.19m. Determine the following for a pair of standard spur gears in mesh: (*a*) an equation for the center distance *C* as a function of the numbers of teeth and module; (*b*) the various combinations of 20° gears that can be used to operate at a center distance of 120 mm with an angular velocity ratio of 3:1. The module is not to be less than 2, and the gears are not to be undercut. The gears are to be hobbed.

4.20m. A 4-module, 20° pinion with 30 teeth drives a rack. Calculate the length of action and the contact ratio.

4.21m. A 12-module, 20° pinion with 24 teeth drives a rack. If the pinion rotates counterclockwise at 360 rpm, determine graphically the velocity of sliding between the pinion tooth and the rack tooth at the beginning of contact, pitch point, and at the end of contact.

4.22m. Two shafts whose axes are 216 mm apart are to be coupled together by standard spur gears with an angular velocity ratio of 1.5:1. Using a module of 4, select two pairs of gears to best fit the above requirements. What change in the given data would have to be allowed if each set were to be used?

4.23m. A 3-module, 20° hob is used to cut a spur gear. The hob is right-handed with a lead angle of 2°40′, a length of 75 mm, and an outside diameter of 75 mm. Make a full-size sketch of the hob cutting a 48-tooth spur gear. The gear blank is 38 mm wide. Show the pitch cylinder of the hob on top of the gear blank with the pitch helix of the hob in correct relation to the pitch element of the gear tooth. Show three tooth elements on the gear and $1\frac{1}{2}$ turns of the thread on the hob; position these elements by means of the normal circular pitch. Label the axis of the hob and gear blank, the lead angle of hob, and the direction of rotation of the hob and gear blank.

4.24m. For a pressure angle of 22.5°, calculate the minimum number of teeth in a pinion to mesh with a rack without involute interference. Also calculate the number of teeth in a pinion to mesh with a gear of equal size without involute interference. The addendum equals the module.

4.25m. A 3-module, 20° pinion with 24 teeth drives a 56-tooth gear. Determine the outside radii so that the addendum circle of each gear passes through the interference point of the other. Calculate the value of *k* for each gear.

4.26m. Two equal 5-module, 20° gears mesh together such that the addendum circle of each gear passes through the interference point of the other. If the contact ratio is 1.622, calculate the number of teeth and the outside radius on each gear.

4.27m. Two equal 20° involute gears are in mesh at the standard center distance. The addendum circle of each gear passes through the interference point of the other. Derive an equation for *k* as a function of *N*, where *N* is the number of teeth and *k* is a constant which, when multiplied by the module, gives the addendum.

4.28m. In the sketch of a standard gear shown in Fig. 4.27, the teeth are 20° full depth. If the pitch diameter is 120 mm and the module is 5, calculate the radius of the pin which contacts the profile at the pitch point. Compute the diameter D_M measured over two opposite pins.

4.29m. A 2.5-module, 20° pinion with 40 teeth meshes with a rack with no backlash. If the rack is pulled out 1.27 mm, calculate the backlash produced.

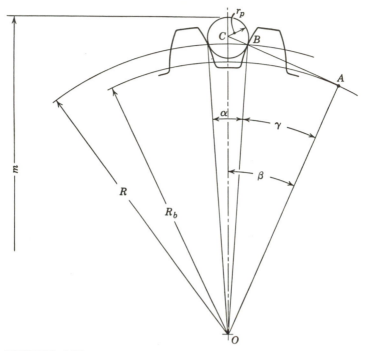

FIGURE 4.27

4.30m. A 2-module, 20° pinion of 18 teeth drives a gear of 54 teeth. If the center distance at which the gears operate is 73.27 mm, calculate the operating pressure angle.

4.31m. A 2.5-module, 20° pinion with 36 teeth drives a gear with 60 teeth. If the center distance is increased by 0.650 mm, calculate (a) the radii of the operating pitch circles, (b) the operating pressure angle, and (c) the backlash produced.

4.32m. A 6-module, 20° pinion of 24 teeth drives a gear of 40 teeth. Calculate (a) the maximum theoretical distance that these gears can be drawn apart and still mesh together with continuous driving, and (b) the backlash on the new pitch circles when the gears are drawn apart the amount calculated in part a.

4.33m. A pinion with 25 teeth has a tooth thickness of 6.477 mm at a cutting pitch radius of 37.50 mm and a pressure angle of 20°. A gear having 42 teeth has a tooth thickness of 5.842 mm at a cutting pitch radius of 63.00 mm and a pressure angle of 20°. Calculate the pressure angle and center distance if these gears are meshed together without backlash.

4.34m. A 2.5-module, 20° pinion of 20 teeth drives a gear of 45 teeth. Using a computer, calculate the backlash produced when the center distance is increased from 81.25 mm to 82.00 by increments of 0.025 mm.

4.35m. A 0.3-module pinion of 34 teeth drives a gear with 60 teeth. If the center distance is increased by 0.127 mm, compare the backlash produced with pressure angles of $14\frac{1}{2}°$, 20°, and 25°.

Chapter Five

Nonstandard Spur Gears

Although many textbooks on kinematics avoid the subject of nonstandard gears altogether, it is a natural and important extension of standard gear theory. Most of the gears used in automobiles and aircraft are nonstandard. In many design situations, an exact speed ratio can only be obtained by using nonstandard gears. In other cases, it will be possible to improve the smoothness of operation or to increase the load-carrying capacity of the gear set by using nonstandard gears. A basic knowledge of nonstandard gear theory will often enhance the designer's ability to produce a superior design.

5.1 THEORY OF NONSTANDARD SPUR GEARS

The most serious defect of the involute system of gearing is the possibility of interference between the tip of the gear tooth and the flank of the pinion tooth when the number of teeth in the pinion is reduced below the minimum for that system of gearing.

When interference does occur, the interfering metal is removed from the flank of the pinion tooth by the cutter when the teeth are generated. This removal of metal by the cutter is known as *undercutting* and would normally occur unless steps are taken to prevent it. If the cutter did not remove this metal, the two gears would not rotate when meshed together because the gear causing the interference would jam on the flank of the pinion. Actually, however, the gears will be able to rotate freely together because the flank of the pinion has been undercut. This undercutting, however, not only weakens the pinion tooth but may also remove a small portion of the involute adjacent to the base circle, which may cause a serious reduction in the length of action.

The attempt to eliminate interference and its consequent undercutting has led to the development of several nonstandard systems of gearing, some of which require special cutters. However, two of these systems have been successful and find wide application because standard cutters can be used to generate the teeth. In the first method, when the pinion is being cut, the cutter is withdrawn a certain amount from the blank so that the addendum of the basic rack passes through the interference point of the pinion. This will eliminate the undercutting, but the width of the tooth will be increased with a corresponding decrease in the tooth space. This is illustrated in Fig. 5.1, where (a) shows undercut teeth, and (b) shows the teeth resulting when the cutter has been withdrawn. When this pinion (Fig. 5.1b) is now mated with its gear, it will be found that the center distance has been increased because of the decreased tooth space. It can no longer be calculated directly from a diametral pitch and numbers of teeth and is therefore not considered standard. The pressure angle at which the gears operate also increases. This method of eliminating interference is known as the *extended center distance system.*

The withdrawal of the cutter need not be limited to the pinion blank alone but may be applied to both the pinion and gear if conditions warrant.

A variation of the extended center distance system is the practice of advancing the cutter into the gear blank the same amount that it will be withdrawn from the pinion blank. This will result in an increased addendum for the pinion and a decreased addendum for the gear; the increase in pinion addendum will equal the decrease in gear addendum. The dedendums will also change on both the pinion and gear so that the working depth will be the same as if the gears were standard. The center distance remains standard as well as the pressure angle. This system is known as the *long and short addendum system.*

Because of the change in the tooth proportions, the thickness of the gear tooth on the pitch circle is reduced and that of the pinion increased. Because of the fact that pinion teeth are weaker than gear teeth when both are made of the same material, the long and short addendum system will tend to equalize the tooth strengths. The long and short addendum system can be applied only when the interference occurs on one gear of a pair in mesh. Also, this system cannot be applied when the gears are equal or nearly equal in size because, although it would eliminate the interference on one gear, it would accentuate it on the other gear.

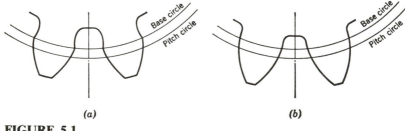

(a)   (b)

FIGURE 5.1

These two methods were developed primarily as a means of eliminating interference. However, they are also used extensively for improving contact ratio, for changing tooth shape to increase the strength of a tooth even though interference may not be present, and for fitting gears to nonstandard center distances.

The two systems can be applied to spur, helical, and bevel gearing. In fact, the standard tooth system for bevel gears is a long and short addendum system.

Formulas for the application of these two systems to spur gears cut with a hob will now be developed.

5.2   EXTENDED CENTER DISTANCE SYSTEM

Figure 5.2a in solid line shows a rack cutting a pinion where the pinion has fewer teeth than the minimum allowed to prevent interference. The rack and pinion are meshing at the standard center distance, with the standard pitch line of the rack tangent to the standard or cutting pitch circle of the pinion. The addendum line of the rack falls above the interference point E of the pinion so that the flanks of the pinion teeth are undercut as shown. For the rack tooth to cut the necessary clearance at the root of the pinion tooth, its height would have to be increased. To simplify the sketch, this additional height is shown (dotted) on only one tooth. The same layout may be used to illustrate the action of a hob cutting the pinion because kinematically a rack tooth and a hob tooth are identical.

To avoid undercutting, the rack is withdrawn a distance e so that the addendum line of the rack passes through the interference point E. This condition is shown dotted in Fig. 5.2a and results in the rack cutting a pinion with a wider

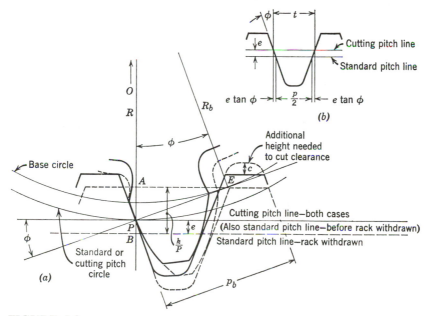

FIGURE 5.2

tooth than before. As the rack is withdrawn, the outside radius of the pinion must also be increased (by machining a larger pinion blank) to allow the clearance between the tip of the pinion tooth and the root of the rack tooth to remain the same. The same clearance is used regardless of whether a gear is standard or nonstandard. To show the change in the pinion tooth more clearly, the rack in Fig. 5.2a was withdrawn downward to the right to keep the left side of the tooth profile the same in both cases. When two gears, one or both of which have been generated with the cutter offset, are meshed together, the center distance will be greater than the standard center distance. In addition, the pressure angle at which these gears operate will be greater than the cutting pressure angle.

As mentioned previously, when a standard pinion is generated by the rack, the standard pitch line of the rack is tangent to the cutting pitch circle of the pinion. In this case, the standard pitch line is also the cutting pitch line. When the rack is withdrawn a distance e, however, the standard pitch line is no longer tangent to the cutting pitch circle; therefore, it cannot serve as the cutting pitch line. A new line on the rack will therefore act as a cutting pitch line. Figure 5.2b shows more clearly the two pitch lines on the rack when it is cutting a nonstandard tooth. From Fig. 5.2a, it can be seen that the cutting pitch circle on the pinion remains the same regardless of whether the pinion being cut is standard or nonstandard.

The width of the enlarged pinion tooth on its cutting pitch circle can be determined from the tooth space of the rack on its cutting pitch line. From Fig. 5.2b, this thickness can be expressed by the following equation:

$$t = 2e \tan \phi + \frac{p}{2} \tag{5.1}$$

Equation 5.1 can therefore be used to calculate the tooth thickness on the standard or cutting pitch circle of a gear generated by a standard cutter offset an amount e; e will be negative if the hob is advanced into the gear blank. This equation can also be used to determine the amount a cutter must be fed into a gear blank to give a specified amount of backlash.

In Fig. 5.2, the rack was withdrawn just enough so that the addendum line passed through the interference point of the pinion. It is possible to develop an equation so that e can be determined to satisfy this condition:

$$e = AB + OA - OP$$

$$= \frac{k}{P_d} + R_b \cos \phi - R$$

$$R_b = R \cos \phi$$

$$R = \frac{N}{2P_d}$$

Therefore,

$$e = \frac{k}{P_d} - R(1 - \cos^2 \phi)$$

$$e = \frac{1}{P_d}\left(k - \frac{N}{2}\sin^2 \phi\right) \qquad \text{(U.S.)} \qquad \text{(5.2)}$$

$$e = m\left(1.000 - \frac{N}{2}\sin^2 \phi\right) \qquad \text{(metric)}$$

Two equations that were developed in the section on involutometry (Chapter 4) find particular application in the study of nonstandard gearing:

$$\cos \phi_B = \frac{R_A}{R_B} \cos \phi_A \qquad \text{(5.3)}$$

$$t_B = 2R_B\left[\frac{t_A}{2R_A} + \text{inv } \phi_A - \text{inv } \phi_B\right] \qquad \text{(5.4)}$$

By means of these equations, it is possible to determine the pressure angle and tooth thickness at any radius R_B if the pressure angle and tooth thickness are known at some other radius R_A. For nonstandard gears, the reference thickness that corresponds to the thickness t_A in Eq. 5.4 is the tooth thickness on the cutting pitch circle, which can easily be calculated for any cutter offset by Eq. 5.1. The reference pressure angle that corresponds to ϕ_A is the pressure angle of the cutter. The radius at this pressure angle is the radius of the cutting pitch circle.

When two gears, gear 1 and gear 2, which have been cut with a hob offset e_1 and e_2, respectively, are meshed together, they will operate on new pitch circles of radii R_1' and R_2' and at a new pressure angle ϕ'. The thickness of the teeth on the operating pitch circles can be expressed as t_1' and t_2' and can easily be calculated from Eq. 5.4. These dimensions are shown in Fig. 5.3 together with the thickness of the teeth t_1 and t_2 on the cutting pitch circles of radii R_1 and R_2.

An equation will now be developed for determining the pressure angle ϕ' at which these two gears will operate:

$$\frac{\omega_2}{\omega_1} = \frac{N_1}{N_2} = \frac{R_1'}{R_2'} \qquad \text{(5.5)}$$

and

$$t_1' + t_2' = \frac{2\pi R_1'}{N_1} = \frac{2\pi R_2'}{N_2} \qquad \text{(5.6)}$$

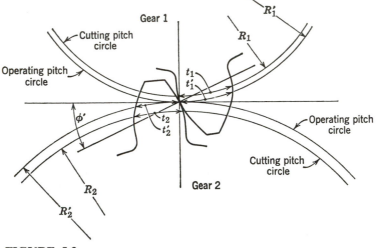

FIGURE 5.3

Substituting Eq. 5.4 for t_1' and t_2',

$$2R_1'\left[\frac{t_1}{2R_1} + (\text{inv }\phi - \text{inv }\phi')\right] + 2R_2'\left[\frac{t_2}{2R_2} + (\text{inv }\phi - \text{inv }\phi')\right] = \frac{2\pi R_1'}{N_1}$$

Dividing by $2R_1'$,

$$\left[\frac{t_1}{2R_1} + (\text{inv }\phi - \text{inv }\phi')\right] + \frac{R_2'}{R_1'}\left[\frac{t_2}{2R_2} + (\text{inv }\phi - \text{inv }\phi')\right] = \frac{\pi}{N_1}$$

$$\frac{t_1}{2R_1} + \frac{R_2'}{R_1'}\frac{t_2}{2R_2} = \frac{\pi}{N_1} + \left(1 + \frac{R_2'}{R_1'}\right)(\text{inv }\phi' - \text{inv }\phi)$$

By substituting Eq. 5.5 and $2R = N/P_d$,

$$\frac{t_1 P_d}{N_1} + \frac{N_2}{N_1}\frac{t_2 P_d}{N_2} = \frac{\pi}{N_1} + \frac{N_1 + N_2}{N_1}(\text{inv }\phi' - \text{inv }\phi)$$

By multiplying by N_1/P_d,

$$t_1 + t_2 = \frac{\pi}{P_d} + \frac{N_1 + N_2}{P_d}(\text{inv }\phi' - \text{inv }\phi)$$

By substituting Eq. 5.1 for t_1 and t_2,

$$2e_1 \tan \phi + \frac{p}{2} + 2e_2 \tan \phi + \frac{p}{2} = \frac{\pi}{P_d} + \frac{N_1 + N_2}{P_d}(\text{inv }\phi' - \text{inv }\phi)$$

$$2 \tan \phi(e_1 + e_2) + p = \frac{\pi}{P_d} + \frac{N_1 + N_2}{P_d}(\text{inv }\phi' - \text{inv }\phi)$$

By substituting $p = \pi/P_d$ and solving for inv ϕ',

$$\text{inv } \phi' = \text{inv } \phi + \frac{2P_d(e_1 + e_2)\tan\phi}{N_1 + N_2} \tag{5.7}$$

or

$$e_1 + e_2 = \frac{(N_1 + N_2)(\text{inv } \phi' - \text{inv } \phi)}{2P_d\tan\phi} \qquad \text{(U.S.)} \tag{5.7a}$$

$$e_1 + e_2 = \frac{m(N_1 + N_2)(\text{inv } \phi' - \text{inv } \phi)}{2\tan\phi} \qquad \text{(metric)}$$

By using Eq. 5.7, it is possible to determine the pressure angle ϕ' at which the two gears will operate after having been cut with a hob offset e_1 and e_2, respectively. To calculate the increase in the center distance (over the standard center distance C) due to the increased pressure angle, Eq. 4.18 can be used and is repeated here:

$$\Delta C = C\left[\frac{\cos\phi}{\cos\phi'} - 1\right] \tag{5.8}$$

Very often, it is necessary to design gears to fit a predetermined center distance. In this case, the pressure angle is fixed by the conditions of the problem, and it is necessary to determine the hob offsets e_1 and e_2. The sum $e_1 + e_2$ can be determined from Eq. 5.7a. However, it should be noted that the sum of e_1 and e_2 does not equal the increase in the center distance over the standard center distance. Unfortunately, there is no way of rationally determining e_1 and e_2 independently of each other. Because of this, the values are usually selected by assuming one of them or by using some empirical relation such as letting e_1 and e_2 vary inversely (or directly if $e_1 + e_2$ is negative) with the numbers of teeth in the gears in an attempt to strengthen the pinion teeth. However, this method of selecting e_1 and e_2 generally does not yield pinion and gear teeth of anywhere near equal strength. In an attempt to correct this situation, a method was developed by Walsh and Mabie[1] for determining the hob offset e_1 from the value of $e_1 + e_2$ for a pair of spur gears designed to operate at a nonstandard center distance. By using a digital computer, it was possible to adjust e_1 and e_2 for various velocity ratios and changes in center distance so that the strength of the pinion teeth was approximately equal to that of the gear teeth.

Because of the complexity of the problem, the results had to be given in the form of design charts. These show curves of $e_1/(e_1 + e_2)$ versus $N_2/(N_1 + N_2)$ for various changes in center distance. These design charts were developed for a cutting pressure angle ϕ of 20°, full-depth teeth ($k = 1$), and coarse pitch.

---

[1]E. J. Walsh and H. H. Mabie, "A Simplified Method for Determining Hob Offset Values in the Design of Nonstandard Spur Gears," *Proceedings*, Second OSU Applied Mechanism Conference, Stillwater, Oklahoma, October 1971.

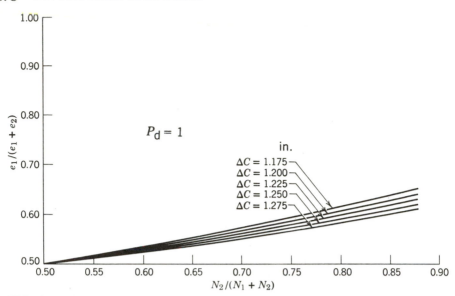

FIGURE 5.4

Although the charts were plotted for data based on a diametral pitch of 1, they can be used for any diametral pitch up to 19.99 (end of coarse pitch). The charts were also plotted for $N_1 = 18$ and N_2 from 18 to 130 teeth. Where N_1 takes on other values, a very slight error (less than 4%) is introduced. A sample chart is shown in Fig. 5.4 for changes in center distance $\Delta C = 1.175$ to 1.275 in. for $P_d = 1$.

Example 5.1. A pinion and gear of 20 and 30 teeth, respectively, are to be cut by a 5-pitch, 20° full-depth hob to operate on a center distance of 5.25 in. with no backlash. Determine the value of e_1 and e_2 to give teeth of the proper thickness so that the strengths of the pinion teeth and gear teeth will be approximately equal. The standard center distance is given by

$$C = \frac{N_1 + N_2}{2P_d} = \frac{20 + 30}{2 \times 5}$$

$$= 5.00 \text{ in.}$$

Operating pressure angle:

$$\cos \phi' = \frac{C}{C'} \cos \phi = \frac{5.0}{5.25} \cos 20°$$

$$\phi' = 26.50°$$

Change in center distance:

$$\nabla C = C' - C = 5.25 - 5.00$$

$$= +0.25 \text{ in.}$$

The value of ∇C must be multiplied by the diametral pitch because the charts are based on $P_d = 1$:

$$\Delta C = \nabla C \times P_d = 0.25 \times 5$$

$$= 1.25 \text{ in.}$$

Also

$$\frac{N_2}{N_1 + N_2} = \frac{30}{20 + 30} = 0.60$$

Therefore, from Fig. 5.4,

$$\frac{e_1}{e_1 + e_2} = 0.543$$

By calculating the value of $e_1 + e_2$ from Eq. 5.7a,

$$e_1 + e_2 = \frac{(N_1 + N_2)(\text{inv } \phi' - \text{inv } \phi)}{2P_d \tan \phi}$$

$$= \frac{(20 + 30)(\text{inv } 26.5° - \text{inv } 20°)}{2 \times 5 \tan 20°}$$

$$= 0.29073 \text{ in.}$$

By combining these results,

$$e_1 = 0.543(e_1 + e_2)$$

$$= 0.543(0.29073)$$

$$= 0.15787 \text{ in.}$$

and

$$e_2 = 0.13286 \text{ in.}$$

Although it is not practical to give all the calculations necessary to determine the stresses in the pinion and gear teeth, it can be shown that

$$S_1 = \frac{9.952 W_n}{F} \quad \text{lb/in.}^2$$

and

$$S_2 = \frac{10.18 W_n}{F} \quad \text{lb/in.}^2$$

where

W_n = normal load at tooth tip (lb)

F = tooth face width (in.)

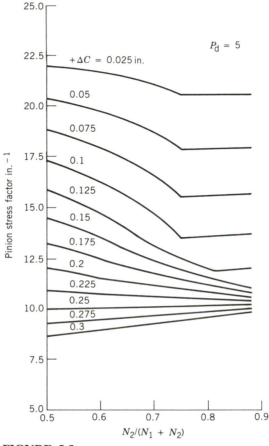

FIGURE 5.5

In addition to the charts for positive changes in center distance, as illustrated in Fig. 5.4, a series of charts is also given in the paper by Walsh and Mabie for negative changes in center distance.

It is a laborious job to calculate tooth stresses in nonstandard spur gears because of the change in standard tooth dimensions caused by the hob offsets e_1 and e_2. In view of this, curves were developed to give stress factors (SF/W_n) as a function of the ratio $N_2/(N_1 + N_2)$ for various changes in center distance and diametral pitch.[2] It was not possible, however, to develop charts for $P_d = 1$ as was done in the case for the hob offset charts shown in Fig. 5.4. Figures 5.5 and 5.6 show stress factor curves for the pinion and gear for $P_d = 5$ of Example 5.1. A comparison of the stress factors for the data of Example 5.1 is shown in Table 5.1 obtained from detailed calculations given in reference 2 and from the curves of Figs. 5.5 and 5.6.

---

[2]H. H. Mabie, E. J. Walsh, and V. I. Bateman, "Determination of Hob Offset Required to Generate Nonstandard Spur Gears with Teeth of Equal Strength," *Mechanism and Machine Theory*, **18** (3), 1983, pp. 181–192.

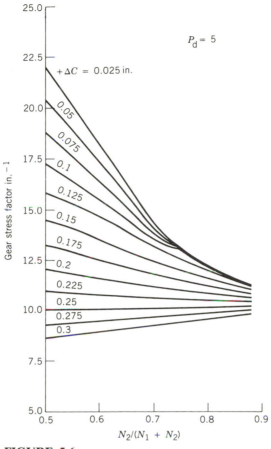

FIGURE 5.6

Another approach to the solution of the problem of determining e_1 and e_2 has been developed by Siegel and Mabie.[3] By this method, e_1 and e_2 are selected for a particular application to give tooth proportions which yield a maximum ratio of recess action to approach action and, at the same time, give a contact ratio m_p of 1.20 or higher. This system is based on the fact that a pair of gears mesh more smoothly coming out of contact than they do going into contact.

TABLE 5.1 Stress Factors (Example 5.1)

	Manual Calculations	Design Charts
Pinion	9.952 in.$^{-1}$	10.0 in.$^{-1}$ (Fig. 5.5)
Gear	10.18 in.$^{-1}$	10.0 in.$^{-1}$ (Fig. 5.6)

[3] R. E. Siegel and H. H. Mabie, "Determination of Hob Offset Values for Nonstandard Gears Based on Maximum Ratio of Recess to Approach Action," *Proceedings,* Third OSU Applied Mechanism Conference, Stillwater, Oklahoma, November 1973.

Therefore, it is advantageous to have as high a ratio of recess to approach action as possible, especially for gears for instrument application.

It is not possible to calculate the addendum and dedendum of an extended center distance gear unless information is available regarding the gear with which it is to mesh. Figure 5.7 shows two gears that are to mesh at a given center distance C'. The gears are to be cut with a hob which is offset e_1 on the pinion and e_2 on the gear. It is necessary to calculate the outside diameter of each gear and the depth of cut. The centerline of gear 2 has been moved to the right so that a cutter tooth may be shown engaged with each blank. If one knows the center distance, the cutting pitch radii, the hob offsets, and the tooth form and diametral pitch of the hob, it is possible to write equations for the outside radii as follows:

$$R_{o_1} = C' - R_2 - e_2 + \frac{k}{P_d} \quad \text{(U.S.)}$$

$$R_{o_1} = C' - R_2 - e_2 + m \quad \text{(metric)}$$

$$R_{o_2} = C' - R_1 - e_1 + \frac{k}{P_d} \quad \text{(U.S.)}$$

$$R_{o_2} = C' - R_1 - e_1 + m \quad \text{(metric)}$$

(5.9)

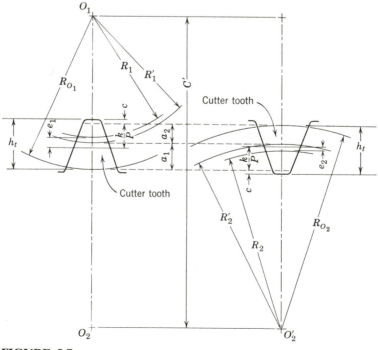

FIGURE 5.7

It should be noted from the sketch that the addendums of the two gears are not equal to each other, nor is either one equal to the k/P_d of the hob.

An equation for the depth of cut can also be easily developed from the sketch:

$$h_t = R_{o_1} + R_{o_2} - C' + c \tag{5.10}$$

where c is obtained from Table 4.1 or 4.2.

5.3 LONG AND SHORT ADDENDUM SYSTEM

If the cutter is advanced into the gear blank the same amount that it is withdrawn from the pinion, then $e_2 = -e_1$ and, from Eq. 5.7, $\phi' = \phi$. Therefore, the pressure angle at which the gears will operate is the same as the pressure angle at which they were cut. Because there is no change in the pressure angle, $R_1' = R_1$ and $R_2' = R_2$, and the gears will operate at the standard center distance.

The addendum of the pinion is increased to $k/P_d + e$ and the addendum of the gear reduced to $k/P_d - e$. The tooth thicknesses on the cutting pitch circles can be readily calculated from Eq. 5.1, keeping in mind that the gear tooth thickness decreases the same amount that the pinion tooth increases. As has been mentioned previously, there are conditions under which the long and short addendum system will not work properly. For the long and short addendum system to be successful, Professor M. F. Spotts of Northwestern University worked out that for $14\frac{1}{2}°$ gears, the sum of the teeth in the gears must be at least 64; and for $20°$ gears, the sum of the teeth should be at least 34. For $25°$ gears, the sum of the teeth should not be less than 24.

The proportions of gears cut by a pinion cutter for either of these two systems will not be the same as when cut by a hob. The preceding formulas apply only to gears cut by a hob or by a rack-type cutter. However, formulas for gears cut by pinion cutters can be developed using the principles above, and these are presented in a later section.

Example 5.2. Two spur gears of 12 and 15 teeth, respectively, are to be cut by a $20°$ full-depth 6-pitch hob. Determine the center distance at which to operate the gears to avoid undercutting.

$$e_1 = \frac{1}{P_d}\left(k - \frac{N}{2}\sin^2\phi\right)$$

$$= \tfrac{1}{6}(1.00 - \tfrac{12}{2}\sin^2 20)$$

$$= 0.04968 \text{ in.}$$

$$e_2 = \tfrac{1}{6}(1.00 - \tfrac{15}{2}\sin^2 20)$$

$$= 0.02045 \text{ in.}$$

$$\text{inv } \phi' = \text{inv } \phi + \frac{2P_d(e_1 + e_2)\tan\phi}{N_1 + N_2}$$

$$= 0.01490 + \frac{2 \times 6(0.04968 + 0.02045) \tan 20°}{12 + 15}$$

$$= 0.01490 + 0.01134$$

$$= 0.02624$$

From the table of involute functions,

$$\phi' = 23.97°$$

$$R'_1 = \frac{R_1 \cos \phi}{\cos \phi'} = \frac{1 \times 0.9397}{0.9135} = 1.0286 \text{ in.}$$

$$R'_2 = \frac{R_2 \cos \phi}{\cos \phi'} = \frac{1.25 \times 0.9397}{0.9135} = 1.2858 \text{ in.}$$

and

$$C' = R'_1 + R'_2 = 2.3144 \text{ in.}$$

Example 5.3. Two 3-module, 20° spur gears of 32 and 48 teeth, respectively, are operating together on the standard center distance of 120.00 mm. To affect a change in the speed ratio, it is desired to replace the 32-tooth gear with one of 31 teeth. The tooth thickness on the cutting pitch circle of the 48-tooth gear and the 120.00 mm center distance are to remain unchanged.

Determine the value of e_1 that will give teeth of the proper thickness to mesh with the 48-tooth gear.

$$R_1 = \frac{N_1 m}{2} = \frac{31 \times 3}{2} = 46.500 \text{ mm}$$

$$R_2 = \frac{N_2 m}{2} = \frac{48 \times 3}{2} = 72.000 \text{ mm}$$

$$R'_1 = \left(\frac{N_1}{N_1 + N_2}\right) C' = \frac{31}{31 + 48} \times 120.00 = 47.089 \text{ mm}$$

$$R'_2 = \left(\frac{N_2}{N_1 + N_2}\right) C' = \frac{48}{31 + 48} \times 120.00 = 72.911 \text{ mm}$$

$$\cos \phi' = \frac{R_1 \cos \phi}{R'_1} = \frac{46.500 \cos 20°}{47.089}$$

$$\phi' = 21.88°$$

$$e_1 + e_2 = \frac{m(N_1 + N_2)(\text{inv } \phi' - \text{inv } \phi)}{2 \tan \phi}$$

$$= \frac{3(31 + 48)(\text{inv } 21.88° - \text{inv } 20°)}{2 \tan 20°}$$

$$= 1.5660 \text{ mm}$$

For $e_2 = 0$, $e_1 = 1.5660$ mm.

5.4 RECESS ACTION GEARS

Another interesting type of nonstandard gears is recess action gears, so called because most or all the action between teeth takes place during the recess portion of contact. The long and short addendum system is a form of recess action gears. It is known that the recess portion of contact of a pair of gears is much smoother than the approach portion. It was on this basis that recess action gears were developed, and it has been found that these gears wear longer and operate with less friction, vibration, and noise than gears with teeth of standard proportions.

Recess action gears can be machined using standard hobs and cutters. The tooth form of such gears is the same as that of standard gears, and they mesh at the same center distance. Therefore, a pair of recess action gears can be used to replace a pair of standard spur gears with no change in center distance.

The strength of recess action gears is approximately the same as for standard gears. However, a recess action gear must be designed to operate either as a driver or as a follower; it cannot be designed to operate as both. A recess action pinion, however, can drive a follower in either direction, that is, it can change direction of rotation during an operating cycle. Also, the gears can be used for a speed-increasing drive as well as for a speed-reducing drive, but the power flow must always be in the same direction. If the power flow changes direction during operation, binding in the tooth contact area will occur with resulting high friction and wear. Because of these limitations, recess action gears cannot be used as idlers operating at standard center distances.

There are two types of recess action gears: (a) full recess action where all the contact is recess, and (b) semi-recess action. In order for a pair of recess action gears to have adequate contact ratio, little or no undercut, and teeth not pointed, full recess action gears cannot have less than 20 teeth in the driver nor less than 27 teeth in the follower. For semi-recess action gears, however, the minimum number of teeth in the driver is reduced to 10 and that in the follower

TABLE 5.2 Tooth Proportions, Recess Action Gears
(Pressure Angle $\phi = 20°$)

	Semi-Recess Action		Full Recess Action	
	Driver	**Follower**	**Driver**	**Follower**
Addendum (a)	$\dfrac{1.500}{P_d}$	$\dfrac{0.500}{P_d}$	$\dfrac{2.000}{P_d}$	0
Dedendum (b)	$\dfrac{0.796}{P_d}$	$\dfrac{1.796}{P_d}$	$\dfrac{0.296}{P_d}$	$\dfrac{2.296}{P_d}$
Pitch diameter (D)	$\dfrac{N}{P_d}$	$\dfrac{N}{P_d}$	$\dfrac{N}{P_d}$	$\dfrac{N}{P_d}$
Outside radius (R_o)	$\dfrac{N+3}{2P_d}$	$\dfrac{N+1}{2P_d}$	$\dfrac{N+4}{2P_d}$	$\dfrac{N}{2P_d}$
Tooth thickness (t)	$\dfrac{1.9348}{P_d}$	$\dfrac{1.2068}{P_d}$	$\dfrac{2.2987}{P_d}$	$\dfrac{0.8429}{P_d}$

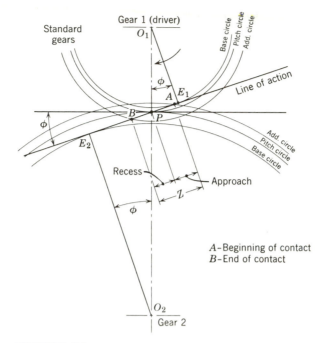

FIGURE 5.8*a*

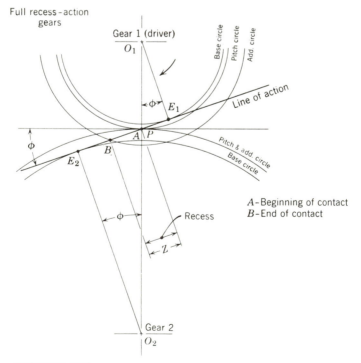

FIGURE 5.8*b*

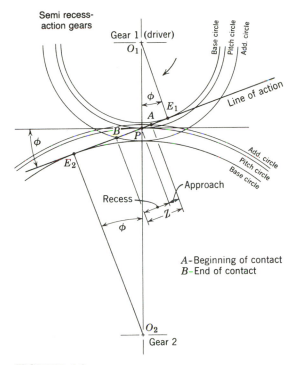

FIGURE 5.8c

to 20. Full recess action gears are to be preferred because all of the action is in the recess portion. Many times, however, the large number of teeth required for full recess action limits their use, and semi-recess action gears must be used instead.

Table 5.2 shows the proportions for the two systems of recess action gears. To give a comparison between these gears and standard gears, Fig. 5.8 shows the addendum, pitch, and base circles and length of action of (a) standard gears, (b) full recess action gears, and (c) semi-recess action gears in mesh. In Fig. 5.8b, for the full recess action system, the pitch circle of the follower (gear 2) becomes the addendum circles because the addendum is zero. Therefore, the approach portion of tooth contact is zero, and all the length of action is in the recess portion. Figure 5.8c, for semi-recess action, shows the recess portion considerably larger than the approach portion for this system.

5.5 NONSTANDARD SPUR GEARS CUT BY A PINION CUTTER

The theory of producing spur gears for an extended center distance application when cut by a pinion cutter is much more complicated than when the gears are cut by a hob or by a rack. When a hob is used to cut a nonstandard gear, the cutting pitch circle of the gear being cut and the cutting pressure angle are the same as if the gear were a standard gear. This fact greatly simplifies the analysis as has been seen in previous sections. When cutting with a pinion cutter, however, and the cutter is withdrawn an amount e, a new cutting pitch circle is defined on the gear and on the pinion cutter. In addition, a new cutting pressure angle develops. These changes make the analysis much more complex. This condition is shown in Fig. 5.9.

Figure 5.9a shows the case of a pinion cutter generating a gear at a standard center distance. The equation for the standard cutting center distance is

$$C_{std} = \frac{N + N_c}{2P_d} \qquad \text{(U.S.)} \qquad \textbf{(5.11)}$$

$$C_{std} = \frac{(N + N_c)m}{2} \qquad \text{(metric)}$$

where

N = number of teeth in gear to be cut

N_c = number of teeth in cutter

P_d = diametral pitch

m = module

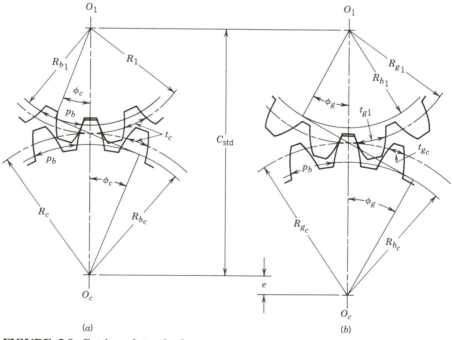

(a)   (b)

FIGURE 5.9 Cutting of standard versus nonstandard gear.

This equation can also be expressed as

$$C_{std} = \frac{(N + N_c)p_b}{2\pi \cos \phi_c} \tag{5.12}$$

where

p_b = base pitch

ϕ_c = standard pressure angle of cutter

Figure 5.9b shows the case where the cutting center distance is increased an amount e. Because the base circle radii remain unchanged, the generating pressure angle ϕ_g is given by

$$\cos \phi_g = \frac{(N + N_c)p_b}{2\pi(C_{std} + e)} \tag{5.13}$$

where

e = pinion cutter offset

The equation from involutometry for the thickness t_B of an involute gear tooth at various radii and corresponding involute pressure angles is given by

$$t_B = 2R_B\left[\frac{t_A}{2R_A} + \text{inv } \phi_A - \text{inv } \phi_B\right] \tag{5.14}$$

where

ϕ_A = involute pressure angle at radius R_A

ϕ_B = involute pressure angle at radius R_B

This equation may also be expressed as

$$t_B = t_A \frac{\cos \phi_A}{\cos \phi_B} + \frac{2R_b}{\cos \phi_B}(\text{inv } \phi_A - \text{inv } \phi_B) \tag{5.15}$$

where

R_b = radius of base circle = $R_A \cos \phi_A = R_B \cos \phi_B$

From Eq. 5.15, it is possible to write an equation for the tooth thickness of a cutter t_{g_c} on its generating pitch circle:

$$t_{g_c} = \frac{t_c \cos \phi_c}{\cos \phi_g} + \frac{2R_{b_c}}{\cos \phi_g}(\text{inv } \phi_c - \text{inv } \phi_g) \tag{5.16}$$

where

t_c = thickness of cutter tooth on standard pitch circle $p_c/2$, p_c being the circular pitch of standard tooth

R_{b_c} = radius of base circle of cutter

The tooth thickness of the cutter as given by Eq. 5.16 is equal to the space width of the gear on its generating pitch circle. The tooth thickness t_g of the gear on its generating pitch circle is therefore equal to the circular pitch on that circle minus the space width. This is given by the equation

$$t_g = \frac{p_b}{\cos \phi_g} - \frac{t_c \cos \phi_c}{\cos \phi_g} - \frac{2R_{b_c}}{\cos \phi_g} (\text{inv } \phi_c - \text{inv}\phi_g) \tag{5.17}$$

When the gear is meshed with a second gear, a running pitch circle is obtained. By using Eq. 5.15, the tooth thickness t_r of a gear on a running pitch circle is determined by

$$t_r = t_g \frac{\cos \phi_g}{\cos \phi_r} + \frac{2R_b}{\cos \phi_r} (\text{inv } \phi_g - \text{inv } \phi_r) \tag{5.18}$$

where

ϕ_r = running pressure angle

R_b = base radius of gear

Substituting Eq. 5.17 into Eq. 5.18 gives

$$t_r = \frac{p_b}{\cos \phi_r} - \frac{t_c \cos \phi_c}{\cos \phi_r} - \frac{2R_{b_c}}{\cos \phi_r} (\text{inv } \phi_c - \text{inv } \phi_g)$$

$$+ \frac{2R_b}{\cos \phi_r} (\text{inv } \phi_g - \text{inv } \phi_r) \tag{5.19}$$

and

$$\text{inv } \phi_g = \frac{t_r \cos \phi_r - p_b + t_c \cos \phi_c + 2R_{b_c} \text{ inv } \phi_c + 2R_b \text{ inv } \phi_r}{2(R_{b_c} + R_b)} \tag{5.20}$$

The generating pressure angle at which a gear must be cut to give a specified tooth thickness at a given running pressure angle can be calculated from Eq. 5.20. The required cutter offset to give this pressure angle may then be calculated from Eq. 5.13.

When gears 1 and 2 have been cut with a pinion cutter to mesh at an extended center distance, equations can be written from Eq. 5.19 to give the tooth thickness of each gear on its running pitch circle:

$$t_{r_1} = \frac{p_b - t_c \cos \phi_c - 2R_{b_c}(\text{inv } \phi_c - \text{inv } \phi_{g_1}) + 2R_{b_1}(\text{inv } \phi_{g_1} - \text{inv } \phi_r)}{\cos \phi_r}$$

$$t_{r_2} = \frac{p_b - t_c \cos \phi_c - 2R_{b_c}(\text{inv } \phi_c - \text{inv}\phi_{g_2}) + 2R_{b_2}(\text{inv } \phi_{g_2} - \text{inv } \phi_r)}{\cos \phi_r}$$

$$\tag{5.21}$$

The running pitch circle diameter of gear 1 is

$$D_{r_1} = \frac{2N_1}{N_1 + N_2}(C + \Delta) \tag{5.22}$$

where Δ is the increase in standard center distance between the two gears. Therefore, the circular pitch on the running pitch circle is

$$\frac{\pi D_{r_1}}{N_1} = \frac{2\pi(C + \Delta)}{N_1 + N_2} \tag{5.23}$$

and

$$\frac{2\pi(C + \Delta)}{N_1 + N_2} = t_{r_1} + t_{r_2} \tag{5.24}$$

for gear pairs with zero backlash. Substituting Eq. 5.21 for t_{r_1} and t_{r_2} into Eq. 5.24 and simplifying gives

$$\text{inv } \phi_{g_1}(R_{b_c} + R_{b_1}) + \text{inv } \phi_{g_2}(R_{b_c} + R_{b_2})$$
$$= 2R_{b_c} \text{ inv } \phi_c + \text{inv } \phi_r(R_{b_1} + R_{b_2}) \tag{5.25}$$

Equation 5.25 can be further simplified by expressing R_{b_c}, R_{b_1}, and R_{b_2} in terms of number of teeth, cutter pressure angle, and diametral pitch to give

$$(N_1 + N_c) \text{ inv } \phi_{g_1} + (N_2 + N_c) \text{ inv } \phi_{g_2}$$
$$= 2N_c \text{ inv } \phi_c + (N_1 + N_2) \text{ inv } \phi_r \tag{5.26}$$

From Eq. 5.26, it can be seen that there is no way of determining ϕ_{g_1} and ϕ_{g_2} independently of each other; therefore, e_1 and e_2 cannot be calculated directly from Eq. 5.13. To overcome this difficulty, a second relationship between e_1 and e_2 was developed by equalizing the static bending stresses in the two gears.[4]

A computer program was written to balance the stresses in the teeth by adjusting the offsets of the pinion cutter used to cut the gears. The gear system was defined by the numbers of teeth in the pinion and gear, the pitch diameter and diametral pitch of the cutter, and the center distance at which the gears were to operate. This system was used to produce a set of design charts for determining values of e_1 and e_2 as a function of N_2/N_1 for various values of ΔC. The charts were based on a 20° cutter pressure angle and a pinion of 20 teeth. Unfortunately,

---

[4]R. N. Green and H. H. Mabie, "Determination of Pinion-Cutter Offsets Required to Produce Nonstandard Spur Gears With Teeth of Equal Strength," *Mechanism and Machine Theory*, **15** (6), 1980, pp. 491–506.

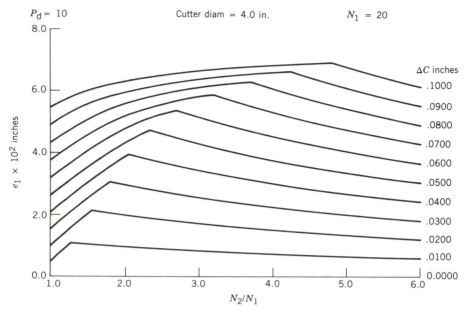

FIGURE 5.10 Cutter offset.

it is not possible to use the charts to obtain offsets for gear sets with pinions containing significantly more or less than the 20 teeth assumed to generate the figures. Sample charts are shown in Figs. 5.10 and 5.11 for determining e_1 and e_2, respectively, for a 10-pitch, 4-in.-diameter cutter for changes in center distances $\Delta C = 0.010$ to 0.100 in. Table 5.3 shows the range of diametral pitches used in the development of the charts.

From the charts, it can be observed that the curves for each value of ΔC,

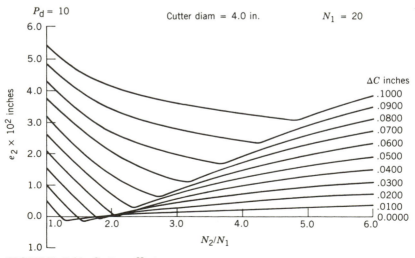

FIGURE 5.11 Cutter offset.

TABLE 5.3 Range of Design Charts[a]

Cutter Pitch Diameter, in.	Diametral Pitch				
	4	6	8	10	12
3			X	X	X
4	X	X	X	X	X
5	X	X	X	X	X
6	X	X	X		

[a]$N_1 = 20$, $\phi = 20°$, $1 \leqq N_2/N_1 \leqq 6$.

except for $\Delta C = 0$, have a slope discontinuity at some point along their length. The slope change marks the point where the design of the gear teeth ceases to be the result of balancing tooth stresses and the design becomes governed by the need to avoid undercutting. This is accomplished by limiting the depth of cut made on the gears to the allowable depth of cut for a standard pinion cutter. The segment to the left of the discontinuity represents the range over which the tooth stresses have been balanced.

Several other equations are required to complete the definition of the gear system geometry. From Fig. 5.12, the outside radii of the two gears are

$$R_{o_1} = C' - (C_{std_2} + e_2) + R_{o_c} - c \quad (c = \text{tooth clearance}) \quad (5.27)$$

$$R_{o_2} = C' - (C_{std_1} + e_1) + R_{o_c} - c \quad (5.28)$$

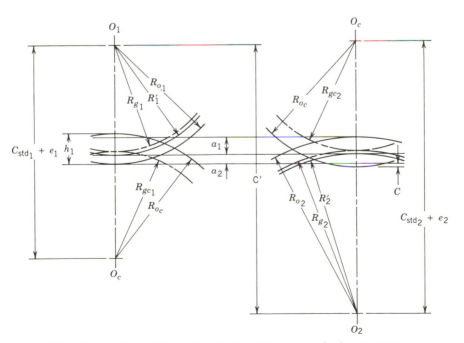

FIGURE 5.12 Outside radii and depth of cut for nonstandard spur gears.

The depth of cut required is

$$h_t = a_1 + a_2 + c$$

or

$$h_t = R_{o_1} + R_{o_2} - C' + c \qquad (5.29)$$

The equation for the outside radii may be further reduced by recognizing that

$$R_{o_c} = R_c + \frac{k}{P_d} + c \qquad \text{(U.S.)}$$

$$R_{o_c} = R_c + m + c \qquad \text{(metric)}$$

Thus,

$$R_{o_1} = C' - R_2 - e_2 + \frac{k}{P_d} \qquad \text{(U.S.)} \qquad (5.30)$$

$$R_{o_1} = C' - R_2 - e_2 + m \qquad \text{(metric)}$$

$$R_{o_2} = C' - R_1 - e_1 + \frac{k}{P_d} \qquad \text{(U.S.)} \qquad (5.31)$$

$$R_{o_2} = C' - R_1 - e_1 + m \qquad \text{(metric)}$$

and so the equations for outside radius and depth of cut for nonstandard gears cut by pinion cutter can be put in the same form as the corresponding equations for nonstandard gears cut by a hob.

Finally, the dedendum radii for the nonstandard gears are given by the following equations:

$$R_{d_1} = R_{o_1} - h_t \qquad (5.32)$$

$$R_{d_2} = R_{o_2} - h_t \qquad (5.33)$$

An equation can be developed to determine the cutter offset that will mark the onset of undercutting. By using the law of cosines, it may be noted from triangle $O_c E_1 O_1$ of Fig. 5.13 that

$$(R_{o_c})^2 = (C_{\text{std}} + e)^2 + (R_{b_1})^2 - 2R_{b_1}(C_{\text{std}} + e) \cos \phi_g \qquad (5.34)$$

at the onset of undercutting. From Eq. 5.13,

$$(C_{\text{std}} + e) \cos \phi_g = \frac{(N_1 + N_c)p_b}{2\pi}$$

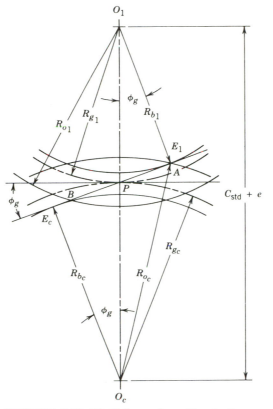

FIGURE 5.13 Limit for undercutting teeth.

which, when substituted into Eq. 5.34, gives

$$e^* = \sqrt{(R_{o_c})^2 - (R_{b_1})^2 + R_{b_1}\frac{(N_1 + N_c)p_b}{\pi}} - C_{std} \qquad (5.35)$$

where e^* is the minimum offset that will prevent undercutting.

For the special case of long and short addendum gears, the change in center distance ΔC equals zero. For hobbed gears, it was seen that $e_1 = -e_2$ from Eq. 5.7a. For nonstandard gears cut by a pinion cutter, no such simplification results, and the relationship between e_1 and e_2 remains highly nonlinear, with e_1 not equal to the negative of e_2. Therefore, long and short addendum gears cannot be cut by standard pinion cutters.

Example 5.4. It is necessary to design a 20-tooth pinion and a 40-tooth gear to operate at a center distance of 3.100 in. with no backlash. The gears are to be cut with a 10-pitch,

20° pinion cutter, with 4-in. pitch diameter. Determine the value of e_1 and e_2 to approximately balance the bending stresses in the teeth of the pinion and gear.

$$C = \frac{N_1 + N_2}{2P_d} = \frac{20 + 40}{2(10)} = 3.000 \text{ in.}$$

$$\Delta C = C' - C = 3.100 - 3.000 = 0.100 \text{ in.}$$

$$\frac{N_2}{N_1} = \frac{40}{20} = 2$$

Therefore, from Fig. 5.10,

$$e_1 = 0.063 \text{ in.}$$

and from Fig. 5.11,

$$e_2 = 0.042 \text{ in.}$$

The stresses were calculated to be

$$S_1 = 22.85 \frac{W_n}{F} \text{ lb/in.}^2$$

and

$$S_2 = 22.87 \frac{W_n}{F} \text{ lb/in.}^2$$

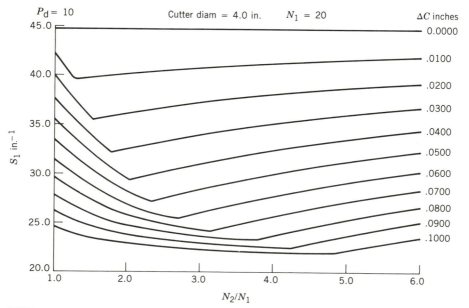

FIGURE 5.14 Tooth stress factor.

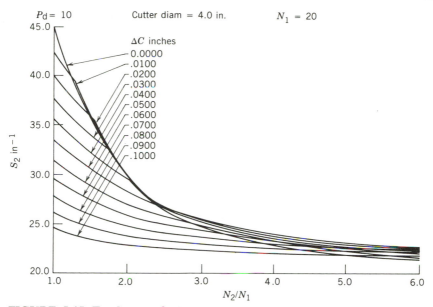

$P_d = 10$ Cutter diam = 4.0 in. $N_1 = 20$

ΔC inches

- 0.0000
- .0100
- .0200
- .0300
- .0400
- .0500
- .0600
- .0700
- .0800
- .0900
- .1000

S_2 in^{-1}

N_2/N_1

FIGURE 5.15 Tooth stress factor.

where

    W_n = normal load at tooth tip (lb)

    F = tooth face width (in.)

As in the case of nonstandard spur gears cut by hobbing, it is also a very tedious job to calculate tooth stresses in nonstandard spur gears cut by pinion cutters. Because of this, curves were developed to give stress factors (SF/W_n) as a function of N_2/N_1 for various changes in center distance.[5] Figures 5.14 and 5.15 show stress factor curves for the pinion and gear for $P_d = 10$ of Example 5.4. A comparison of the stress factors for the data of Example 5.4 is shown in Table 5.4 obtained from detailed calculations given in reference 5 and from the curves of Figs. 5.14 and 5.15.

TABLE 5.4 Stress Factors (Example 5.4)

	Manual Calculations	Design Charts
Pinion	22.85 in.$^{-1}$	22.90 in.$^{-1}$ (Fig. 5.14)
Gear	22.87 in.$^{-1}$	22.90 in.$^{-1}$ (Fig. 5.15)

---

[5]R. N. Green and H. H. Mabie, "Determination of Static Tooth Stresses in Nonstandard Spur Gears Cut by Pinion Cutter," *Mechanism and Machine Theory*, **15** (6), 1980, pp. 507–514.

Example 5.5. Two spur gears of 32 and 48 teeth cut by an 8-pitch, 20° pinion cutter mesh together without backlash at the standard center distance of 5 in. To change the speed ratio, it is necessary to replace the 32-tooth pinion with one of 31 teeth. The tooth thickness on the cutting pitch circle of the 48-tooth gear and the 5-in. center distance are to remain unchanged. Determine the value of e_1 that will give teeth of the proper thickness to mesh with the 48-tooth gear. The pitch diameter of the pinion cutter D_c is 3.000 in., and the number of teeth in the cutter N_c is 24.

$$R_1 = \frac{N_1}{2P_d} = \frac{31}{2(8)} = 1.9375 \text{ in.}$$

$$R_2 = \frac{N_2}{2P_d} = \frac{48}{2(8)} = 3.000 \text{ in.}$$

$$C = \frac{N_1 + N_2}{2P_d} = \frac{31 + 48}{2(8)} = 4.938 \text{ in.}$$

$$C' = 5.000 \text{ in.}$$

$$\cos \phi' = \frac{C \cos \phi_c}{C'} = \frac{4.938 \cos 20°}{5.000}$$

$$\phi' = 21.87° = \phi_r$$

Because $e_2 = 0$, the generating pressure angle of the gear $\phi_{g_2} = 20°$, and Eq. 5.26 can easily be solved for ϕ_{g_1}:

$$(N_1 + N_c) \text{ inv } \phi_{g_1} + (N_2 + N_c) \text{ inv } \phi_{g_2} = 2N_c \text{ inv } \phi_c + (N_1 + N_2) \text{ inv } \phi_r$$

$$(31 + 24) \text{ inv } \phi_{g_1} + (48 + 24) \text{ inv } 20° = 2(24) \text{ inv } 20° + (31 + 48) \text{ inv } 21.87°$$

Therefore,

$$\text{inv } \phi_{g_1} = 0.021773$$

and

$$\phi_{g_1} = 22.59°$$

From Eq. 5.13,

$$e_1 = \frac{(N_1 + N_c)p_b}{2\pi \cos \phi_{g_1}} - C_{\text{std}}$$

(C_{std} is the standard center distance between gear 1 and the cutter.)

$$p_b = p \cos \phi_c = \frac{\pi}{8} \cos 20° = 0.3690 \text{ in.}$$

$$C_{\text{std}} = \frac{N_1 + N_c}{2P_d} = \frac{31 + 24}{2(8)} = 3.4375 \text{ in.}$$

$$e_1 = \frac{(31 + 24)(0.3690)}{2\pi \cos 22.59°} - 3.4375$$

$$= 0.06096 \text{ in.}$$

Problems—U.S. Standard

5.1. A 12-tooth pinion is to be cut with a 2-pitch, 20° hob. Make a layout of the theoretical rack and pinion tooth at the standard setting as shown in Fig. 5.2a. Draw the pinion involute using the equations of involutometry. Show the effect on the pinion tooth of withdrawing the basic rack until its addendum line just passes through the interference point. This layout should be shown dotted and superimposed upon the first sketch with the side of the rack tooth passing through the pitch point. Label the base circle, cutting pitch circle, hob offset, pressure angle, and pitch lines (cutting and standard) of the rack.

5.2. A 24-tooth pinion is to be cut with a 10-pitch, $14\frac{1}{2}$° hob. Calculate the minimum distance the hob will have to be withdrawn to avoid undercutting. Calculate the radius of the cutting pitch circle and the tooth thickness on the cutting pitch circle.

5.3. A 26-tooth gear is to be cut with a 7-pitch, 20° hob. Calculate the maximum distance the hob can be advanced into the gear blank without causing undercutting. Calculate the radius of the cutting pitch circle and the tooth thickness on the cutting pitch circle.

5.4. A 20-tooth gear is cut by a 4-pitch, $14\frac{1}{2}$° hob that has been withdrawn 0.10 in. Determine if this hob offset is enough to eliminate undercutting. If so, calculate the tooth thickness on the cutting pitch circle and on the base circle.

5.5. A 35-tooth gear is to be cut with a 4-pitch, $14\frac{1}{2}$° hob. Calculate the change in cutter setting from its standard position to give a tooth thickness of 0.400 in. on a circle for which the pressure angle is 20°.

5.6. A 20-tooth pinion is to be cut with a 6-pitch, 20° hob. What would be the change in cutter setting from its standard position to give a tooth thickness of 0.274 in. on a circle for which the pressure angle is $14\frac{1}{2}$°?

5.7. A 20-tooth pinion is to be cut with a 6-pitch, 20° hob. Calculate the minimum tooth width that can be produced on a circle for which the pressure angle is $14\frac{1}{2}$°. The tooth is not to be undercut.

5.8. A pinion of 11 teeth and a gear of 14 teeth were cut with an 8-pitch, 20° hob. To avoid undercutting, the hob was withdrawn 0.0446 in. on the pinion and 0.0227 in. on the gear. Calculate the pressure angle and the center distance at which these gears will operate when meshed together. Determine the difference between the center distance calculated above and the standard center distance, and compare with $e_1 + e_2$.

5.9. Prove that

$$(e_1 + e_2) > \Delta C \qquad \text{for} \qquad \phi' > \phi$$

and that

$$(e_1 + e_2) < \Delta C \qquad \text{for} \qquad \phi' < \phi$$

5.10. A pinion of 15 teeth and a gear of 21 teeth are to be cut with a 6-pitch, $14\frac{1}{2}$° hob to operate on a center distance of 3.20 in. Determine whether these gears can be cut without undercutting to operate at this center distance.

5.11. Using the data from Example 5.2, calculate the outside radii of the gear blanks, the depth of cut, and the contact ratio.

5.12. A pinion and gear of 13 and 24 teeth, respectively, are to be cut by a 4-pitch, 20° hob to operate at a center distance of 4.83 in. Calculate the pressure angle at which the gears will operate and the value of e_1 and e_2. Let e_1 and e_2 vary inversely as the number of teeth. Check e_1 to see if it is large enough to prevent undercutting. Determine the outside radii of the gear blanks, the depth of cut, and the contact ratio.

5.13. A 12-tooth pinion has a tooth thickness of 0.2608 in. on its cutting pitch circle. A 32-tooth gear that meshes with the pinion has a tooth thickness of 0.1880 in. on its cutting pitch circle. If both gears have been cut by a 7-pitch, 20 hob, calculate the hob offset e used in cutting each gear and the pressure angle at which the gears operate.

5.14. A nonstandard 35-tooth pinion has a tooth thickness of 0.188 in. at a radius of 2.50 in. and a pressure angle of 20°. The pinion meshes with a rack at the 2.50-in. radius with zero backlash. If the rack is 7-pitch, 20°, calculate the distance from the center of the pinion to the standard pitch line of the rack.

5.15. An 11-tooth pinion is to drive a 23-tooth gear at a center distance of 2.00 in. If the gears are to be cut by a 9-pitch, 20° hob, calculate the value of e_1 and e_2 so that the beginning of contact during cutting of the pinion occurs at the interference point of the pinion.

5.16. A 20-tooth pinion cut with a 10-pitch, 20° hob drives a 30-tooth gear at a center distance of 2.50 in. It is necessary to replace these gears with a pair that will give a velocity ratio of $1\frac{1}{3}:1$ and yet maintain the same center distance. Using the same diametral pitch hob as the original gears, select a pair of gears for the job which vary as little as possible from standard gears. Determine the hob offsets, the outside radii, and the depth of cut.

5.17. It is necessary to connect two shafts whose center distance is 3.90 in. with a pair of spur gears having a velocity ratio of 1.25:1. Using a 10-pitch, $14\frac{1}{2}°$ hob, recommend a pair of gears for the job whose angular velocity ratio will approach 1.25:1 as closely as possible and not be undercut. Calculate the hob offsets, the outside diameters, depth of cut, and the contact ratio.

5.18. A pinion and gear of 27 and 39 teeth, respectively, are to be cut with a 6-pitch, $14\frac{1}{2}°$ hob to give long and short addendum teeth. The hob is offset 0.03 in. Determine for each gear the pitch diameter, the outside diameter, the depth of cut, and the tooth thickness on the pitch circle.

5.19. A pair of long and short addendum gears of 18 and 28 teeth, respectively, are cut with a 4-pitch, 20° hob that has been offset 0.060 in. Compare the contact ratio of these gears with the contact ratio of a pair of standard gears of the same pitch and numbers of teeth.

5.20. A 30-tooth pinion cut with a 20-pitch, 20° hob drives a 40-tooth gear at the standard center distance. If 0.004 in. backlash is required, calculate the amount the hob must be fed into the pinion and into the gear to give this backlash. Assume both gears to be thinned the same amount.

5.21. A 20-tooth pinion cut with an 8-pitch, 25° hob is to mesh with a 40-tooth gear at a center distance of 3.80 in. If the hob is pulled out 0.0352 in. when cutting the pinion and 0.0165 in. when cutting the gear, calculate the backlash produced.

5.22. A pair of long and short addendum gears of 18 and 30 teeth, respectively, cut with a 6-pitch, 25° hob are designed to give zero backlash when the hob is offset 0.05 in. Calculate the value of e_1 and e_2 if these gears are modified to give a backlash of 0.005 in. assuming that both gears are thinned the same amount.

5.23. An 18-tooth pinion cut with a 12-pitch, 20° hob drives a 42-tooth gear. If these gears are semi-recess action gears, calculate the ratio of the recess action to the approach action.

5.24. A pair of semi-recess action gears mesh together without backlash. The pinion has 20 teeth and the gear 48 teeth. If the gears are cut with a 10-pitch, 20° hob, calculate the contact ratio.

5.25. A pair of recess action gears are to be designed to mesh together without backlash. The pinion is to have 20 teeth and the gear 44 teeth, and the gears are to be cut with an 8-pitch, 20° hob. Calculate whether a contact ratio of 1.40 can be attained using semi- or full recess action gears or both.

5.26. A 24-tooth pinion cut with a 10-pitch, 20° hob drives a 40-tooth gear. The gears have semi-recess action, and the length of action $Z = 0.4680$ in. Calculate the ratio of recess action to approach action.

5.27. A 24-tooth pinion is to be cut with a 10-pitch, $14\frac{1}{2}°$ pinion cutter, $N_c = 30$ and $D_c = 3$ in. Calculate the minimum distance the cutter will have to be withdrawn to avoid undercutting. Calculate the radius of the cutting pitch circle and the tooth thickness on the cutting pitch circle.

5.28. A 26-tooth gear is to be cut with an 8-pitch, 20° pinion cutter, $N_c = 24$ and $D_c = 3$ in. Calculate the maximum distance the cutter can be advanced into the gear blank without causing undercutting. Calculate the radius of the cutting pitch circle and the tooth thickness on the cutting pitch circle.

5.29. A 20-tooth gear is cut by a 4-pitch, $14\frac{1}{2}°$ pinion cutter ($N_c = 16$ and $D_c = 4$ in.) that has been withdrawn 0.100 in. Determine if this offset is enough to eliminate undercutting. If so, calculate the tooth thickness on the cutting pitch circle and on the base circle.

5.30. A 35-tooth gear is to be cut by a 4-pitch, $14\frac{1}{2}°$ pinion cutter, $N_c = 20$ and $D_c = 5$ in. Calculate the change in cutter setting from its standard position to give a tooth thickness of 0.400 in. on a circle for which the pressure angle is 20°.

5.31. A 20-tooth pinion is to be cut with a 6-pitch, 20° pinion cutter, $N_c = 36$ and $D_c = 6$ in. What would be the change in cutter setting from its standard position to give a tooth thickness of 0.274 in. on a circle for which the pressure angle is $14\frac{1}{2}°$?

5.32. A 20-tooth pinion is to be cut with a 6-pitch, 20° pinion cutter, $N_c = 30$ and $D_c = 5$ in. Calculate the minimum tooth width that can be produced on a circle for which the pressure angle is $14\frac{1}{2}°$. The tooth is not to be undercut.

5.33. A pinion of 11 teeth and a gear of 14 teeth were cut with an 8-pitch, 20° pinion cutter, $N_c = 24$ and $D_c = 3$ in. To avoid undercutting, the cutter was withdrawn 0.0446 in. on the pinion and 0.0227 in. on the gear. Calculate the pressure angle and the center distance at which these gears will operate when meshed together. Determine the difference between the center distance calculated above and the standard center distance, and compare with $e_1 + e_2$.

5.34. A pinion of 15 teeth and a gear of 21 teeth are to be cut with a 6-pitch, $14\frac{1}{2}°$ pinion cutter ($N_c = 24$ and $D_c = 4$ in.) to operate on a center distance of 3.200 in. Determine whether these gears can be cut without undercutting to operate at this center distance.

5.35. Two spur gears of 12 and 15 teeth, respectively, are to be cut with a 3-pitch, 20° pinion cutter, $N_c = 12$ and $D_c = 4$ in. Determine the center distance at which to operate the gears to avoid undercutting. Calculate the outside radii of the gear blanks, the depth of cut, and the contact ratio.

5.36. A 12-tooth pinion has a tooth thickness of 0.2608 in. on its cutting pitch circle. A 32-tooth gear that meshes with the pinion has a tooth thickness of 0.1880 in. on its cutting pitch circle. If both gears have been cut by an 8-pitch, 20° pinion cutter ($N_c = 24$ and $D_c = 3$ in.), calculate the offset e used in cutting each gear and the pressure angle at which the gears operate.

5.37. An 11-tooth pinion is to drive a 23-tooth gear at a center distance of 2.000 in. If the gears are to be cut by a 10-pitch, 20° pinion cutter ($N_c = 40$ and $D_c = 4$ in.), calculate the value of e_1 and e_2 so that the beginning of contact during cutting of the pinion occurs at the interference point of the pinion.

5.38. A 20-tooth pinion cut with a 10-pitch, 20° pinion cutter ($N_c = 40$ and $D_c = 4$ in.) drives a 30-tooth gear at a center distance of 2.500 in. It is necessary to replace these gears with a pair that will give a velocity ratio of $1\frac{1}{3}$:1 and yet maintain the same center distance. Using the same cutter as the original gears, select a pair of gears for the job that vary as little as possible from standard gears. Determine the offsets, the outside radii, and the depth of cut.

5.39. It is necessary to connect two shafts whose center distance is 3.900 in. with a pair of spur gears having a velocity ratio of 1.25:1. Using a 10-pitch, $14\frac{1}{2}$° pinion cutter ($N_c = 30$ and $D_c = 3$ in.), recommend a pair of gears for the job whose angular velocity ratio will approach 1.25:1 as closely as possible and not be undercut. Calculate the cutter offsets, the outside diameters, the depth of cut, and the contact ratio.

5.40. A 30-tooth pinion cut with a 20-pitch, 20° pinion cutter ($N_c = 60$ and $D_c = 3$ in.) is to mesh with a 40-tooth gear at the standard center distance. If 0.004 in. backlash is required, calculate the amount the cutter must be fed into the pinion and into the gear to give this backlash. Assume both gears to be thinned the same amount.

5.41. A 20-tooth pinion cut with an 8-pitch, 20° pinion cutter ($N_c = 48$ and $D_c = 6$ in.) is to mesh with a 40-tooth gear at a center distance of 3.800 in. If the cutter is pulled out 0.0352 in. when cutting the pinion and 0.0165 in. when cutting the gear, calculate the backlash produced.

Problems—Metric

5.1m. A 12-tooth pinion is to be cut with a 12-module, 20° hob. Make a layout of the theoretical rack and pinion tooth at the standard setting as shown in Fig. 5.2a. Draw the pinion involute using the equations of involutometry. Show the effect on the pinion tooth of withdrawing the basic rack until its addendum line just passes through the interference point. This layout should be dotted and superimposed upon the first sketch with the side of the rack tooth passing through the pitch point. Label the base circle, cutting pitch circle, hob offset, pressure angle, and pitch lines (cutting and standard) of the rack.

5.2m. A 16-tooth pinion is to be cut with a 2.5-module, 20° hob. Calculate the minimum distance the hob will have to be withdrawn to avoid undercutting. Calculate the radius of the cutting pitch circle and the tooth thickness on the cutting pitch circle.

5.3m. A 26-tooth gear is to be cut with a 3.5-module, 20° hob. Calculate the maximum distance the hob can be advanced into the gear blank without causing undercutting. Calculate the radius of the cutting pitch circle and the tooth thickness on the cutting pitch circle.

5.4m. A 16-tooth gear is cut by a 6-module, 20° hob that has been withdrawn 0.5000 mm. Determine if this hob offset is enough to eliminate undercutting. If so, calculate the tooth thickness on the cutting pitch circle and on the base circle.

5.5m. A 35-tooth gear is to be cut with a 6-module, 20° hob. Calculate the change in cutter setting from its standard position to give a tooth thickness of 10.2 mm on a circle for which the pressure angle is 20°.

5.6m. A 20-tooth pinion is to be cut with a 4-module, 20° hob. What would be the change in cutter setting from its standard position to give a tooth thickness of 6.960 mm on a circle for which the pressure angle is $14\frac{1}{2}$°?

5.7m. A 20-tooth pinion is to be cut with a 4-module, 20° hob. Calculate the minimum tooth width that can be produced on a circle for which the pressure angle is $14\frac{1}{2}°$. The tooth is not to be undercut.

5.8m. A pinion of 11 teeth and a gear of 14 teeth were cut with a 3-module, 20° hob. To avoid undercutting, the hob was withdrawn 1.0698 mm on the pinion and 0.5434 mm on the gear. Calculate the pressure angle and the center distance at which these gears will operate when meshed together. Determine the difference between the center distance calculated above and the standard center distance, and compare with $e_1 + e_2$.

5.9m. Prove that

$$(e_1 + e_2) > \Delta c \quad \text{for} \quad \phi' > \phi$$

and that

$$(e_1 + e_2) < \Delta C \quad \text{for} \quad \phi' < \phi$$

5.10m. A pinion of 12 teeth and a gear of 15 teeth are to be cut with a 6-module, 20° hob to operate at a center distance of 83.50 mm. Determine whether these gears can be cut without undercutting to operate at this center distance.

5.11m. A pinion and gear of 13 and 24 teeth, respectively, are to be cut by a 6-module, 20° hob to operate at a center distance of 115.9 mm. Calculate the pressure angle at which the gears will operate and the value of e_1 and e_2. Let e_1 and e_2 vary inversely as the number of teeth. Check e_1 to see if it is large enough to prevent undercutting. Determine the outside radii of the gear blanks, the depth of cut, and the contact ratio.

5.12m. Using the data from Example 5.3, check to see if the value of e_1 is large enough to avoid undercutting. Calculate the outside radii of the gear blanks, the depth of cut, and the contact ratio.

5.13m. A 12-tooth pinion has a tooth thickness of 6.624 mm on its cutting pitch circle. A 32-tooth gear that meshes with the pinion has a tooth thickness of 4.372 mm on its cutting pitch circle. If both gears have been cut by a 3.5-module, 20° hob, calculate the hob offset e used in cutting each gear and the pressure angle at which the gears operate.

5.14m. A nonstandard 35-tooth pinion has a tooth thickness of 4.604 mm at a radius of 61.25 mm and a pressure angle of 20°. The pinion meshes with a rack at the 61.25 mm radius with zero backlash. If the rack is 3.5-module, 20°, calculate the distance from the center of the pinion to the standard pitch line of the rack.

5.15m. An 11-tooth pinion is to drive a 23-tooth gear at a center distance of 54.0 mm. If the gears are to be cut by a 3-module, 20° hob, calculate the value of e_1 and e_2 so that the beginning of contact during cutting of the pinion occurs at the interference point of the pinion.

5.16m. A 2.5-module, 20° pinion with 20 teeth drives a gear with 30 teeth at a center distance of 62.50 mm. It is necessary to replace these gears with a pair that will give a velocity ratio of $1\frac{1}{3}:1$ and yet maintain the same center distance. Using the same hob as the original gears, select a pair of gears for the job which vary as little as possible from standard gears. Determine the hob offsets, outside radii, and the depth of cut.

5.17m. It is necessary to connect two shafts whose center distance is 99.06 mm with a pair of spur gears having a velocity ratio of 1.25:1. Using a 2.5-module, 20° hob, recommend a pair of gears for the job whose angular velocity ratio will approach 1.25:1 as closely as possible and not be undercut. Calculate the hob offsets, the outside diameters, depth of cut, and the contact ratio.

5.18m. A pinion and gear of 27 and 39 teeth, respectively, are to be cut with a 4-module, 20° hob to give long and short addendum teeth. The hob is offset 0.720 mm. Determine for each gear the pitch diameter, the outside diameter, the depth of cut, and the tooth thickness on the pitch circle.

5.19m. A pair of long and short addendum gears of 18 and 28 teeth, respectively, are cut with a 6-module, 20° hob that has been offset 1.524 mm. Compare the contact ratio of these gears with the contact ratio of a pair of standard gears of the same module and numbers of teeth.

5.20m. A 1.25-module, 20° pinion with 30 teeth is to mesh with a 40-tooth gear at the standard center distance. If 0.1016 mm backlash is required, calculate the amount the hob must be fed into the pinion and into the gear to give the backlash. Assume both gears to be thinned the same amount.

5.21m. A 3-module, 20° pinion with 20 teeth is to mesh with a 40-tooth gear at a center distance of 90.52 mm. If the hob is pulled out 0.2271 mm when cutting the pinion and 0.1096 mm when cutting the gear, calculate the backlash produced.

5.22m. A pair of long and short addendum gears of 18 and 30 teeth, respectively, cut with a 4-module, 20° hob are designed to give zero backlash when the hob is offset 1.2700 mm. Calculate the value of e_1 and e_2 if these gears are modified to give a backlash of 0.1270 mm, assuming that both gears are thinned the same amount.

5.23m. A 2-module, 20° pinion of 18 teeth drives a gear of 42 teeth. If these gears are semi-recess action gears, calculate the ratio of the recess action to the approach action.

5.24m. A pair of semi-recess action gears mesh together without backlash. The pinion has 20 teeth and the gear 48 teeth. If the gears are cut with a 2.5-module, 20° hob, calculate the contact ratio.

5.25m. A pair of recess action gears are to be designed to mesh together without backlash. The pinion is to have 20 teeth and the gear 44 teeth, and the gears are to be cut with a 3-module, 20° hob. Calculate whether a contact ratio of 1.40 can be attained using semi- or full recess action gears or both.

5.26m. A 24-tooth pinion cut with a 2.5 module hob drives a 40-tooth gear. The gears have semi-recess action, and the length of action Z is 11.663 mm. Calculate the ratio of recess action to approach action.

5.27m. A 24-tooth pinion is to be cut with a 2.5-module, 20° pinion cutter, $N_c = 30$ and $D_c = 75$ mm. Calculate the minimum distance the cutter will have to be withdrawn to avoid undercutting. Calculate the radius of the cutting pitch circle and the tooth thickness on the cutting pitch circle.

5.28m. A 26-tooth gear is to be cut with a 3-module, 20° pinion cutter, $N_c = 26$ and $D_c = 78$ mm. Calculate the maximum distance the cutter can be advanced into the gear blank without causing undercutting. Calculate the radius of the cutting pitch circle and the tooth thickness on the cutting pitch circle.

5.29m. A 20-tooth gear is to be cut with a 6-module, 20° pinion cutter ($N_c = 18$ and $D_c = 108$ mm) that has been withdrawn 2.54 mm. Determine if this cutter offset is enough to eliminate undercutting. If so, calculate the tooth thickness on the cutting pitch circle and on the base circle.

5.30m. A 35-tooth gear is to be cut with a 6-module, 20° pinion cutter, $N_c = 18$ and $D_c = 108$ mm. Calculate the change in cutter setting from its standard position to give a tooth thickness of 10.2 mm on a circle for which the pressure angle is 20°.

5.31m. A 20-tooth pinion is to be cut with a 4-module, 20° pinion cutter, $N_c = 26$ and

D_c = 104 mm. What would be the change in cutter setting from its standard position to give a tooth thickness of 6.960 mm on a circle for which the pressure angle is $14\frac{1}{2}°$?

5.32m. A 20-tooth pinion is to be cut with a 4-module, 20° pinion cutter, N_c = 26 and D_c = 104 mm. Calculate the minimum tooth width that can be produced on a circle for which the pressure angle is $14\frac{1}{2}°$. The tooth is not to be undercut.

5.33m. A pinion of 11 teeth and a gear of 14 teeth were cut with a 3-module, 20° pinion cutter, N_c = 26 and D_c = 78 mm. To avoid undercutting, the cutter was withdrawn 1.0698 mm on the pinion and 0.5434 mm on the gear. Calculate the pressure angle and the center distance at which these gears will operate when meshed together. Determine the difference between the center distance calculated above and the standard center distance, and compare with $e_1 + e_2$.

5.34m. A pinion of 12 teeth and a gear of 15 teeth are to be cut with a 6-module, 20° pinion cutter (N_c = 18 and D_c = 108 mm) to operate at a center distance of 83.50 mm. Determine whether these gears can be cut without undercutting to operate at this center distance.

5.35m. Two spur gears of 12 and 15 teeth are to be cut with a 4-module, 20° pinion cutter, N_c = 26 and D_c = 104 mm. Determine the center distance at which to operate the gears to avoid undercutting. Calculate the outside radii of the gear blanks, the depth of cut, and the contact ratio.

5.36m. A 12-tooth pinion has a tooth thickness of 6.624 mm on its cutting pitch circle. A 32-tooth gear that meshes with the pinion has a tooth thickness of 4.372 mm on its cutting pitch circle. If both gears have been cut by a 3.5-module, 20° pinion cutter (N_c = 22 and D_c = 77 mm), calculate the offset e used in cutting each gear and the pressure angle at which the gears operate. Can these gears be cut with these tooth thicknesses without undercutting?

5.37m. An 11-tooth pinion is to drive a 23-tooth gear at a center distance of 54.0 mm. If the gears are to be cut by a 2.5-module, 20° pinion cutter (N_c = 30 and D_c = 75 mm), calculate the value of e_1 and e_2 so that the beginning of contact during cutting of the pinion occurs at the interference point of the pinion.

5.38m. A 20-tooth pinion cut with a 2.5-module, 20° pinion cutter (N_c = 30 and D_c = 75 mm) drives a 30-tooth gear at a center distance of 62.50 mm. It is necessary to replace these gears with a pair that will give a velocity ratio of $1\frac{1}{3}$:1 and yet maintain the same center distance. Using the same cutter as the original gears, select a pair of gears for the job that vary as little as possible from standard gears. Determine the cutter offsets, the gear outside radii, and the gear depth of cut.

5.39m. It is necessary to connect two shafts whose center distance is 99.06 mm with a pair of spur gears having a velocity ratio of 1.25:1. Using a 2.5-module, 20° pinion cutter (N_c = 30 and D_c = 75 mm), recommend a pair of gears for the job whose angular-velocity ratio will approach 1.25:1 as closely as possible and not be undercut. Calculate the cutter offsets, the gear outside diameters, the gear depth of cut, and the contact ratio.

5.40m. A 30-tooth pinion cut with a 1-module, 20° pinion cutter (N_c = 76 and D_c = 76 mm) drives a 40-tooth gear at the standard center distance. If 0.1016 mm backlash is required, calculate the amount the cutter must be fed into the pinion and into the gear to give the backlash. Assume both gears to be thinned the same amount.

Chapter Six

Bevel, Helical,
and Worm Gearing

6.1 THEORY OF BEVEL GEARS

Bevel gears (Fig. 6.1) are used to connect shafts whose axes intersect. The *shaft angle* is defined as the angle between the centerlines which contains the engaging teeth. Although the shaft angle is usually 90°, there are many bevel gear applications that require shaft angles larger or smaller than this amount.

The pitch surface of a bevel gear is a cone. When two bevel gears mesh

FIGURE 6.1 Straight bevel gears. (Courtesy of Gleason Works.)

together, their cones contact along a common element and have a common apex where the shaft centerlines intersect. The cones roll together without slipping and have spherical motion. Each point in a bevel gear remains at a constant distance from the common apex.

Figure 6.2 shows an axial section of a pair of bevel gears in mesh with the shafts at right angles. Because the pitch cones roll together without slipping, the angular velocity ratio is inversely proportional to the diameters of the bases of the cones. These cone diameters become the pitch diameters of the gears. The angular-velocity ratio can then be expressed as $\omega_1/\omega_2 = D_2/D_1 = N_2/N_1$ as in the case of spur gears. The relation $P_d = N/D$ also holds as in spur gears.

In making a sketch of a pair of spur gears in mesh, it was a simple matter, knowing the pitch diameters, to draw the pitch circles in their correct position. In the case of bevel gears, however, pitch angles as well as pitch diameters have to be considered. Equations for the pitch angles are derived below; A_O is the length of the pitch cone element:

$$\sin \Gamma_1 = \frac{D_1}{2A_O} = \sin (\Sigma - \Gamma_2)$$

$$\sin \Gamma_1 = \sin \Sigma \cos \Gamma_2 - \cos \Sigma \sin \Gamma_2$$

$$\frac{\sin \Gamma_1}{\sin \Sigma \sin \Gamma_2} = \frac{\cos \Gamma_2}{\sin \Gamma_2} - \frac{\cos \Sigma}{\sin \Sigma}$$

$$\frac{1}{\sin \Sigma} \left[\frac{\sin \Gamma_1}{\sin \Gamma_2} + \cos \Sigma \right] = \frac{1}{\tan \Gamma_2}$$

Also,

$$\frac{\sin \Gamma_1}{\sin \Gamma_2} = \frac{D_1}{D_2}$$

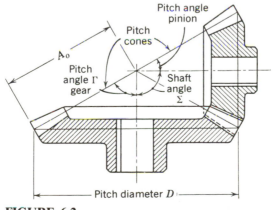

FIGURE 6.2

Therefore,

$$\tan \Gamma_2 = \frac{\sin \Sigma}{\cos \Sigma + (D_1/D_2)} = \frac{\sin \Sigma}{\cos \Sigma + (N_1/N_2)} \qquad (6.1)$$

Similarly,

$$\tan \Gamma_1 = \frac{\sin \Sigma}{\cos \Sigma + (N_2/N_1)} \qquad (6.2)$$

Although Eqs. 6.1 and 6.2 were derived for gears with shafts at right angles, these equations also apply to bevel gears with any shaft angle.

In making a layout of a pair of bevel gears in mesh, the position of the common pitch cone element can be determined graphically if the angular-velocity ratio and the shaft angle are known.

As has been mentioned, the pitch cones of a pair of bevel gears have spherical motion. Therefore, to have the large ends of the bevel gear teeth match perfectly when in mesh, they should lie in the surface of a sphere whose center is the apex of the pitch cones and whose radius is the common pitch cone element. It is not customary, however, to make the back of a bevel gear spherical, so it is made conical as shown in Fig. 6.3. This cone is known as the *back cone* and is tangent to the theoretical sphere at the pitch diameter. The elements of the

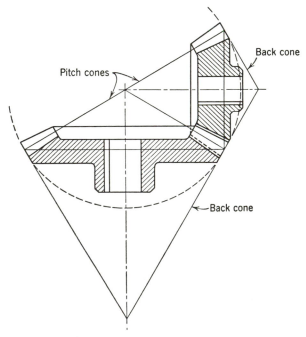

Back cone

Pitch cones

Back cone

FIGURE 6.3

back cone are therefore perpendicular to those of the pitch cone. For all practical purposes, the surface of the back cone and the surface of the sphere are identical in the region of the ends of the bevel gear teeth. The distances from the apex of the pitch cones to the outer ends of the teeth at any point except the pitch point are not equal, so the end surfaces of meshing teeth will not be quite flush. However, this variation is slight and does not affect tooth action.

All the proportions of the tooth of a bevel gear are figured at the large end of the tooth. This will be amplified in a later section. When it is necessary to show the outline of the large end of the tooth, use is made of the fact that the profile of the bevel gear tooth closely corresponds to that of a spur gear tooth having a pitch radius equal to the back cone element and a diametral pitch equal to that of the bevel gear. This spur gear is called the *equivalent spur gear,* and this section through the bevel gear is known as the *transverse section.*

In addition to the general type of bevel gears shown in Fig. 6.2, there are the following three special types:

1. *Miter Gears.* The gears are of equal size and the shaft angle is 90°.
2. *Angular Bevel Gears.* The shaft angle is greater or less than 90°. A sketch is shown in Fig. 6.4.
3. *Crown Gear.* The pitch angle equals 90°, and the pitch surface becomes a plane. A sketch is shown in Fig. 6.5.

Up to the present time, the discussion has dealt primarily with the general theory and types of bevel gears. We are now ready to consider the form of the bevel gear tooth.

As was seen from the study of Chapter 4, the involute profile of a spur gear was easily generated from a base circle and took the form of a cylindrical involute when the thickness of the gear was considered. The involute form is not used for bevel gears, however, because the base surface would be a cone. This means that, when a plane is rolled on this base cone, a line in the plane generates a spherical involute. A spherical involute is impractical to manufacture.

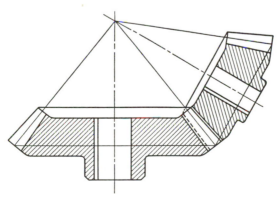

FIGURE 6.4

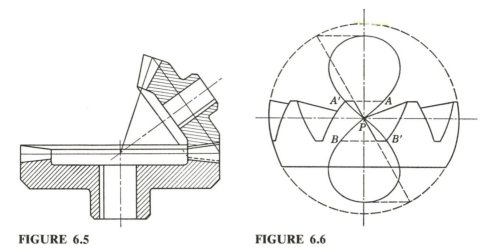

FIGURE 6.5   **FIGURE 6.6**

The bevel gear system that has been developed is one in which the teeth are generated conjugate to a crown gear having teeth with flat sides. The crown gear therefore bears the same relation to bevel gears as a rack does to spur gears. Figure 6.6 shows the sketch of a theoretical crown gear. The sides of the teeth lie in planes which pass through the center of the sphere. When the crown gear is meshed with a conjugate gear, the complete path of contact in the surface of the sphere is in the form of a figure 8. Because of this, the teeth in the crown gear and in the conjugate gear are called *octoid teeth*. Only a portion of the path is used, and for teeth of the height shown the path of contact is either APB or $A'PB'$.

6.2 BEVEL GEAR DETAILS

For considering the details of a bevel gear, an axial section of a pair of Gleason straight-tooth bevel gears is shown in Fig. 6.7a. The Gleason system has been adopted as the standard for bevel gears. As seen in the sketch, the dedendum elements are drawn toward the apex of the pitch cones. The addendum elements, however, are drawn parallel to the dedendum elements of the mating member, thus giving a constant clearance and eliminating possible fillet interference at the small ends of the teeth. Elimination of this possible interference allows larger edge radii to be used on the generating tools, which will increase tooth strength through increased fillets. The large ends of the teeth are proportioned according to the long and short addendum system discussed in Chapter 5 so that the addendum on the pinion will be greater than that on the gear. Long addendums are used on the pinion primarily to avoid undercutting, to balance tooth wear, and to increase tooth strength. The Gleason standard for the proportions of straight bevel gear teeth is given in a later section. Figure 6.7b is the transverse section $A–A$ showing the tooth profiles.

The addendum and dedendum are measured perpendicularly to the pitch

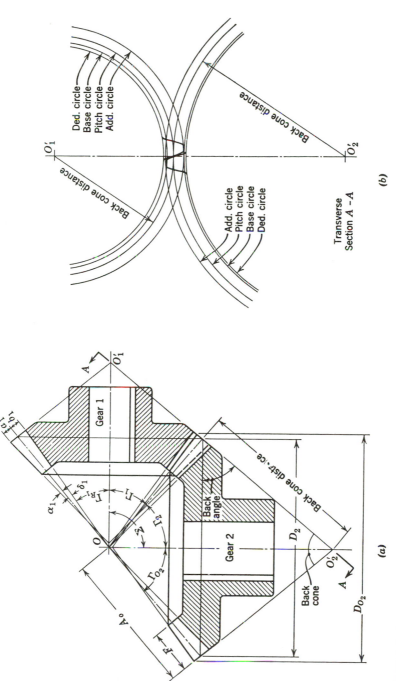

FIGURE 6.7 D = Pitch diameter, D_o = outside diameter, A_o = pitch cone distance, F = face width, a = addendum, b = dedendum, Σ = shaft angle, Γ = pitch angle, α = addendum angle, δ = dedendum angle, Γ_O = face angle, Γ_R = foot angle.

cone element at the outside of the gear; therefore, the dedendum angle is given by

$$\tan \delta = \frac{b}{A_O} \tag{6.3}$$

Because the addendum element is not drawn toward the apex of the pitch cones, the addendum angle α must be determined indirectly. It can be shown that the addendum angle of the pinion will equal the dedendum angle of the gear. Likewise, the addendum angle of the gear will equal the dedendum angle of the pinion. The face angle and the root angle are therefore

$$\Gamma_O = \Gamma + \alpha \tag{6.4}$$

$$\Gamma_R = \Gamma - \delta \tag{6.5}$$

Because the back angle is equal to the pitch angle, the outside diameter of a bevel gear is

$$D_O = D + 2a \cos \Gamma \tag{6.6}$$

The face width of a bevel gear is not determined by the kinematics of tooth action but by requirements of manufacture and load capacity. Manufacturing difficulties are encountered if the face width of the gear is too large a proportion of the cone distance A_O. Therefore, the face width is limited as follows:

$$F < 0.3A_O \quad \text{or} \quad \frac{10.0}{P_d} \text{ (or } 10.0\,m) \tag{6.7}$$

whichever is smaller.

Although integral diametral pitches or modules are frequently used on bevel gears, there is not the same necessity for restricting designs on this account since tooling for bevel gears is not limited to standard pitches or modules as in the case of spur gearing. Table 6.1 shows the symbols for bevel gears for the AGMA

TABLE 6.1 Bevel Gear Symbols

	AGMA	ISO 701
Cone distance	A_O	R
Pitch angle	Γ	δ'
Shaft angle	Σ	Σ
Addendum angle	α	θ_a
Dedendum angle	δ	θ_f
Face angle	Γ_O	δ_a
Root angle	Γ_R	δ_t

and for the proposed International Standard ISO 701. Symbols that are the same as those for spur gears (Table 4.6) are not included.

6.3 GLEASON STRAIGHT BEVEL GEAR TOOTH PROPORTIONS

(For straight bevel gears with axes at right angles and 13 or more pinion teeth.)

1. Number of teeth:

> 16 or more teeth in the pinion
> 15 teeth in pinion and 17 or more teeth in gear
> 14 teeth in pinion and 20 or more teeth in gear
> 13 teeth in pinion and 30 or more teeth in gear

2. Pressure angle, $\phi = 20°$

3. Working depth, $h_k = 2.000/P_d$ (U.S.)

$$= 2.000m \qquad \text{(metric)}$$

4. Whole depth, $h_t = 2.188/P_d + 0.002$ (U.S.)

$$= 2.188m + 0.05 \qquad \text{(metric)}$$

5. Addendum,

Gear: $a_G = \dfrac{0.540}{P_d} + \dfrac{0.460}{P_d(N_2/N_1)^2}$ (U.S.)

$$a_G = 0.540m + \frac{0.460m}{(N_2/N_1)^2} \qquad \text{(metric)}$$

Pinion: $a_P = \dfrac{2.000}{P_d} - a_G$ (U.S.)

$$a_P = 2.000m - a_G \qquad \text{(metric)}$$

6. Dedendum,

Gear: $b_G = \dfrac{2.188}{P_d} + 0.002 - a_G$ (U.S.)

$$b_G = 2.188m + 0.05 - a_G \qquad \text{(metric)}$$

Pinion: $b_P = \dfrac{2.188}{P_d} + 0.002 - a_P$ (U.S.)

$$b_P = 2.188m + 0.05 - a_P \qquad \text{(metric)}$$

7. Circular thickness (tooth thickness on pitch circle),

Gear: $t_G = \dfrac{p}{2} - (a_P - a_G) \tan \phi$ (approximately)[1]

Pinion: $t_P = p - t_G$

where p is the circular pitch.

6.4   ANGULAR STRAIGHT BEVEL GEARS

The proportions of angular straight bevel gears can be determined from the same relations as given for bevel gears at right angles with the following exceptions:

1. The limiting numbers of teeth cannot be taken from item one in section 6.3. Each application must be examined separately for undercutting with the aid of a chart in Gleason's *Design Manual*. This chart shows a plot of maximum pinion dedendum angle for no undercut versus pitch angle. Curves are given for several pressure angles.

2. The pressure angle is determined in conjunction with the preceding item.

3. In determining the gear addendum from item five in section 6.3, it is necessary to use an equivalent 90° bevel gear ratio for the ratio N_2/N_1.

$$\text{Equivalent 90° ratio} = \sqrt{\frac{N_2 \cos \Gamma_1}{N_1 \cos \Gamma_2}}$$

For a crown gear ($\Gamma = 90°$), this ratio equals infinity.

For angular bevel gears where the shaft angle is greater than 90° and the pitch angle of the gear is also greater than 90°, an internal bevel gear results. In this case, the calculations should be referred to the Gleason Works to determine whether the gears can be cut.

6.5   ZEROL BEVEL GEARS

In addition to straight bevel gears, there are two other types of bevel gears, one of which is the Zerol bevel. Zerol bevel gears have curved teeth with zero spiral angle at the middle of the face width, as shown in Fig. 6.8, and have the same thrust and tooth action as straight bevel gears. They may therefore be used in the same mountings. The advantage of the Zerol gear over the straight bevel is

---

[1]To obtain the exact value, a set of curves is necessary which is not suitable for inclusion here. See Gleason, *Design Manual*.

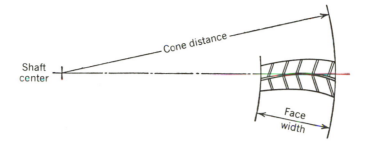

FIGURE 6.8 Diagram showing section through zerol pinion and gear teeth in mesh.

that it can have its tooth surfaces ground. Also, the Zerol gear has localized tooth contact, that is, contact over only the central portion of the tooth instead of along the entire tooth, whereas the straight bevel may or may not, depending upon the bevel gear generator used. Modern straight bevel gear generators produce a tooth with localized bearing by curving the teeth along their length ever so slightly. Mating teeth are therefore slightly convex so that contact takes place near the middle of the tooth. A straight bevel gear with this feature is known as a Coniflex gear. The localized contact allows a slight amount of adjustment during assembly and some displacement due to deflection under operating loads without concentrating the load on the ends of the teeth. Photographs of Coniflex and Zerol bevel gears appear in Fig. 6.9. Figure 6.10 shows the machining of a Coniflex bevel gear.

6.6 SPIRAL BEVEL GEARS

The second type is the spiral bevel gear, which has obliquely curved teeth. Figure 6.11a shows a section of a pair of teeth in contact, and Fig. 6.11b shows the tooth spiral of one gear. The teeth are given a spiral angle such that the face advance (Fig. 6.11b) is greater than the circular pitch, which results in continuous pitch line contact in the plane of the axes of the gears. This makes it possible to obtain smooth operation with a smaller number of teeth in the pinion than with straight or Zerol bevel gears, which do not have continuous pitch line contact. Also, in spiral bevel gears, contact between the teeth begins at one end of the tooth and progresses obliquely across the face of the tooth. This is in contrast to the tooth action of straight or Zerol bevel gears, where contact takes place all at once across the entire width of face. For these reasons, spiral bevel gears have smoother action than either straight or Zerol bevel gears and are especially suitable for high-speed work. As shown in Fig. 6.11a, spiral bevel gears have localized tooth contact, which is easily controlled by varying the radii of curvature of mating teeth. Spiral bevel gears can also have their tooth surfaces ground. Figure 6.12 shows a pair of spiral bevel gears in mesh. Figure 6.13 illustrates the production of a spiral bevel gear pinion.

FIGURE 6.9   (*a*) Coniflex bevel gears showing localized contact. (*b*) Zerol bevel gears showing localized contact. (Courtesy of Gleason Works.)

FIGURE 6.10 Cutting coniflex bevel gear. (Courtesy of Gleason Works.)

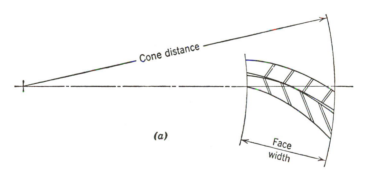

(a)

FIGURE 6.11 (*a*) Diagram showing section through spiral bevel pinion and gear teeth in mesh.

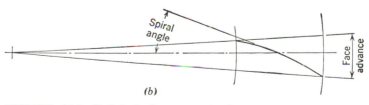

(b)

FIGURE 6.11 (*b*) Spiral of one gear tooth.

FIGURE 6.12 Spiral bevel gears. (Courtesy of Gleason Works.)

FIGURE 6.13 Generating spiral bevel pinion. (Courtesy of Gleason Works.)

6.7 HYPOID GEARS

At one time, spiral bevel gears were used exclusively for automotive rear axle drive gears (ring gear and pinion). In 1925, Gleason introduced the hypoid gear, which has replaced the spiral bevel for this application. Hypoid gears are similar in appearance to spiral bevel gears, with the exception that the axis of the pinion is offset from that of the gear so that the axes no longer intersect. See Fig. 6.14. To take care of this offset and still maintain line contact, the pitch surface of a hypoid gear approaches a hyperboloid of revolution rather than a cone as in bevel gears. In automotive applications, the offset is advantageous because it allows the drive shaft to be lowered, resulting in a lower slung body. In addition, hypoid pinions are stronger than spiral bevel pinions. The reason for this is that hypoid gears can be designed so that the spiral angle of the pinion is larger than that of the gear. This results in a larger, and hence stronger, pinion than the corresponding spiral bevel pinion. Another difference is that hypoid gears have sliding action along the teeth, whereas spiral bevel gears do not. Hypoid gears operate more quietly and can be used for higher reduction ratios than spiral bevel gears. Hypoid gears can also be ground.

The tooth form for Zerol bevel, spiral bevel, and hypoid gears is the long and short addendum system, except when both gears have the same number of teeth. Standards similar to that given for straight bevel gears have been developed for these systems and may be found in the Gleason *Design Manual* for bevel and hypoid gears.

FIGURE 6.14 Hypoid gears. (Courtesy of Gleason Works.)

6.8 THEORY OF HELICAL GEARS

If a plane is rolled on a base cylinder, a line in the plane parallel to the axis of the cylinder will generate the surface of an involute spur gear tooth. If the generating line is inclined to the axis, however, the surface of a helical gear tooth will be generated. These two conditions are shown in Figs 6.15a and 6.15b, respectively.

Helical gears are used to connect parallel shafts and nonparallel, nonintersecting shafts. The former are known as *parallel helical gears* and the latter as *crossed helical gears*. See Figs. 6.16a and 6.16b.

In determining the tooth proportions of a helical gear for either crossed or parallel shafts, it is necessary to consider the manner in which the teeth are to be cut. If the gear is to be hobbed, all dimensions are figured in a plane that is normal to the tooth pitch element, and the diametral pitch and the pressure angle are standard values in that plane. Because the cutting action of a hob occurs in the normal plane, it is possible to use the same hob to cut both helical gears and

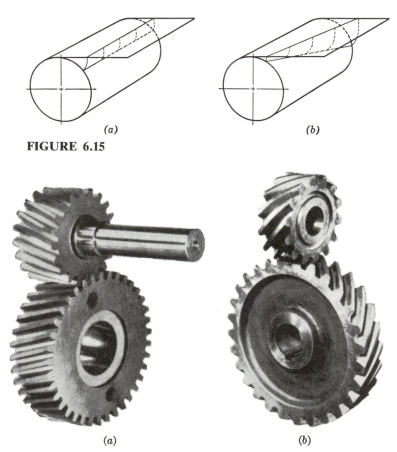

(a)   *(b)*

FIGURE 6.15

(a)   *(b)*

FIGURE 6.16 Helical gears *(a)* for parallel shafts and *(b)* for crossed shafts. (Courtesy of D. O. James Gear Manufacturing Company.)

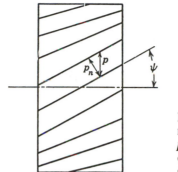

FIGURE 6.17 p_n = normal circular pitch, p = circular pitch in plane of rotation, ψ = helix angle.

spur gears of a given pitch; in a spur gear the normal plane and the plane of rotation are identical. Figure 6.17 shows a sketch of a helical gear with the circular pitch measured in the normal plane and in the plane of rotation. From the sketch,

$$p_n = p \cos \psi = \frac{\pi \cos \psi}{P_d} \qquad \text{(U.S.)}$$

$$p_n = \pi m \cos \psi \qquad \text{(metric)}$$

(6.8)

where P_d = diametral pitch in plane of rotation (also known as *transverse diametral pitch*). Figure 6.18 shows hobbing of helical gears.

When a helical gear is cut by a hob, the normal circular pitch p_n of Fig. 6.17 becomes equal to the circular pitch of the hob. From this and from the fact that $p = \pi/P_d$ the following relation can be written:

$$p_n = \frac{\pi}{P_{nd}} \qquad \text{(U.S.)}$$

$$p_n = \pi m_n \qquad \text{(metric)}$$

where P_{nd} is the normal diametral pitch and is equal to the diametral pitch of the hob, and m_n is the normal module. Substituting for p_n in Eq. 6.8,

$$P_d = P_{nd} \cos \psi \qquad \text{(U.S.)}$$

$$m = \frac{m_n}{\cos \psi} \qquad \text{(metric)}$$

(6.9)

Also, by substituting $P_d = N/D$ in Eq. 6.9,

$$D = \frac{N}{P_{nd} \cos \psi} \qquad \text{(U.S.)}$$

$$D = \frac{N m_n}{\cos \psi} \qquad \text{(metric)}$$

(6.10)

FIGURE 6.18 Hobbing a helical gear. (Courtesy of Falk Gear Company.)

While it is not our intent to go into detail concerning the forces acting on a helical gear, it is necessary to consider them in determining the relation between pressure angle in the plane of rotation ϕ and the normal pressure angle ϕ_n and the helix angle ψ. From Fig. 6.19 showing these forces,

$$\tan \phi = \frac{F_s}{F_t} \text{ (plane } OABH)$$

$$\tan \phi_n = \frac{F_s}{OD} \text{ (plane } ODC)$$

$$OD = \frac{F_t}{\cos \psi} \text{ (plane } OADG)$$

Therefore,

$$\tan \phi_n = \frac{F_s \cos \psi}{F_t}$$

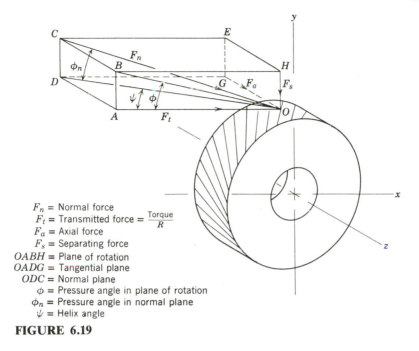

F_n = Normal force
F_t = Transmitted force = $\dfrac{\text{Torque}}{R}$
F_a = Axial force
F_s = Separating force
$OABH$ = Plane of rotation
$OADG$ = Tangential plane
ODC = Normal plane
ϕ = Pressure angle in plane of rotation
ϕ_n = Pressure angle in normal plane
ψ = Helix angle

FIGURE 6.19

and

$$\tan \phi = \frac{\tan \phi_n}{\cos \psi} \tag{6.11}$$

It is also interesting to consider the effect of the helix angle on the number of teeth that can be cut by a hob on a helical gear without undercutting. By referring to Fig. 4.17 (spur gears), an equation can be developed for the minimum number of teeth for helical gears cut by a hob as was done for spur gears in Chapter 4. (See Eq. 4.12.) This results in

$$N = \frac{2k \cos \psi}{\sin^2 \phi} \tag{6.12}$$

A table has been compiled by AGMA (207.05, June 1971) to give the minimum number of teeth that can be hobbed on a helical gear without undercutting. These are given in Table 6.2 as a function of helix angle ψ and normal pressure angle ϕ_n full-depth teeth.

If it is necessary to use a pinion smaller than those given in Table 6.2, the pinion may be cut without undercutting by withdrawing the hob in a manner similar to that shown for spur gears in Chapter 5. An equation which is equivalent

TABLE 6.2 Minimum Number of Teeth for Helical Gears Without Undercutting[a]

	ϕ_n		
ψ	$14\frac{1}{2}°$	20°	25°
0° (spur gears)	32	18	12
5°	32	17	12
10°	31	17	11
15°	29	16	11
20°	27	15	10
23°	26	14	10
25°	25	14	9
30°	22	12	8
35°	19	10	7
40°	15	9	6
45°	12	7	5

[a]Extracted from AGMA Standard System—Tooth Proportions for Fine-Pitch Spur and Helical Gears (AGMA 207.06), with the permission of the publisher, the American Gear Manufacturers Association, 1500 King Street, Alexandria, VA 22314.

to Eq. 5.2 for spur gears can be derived for helical gears as

$$e = \frac{1}{P_{nd}} \left[k - \frac{N \sin^2 \phi}{2 \cos \psi} \right] \qquad \text{(U.S.)}$$

$$e = m_n \left[1.000 - \frac{N \sin^2 \phi}{2 \cos \psi} \right] \qquad \text{(metric)}$$

(6.13)

The value of e given by Eq. 6.13 is the amount the hob will have to be withdrawn in order to have the addendum line of the rack or hob just pass through the interference point of the pinion being cut.

Although most hobs are designed to have the diametral pitch a standard value in the normal plane, some hobs are produced that have the diametral pitch a standard value in the plane of rotation. These hobs are known as *transversal hobs,* and the pitch in the plane of rotation is known as the *transverse diametral pitch.*

If the gear is to be cut by the Fellows gear shaper method, the dimensions are considered in the plane of rotation, and the diametral pitch and the pressure angle are standard values in that plane. When a helical gear is cut by a Fellows cutter, the circular pitch p of Fig. 6.17 becomes equal to the circular pitch of the

FIGURE 6.20 Pinion cutters for spur and helical gears. (Courtesy of Fellows Corporation.)

cutter so that the following relations apply:

$$p = \frac{\pi D}{N} = \frac{\pi}{P_d} \quad \text{(U.S.)}$$

$$p = \pi m \quad \text{(metric)}$$

(6.14)

and

$$P_d = \frac{N}{D} \quad \text{(U.S.)}$$

$$m = \frac{D}{N} \quad \text{(metric)}$$

(6.15)

In the Fellows method, the same cutter cannot be used to cut both helical and spur gears. Figure 6.20 shows pinion cutters for spur and helical gears.

The features discussed apply to helical gears with parallel shafts and with crossed shafts. The two types will now be considered separately.

Table 6.3 shows the symbols for helical gears for the AGMA and for the

TABLE 6.3 Helical Gear Symbols

	AGMA	ISO 701
Circular pitch (plane of rotation)	p	p
Circular pitch (normal plane)	p_n	p_n
Pressure angle (plane of rotation)	ϕ	α
Pressure angle (normal plane)	ϕ_n	α_n
Helix angle	ψ	β
Shaft angle	Σ	Σ

proposed International Standard ISO 701. Symbols which are the same as those for spur gears (Table 4.8) are not included.

6.9 PARALLEL HELICAL GEARS

For parallel helical gears to mesh properly, the following conditions must be satisfied:

1. Equal helix angles.
2. Equal pitches or modules.
3. Opposite hand, that is, one gear with a left-hand helix and the other with a right-hand helix.

From Eq. 6.10, the velocity ratio can be expressed as

$$\frac{\omega_1}{\omega_2} = \frac{N_2}{N_1} = \frac{P_{nd} \, D_2 \cos \psi_2}{P_{nd} \, D_1 \cos \psi_1} = \frac{D_2}{D_1} \tag{6.16}$$

The spur gear equation for center distance,

$$C = \frac{(N_1 + N_2)}{2P_d} \qquad \text{(U.S.)}$$

$$C = \frac{(N_1 + N_2)m}{2} \qquad \text{(metric)}$$

can also be used for parallel helical gears provided P_d is the diametral pitch in the plane of rotation.

In a parallel helical gear, the face width is made large enough so that for a given helix angle ψ, the face advance is greater than the circular pitch, as illustrated in Fig. 6.21. It will give continuous contact in the axial plane as the gears rotate. This ratio (face advance to circular pitch) may be considered as a contact ratio. From Fig. 6.21, it can be seen that to have the face advance just equal the circular pitch, the face width would have to equal $p/\tan \psi$. To provide a margin of safety, the AGMA recommends that this limiting face width be increased by at least 15%, which results in the following equation:

$$F > \frac{1.15p}{\tan \psi} \tag{6.17}$$

In addition to the contact ratio resulting from the twist of the teeth, parallel helical gears will also have a contact ratio in the plane of rotation the same as spur gears. The total contact ratio will therefore be the sum of these two values and is greater than that for spur gears.

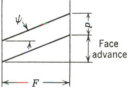

FIGURE 6.21

FIGURE 6.22 Herringbone gears. (Courtesy of D. O. James Gear Manufacturing Company.)

Helical gears connecting parallel shafts have line contact similar to spur gears. However, in spur gears, the contact line is parallel to the axis, whereas in helical gears, it runs diagonally across the face of the tooth. Parallel helical gears have smoother action and hence less noise and vibration than spur gears and are therefore to be preferred for high-speed work. The reason for the smoother action is that the teeth come into contact gradually beginning at one end of the tooth and progressing across the tooth surface, whereas in spur gears, contact takes place simultaneously over the entire face width. The disadvantage of parallel helical gears is in the end thrust produced by the tooth helix. If this end thrust is so large that it cannot be conveniently carried by the bearings, it may be counterbalanced by using two helical gears of opposite hand or by using a herringbone gear which is in effect a double helical gear cut on one blank. Figure 6.22 shows a photograph of a herringbone gear.

Example 6.1. As an example of parallel helical gears, consider that to reduce the noise in a gear drive, two 16-pitch spur gears of 30 and 80 teeth, respectively, are to be replaced by helical gears. The center distance and the angular-velocity ratio must remain the same.

Determine the helix angle, the outside diameters, and the face width of the new gears. Assume the helical gears to be cut by a 16-pitch, 20° full-depth hob.

From spur gear data,

$$C = \frac{N_1 + N_2}{2P_d} = \frac{30 + 80}{2 \times 16} = 3.4375 \text{ in.}$$

$$\frac{\omega_1}{\omega_2} = \frac{N_2}{N_1} = \frac{80}{30} = \frac{8}{3}$$

For the helical gears,

$$P_{nd} = 16 \qquad C = \frac{N_1 + N_2}{2P_d} \qquad \text{or} \qquad P_d = \frac{N_1 + N_2}{2C}$$

$$P_d < 16$$

$$N_2 = \frac{8}{3} N_1 \qquad P_d = \frac{(\frac{11}{3})N_1}{2(\frac{55}{16})} = \frac{8}{15} N_1$$

$$C = 3.4375 \text{ in. } (3\tfrac{7}{16} \text{ in.})$$

By trial, find numbers of teeth:

N_1	N_2	P_d	REMARKS
30	80	16	Original spur gears
29	77.33	15.47	N_2 not whole number
28	74.67	14.93	N_2 not whole number
27	72	14.40	Satisfactory to use

Therefore, let

$$N_1 = 27$$

$$N_2 = 72$$

$$\cos \psi = \frac{P_d}{P_{nd}} = \frac{14.40}{16} = 0.9000$$

$$\psi = 25.84°$$

There are other combinations of numbers of teeth and helix angle that will satisfy the conditions, but the one listed should be selected because it will give the smallest helix angle.

The outside diameters of the two gears are

$$D_{O_1} = D_1 + 2a = \frac{N_1}{P_d} + 2\left(\frac{k}{P_{nd}}\right) = \frac{27}{14.4} + 2\left(\frac{1}{16}\right) = 2.000 \text{ in.}$$

$$D_{O_2} = D_2 + 2a = \frac{N_2}{P_d} + 2\left(\frac{k}{P_{nd}}\right) = \frac{72}{14.4} + 2\left(\frac{1}{16}\right) = 5.125 \text{ in.}$$

Note that the addendum was calculated using the diametral pitch of the hob (P_{nd}).
The face width is

$$F > \frac{1.15p}{\tan \psi}$$

$$p = \frac{\pi}{P_d} = \frac{\pi}{14.4} = 0.2185 \text{ in.}$$

Therefore,

$$F > \frac{(1.15)(0.2185)}{\tan 25.84°} > 0.5189 \text{ in.}$$

Use

$$F = \frac{9}{16} \text{ in.}$$

6.10   CROSSED HELICAL GEARS

For crossed helical gears to mesh properly, there is only one requirement, that
is, they must have common normal pitches or modules. Their pitches in the plane
of rotation are not necessarily and not usually equal. Their helix angles may or
may not be equal, and the gears may be of the same or of opposite hand.

From Eq. 6.10, the velocity ratio becomes

$$\frac{\omega_1}{\omega_2} = \frac{N_2}{N_1} = \frac{P_{nd} D_2 \cos \psi_2}{P_{nd} D_1 \cos \psi_1} = \frac{D_2 \cos \psi_2}{D_1 \cos \psi_1} \qquad \textbf{(6.18)}$$

If Σ is the angle between two shafts connected by crossed helical gears and ψ_1
and ψ_2 are the helix angles of the gears, then

$$\Sigma = \psi_1 \pm \psi_2 \qquad \textbf{(6.19)}$$

The plus and minus signs apply, respectively, when the gears have the same or
the opposite hand. Equation 6.19 is illustrated in Fig. 6.23 showing pairs of crossed
helical gears in and out of mesh.

The action of crossed helical gears is quite different from that of parallel
helical gears. Crossed helical gears have point contact. In addition, sliding action
takes place along the tooth, which is not present in parallel helical gears. For
these reasons, crossed helical gears are used to transmit only small amounts of
power. An application of these gears is on the distributor drive on an automotive
engine.

Using the principle of the velocity of sliding developed in Chapter 1, it is

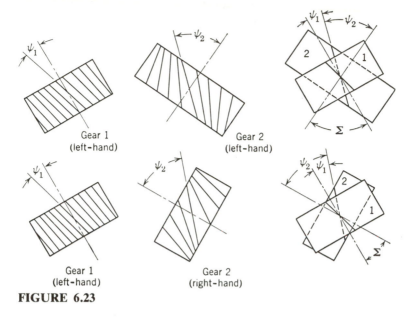

Gear 1
(left-hand)

Gear 2
(left-hand)

Gear 1
(left-hand)

Gear 2
(right-hand)

FIGURE 6.23

possible to determine the tooth helices across the faces of two crossed helical gears provided the peripheral velocity of the pitch point of each gear is known. Figure 6.24 shows this construction, where V_1 and V_2 are known, and it is required to find the tooth helices and helix angles for these velocities and given shaft angle. The two helices in contact at point P are parallel to the line M_1M_2. This contact occurs on the bottom of gear 1 and on the top of gear 2.

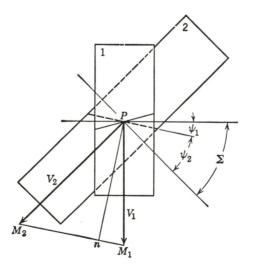

FIGURE 6.24

Example 6.2. To illustrate crossed helical gears, consider a pair of gears connecting two shafts at an angle of 60° with a velocity ratio of 1.5:1. The pinion has a normal diametral pitch of 6, a pitch diameter of 7.75 in., and a helix angle of 35°. Determine the helix angle and the pitch diameter of the gear and the numbers of teeth on both the pinion and the gear.

To find the helix angle of the gear, assume both gears have the same hand. Then,

$$\Sigma = \psi_1 + \psi_2$$

where $\Sigma = 60°$ and $\psi_1 = 35°$. Therefore,

$$\psi_2 = 25°$$

The pitch diameter of the gear can be determined as follows:

$$\frac{\omega_1}{\omega_2} = \frac{N_2}{N_1} = \frac{D_2 \cos \psi_2}{D_1 \cos \psi_1}$$

$$D_2 = \frac{D_1 \cos \psi_1}{\cos \psi_2} \times \frac{\omega_1}{\omega_2} = \frac{(7.750)(\cos 35°)(1.5)}{\cos 25°}$$

$$D_2 = 10.50 \text{ in.}$$

The numbers of teeth on the pinion and on the gear are

$$N_1 = P_{nd} D_1 \cos \psi_1 = (6)(7.750)(\cos 35°)$$

$$N_1 = 38$$

$$N_2 = N_1 \frac{\omega_1}{\omega_2} = (38)(1.5)$$

$$N_2 = 57$$

6.11 WORM GEARING

If a tooth on a helical gear makes a complete revolution on the pitch cylinder, the resulting gear is known as a *worm.* The mating gear for a worm is designated as a *worm gear,* or *worm wheel*; however, the worm gear is not a helical gear. A worm and worm gear are used to connect nonparallel, nonintersecting shafts usually at right angles. See Fig. 6.25. The gear reduction is generally quite large. The relation between a spur or helical gear and its hob during cutting is similar to the relation between a worm and a worm gear. Worms that are true involute helical gears may be used to drive spur or helical gears, but point contact obviously results, which is unsatisfactory from the standpoint of wear. It is possible, however, to secure line contact by mating the worm with a worm gear that has been cut with a hob having the same diameter and same form of tooth as the worm. If this is done, the worm and worm gear will be conjugate, but the worm will

(a)

(b)

FIGURE 6.25 (a) Worm and worm gear. (Courtesy of Foote Brothers Gear & Manufacturing Corp.) (b) Hourglass worm and worm gear. (Courtesy of Cone Drive Gears, Division of Michigan Tool Company.)

not have involute teeth. Figure 6.26a shows a sketch of a worm, where λ is the lead angle, ψ the helix angle, p_x the axial pitch, and D the pitch diameter. The axial pitch of the worm is the distance between corresponding points of adjacent threads measured parallel to the axis.

In considering the characteristics of a worm, the lead is of primary importance and may be defined as the axial distance that a point on the helix of the worm will move in one revolution of the worm. The relation between the lead and axial pitch is

$$L = p_x N_1 \qquad \qquad \textbf{(6.20)}$$

where N_1 is the number of threads (or teeth) wrapped on the pitch cylinder of the worm. A worm may be obtained with one to ten threads.

If a complete revolution of a thread on a worm is unwrapped, a triangle results as shown in Fig. 6.26b. From the figure, it can be seen that

$$\tan \lambda = \frac{L}{\pi D_1} \qquad \qquad \textbf{(6.21)}$$

where D_1 is the diameter of the worm.

The diameter of a worm gear can be calculated from

$$D_2 = \frac{p N_2}{\pi} \qquad \qquad \textbf{(6.22)}$$

where N_2 is the number of teeth in the worm gear. The velocity ratio is

$$\frac{\omega_1}{\omega_2} = \frac{N_2}{N_1} = \frac{D_2 \cos \psi_2}{D_1 \cos \psi_1} \qquad \qquad \textbf{(6.23)}$$

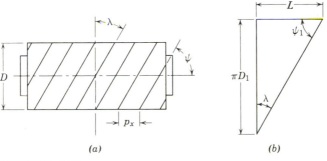

(a)   (b)

FIGURE 6.26

TABLE 6.4 Worm Gear Symbols

	AGMA	ISO 701
Circular pitch (worm gear)	p	p
Axial pitch (worm)	p_x	p_x
Lead	L	p_z
Helix angle	ψ	β
Lead angle	λ	γ

and

$$\frac{\omega_1}{\omega_2} = \frac{\pi D_2}{L}$$

(6.24)

for shafts at right angles.

For a worm and worm gear with shafts *at right angles* to mesh properly, the following conditions must be satisfied:

1. Lead angle of worm = helix angle of worm gear
2. Axial pitch of worm = circular pitch of worm gear

A worm and worm gear drive may or may not be reversible depending on the application. When used as a drive for a hoist, it is necessary that the unit be self-locking and driven only by the worm. However, if a worm drive is used for an automotive drive, it is necessary that the drive be reversible and that the worm gear be able to drive the worm. If the lead angle of the worm is greater than the friction angle of the surfaces in contact, the drive will be reversible. The coefficient of friction μ and the friction angle ϕ are related by the equation $\mu = \tan \phi$. A worm and worm gear are considered self-locking when the lead angle of the worm is less than 5°.

Table 6.4 shows the symbols for worm gears for the AGMA and for the proposed International Standard ISO 701. Symbols which are the same as those for spur gears (Table 4.8) are not included.

Example 6.3. As an example of worm gearing, consider a triple-threaded worm driving a worm gear of 60 teeth; the shaft angle is 90° as shown in Fig. 6.27. The circular pitch of the worm gear is $1\frac{1}{4}$ in., and the pitch diameter of the worm is 3.80 in. Determine the

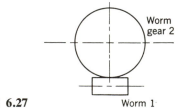

FIGURE 6.27

lead angle of the worm, the helix angle of the worm gear, and the distance between shaft centers.

The lead angle of the worm can be found from

$$L = p_x N_1 \quad \text{where} \quad p_x = p$$

$$L = 1.250 \times 3 = 3.750 \text{ in.}$$

$$\tan \lambda = \frac{L}{\pi D_1} = \frac{3.750}{\pi \times 3.800} = 0.314$$

Therefore,

$$\lambda = 17.4°$$

The helix angle of the worm gear equals the lead angle of the worm. Therefore,

$$\psi_2 = 17.4°$$

The center distance is found by

$$D_2 = \frac{p N_2}{\pi} = \frac{(1.250)(60)}{\pi} = 23.90 \text{ in.}$$

$$C = \frac{D_1 + D_2}{2} = \frac{3.800 + 23.90}{2} = 13.85 \text{ in.}$$

Problems—U.S. Standard

6.1. A pair of bevel gears have a velocity ratio of ω_1/ω_2, and the shaft centerlines intersect at an angle Σ. If distances x and y are laid off from the intersection point along the shaft axes in the ratio ω_1/ω_2, prove that the diagonal of a parallelogram with sides x and y will be the common pitch cone element of the bevel gears.

6.2. A Gleason crown bevel gear of 24 teeth and a diametral pitch of 5 is driven by a 16-tooth pinion. Calculate the pitch diameter and pitch angle of the pinion, the addendum and dedendum, the face width, and the pitch diameter of the gear. Make a full-size axial sketch of the pinion and gear in mesh using reasonable proportions for the hubs and webs as shown in Fig. 6.7a.

6.3. A Gleason crown bevel gear of 48 teeth and a diametral pitch of 12 is driven by a 24-tooth pinion. (a) Calculate the pitch angle of the pinion and the shaft angle. (b) Make a sketch (to scale) of the pitch cones of the two gears in mesh. Show the back cones of each gear and label the pitch cones and the back cones.

6.4. A pair of Gleason miter gears have 20 teeth and a diametral pitch of 4. Calculate the pitch diameter, the addendum and dedendum, the face width, the pitch cone distance, the face angle, the root angle, and the outside diameter. Make a full-size axial sketch of the gears in mesh using reasonable proportions for the hub and web as shown in Fig. 6.7a. Dimension the drawing with the values calculated.

6.5. A Gleason 6-pitch, straight bevel pinion of 21 teeth drives a gear of 27 teeth. The shaft angle is 90°. Calculate the pitch angles, the addendums and dedendums, and the face width of each gear. Make a full-size axial sketch of the gears in mesh using reasonable proportions for the hubs and webs as shown in Fig. 6.7a.

6.6. A Gleason 4-pitch, straight bevel pinion of 14 teeth drives a gear of 20 teeth. The shaft angle is 90°. Calculate the addendum and dedendum, circular tooth thickness for each gear, and the pitch and base radii of the equivalent spur gears. Make a full-size sketch of the equivalent gears showing two teeth in contact as in Fig. 6.7b.

6.7. A Gleason 5-pitch, straight bevel pinion of 16 teeth drives a gear of 24 teeth. The shaft angle is 45°. After making the necessary calculations, lay out a full-size axial sketch of the pinion and gear in mesh using reasonable proportions for the hubs and webs as shown in Fig. 6.7a.

6.8. A pair of Gleason bevel gears mesh with a shaft angle of 75°. The diametral pitch is 10, and the numbers of teeth in the pinion and gear are 30 and 40, respectively. (a) Calculate the pitch angles and the addendums and dedendums of the pinion and gear. (b) Make a full-size sketch of the pitch cones and the back cones of the two gears in mesh. Label the pitch cones, back cones, and pitch angles of both gears. (c) Mark off (double size) the addendum and dedendum of the pinion on the sketch and clearly label them.

6.9. Prove with the aid of a suitable sketch that in a Gleason straight tooth bevel gear the addendum angle of the pinion equals the dedendum angle of the gear and that $\Gamma_o = \Gamma + \alpha$.

6.10. A 14-tooth helical gear is to be cut by a 10-pitch, 20° hob. Calculate the following: (a) the minimum helix angle which this gear must have to be cut at the standard setting without undercutting. (b) the amount the hob will have to be withdrawn to avoid undercutting if the helix angle of the gear is made 20°.

6.11. A 12-tooth helical pinion is to be cut with an 8-pitch, 20° hob. If the helix angle is to be 20°, calculate the amount the hob will have to be withdrawn to avoid undercutting.

6.12. Two equal spur gears of 48 teeth, 1 in. face width, and 6 diametral pitch mesh together in the drive of a fatigue tester. Calculate the helix angle of a pair of helical gears to replace the spur gears if the face width, center distance, and velocity ratio are to remain the same. Use the following cutters: (a) Fellows of 6 diametral pitch, (b) hob of 6 normal diametral pitch.

6.13. Two standard spur gears were cut with a 10-pitch, 20° hob to give a velocity ratio of 3.5:1 and center distance of 6.75 in. Helical gears are to be cut with the same hob to replace the spur gears keeping the center distance and angular-velocity ratio the same. Determine the helix angle, numbers of teeth, and face width of the new gears keeping the helix angle to a minimum.

6.14. Two standard spur gears are to be replaced by helical gears. The spur gears were cut by an 8-pitch, 20° hob, the velocity ratio is 1.75:1, and the center distance is 5.5 in. The helical gears are to be cut with the same hob and maintain the same center distance. The helix angle is to be between 15° and 20° and the velocity ratio between 1.70 and 1.75. Find the numbers of teeth, helix angle, and velocity ratio.

6.15. In a proposed gear drive, two standard spur gears (16 diametral pitch) with 36 and 100 teeth, respectively, are meshed at the standard center distance. It is decided to replace these spur gears with helical gears having a helix angle of 22° and the same number of teeth. Determine the change in center distance required if the helical gears are cut (a) with a 16-pitch, 20° hob, (b) with a 16-pitch, 20° Fellows cutter.

6.16. A pair of helical gears for parallel shafts are to be cut with an 8-pitch, 25° hob. The helix angle is to be 20° and the center distance between 6.00 and 6.25 in. The angular-velocity ratio is to approach as closely as possible 2:1. Calculate the circular pitch and the diametral pitch in the plane of rotation. Determine the numbers of teeth, pitch diameters, and center distance to satisfy the above conditions.

6.17. A 10-pitch, 20-tooth spur pinion drives two gears, one of 36 teeth and the other of 48 teeth. It is desired to replace all three gears with helical gears and to change the velocity ratio between the 20-tooth gear shaft and the 48-tooth gear shaft to 2:1. The velocity ratio and the center distance between the 20-tooth gear shaft and the 36-tooth gear shaft is to remain the same. Using an 8-pitch, 20° stub hob and keeping the helix angle as low as possible, determine the number of teeth, helix angle and hand, face width, and outside diameter for each gear. Calculate the change in center distance between the shafts that originally mounted the 20- and 48-tooth gears.

6.18. A 12-pitch, 24-tooth spur pinion drives two gears, one of 36 teeth and the other of 60 teeth. It is necessary to replace all three gears with helical gears keeping the same velocity ratios and center distances. Using a 16-pitch, 20° stub hob and keeping the helix angle as low as possible, determine the number of teeth, helix angle and hand, face width, and outside diameter for each gear.

6.19. Two parallel shafts are to be connected by a pair of helical gears (gears 1 and 2). The angular-velocity ratio is to be 1.25:1 and the center distance 4.5 in. In addition, gear 2 is to drive a helical gear 3 whose shaft is at right angles to shaft 2. The angular-velocity ratio between gears 2 and 3 is to be 2:1. Using a 9-pitch, 20° hob, determine the number of teeth, helix angle, and pitch diameter of each gear and find center distance C_{23}.

6.20. Two parallel shafts are to be connected by a pair of helical gears (gears 1 and 2). The angular velocity ratio is to be 1.75:1 and the center distance 2.75 in. In addition, gear 2 is to drive a third helical gear (gear 3) with an angular velocity ratio of 2:1. Three hobs are available for cutting the gears: hob A (7 pitch, 20°), hob B (9 pitch, 20°), and hob C (12 pitch, 20°). (*a*) Choose the hob that will result in the smallest helix angle ψ. (*b*) Which hob will permit the shortest center distance C_{23} between shafts 2 and 3 while maintaining a helix angle *less than* 35°?

6.21. The formula for the center distance between two spur or parallel helical gears is given by $C = (N_1 + N_2)/2P_d$, where C is dependent upon the number of gear teeth N_1 and N_2 and the diametral pitch P_d. Show that C_{23} is independent of P_d for three gears (spur or parallel helical) in mesh whose center distance C_{12} and angular-velocity ratios ω_1/ω_2 and ω_2/ω_3 are known.

6.22. Two 18-pitch spur gears of 36 and 90 teeth, respectively, are to be replaced by helical gears. The center distance and the velocity ratio are to remain the same. If the width of the gears cannot exceed $\frac{1}{2}$ in. because of space limitations, determine a pair of helical gears for this job keeping the helix angle as small as possible. Use an 18-pitch, 20° hob, and determine the numbers of teeth, helix angle, face width, and outside diameters.

6.23. Two 18-pitch spur gears of 32 and 64 teeth, respectively, are to be replaced by helical gears. The center distance and velocity ratio are to remain the same. If the width of the gears cannot exceed $\frac{7}{16}$ in. because of space limitations, determine which of the following hobs should be used keeping the helix angle as small as possible: hob A (18-pitch, 20°) or hob B (20-pitch, 20°). In addition, determine the numbers of teeth, helix angle, face width, and outside diameters.

6.24. Two parallel shafts are to be connected by a pair of helical gears (gears 1 and 2). The angular velocity ratio is to be $1\frac{1}{3}$:1 and the center distance 3.50 in. Considering that hobs are available from 6 to 12 pitch (inclusive), tabulate the numbers of teeth, helix angle, and face width for the various combinations (of N_1 and N_2) that will satisfy the given conditions. What is the best selection for this drive? Why? Let 15 be the lowest number of teeth for the smaller gear at $P_{nd} = 6$.

6.25. Two shafts crossing at right angles are to be connected by helical gears. The angular-velocity ratio is to be $1\frac{1}{2}$:1 and the center distance 5.00 in. Assuming the gears to have equal helix angles, calculate the diametral pitch of a cutter to generate 20 teeth on the pinion if the cutter is (a) a hob and (b) a Fellows cutter.

6.26. The following helical gears, cut with a 12-pitch, 20° hob, are meshed without backlash:

Gear 1—36 teeth, right-hand, 30° helix angle
Gear 2—72 teeth, left-hand, 40° helix angle

Determine the shaft angle, the angular velocity ratio, and the center distance.

6.27. Two shafts crossed at right angles are connected by helical gears (gears 1 and 2) cut with a 12-pitch, 20° full-depth hob. Both gears are right-hand and the angular velocity ratio is 15:1; $D_2 = 5.196$ in. and $\psi_1 = 60°$. A design modification requires a reduction of the outside diameter (o.d.) of gear 1 by 0.25 in. to provide clearance for a new component. Assuming that the same hob must be used for cutting any new gears, show that the o.d. of gear 1 can be reduced without changing the velocity ratio, the shaft angle, and the numbers of gear teeth N_1 and N_2. The o.d. of gear 2 and the center distance may be altered if necessary. In the analysis, calculate and compare the following data for both the original and the new gears: C_{12}, D_1, D_2, N_1, N_2, ψ_1, ψ_2.

6.28. A 21-tooth helical gear of 6 normal diametral pitch is to drive a spur gear. The angular velocity ratio is to be 2:1 and the angle between the shafts 45°. Determine the pitch diameters for the two gears and the helix angle for the helical gear. Make a full-size sketch of the two gears (pitch cylinders) in contact similar to Fig. 6.24 with the pinion on top; the width of the gears is to be 1 in. Show the tooth elements in contact and also a tooth element on top of the pinion. Label and dimension the helix angle and the shaft angle.

6.29. Two crossed shafts are to be connected by helical gears. The angular velocity ratio is to be $1\frac{1}{2}$:1 and the center distance 8.50 in. If one gear is available from a previous job with 30 teeth, 30° helix angle, and a normal diametral pitch of 5, calculate the shaft angle that must be used. Let both gears be of the same hand and let the 30-tooth gear be the pinion.

6.30. Two crossed shafts are connected by helical gears. The velocity ratio is 1.8:1 and the shaft angle 45°. If $D_1 = 2.31$ in. and $D_2 = 3.73$ in., calculate the helix angles if both gears have the same hand.

6.31. Two shafts crossed at right angles are to be connected by helical gears. The angular-velocity ratio is to be $1\frac{1}{2}$:1 and the center distance 5.00 in. Select a pair of gears for this application to be cut by the Fellows method.

6.32. Two crossed shafts are connected by helical gears. The velocity ratio is 3:1, the shaft angle 60°, and the center distance 10.00 in. If the pinion has 35 teeth and a normal diametral pitch of 8, calculate the helix angles and pitch diameters if the gears are of the same hand.

6.33. A helical pinion of 2.00 in. pitch diameter drives a helical gear of 3.25 in. as shown in Fig. 6.24, $\Sigma = 30°$. Let the velocity of the pitch point of gear 1 be represented by a vector 2 in. long and that of gear 2 by a vector 3 in. long. Using a face width of 1 in. for the gears, graphically determine the tooth element on the top of each gear, the helix angle and the hand of each gear, and the velocity of sliding.

6.34. An 8-pitch, $14\frac{1}{2}°$ hob is used to cut a helical gear. The hob is right-hand with a lead angle of 2° 40', a length of 3.00 in., and an outside diameter of 3.00 in. Make a full-size sketch of the hob cutting a 47-tooth right-hand helical gear with a helix angle of 20°. The gear blank is $1\frac{1}{2}$ in. wide. Show the pitch cylinder of the hob on top of the gear blank with the pitch helix of the hob in correct relation to the pitch element of the gear tooth. Show three tooth elements on the gear and $1\frac{1}{2}$ turns of the thread on the hob; position these elements by means of the normal circular pitch. Label the axis of the hob and gear blank, the lead angle of hob, the helix angle of the gear, and the direction of rotation of the hob and gear blank.

6.35. Repeat Problem 6.34 with a left-hand helical gear.

6.36. A double-threaded worm having a lead of 2.00 in. drives a worm gear with a velocity ratio of 20:1; the angle between the shafts is 90°. If the center distance is 9.00 in., determine the pitch diameter of the worm and worm gear.

6.37. A worm and worm gear with shafts at 90° and a center distance of 7.00 in. are to have a velocity ratio of 18:1. If the axial pitch of the worm is to be $\frac{1}{2}$ in., determine the maximum number of teeth in the worm and worm gear that can be used for the drive and their corresponding pitch diameters.

6.38. A worm and worm gear connect shafts at 90°. Derive equations for the diameters of the worm and worm gear in terms of the center distance C, velocity ratio ω_1/ω_2, and lead angle λ.

6.39. A worm and worm gear with shafts at 90° and a center distance of 6.00 in. are to have a velocity ratio of 20:1. If the axial pitch of the worm is to be $\frac{1}{2}$ in., determine the smallest diameter worm that can be used for the drive.

6.40. A four-threaded worm drives a 60-tooth worm gear with shafts at 90°. If the center distance is 8.00 in. and the lead angle of the worm 20°, calculate the axial pitch of the worm and the pitch diameters of the two gears.

6.41. A four-threaded worm drives a 48-tooth worm gear having a pitch diameter of 7.64 in. and a helix angle of 20°. If the shafts are at right angles, calculate the lead and the pitch diameter of the worm.

6.42. A six-threaded worm drives a worm gear with an angular velocity ratio of 8:1 and a shaft angle of 80°. The axial pitch of the worm is $\frac{1}{2}$ in. and the lead angle 20°. Calculate the pitch diameters of the worm and worm gear and the circular pitch of the gear.

6.43. A five-threaded worm drives a 33-tooth worm gear with a shaft angle of 90°. The center distance is 2.75 in. and the lead angle 20°. Calculate the pitch diameters, the lead, and the axial pitch of the worm.

6.44. A worm and worm gear with shafts at 90° and a center distance of 3.10 in. are to have a velocity ratio of 7:1. Using a lead angle of 20°, determine the pitch diameters and numbers of teeth for the gears. Make the axial pitch a simple fraction.

6.45. A worm and worm gear with shafts at 90° and a center distance of 3.00 in. are to have a velocity ratio of 30:1. Determine a pair of gears for the job, and specify the numbers of teeth, pitch diameters, and lead angle. Make the axial pitch a simple fraction.

Problems—Metric

6.1m. A pair of bevel gears have a velocity ratio of ω_1/ω_2, and the shaft centerlines intersect at an angle Σ. If distances x and y are laid off from the intersection point along the shaft axes in the ratio ω_1/ω_2, prove that the diagonal of a parallelogram with sides x and y will be the common pitch cone element of the bevel gears.

6.2m. A Gleason crown bevel gear of 24 teeth and a module of 5.08 is driven by a 16-tooth pinion. Calculate the pitch diameter and pitch angle of the pinion, the addendum and dedendum, the face width, and the pitch diameter of the gear. Make a full-size axial sketch of the pinion and gear in mesh using reasonable proportions for the hubs and webs as shown in Fig. 6.7a.

6.3m. A Gleason crown bevel gear of 48 teeth and a module of 2.12 is driven by a 24-tooth pinion. (a) Calculate the pitch angle of the pinion and the shaft angle. (b) Make a sketch (to scale) of the pitch cones of the two gears in mesh. Show the back cones of each gear and label the pitch cones and the back cones.

6.4m. A pair of Gleason miter gears have 20 teeth and a module of 6.35. Calculate the pitch diameter, the addendum and dedendum, the face width, the pitch cone distance, the face angle, the root angle, and the outside diameter. Make a full-size axial sketch of the gears in mesh using reasonable proportions for the hub and web as shown in Fig. 6.7a. Dimension the drawing with the values calculated.

6.5m. A Gleason 4.23-module, straight bevel pinion of 21 teeth drives a gear of 27 teeth. The shaft angle is 90°. Calculate the pitch angles, the addendums and dedendums, and the face width of each gear. Make a full-size axial sketch of the gears in mesh using reasonable proportions for the hubs and webs as shown in Fig. 6.7a.

6.6m. A Gleason 6.35-module, straight bevel pinion of 14 teeth drives a gear of 20 teeth. The shaft angle is 90°. Calculate the addendum and dedendum, circular tooth thickness for each gear, and the pitch and base radii of the equivalent spur gears. Make a full-size sketch of the equivalent gears showing two teeth in contact as in Fig. 6.7b.

6.7m. A Gleason 5.08-module, straight bevel pinion of 16 teeth drives a gear of 24 teeth. The shaft angle is 45°. After making the necessary calculations, lay out a full-size axial sketch of the pinion and gear in mesh using reasonable proportions for the hubs and webs as shown in Fig. 6.7a.

6.8m. A pair of Gleason bevel gears mesh with a shaft angle of 75°. The module is 2.54, and the numbers of teeth in the pinion and gear are 30 and 40, respectively. (a) Calculate the pitch angles and the addendums and dedendums of the pinion and gear. (b) Make a full-size sketch of the pitch cones and the back cones of the two gears in mesh. Label the pitch cones, back cones, and pitch angles of both gears. (c) Mark off (double size) the addendum and dedendum of the pinion on the sketch and clearly label them.

6.9m. Prove with the aid of a suitable sketch that in a Gleason straight tooth bevel gear the addendum angle of the pinion equals the dedendum angle of the gear and that $\Gamma_o = \Gamma + \alpha$.

6.10m. A 14-tooth helical gear is to be cut by a 2.5-module, 20° hob. Calculate the following: (a) the minimum helix angle which this gear must have to be cut at the standard setting without undercutting; (b) the amount the hob will have to be withdrawn to avoid undercutting if the helix angle of the gear is made 20°.

6.11m. A 12-tooth helical pinion is to be cut with a 3-module, 20° hob. If the helix angle is to be 20°, calculate the amount the hob will have to be withdrawn to avoid undercutting.

6.12m. Two equal spur gears of 48 teeth, 25.4 mm face width, and of a 4 module mesh together in the drive of a fatigue tester. Calculate the helix angle of a pair of helical gears to replace the spur gears if the face width, center distance, and velocity ratio are to remain the same. Use the following cutters: (a) 4-module Fellows, (b) 4-normal-module hob.

6.13m. Two standard spur gears were cut with a 2.5-module, 20° hob to give a velocity ratio of 3.5:1 and center distance of 168.75 mm. Helical gears are to be cut with the same hob to replace the spur gears keeping the center distance and angular-velocity ratio the same. Determine the helix angle, numbers of teeth, and face width of the new gears keeping the helix angle to a minimum.

6.14m. Two standard spur gears are to be replaced by helical gears. The spur gears were cut by a 3-module, 20° hob, the velocity ratio is 1.75:1, and the center distance is 132 mm. The helical gears are to be cut with the same hob and maintain the same center distance. The helix angle is to be between 15° and 20° and the velocity ratio between 1.70 and 1.75. Find the numbers of teeth, helix angle, and velocity ratio.

6.15m. In a proposed gear drive, two standard spur gears (1.5-module) with 36 and 100 teeth, respectively, are meshed at the standard center distance. It is decided to replace these spur gears with helical gears having a helix angle of 22° and the same number of teeth. Determine the change in center distance required if the helical gears are cut (*a*) with a 1.5-module, 20° hob, (*b*) with a 1.5-module, 20° Fellows cutter.

6.16m. A pair of helical gears for parallel shafts are to be cut with a 3-module hob. The helix angle is to be 20° and the center distance between 152.40 and 158.75 mm. The angular velocity ratio is to approach 2:1 as closely as possible. Calculate the circular pitch and the module in the plane of rotation. Determine the numbers of teeth, pitch diameters, and center distance to satisfy the above conditions.

6.17m. A 2.5-module, 20-tooth spur pinion drives two gears, one of 36 teeth and the other of 48 teeth. It is desired to replace all three gears with helical gears and to change the velocity ratio between the 20-tooth gear shaft and the 48-tooth gear shaft to 2:1. The velocity ratio and the center distance between the 20-tooth gear shaft and the 36-tooth gear shaft is to remain the same. Using a 3-module, 20° hob and keeping the helix angle as low as possible, determine the number of teeth, helix angle and hand, face width, and outside diameter for each gear. Calculate the change in center distance between the shafts that originally mounted the 20- and 48-tooth gears.

6.18m. A 2-module, 24-tooth spur pinion drives two gears, one of 36 teeth and the other of 60 teeth. It is necessary to replace all three gears with helical gears keeping the same velocity ratios and center distances. Using a 1.5-module, 20° hob and keeping the helix angle as low as possible, determine the number of teeth, helix angle and hand, face width, and outside diameter for each gear.

6.19m. Two parallel shafts are to be connected by a pair of helical gears (gears 1 and 2). The angular-velocity ratio is to be 1.25:1 and the center distance 114.3 mm. In addition, gear 2 is to drive a helical gear 3 whose shaft is at right angles to shaft 2. The angular-velocity ratio between gears 2 and 3 is to be 2:1. Using a 2.75-module, 20° hob, determine the number of teeth, helix angle, and pitch diameter of each gear and find center distance C_{23}.

6.20m. Two parallel shafts are to be connected by a pair of helical gears (gears 1 and 2). The angular-velocity ratio is to be 1.75:1 and the center distance 69.85 mm. In addition, gear 2 is to drive a third helical gear (gear 3) with an angular-velocity ratio of 2:1. Three hobs are available for cutting the gears: hob *A* (3.5-module), hob *B* (2.75-module), and hob *C* (2-module). (*a*) Choose the hob that will result in the smallest helix angle ψ. (*b*) Which hob will permit the shortest center distance C_{23} between shafts 2 and 3 while maintaining a helix angle *less than* 35°?

6.21m. The formula for the center distance between two spur or parallel helical gears is given by $C = [(N_1 + N_2)/2]m$, where C is dependent upon the number of gear teeth N_1 and N_2 and the module m. Show that C_{23} is independent of m for three gears (spur or parallel helical) in mesh whose center distance C_{12} and angular-velocity ratios ω_1/ω_2 and ω_2/ω_3 are known.

6.22m. Two 1.5-module, 20° spur gears of 36 and 90 teeth, respectively, are to be replaced by helical gears. The center distance and the velocity ratio are to remain the same. If the

242 BEVEL, HELICAL, AND WORM GEARING

width of the gears cannot exceed 12.70 mm because of space limitations, determine a pair of helical gears for this job keeping the helix angle as small as possible. Use a 1.5-module hob, and determine the numbers of teeth, helix angle, face width, and outside diameters.

6.23m. Two 1.5-module spur gears of 32 and 64 teeth, respectively, are to be replaced by helical gears. The center distance and velocity ratio are to remain the same. If the width of the gears cannot exceed 11.11 mm because of space limitations, determine which of the following hobs should be used keeping the helix angle as small as possible: hob A (1.5-module) or hob B (1.25-module). In addition, determine the numbers of teeth, helix angle, face width, and outside diameters.

6.24m. Two parallel shafts are to be connected by a pair of helical gears (gears 1 and 2). The angular-velocity ratio is to be $1\frac{1}{3}$:1 and the center distance 88.90 mm. Considering that hobs are available for modules from 2 to 4 inclusive, tabulate the numbers of teeth, helix angle, and face width for the various combinations of N_1 and N_2 that will satisfy the given conditions. What is the best selection for this drive? Why? Let 15 be the lowest number of teeth for the smaller gear at $m_n = 4$.

6.25m. Two shafts crossing at right angles are to be connected by helical gears. The angular-velocity ratio is to be $1\frac{1}{2}$:1 and the center distance 127.0 mm. Assuming the gears to have equal helix angles, calculate the module of a cutter to generate 20 teeth on the pinion if the cutter is (*a*) a hob and (*b*) a Fellows cutter. Determine the change that must be made in center distance in each case in order to use standard cutters.

6.26m. The following helical gears, cut with a 2-module 20° hob, are meshed without backlash:

> Gear 1—36 teeth, right-hand, 30° helix angle
> Gear 2—72 teeth, left-hand, 40° helix angle

Determine the shaft angle, the angular-velocity ratio, and the center distance.

6.27m. Two shafts crossed at right angles are connected by helical gears (gears 1 and 2) cut with a 2-module, 20° hob. Both gears are right-hand and the angular-velocity ratio is 15:1; $D_2 = 131.64$ mm and $\psi = 60°$. A design modification requires a reduction of the outside diameter (o.d.) of gear 1 by 6.35 mm to provide clearance for a new component. Assuming that the same hob must be used for cutting any new gears, show that the o.d. of gear 1 can be reduced without changing the velocity ratio, the shaft angle, and the numbers of gear teeth N_1 and N_2. The o.d. of gear 2 and the center distance may be altered if necessary. In the analysis, calculate and compare the following data for both the original and new gears: C_{12}, D_1, D_2, N_1, N_2, ψ_1, ψ_2.

6.28m. A 21-tooth helical gear of a normal module of 4 is to drive a spur gear. The angular-velocity ratio is to be 2:1 and the angle between the shafts 45°. Determine the pitch diameters for the two gears and the helix angle for the helical gear. Make a full-size sketch of the two gears (pitch cylinders) in contact similar to Fig. 6.24 with the pinion on top; the width of the gears is to be 25.0 mm. Show the tooth elements in contact and also a tooth element on top of the pinion. Label and dimension the helix angle and the shaft angle.

6.29m. Two crossed shafts are to be connected by helical gears. The angular-velocity ratio is to be $1\frac{1}{2}$:1 and the center distance 215.9 mm. If one gear is available from a previous job with 30 teeth, 30° helix angle, and a normal module of 5, calculate the shaft angle that must be used. Let both gears be of the same hand and let the 30-tooth gear be the pinion.

6.30m. Two crossed shafts are connected by helical gears. The velocity ratio is 18.1:1 and the shaft angle 45°. If D_1 = 57.735 mm and D_2 = 93.175 mm, calculate the helix angles if both gears have the same hand.

6.31m. Two shafts crossed at right angles are to be connected by helical gears. The angular-velocity ratio is to be $1\frac{1}{2}$:1 and the center distance 125.0 mm. Select a pair of gears for this application to be cut by the Fellows method.

6.32m. Two crossed shafts are connected by helical gears. The velocity ratio is 3:1, the shaft angle 60°, and the center distance 254.0 mm. If the pinion has 35 teeth and a normal module of 3, calculate the helix angles and pitch diameters if the gears are of the same hand.

6.33m. A helical pinion of 50.0 mm pitch diameter drives a helical gear of 84.0 mm diameter as shown in Fig. 6.24, Σ = 30°. Let the velocity of the pitch point of gear 1 be represented by a vector 50.0 mm long and that of gear 2 by a vector 72.5 mm long. Using a face width of 26.0 mm for the gears, graphically determine the tooth element on the top of each gear, the helix angle and the hand of each gear, and the velocity of sliding.

6.34m. A 3-module, 20° hob is used to cut a helical gear. The hob is right-hand with a lead angle of 2° 40', a length of 75 mm, and an outside diameter of 75 mm. Make a full-size sketch of the hob cutting a 47-tooth right-hand helical gear with a helix angle of 20°. The gear blank is 38 mm wide. Show the pitch cylinder of the hob on top of the gear blank with the pitch helix of the hob in correct relation to the pitch element of the gear tooth. Show three tooth elements on the gear and $1\frac{1}{2}$ turns of the thread on the hob; position these elements by means of the normal circular pitch. Label the axis of the hob and gear blank, the lead angle of hob, the helix angle of the gear, and the direction of rotation of the hob and gear blank.

6.35m. Repeat Problem 6.34m with a left-hand helical gear.

6.36m. A double-threaded worm having a lead of 64.292 mm drives a worm gear with a velocity ratio of $19\frac{1}{2}$:1; the angle between the shafts is 90°. If the center distance is 235.0 mm, determine the pitch diameter of the worm and worm gear.

6.37m. A worm and worm gear with shafts at 90° and a center distance of 178.0 mm are to have a velocity ratio of $17\frac{1}{2}$:1. If the axial pitch of the worm is to be 26.192 mm, determine the maximum number of teeth in the worm and worm gear that can be used for the drive and their corresponding pitch diameters.

6.38m. A worm and worm gear connect shafts at 90°. Derive equations for the diameters of the worm and worm gear in terms of the center distance C, velocity ratio ω_1/ω_2, and lead angle λ.

6.39m. A worm and worm gear with shafts at 90° and a center distance of 152.0 mm are to have a velocity ratio of 20:1. If the axial pitch of the worm is to be 17.463 mm, determine the smallest diameter worm that can be used for the drive.

6.40m. A double-threaded worm drives a 31-tooth worm gear with shafts at 90°. If the center distance is 210.0 mm and the lead angle of the worm 18.83°, calculate the axial pitch of the worm and the pitch diameters of the two gears.

6.41m. A three-threaded worm drives a 35-tooth worm gear having a pitch diameter of 207.8 mm and a helix angle of 21.08°. If the shafts are at right angles, calculate the lead and the pitch diameter of the worm.

6.42m. A four-threaded worm drives a worm gear with an angular-velocity ratio of $8\frac{3}{4}$:1 and a shaft angle of 90°. The axial pitch of the worm is 18.654 mm and the lead angle 27.22°. Calculate the pitch diameters of the worm and worm gear.

6.43m. A six-threaded worm drives a 41-tooth worm gear with a shaft angle of 90°. The center distance is 88.90 mm and the lead angle 26.98°. Calculate the pitch diameters, the lead, and the axial pitch of the worm.

6.44m. A worm and worm gear with shafts at 90° and a center distance of 76.20 mm are to have a velocity ratio of $7\frac{1}{4}:1$. Using a lead angle of 28.88°, determine the pitch diameters. Select numbers of teeth for the gears considering worms with 1 to 10 threads.

6.45m. A worm and worm gear with shafts at 90° and a center distance of 102.0 mm are to have a velocity ratio of $16\frac{1}{2}:1$ and a lead angle of the worm of 13.63°. Determine the various pairs of gears that can be used considering worms with one to ten threads. Specify the numbers of teeth and pitch diameters.

Chapter Seven

Gear Trains

7.1 INTRODUCTION TO GEAR TRAINS

Often, it is necessary to combine several gears and to obtain by so doing what is known as a *gear train*. Given the input angular velocity to a gear train, it is important to be able to determine easily the angular velocity of the output gear and its direction of rotation. The ratio of the input angular velocity to the output angular velocity is known as the *angular-velocity ratio* and is expressed as ω_{in}/ω_{out}.

Figure 7.1 shows a pinion driving an external spur gear and a pinion driving an internal spur gear. In both cases, the angular-velocity ratio is inversely proportional to the number of teeth as indicated. The external gears rotate in opposite directions, and the internal gear rotates in the same direction as its pinion. This is indicated by a minus sign on the velocity ratio in the first case and by a plus sign in the second case. Up to the present time, it has been unnecessary to assign

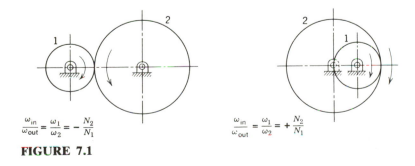

$$\frac{\omega_{in}}{\omega_{out}} = \frac{\omega_1}{\omega_2} = -\frac{N_2}{N_1}$$

$$\frac{\omega_{in}}{\omega_{out}} = \frac{\omega_1}{\omega_2} = +\frac{N_2}{N_1}$$

FIGURE 7.1

245

an algebraic sign to the angular-velocity ratio of a pair of gears. However, when gears are combined to give a gear train, it is important to consider the sign because it indicates direction of rotation. This is especially true in the analysis of planetary gear trains.

Occasionally, it is necessary to change the direction of rotation of a gear without changing its angular velocity. This can be done by placing an *idler gear* between the driven gear and the driver gear. When an idler gear is used, the direction of rotation is changed but the velocity ratio remains the same.

It can be shown that the angular-velocity ratio of a gear train where all gears have fixed axes of rotation is the product of the number of teeth of all the driven gears divided by the product of the number of teeth of all the driving gears. This relation is given in equation form by

$$\frac{\omega_{in}}{\omega_{out}} = \frac{\omega_{driver}}{\omega_{driven}} = \frac{\text{Product of teeth of driven gears}}{\text{Product of teeth of driving gears}} \tag{7.1}$$

To illustrate the use of Eq. 7.1, consider the gear train of Fig. 7.2, where gear 2 and gear 3 are mounted on the same shaft. The angular-velocity ratio is given by

$$\frac{\omega_{in}}{\omega_{out}} = \frac{\omega_1}{\omega_4} = +\frac{N_2 \times N_4}{N_1 \times N_3}$$

The plus sign is determined by observation. That the preceding equation is correct

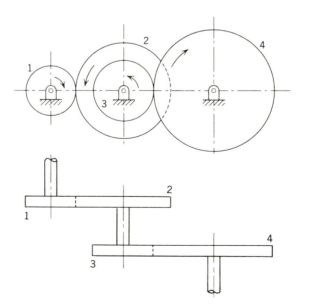

FIGURE 7.2

can be easily shown:

$$\frac{\omega_1}{\omega_2} = -\frac{N_2}{N_1} \quad \text{and} \quad \frac{\omega_3}{\omega_4} = -\frac{N_4}{N_3}$$

$$\frac{\omega_1}{\omega_2} \times \frac{\omega_3}{\omega_4} = +\frac{N_2}{N_1} \times \frac{N_4}{N_3}$$

But

$$\omega_2 = \omega_3$$

Therefore,

$$\frac{\omega_1}{\omega_4} = +\frac{N_2}{N_1} \times \frac{N_4}{N_3}$$

FIGURE 7.3 Triple-reduction speed reducer. (Courtesy of Jones Machinery, Division of Hewitt-Robins, Inc.)

When two gears are fixed to the same shaft as gears 2 and 3 in Fig. 7.2, the gears form a *compound gear*.

Although the angular-velocity ratio is used for calculations involving just one pair of gears, it is more convenient when working with a gear train to use the reciprocal of the angular-velocity ratio. The reason for this is that the angular velocity of the driver will be known from the speed of the motor, and it is necessary only to multiply the speed of the driver by a factor to find the speed of the last gear in the train. This reciprocal is known as the *train value* and is given in equation form by

$$\frac{\omega_{driven}}{\omega_{driver}} = \frac{\text{Product of teeth of driving gears}}{\text{Product of teeth of driven gears}} \qquad (7.2)$$

In general, gear velocities step down so that this value will be less than 1.00. A typical gear train is illustrated in the triple-reduction speed reducer shown in Fig. 7.3.

7.2 PLANETARY GEAR TRAINS

To obtain a desired gear ratio, it is often advantageous to design a gear train so that one of the gears will have planetary motion. With this motion, a gear will be so driven that it not only rotates about its own center but at the same time its center rotates about another center. Figures 7.4a and 7.4b show two planetary gear trains, where gear 1 is often referred to as the *sun* gear and gear 2 as the *planet* gear. In Fig. 7.4a, arm 3 drives gear 2 about gear 1, which is a fixed external gear. As can be seen, gear 2 rotates about its center *B* while this center rotates about center *A*. As gear 2 rolls on the outside of gear 1, a point on its surface will generate an epicycloid. Figure 7.4b shows the case where gear 1 is an internal gear. In this case, a hypocycloid will be generated by a point on the surface of gear 2. Because of the curves generated, a planetary gear train is often referred to as an *epicyclic,* or *cyclic,* gear train.

It is more difficult to determine the angular-velocity ratio of a planetary gear train than that of an ordinary train because of the double rotation of the

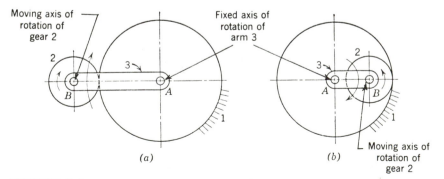

FIGURE 7.4

planet. The angular-velocity ratio may be obtained by the instantaneous center method, the formula method, or the tabulation method. The instantaneous center method will be reserved for Chapter 8, with the other two methods presented here. The formula method will be treated first.

In Fig. 7.4, let it be required to determine ω_{21} given ω_{31}. It should be noted that ω_{21} is defined as the angular velocity of gear 2 relative to gear 1 and ω_{31} is defined as the angular velocity of arm 3 relative to gear 1. Because gear 1 is fixed, this is the same as the angular velocity of gear 2 and of arm 3 relative to the ground. In the solution of the problem, ω_{23}/ω_{13} plays an important part.

Consider the gear train in Fig. 7.4a to be changed so that arm 3 is stationary instead of gear 1. Arm 3 then becomes the ground, and an ordinary gear train results. The ratio ω_{23}/ω_{13} can therefore be evaluated as $-N_1/N_2$. If the mechanism is now returned to its original condition, that is, arm 3 moving and gear 1 fixed, the ratio ω_{23}/ω_{13} will still be $-N_1/N_2$. The reason for this is that when a mechanism is inverted, the relative motion between links does not change. A solution for ω_{21} in terms of the known quantities ω_{31} and ω_{23}/ω_{13} can now be affected by writing an equation of ω_{21} and dividing by ω_{31} as follows:

$$\omega_{21} = \omega_{31} + \omega_{23}$$

$$\frac{\omega_{21}}{\omega_{31}} = 1 + \frac{\omega_{23}}{\omega_{31}} = 1 - \frac{\omega_{23}}{\omega_{13}}$$

Therefore,

$$\omega_{21} = \omega_{31}\left(1 - \frac{\omega_{23}}{\omega_{13}}\right) \tag{7.3}$$

For Fig. 7.4a,

$$\frac{\omega_{23}}{\omega_{13}} = -\frac{N_1}{N_2}$$

and

$$\omega_{21} = \omega_{31}\left(1 + \frac{N_1}{N_2}\right) \tag{7.3a}$$

For Fig. 7.4b,

$$\frac{\omega_{23}}{\omega_{13}} = +\frac{N_1}{N_2}$$

and

$$\omega_{21} = \omega_{31}\left(1 - \frac{N_1}{N_2}\right) \tag{7.3b}$$

From comparison of Eqs. 7.3a and 7.3b, it is apparent why it is important that the correct algebraic sign of ω_{23}/ω_{13} be substituted into Eq. 7.3.

Consider next the case where all of the gears rotate as well as the arm. This is illustrated in Fig. 7.5, where ω_{31} and ω_{41} are known, and it is required to find ω_{21}. In solving this problem, ω_{24}/ω_{34} is the key ratio because it is the velocity ratio of the two gears relative to the arm and can be easily evaluated. Equations can be written for ω_{24} and ω_{34} and combined so that the ratio ω_{24}/ω_{34} appears. This is illustrated in the following:

$$\omega_{24} = \omega_{21} - \omega_{41}$$

$$\omega_{34} = \omega_{31} - \omega_{41}$$

By dividing the first equation by the second,

$$\frac{\omega_{24}}{\omega_{34}} = \frac{\omega_{21} - \omega_{41}}{\omega_{31} - \omega_{41}}$$

$$\frac{\omega_{24}}{\omega_{34}}(\omega_{31} - \omega_{41}) = \omega_{21} - \omega_{41}$$

$$\omega_{21} = \left(\frac{\omega_{24}}{\omega_{34}}\right)\omega_{31} + \omega_{41}\left(1 - \frac{\omega_{24}}{\omega_{34}}\right)$$

But

$$\frac{\omega_{24}}{\omega_{34}} = -\frac{N_3}{N_2}$$

Therefore,

$$\omega_{21} = \left(-\frac{N_3}{N_2}\right)\omega_{31} + \omega_{41}\left(1 + \frac{N_3}{N_2}\right) \tag{7.4}$$

In deriving Eqs. 7.3 and 7.4, it was seen that in each case the angular-velocity ratio of the gears relative to the arm was first obtained, and then the

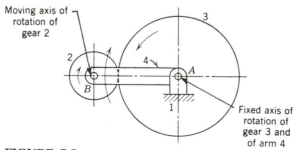

FIGURE 7.5

equations of relative velocity were written and combined to contain this ratio. Although this method is basic, it means that a new equation must be developed for each planetary system encountered. To avoid this repetition, it is possible to derive a general equation which can be applied to any planetary gear train.

Consider Fig. 7.5 again and the equations

$$\omega_{24} = \omega_{21} - \omega_{41}$$

$$\omega_{34} = \omega_{31} - \omega_{41}$$

and

$$\frac{\omega_{24}}{\omega_{34}} = \frac{\omega_{21} - \omega_{41}}{\omega_{31} - \omega_{41}}$$

If in Fig. 7.5, gear 3 is considered the first gear and gear 2 the last gear, the preceding equation may be written as

$$\frac{\omega_{LA}}{\omega_{FA}} = \frac{\omega_L - \omega_A}{\omega_F - \omega_A} \tag{7.5}$$

where

$\dfrac{\omega_{LA}}{\omega_{FA}}$ = velocity ratio of last gear to first gear both relative to arm

ω_L = angular velocity of last gear in train relative to fixed link

ω_A = angular velocity of arm relative to fixed link

ω_F = angular velocity of first gear in train relative to fixed link

In using Eq. 7.5, it must be emphasized that the *first gear* and the *last gear* must be gears that mesh with the gear or gears that have planetary motion. Also, the first gear and the last gear must be on parallel shafts because angular velocities cannot be treated algebraically unless the vectors representing these velocities are parallel.

Equation 7.5 will now be used to write the equation for the gear train in Fig. 7.4a. Let gear 1 be considered the first gear and gear 2 the last gear:

$$\frac{\omega_{LA}}{\omega_{FA}} = \frac{\omega_L - \omega_A}{\omega_F - \omega_A}$$

$$\frac{\omega_{LA}}{\omega_{FA}} = \frac{\omega_{23}}{\omega_{13}} = -\frac{N_1}{N_2}$$

$$\omega_L = \omega_{21}$$

$$\omega_A = \omega_{31}$$

$$\omega_F = \omega_1 = 0$$

Substituting these values gives

$$-\frac{N_1}{N_2} = \frac{\omega_{21} - \omega_{31}}{0 - \omega_{31}}$$

$$\omega_{21} - \omega_{31} = \left(\frac{N_1}{N_2}\right)\omega_{31}$$

and

$$\omega_{21} = \omega_{31}\left(1 + \frac{N_1}{N_2}\right)$$

which agrees with Eq. 7.3*a*. The application of Eq. 7.5 to a more complicated train is given in the following example.

Example 7.1. If arm 6 and gear 5 in Fig. 7.6 are driven clockwise (viewed from the right end) at 150 and 50 rad/s, respectively, determine ω_{21} in magnitude and direction. Use Eq. 7.5 and let gear 5 be the first gear and gear 2 the last gear:

$$\frac{\omega_{LA}}{\omega_{FA}} = \frac{\omega_L - \omega_A}{\omega_F - \omega_A}$$

$$\frac{\omega_{26}}{\omega_{56}} = \frac{\omega_{21} - \omega_{61}}{\omega_{51} - \omega_{61}}$$

$$\frac{\omega_{26}}{\omega_{56}} = \frac{N_5 \times N_3}{N_4 \times N_2} = \frac{20 \times 30}{28 \times 18} = \frac{25}{21}$$

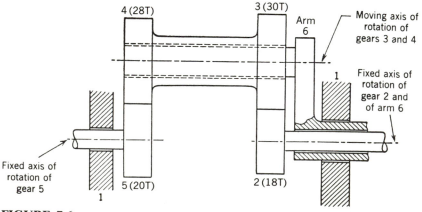

FIGURE 7.6

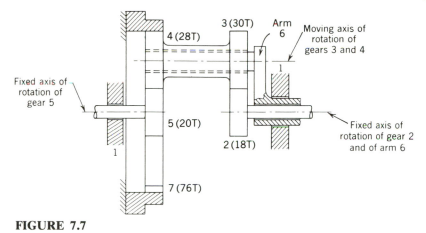

3 (30T) Arm 6 Moving axis of rotation of gears 3 and 4

4 (28T)

Fixed axis of rotation of gear 5

5 (20T)

2 (18T)

Fixed axis of rotation of gear 2 and of arm 6

7 (76T)

FIGURE 7.7

Therefore,

$$\frac{25}{21} = \frac{\omega_{21} - 150}{50 - 150}$$

$$\omega_{21} = \frac{25}{21}(-100) + 150$$

$$= +30.9 \text{ rad/s}$$

Because the sign of ω_{21} is the same as that of ω_{51} and ω_{61}, ω_{21} is in the same direction, namely, clockwise viewed from the right end.

Occasionally, it is necessary to analyze a planetary gear train that cannot be solved by a single application of Eq. 7.5 as was done in Example 7.1. For instance, if a fixed internal gear 7 is added to the train of Fig. 7.6 to mesh with gear 4 as shown in Fig. 7.7 and it is required to calculate ω_{51} given ω_{21}, it will be necessary to use Eq. 7.5 twice to solve the problem. The first application of Eq. 7.5 would consider gears 2, 3, 4, and 5 and the arm 6; and the second application would consider gears 2, 3, 4, and 7 and the arm 6. This will be illustrated in the following example.

Example 7.2. If ω_{21} rotates clockwise (viewed from the right end) at 60 rad/s, determine ω_{51} and its direction of rotation (Fig. 7.7).

Considering first gears 2, 3, 4, and 5 and arm 6, let gear 2 be the first gear and gear 5 the last gear:

$$\frac{\omega_{LA}}{\omega_{FA}} = \frac{\omega_L - \omega_A}{\omega_F - \omega_A}$$

$$\frac{\omega_{56}}{\omega_{26}} = \frac{\omega_{51} - \omega_{61}}{\omega_{21} - \omega_{61}}$$

$$\frac{\omega_{56}}{\omega_{26}} = \frac{N_2 \times N_4}{N_3 \times N_5} = \frac{18 \times 28}{30 \times 20} = \frac{21}{25}$$

Therefore,

$$\frac{21}{25} = \frac{\omega_{51} - \omega_{61}}{\omega_{21} - \omega_{61}} = \frac{\omega_{51} - \omega_{61}}{60 - \omega_{61}} \tag{a}$$

However, Eq. (a) cannot be solved because it contains two unknowns, ω_{51} and ω_{61}. It is necessary therefore to consider gears 2, 3, 4, and 7 and arm 6. Let gear 2 be the first gear and gear 7 the last gear:

$$\frac{\omega_{76}}{\omega_{26}} = \frac{\omega_{71} - \omega_{61}}{\omega_{21} - \omega_{61}}$$

$$\frac{\omega_{76}}{\omega_{26}} = -\frac{N_2 \times N_4}{N_3 \times N_7} = -\frac{18 \times 28}{30 \times 76} = -\frac{21}{95}$$

$$-\frac{21}{95} = \frac{\omega_{71} - \omega_{61}}{\omega_{21} - \omega_{61}} = \frac{0 - \omega_{61}}{60 - \omega_{61}} \tag{b}$$

By solving Eq. (b) for ω_{61},

$$-\frac{21}{95}(60 - \omega_{61}) = 0 - \omega_{61}$$

$$\omega_{61} = \frac{21 \times 15}{29} = +10.86 \text{ rad/s}$$

From Eq. (a),

$$\frac{21}{25}(60 - \omega_{61}) = \omega_{51} - \omega_{61}$$

$$\frac{21}{25}(60 - 10.86) = \omega_{51} - 10.86$$

Therefore, $\omega_{51} = +52.14$ rad/s, with the direction of rotation same as that of ω_{21}.

Example 7.3. Consider that in the differential shown in Fig. 7.8, the angular velocity of shaft A is 350 rad/s in the direction shown and that of shaft B is 2000 rad/s. Determine the angular velocity of shaft C.

Use Eq. 7.5 and remember that the first gear and the last gear selected for the equation must be gears that mesh with the gears that have planetary motion. Therefore, let gear 4 be the first gear and gear 7 the last gear:

$$\frac{\omega_{LA}}{\omega_{FA}} = \frac{\omega_L - \omega_A}{\omega_F - \omega_A}$$

$$\frac{\omega_{78}}{\omega_{48}} = \frac{\omega_{71} - \omega_{81}}{\omega_{41} - \omega_{81}}$$

$$\frac{\omega_{78}}{\omega_{48}} = -\frac{N_4 \times N_6}{N_5 \times N_7} = -\frac{30 \times 24}{64 \times 18} = -\frac{5}{8}$$

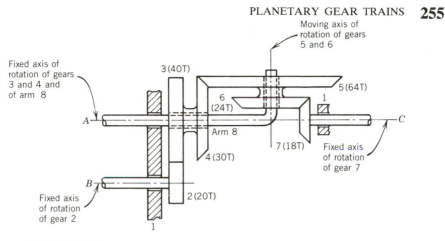

FIGURE 7.8

Also,

$$\omega_{41} = \omega_{31} = \omega_B \times \frac{N_2}{N_3} = 2000 \times \frac{20}{40}$$

$$= +1000 \text{ rad/s and same direction as } \omega_A$$

and

$$\omega_{81} = \omega_A = 350 \text{ rad/s}$$

By making the substitutions,

$$-\frac{5}{8} = \frac{\omega_{71} - 350}{1000 - 350}$$

$$\omega_{71} = -\frac{5}{8}(650) + 350$$

$$= -406.3 + 350$$

$$= -56.3 \text{ rad/s and opposite direction to } \omega_A$$

The tabulation method is another convenient way of solving planetary gear problems. To illustrate its use, consider the gear train of Fig. 7.4a and the following procedure:

1. Disconnect gear 1 from the ground and lock it to arm 3 together with gear 2. There can now be no relative motion among members 1, 2, and 3.
2. Rotate arm 3 (and gears 1 and 2) one positive revolution about center A.
3. Unlock gears 1 and 2 from arm 3. *Hold arm 3 fixed* and rotate gear 1 one negative revolution. Gear 2 therefore rotates $+N_1/N_2$ revolutions.

TABLE 7.1

	Gear 1	Gear 2	Arm 3
Motion with arm relative to frame (item 2)	$+1$	$+1$	$+1$
Motion relative to arm (item 3)	-1	$+\dfrac{N_1}{N_2}$	0
Total motion relative to frame	0	$1 + \dfrac{N_1}{N_2}$	$+1$

The results of steps 2 and 3 are entered in Table 7.1 together with the total number of revolutions made by each member of the train relative to the ground. It can be seen from the "Total" line of Table 7.1 that with gear 1 stationary, gear 2 turns $1 + (N_1/N_2)$ revolutions for one revolution of arm 3. This agrees with Eq. 7.3a.

Two examples will be given to illustrate the use of the tabular method.

Example 7.4. Consider that arm 4 of Fig. 7.9 rotates counterclockwise at 50 rad/s. Determine ω_{21} in magnitude and direction. See Table 7.2.

$$\frac{\omega_{21}}{\omega_{41}} = \frac{1 + (N_1/N_2)}{1}$$

$$\omega_{21} = \omega_{41}\left(1 + \frac{N_1}{N_2}\right) = 50\left(1 + \frac{80}{40}\right)$$

$$= +150 \text{ rad/s and same direction as } \omega_{41}$$

One distinct advantage of the tabular method is that more than one ratio can be obtained from a solution. In Example 7.4, if it had been necessary to

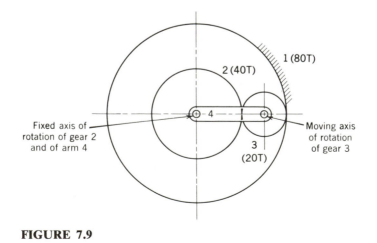

FIGURE 7.9

TABLE 7.2

	Gear 1	Gear 2	Gear 3	Arm 4
Motion with arm relative to frame	+1	+1	+1	+1
Motion relative to arm	-1	$+\dfrac{N_1}{N_2}$	$-\dfrac{N_1}{N_3}$	0
Total motion relative to frame	0	$1+\dfrac{N_1}{N_2}$	$1-\dfrac{N_1}{N_3}$	+1

determine the value of ω_{31}, this could have easily been accomplished from the data in the table.

Example 7.5. Example 7.1 and Fig. 7.6 will now be worked by the tabular method. Because all of the gears in this train rotate, it is easier to work with the actual speeds of gear 5 and arm 6 in the table rather than with one revolution as in Example 7.4. Because arm 6 rotates 150 rad/s, this must be the number of turns to which the entire train is subjected when locked together for line 1 of Table 7.3; note the 0 for arm 6 in line 2. With +150 for gear 5 in line 1, −100 must be inserted for gear 5 in line 2 to give the correct total of +50. With arm 6 stationary in line 2, and gear 5 now rotating a known amount, the rotation of gears 2, 3, and 4 can easily be determined for line 2.

$$\omega_{21} = 150 - 100\left(\frac{N_5 \times N_3}{N_4 \times N_2}\right) = 150 - 100\left(\frac{20 \times 30}{28 \times 18}\right)$$

$$= 150 - 100 \times \frac{25}{21}$$

$$= +30.9 \text{ rad/s and same direction as } \omega_{51} \text{ and } \omega_{61}$$

TABLE 7.3

	Gear 2	Gear 3	Gear 4	Gear 5	Arm 6
Motion with arm relative to frame	+150	+150	+150	+150	+150
Motion relative to arm	$-100\dfrac{N_5 \times N_3}{N_4 \times N_2}$	$+100\dfrac{N_5}{N_4}$	$+100\dfrac{N_5}{N_4}$	-100	0
Total motion relative to frame	$150 - 100\left(\dfrac{N_5 \times N_3}{N_4 \times N_2}\right)$			+50	+150

Example 7.6. Example 7.2 and Fig. 7.7 will now be solved using the tabular method (Table 7.4). In this problem, the angular velocity of the arm is unknown, which requires a variation in the solution from that of Example 7.5.

In Fig. 7.7, with data from Example 7.2, $\omega_{21} = 60$ rad/s with clockwise direction

TABLE 7.4

	Gear 2	Gear 3	Gear 4	Gear 5	Gear 7	Arm 6
Motion with arm relative to frame	x	x	x	x	x	x
Motion relative to arm	$+x\left(\dfrac{N_3 \times N_7}{N_2 \times N_4}\right)$			$+x\left(\dfrac{N_7}{N_5}\right)$	$-x$	0
Total motion relative to frame	60			y	0	x

viewed from the right end. It is required to determine ω_{51} and its direction; ω_{61} is unknown, and $\omega_{71} = 0$.

Let $x = \omega_{61}$ and $y = \omega_{51}$:

$$x + x\left(\frac{N_3 \times N_7}{N_2 \times N_4}\right) = 60$$

$$x\left(1 + \frac{30 \times 76}{18 \times 28}\right) = 60$$

$$x\left(1 + \frac{95}{21}\right) = 60$$

$$x = \frac{21}{116}\,(60) = +10.86 \text{ rad/s}$$

$\therefore \omega_{61} = +10.86$ rad/s and direction of rotation same as that of ω_{21}

$$y = x + x\left(\frac{N_7}{N_5}\right)$$

$$= x\left(1 + \frac{N_7}{N_5}\right)$$

$$= 10.86\left(1 + \frac{76}{20}\right) = +52.13 \text{ rad/s}$$

$\therefore \omega_{51} = +52.13$ rad/s and direction of rotation same as that of ω_{21}

As can be seen, this method of solution is somewhat shorter than that used in the solution of Example 7.2.

7.3 APPLICATIONS OF PLANETARY GEAR TRAINS

Planetary gear trains find many applications in machine tools, hoists, aircraft propeller reduction drives, automobile differentials, automatic transmissions, aircraft servo-drives, and many others. Figure 7.10 shows a diagrammatic sketch of a planetary train used as a reduction between the engine and the propeller in an

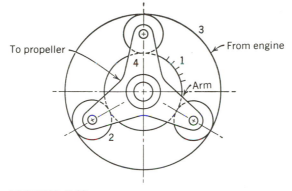

FIGURE 7.10

aircraft power plant. Figure 7.11 shows a photograph of an actual unit. The earlier aircraft engine reduction drives used bevel gears in the planetary train. These were discarded, however, in favor of spur gears because the spur gear planetary drive could transmit more power in a given space.

In Fig. 7.10, the engine drives the internal gear 3. Gear 2 meshes with the fixed gear 1 and with the gear 3 so that it has planetary motion. Arm 4, or planet

FIGURE 7.11 Planetary reduction unit for aircraft propeller drive. (Courtesy of Foote Brothers Gear & Machinery Corp.)

carrier, which is connected to gear 2, drives the propeller at a slower speed than the engine. The equation for the ratio of engine speed ω_{31} to propeller speed ω_{41} can easily be determined from Eq. 7.5 as follows:

$$\frac{\omega_{31}}{\omega_{41}} = 1 - \frac{\omega_{34}}{\omega_{14}}$$

where

$$\frac{\omega_{34}}{\omega_{14}} = -\frac{N_1}{N_3}$$

Therefore,

$$\frac{\omega_{31}}{\omega_{41}} = 1 + \frac{N_1}{N_3}$$

It is interesting to note that it would be impossible to obtain a velocity ratio as high as 2:1 because this would mean that gear 1 would have to have the same number of teeth as gear 3, which is impossible. In determining the limiting ratio for a given drive, it should be noted that all of the gears will have the same diametral pitch.

A planetary gear train used as a differential in an automobile is shown in Fig. 7.12. Figure 7.13 shows a cutaway view of a differential and housing. This mechanism makes it possible for an automobile to turn a corner without the rear wheels slipping. In Fig. 7.12 gear 2 is driven by the engine via the clutch, transmission, and drive shaft. Gear 2 drives gear 3, which is fastened to the carrier 7. If the car is moving straight ahead, gears 4, 5, and 6 turn as a unit with the carrier and there is no relative motion between them. Gears 5 and 6 turn the

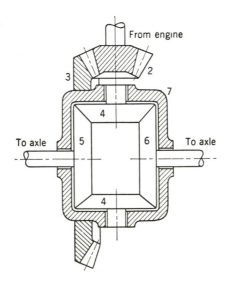

FIGURE 7.12

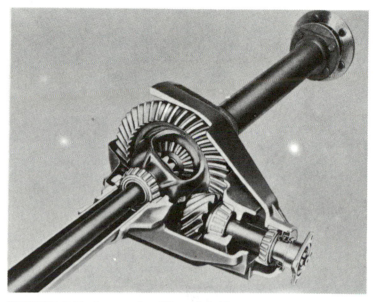

FIGURE 7.13 Automotive differential. (Courtesy of Gleason Works.)

axles. When the car makes a turn, however, gears 5 and 6 no longer rotate at the same speed and gear 4 has to turn about its own axis, as well as being driven by the carrier. It is interesting to note that, if one of the wheels is held stationary and the second is free to rotate, the second wheel will turn at a speed twice that of the carrier. This characteristic is a disadvantage when the car is stuck in snow or mud.

As previously mentioned in Chapter 2 (section 2.13, Computing Elements), electronic computing systems have largely replaced mechanical systems. However, there are applications where mechanical elements are preferable because they do not require electrical power. Figures 7.14a and 7.14b show a bevel gear differential and a spur gear differential, respectively. The bevel gear units are

FIGURE 7.14a Bevel gear differential.

FIGURE 7.14*b* **Spur gear differential. In this differential gears 5 (48T) and 6 (69T) are compounded, and carrier 7 is pinned to shaft 8. The power flows from gear 2 (48T) to planet 3 (18T) to planet 4 (18T) to gear 5, and finally to gear 6.**

available commercially in several stock sizes and are used extensively where a mechanical computing or control system is required.

There are many planetary gear train designs and a wide range of possible ratios. The applications mentioned are only two of a wide variety. In many instances, it is possible to obtain a greater reduction ratio with a smaller drive using a planetary train than when using an ordinary gear train.

7.4 ASSEMBLY OF PLANETARY GEAR TRAINS

When a planetary gear train is designed, the question of assembling the train with equally spaced planets must be considered. With the train illustrated in Fig. 7.15, it is possible that for a given number of teeth in gears 1, 2, and 3 it might not be possible to have three equally spaced planet gears.

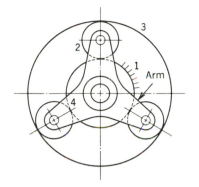

FIGURE 7.15

To determine the number of planets that can be used for a given number of teeth in gears 1, 2, and 3, it is necessary to determine the angle AOB in Fig. 7.16a resulting from gear 3 having been rotated a whole number of tooth spaces with gear 1 stationary. The case must also be investigated where gear 3 is stationary, and gear 1 has been rotated a whole number of spaces. This gives angle AOB' as shown in Fig. 7.16b. The following method was developed by Professor G. B. DuBois of Cornell University.

Let the numbers of teeth in gears 1, 2, and 3 be N_1, N_2, and N_3. If θ_{31} equals the angular motion of gear 3 after it has been rotated one whole tooth space with respect to gear 1, then

$$\theta_{31} = \frac{1}{N_3} \text{ revolutions}$$

The angular motion of arm 4 with respect to gear 1 when gear 3 has been rotated one tooth space is given by

$$\theta_{41} = \theta_{31} \times \frac{\omega_{41}}{\omega_{31}} \text{ revolutions}$$

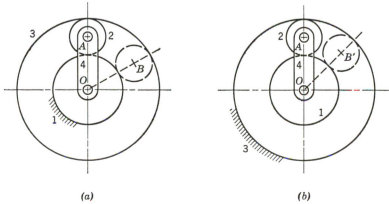

(a)   (b)

FIGURE 7.16

From the velocity analysis of the planetary train of Fig. 7.10, which is identical to the one under consideration,

$$\frac{\omega_{41}}{\omega_{31}} = \frac{N_3}{N_3 + N_1}$$

Therefore,

$$\theta_{41} = \frac{1}{N_3} \times \frac{N_3}{N_3 + N_1} = \frac{1}{N_3 + N_1} \text{ revolutions}$$

Angle AOB is the angle turned through by arm 4 when gear 3 moves relative to gear 1. If gear 3 moves one tooth space, angle AOB equals θ_{41}. This is the smallest possible angle between planet gears if the planets are allowed to overlap. If gear 3 rotates a whole number of tooth spaces c, then

$$\angle AOB = c(\theta_{41}) = \frac{c}{N_3 + N_1} \text{ revolutions} \qquad (7.6)$$

and represents the angle between planets with possible overlapping.

Consider next the case of Fig. 7.16*b* where gear 1 has been rotated one tooth space with gear 3 stationary and it is required to find angle AOB'. If θ_{13} equals the angular motion of gear 1 after it has been rotated one tooth space and θ_{43} equals the resulting motion of arm 4 (both relative to gear 3), then

$$\theta_{13} = \frac{1}{N_1}$$

and

$$\theta_{43} = \theta_{13} \times \frac{\omega_{43}}{\omega_{13}}$$

But it can be easily derived that

$$\frac{\omega_{43}}{\omega_{13}} = \frac{N_1}{N_1 + N_3}$$

Therefore,

$$\theta_{43} = \frac{1}{N_1} \times \frac{N_1}{N_1 + N_3} = \frac{1}{N_1 + N_3}$$

and

$$\angle AOB' = c(\theta_{43}) = \frac{c}{N_1 + N_3} \tag{7.7}$$

By comparing Eqs. 7.6 and 7.7, it can be seen that arm 4 rotates through the same angle regardless of whether gear 3 or gear 1 rotates one or more tooth spaces.

If angle AOB is the fraction of a revolution between planets, then its reciprocal will be the number of planets. By taking the reciprocal of Eq. 7.6, it is possible to obtain an expression for the number of equally spaced planets around gear 1. If n represents the number of planets, then

$$n = \frac{N_3 + N_1}{c} \tag{7.8}$$

These planets may or may not overlap each other, depending on the value of c.

It is now necessary to determine the maximum number of planets n_{max} that can be used without overlapping. In Fig. 7.17, the outside radii R_{o_2} of two planet gears are shown almost touching at point C. From the figure,

$$n_{max} = \frac{360}{\angle AOB} = \frac{180}{\angle AOC}$$

$$\angle AOC = \sin^{-1} \frac{\overline{AC}}{\overline{OA}}$$

where

$$\overline{AC} > R_{o_2}$$

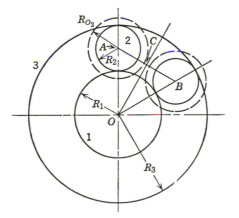

FIGURE 7.17

and

$$\overline{OA} = R_1 + R_2$$

$$R_{o_2} = R_2 + a = \frac{N_2}{2P_d} + \frac{k}{P_d} \qquad (k = 1 \text{ for standard full-depth teeth})$$

or

$$R_{o_2} = \frac{N_2 + 2}{2P_d}$$

and

$$R_1 + R_2 = \frac{N_1 + N_2}{2P_d}$$

Therefore, for standard full-depth teeth,

$$n_{\max} < \frac{180}{\sin^{-1}(N_2 + 2)/(N_1 + N_2)} \qquad (7.9)$$

Also, from the geometry of Fig. 7.16,

$$R_3 = R_1 + 2R_2$$

Because $R = N/2P_d$ for a standard gear and because the diametral pitches of gears 1, 2, and 3 are equal,

$$N_3 = N_1 + 2N_2$$

For nonstandard gears, Eq. 7.9 can be used to give an approximate value of $n_{\max}$. In this case, the fractional value of N_2 resulting from use of the standard equation

$$N_2 = \frac{N_3 - N_1}{2}$$

would be substituted in Eq. 7.9. As a final check, a layout drawing should be made.

Example 7.7. In a planetary gear train similar to that of Fig. 7.15, gear 1 has 50 teeth and gear 3 has 90 teeth. Determine the number of equally spaced planets that can be used without overlapping. The gears are standard:

$$N_2 = \frac{N_3 - N_1}{2} = \frac{90 - 50}{2} = 20$$

$$n_{max} = \frac{180}{\sin^{-1}(N_2 + 2)/(N_1 + N_2)} = \frac{180}{\sin^{-1}(20 + 2)/(50 + 20)} = 9.8 \text{ planets}$$

Therefore, the number of planets in the gear train cannot exceed 9.

$$n = \frac{N_3 + N_1}{c} = \frac{90 + 50}{c} = \frac{140}{c}$$

The value of c must be a whole number of tooth spaces between planets that, when divided into 140, will give a whole number n. For this case, c can equal 140, 70, 35, 28, or 20. Therefore,

$n = 1, 2, 4, 5,$ or 7 equally spaced planets

7.5   CIRCULATING POWER IN CONTROLLED PLANETARY GEAR SYSTEMS

As has been seen in the previous work, a planetary gear train designed as a differential consists primarily of three rotating elements, with the speed of any one element depending upon the speeds of the other two. In all the designs considered so far, there have been two independent inputs. It is possible, however, to design a planetary differential where the speed of one element is controlled by a gear train connected to either of the other two elements. This addition is known as a *branch control circuit* and may contain a fixed-speed or a variable-speed gear train.

Although the angular-velocity ratios for a planetary system with a control circuit are easily calculated, it is difficult to determine the circulating power load within the system. It is very important that the amount of circulating power be considered in the design of the unit, or low efficiencies may result.

The design of a gear train on the basis of strength is beyond the scope of kinematics. However, the calculation of circulating power is so directly connected to kinematic design that we thought it advisable to include a brief treatment of the subject as presented by Laughlin, Holowenko, and Hall[1] and by Sanger.[2]

Figure 7.18a shows a sketch of a planetary gear train with a branch control circuit consisting of gears 2, 3, 4, and 5. Shaft A is the input shaft which powers gear 2 and arm 10. Gear 6 is powered by shaft A through the control circuit. Shaft B, which is powered by gear 9, is the output shaft.

[1]H. G. Laughlin, A. R. Holowenko, and A. S. Hall, "How to Determine Circulating Power in Controlled Epicyclic Gear Systems," *Machine Design,* **28** (6), March 22, 1956.

[2]D. J. Sanger, "The Determination of Power Flow in Multiple-Path Transmission Systems," *Mechanism and Machine Theory,* **7** (1), Spring 1972.

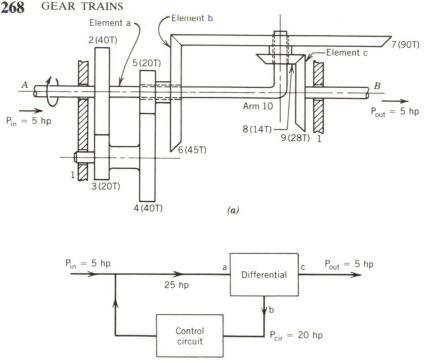

FIGURE 7.18

In analyzing the power flow in a planetary differential, it is necessary to label the three basic rotating elements of the system as element a, element b, and element c. One of these elements is always an arm that carries the planetary gears, and the other two elements are gears on independent axes. Element a will always be that member projecting from the differential to outside the system and connecting to element b through the branch control circuit. Element b will always be that member which transmits power to or from the differential to the branch control circuit but does not transmit power directly to or from outside of the system. Element c will always be that member projecting from the differential directly to outside the system but having no connection with the control circuit. In Fig. 7.18a, arm 10 is element a, since it takes power from outside the differential and is connected to gear 6 through the control circuit. Gear 6 is the controlled element and has no direct connection to outside the system and is therefore element b. Gear 9 transmits power directly to outside the system and has no connection with the control circuit and is therefore element c. It should be mentioned that if the power input were through shaft B and gear 9 which would result in the output through arm 10 and shaft A, the same notation would apply. It can be seen therefore that the notation depends only on the configuration of the differential and not on the direction of power flow.

The planetary system of Fig. 7.18a is shown schematically in Fig. 7.18b, with the three rotating elements labeled. From Fig. 7.18b, an expression for the amount of power circulating through the branch circuit can easily be developed from the general torque and energy relationships between the three elements. If the power flowing through the branch control circuit is defined as the circulating power P_{cir} and the power flowing through element c is labeled P_c, a ratio γ of these powers is

$$\gamma = \frac{P_{cir}}{P_c} = \frac{T_b \omega_b}{T_c \omega_c} \tag{7.10}$$

Also, by considering the differential as an isolated unit,

$$\Sigma T = T_a + T_b + T_c = 0$$

$$\Sigma P = T_a \omega_a + T_b \omega_b + T_c \omega_c = 0$$

Solving the preceding equations simultaneously gives

$$\gamma = \frac{\omega_b(\omega_c - \omega_a)}{\omega_c(\omega_a - \omega_b)}$$

or

$$\gamma = \frac{r(1 - R)}{1 - r} \tag{7.11}$$

where

$$r = \frac{\omega_b}{\omega_a}$$

$$R = \frac{\omega_a}{\omega_c}$$

If γ is positive, it means that power flows into or out of the differential through both elements b and c. If γ is negative, power flows into the differential through one of these elements (b or c) and out through the other.

Example 7.8. Consider that in the differential shown in Fig. 7.18a, $\omega_A = 3600$ rpm in the direction shown, and the power input is 5 hp. Determine the power circulating in the branch control circuit.

$$\omega_a = \omega_{10/1} = 3600 \text{ rpm}$$

$$\omega_b = \omega_{61} = \omega_{10} \frac{N_2 \times N_4}{N_3 \times N_5}$$

$$= 3600 \frac{40 \times 40}{20 \times 20} = 14,400 \text{ rpm}$$

By using the relation (Eq. 7.5)

$$\frac{\omega_{LA}}{\omega_{FA}} = \frac{\omega_L - \omega_A}{\omega_F - \omega_A}$$

the angular velocity of gear 9 can be found as follows:

Let gear 6 be the first gear and gear 9 the last gear. By making the substitution of these gear numbers into Eq. 7.5, the following equation results:

$$\frac{\omega_{9/10}}{\omega_{6/10}} = \frac{\omega_{91} - \omega_{10/1}}{\omega_{61} - \omega_{10/1}}$$

$$\frac{\omega_{9/10}}{\omega_{6/10}} = -\frac{N_6 \times N_8}{N_7 \times N_9} = -\frac{45 \times 14}{90 \times 28} = -\frac{1}{4}$$

$$-\frac{1}{4} = \frac{\omega_{91} - 3600}{14,400 - 3600}$$

$$\omega_{91} = -\frac{1}{4}(10,800) + 3600$$

$$= 900 \text{ rpm} \qquad (\text{same direction of rotation as } \omega_{10/1})$$

Therefore,

$$\omega_c = \omega_{91} = 900 \text{ rpm}$$

By substituting in Eq. 7.11,

$$\gamma = \frac{r(1 - R)}{1 - r}$$

$$r = \frac{\omega_b}{\omega_a} = \frac{\omega_{61}}{\omega_{10/1}} = \frac{14,400}{3600} = 4$$

$$R = \frac{\omega_a}{\omega_c} = \frac{\omega_{10/1}}{\omega_{91}} = \frac{3600}{900} = 4$$

$$\gamma = \frac{4(1 - 4)}{1 - 4} = +4$$

Therefore, from Eq. 7.10,

$$P_{\text{cir}} = \gamma P_c$$

$$= 4 \times 5 = 20 \text{ hp}$$

The positive value of γ indicates that power flows through elements b and c in the same direction (out of or into the differential). Because power flow was given from the differential through element c, the circulating power flows from the differential through element b. The magnitudes and directions of the flow of power are shown on Fig. 7.18b.

Example 7.9. Consider the differential shown in Fig. 7.19a, with $\omega = 100$ rpm in the direction shown and a power input of 20 hp. Determine the power circulating in the branch control circuit.

From the given data,

$$\omega_{21} = 100 \text{ rpm}$$

$$\omega_{31} = -100\left(\frac{N_2}{N_3}\right) = -100\left(\frac{40}{30}\right) = -133.33 \text{ rpm}$$

$$\omega_{41} = \omega_{31} = -133.33 \text{ rpm}$$

$$\omega_{91} = 100 \text{ rpm}$$

$$\omega_{81} = -100\left(\frac{N_9}{N_8}\right) = -100\left(\frac{60}{30}\right) = -200 \text{ rpm}$$

$$\omega_{71} = \omega_{81} = -200 \text{ rpm}$$

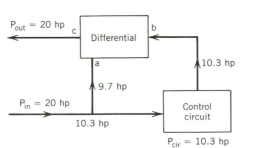

FIGURE 7.19

By using the relation (Eq. 7.5)

$$\frac{\omega_{LA}}{\omega_{FA}} = \frac{\omega_L - \omega_A}{\omega_F - \omega_A}$$

the angular velocity of arm 10 can be determined as follows:

Let gear 4 be the first gear and gear 7 the last gear. By substituting these gear numbers into the above equation, the following relation is obtained:

$$\frac{\omega_{7/10}}{\omega_{4/10}} = \frac{\omega_{71} - \omega_{10/1}}{\omega_{41} - \omega_{10/1}}$$

where

$$\frac{\omega_{7/10}}{\omega_{4/10}} = -\frac{N_6 \times N_4}{N_7 \times N_5} = -\frac{64 \times 38}{59 \times 36} = -1.145$$

Therefore,

$$-1.145 = \frac{-200 - \omega_{10/1}}{-166.67 - \omega_{10/1}}$$

and

$$\omega_{10/1} = -182.2 \text{ rpm} \qquad \text{(direction opposite to } \omega_{21}\text{)}$$

From the analysis given above, the angular velocities of elements a, b, and c shown in Fig. 7.19a are

$$\omega_a = \omega_{41} = -166.67 \text{ rpm}$$

$$\omega_b = \omega_{71} = -200 \text{ rpm}$$

$$\omega_c = \omega_{10/1} = -182.2 \text{ rpm}$$

The circulating power can now be determined from Eq. 7.11:

$$\gamma = \frac{r(1 - R)}{1 - r}$$

where

$$r = \frac{\omega_b}{\omega_a} = \frac{-200}{-166.67} = +1.199$$

$$R = \frac{\omega_a}{\omega_c} = \frac{-166.67}{-182.2} = +0.915$$

Therefore,

$$\gamma = \frac{1.199(1 - 0.915)}{1 - 1.199} = -0.513$$

and

$$P_{\text{cir}} = \gamma P_c = -0.513(20) = -10.3 \text{ hp}$$

The negative value of γ indicates that the power flow through elements b and c will be in opposite directions relative to the differential. That is, since power flow was given from the differential through element c, the circulating power flows through element b into the differential. The magnitudes and directions of the flow of power are shown in Fig. 7.19b.

7.6 HARMONIC DRIVE GEARING[3]

Harmonic drive gearing is a patented principle based on nonrigid body mechanics. It employs the three concentric components shown in Figs. 7.20a and 7.20b to produce high mechanical advantage and speed reduction. The use of nonrigid body mechanics allows a continuous elliptical deflection wave to be induced in a nonrigid external gear, thereby providing a continuous rolling mesh with a rigid, internal gear.

Since the teeth of the nonrigid flexspline and the rigid circular spline are in continuous engagement and since the flexspline has two fewer teeth than the circular spline, one revolution of the input causes relative motion between the flexspline and the circular spline equal to two teeth. Thus, with the circular spline rotationally fixed, the flexspline will rotate in an opposite direction to the input at a reduction ratio equal to the number of teeth on the flexspline divided by 2.

This relative rotation may be seen by examining the motion of a single flexspline tooth over one-half an input revolution, as shown in Fig. 7.20c.

The tooth is fully engaged when the major axis of the wave generator input is at zero degrees. When the wave generator's major axis rotates to 90°, the tooth is disengaged. Full reengagement occurs in the adjacent circular spline tooth space when the major axis is rotated another 180°. This motion repeats as the major axis rotates another 180° back to zero, thereby producing two tooth advancements per input revolution.

It should be mentioned that any elements of the drive can function as the input, the output, or the fixed member depending on whether the gearing is used for speed reduction, speed increasing, or differential operation.

---

[3]The material in this section is adapted directly from the *Harmonic Drive Designer's Manual* of Emhart Machinery Group, Wakefield, Mass., and is used with permission.

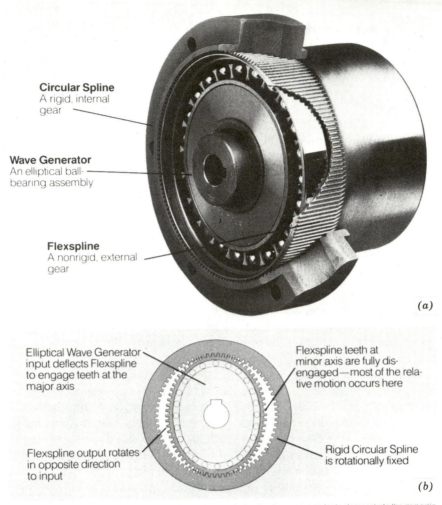

Circular Spline
A rigid, internal gear

Wave Generator
An elliptical ball-bearing assembly

Flexspline
A nonrigid, external gear

(a)

Elliptical Wave Generator input deflects Flexspline to engage teeth at the major axis

Flexspline teeth at minor axis are fully disengaged—most of the relative motion occurs here

Flexspline output rotates in opposite direction to input

Rigid Circular Spline is rotationally fixed

(b)

Note: The amount of Flexspline deflection has been exaggerated in the diagrams in order to demonstrate the principle. Actual deflection is much smaller than shown and is well within the material fatigue limits. Deflection is, therefore, not a factor in the life expectancy of the gearing.

0°

90°

180° *(c)*

FIGURE 7.20

Problems

7.1. In Fig. 7.21, gear 1 rotates in the direction shown at 240 rpm. Determine the speed of pinion 9 (rpm) and the speed (fpm) and direction of rack 10.

7.2. A hoist is operated by a motor driving a 4-threaded worm that engages a 100-tooth worm gear. The worm gear is keyed to a shaft which also contains a 20-tooth spur pinion. The pinion meshes with a 140-tooth spur gear mounted on the end of the hoisting drum. Make a sketch of the unit, and calculate the speed of hoisting (fpm) if the motor operates at 600 rpm and the drum diameter is 12 in.

7.3. Two slitting rolls A and B for cutting sheet metal are driven by means of the gear train shown in Fig. 7.22. The rolls must operate in the direction shown at a peripheral speed of 1150 mm/s. (a) Determine the angular velocity ratio ω_2/ω_3 to drive the rolls at the required speed. Gear 1 runs at 1800 rpm. (b) Determine the direction of rotation of gear 1 and the hand of worm 6 to give the required rotation of the rolls.

7.4. In the sketch of the press shown in Fig. 7.23, 5 and 6 are single-threaded screws of the opposite hand, with 6 threading into 5 as indicated. Gear 4 is fastened to screw 5. Plate B is prevented from turning by a slot in it which engages the frame. If the pitch of 5 is $\frac{1}{4}$ in. and that of 6 is $\frac{1}{8}$ in., determine the direction and the number of turns of shaft A required to lower plate B $\frac{3}{4}$ in.

7.5. The gear train in Fig. 7.24 shows the essential features of the work spindle drive for a gear hobbing machine. Gear blank B and the worm gear 9 are mounted on the same shaft and must rotate together. (a) If the gear blank B is to be driven clockwise, determine the hand of the hob A. (b) Determine the angular-velocity ratio ω_7/ω_5 to cut 72 teeth on the gear blank B.

7.6. A gear train contains shaft A to which is keyed gears 1 and 2, an intermediate shaft B with a sliding compound gear 3, 4, and 5 and shaft C to which are keyed gears 6 and 7. The gears are numbered from left to right and are all spur gears with shaft center distances of 12 in. and a diametral pitch of 5. The compound gear can be shifted to the left to give a velocity ratio of 5:1 through gears 1, 4, 3, and 6, or to the right to give a velocity ratio 25:9 through gears 2, 4, 5, and 7. Draw a sketch of the unit and calculate the number of teeth in each gear if $N_5 = N_2$.

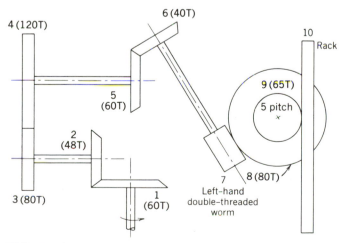

FIGURE 7.21

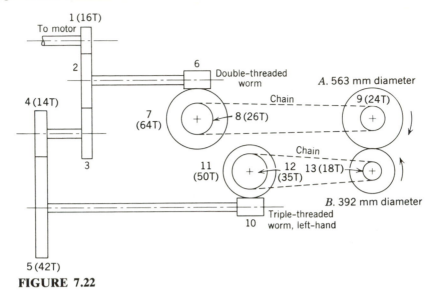

FIGURE 7.22

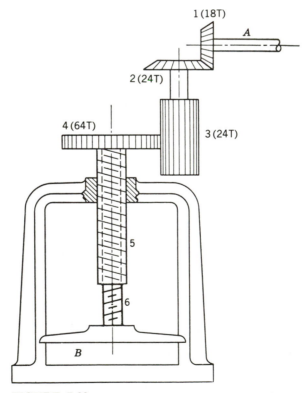

FIGURE 7.23

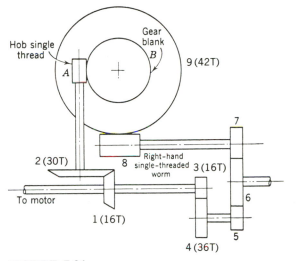

FIGURE 7.24

7.7. In the gear train in Fig. 7.25, screws 5 and 6 are single-threaded of opposite hand as shown. Screw 5 has a pitch of 3 mm and screw 6 a pitch of 2.5 mm. Screw 6 threads into screw 5 and screw 5 threads into the frame. Determine the change in x and y in magnitude and direction for one revolution of the handwheel in the direction shown. Gears 1 and 2 are compounded on the handwheel shaft.

7.8. Figure 7.26 shows part of a gear train for a vertical milling machine. Power input is through the pulley and power output through gear 12. Compound gears 1 and 2, 3 and

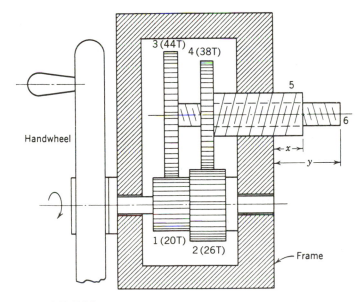

FIGURE 7.25

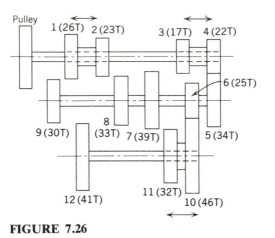

FIGURE 7.26

4, and 10 and 11 can slide as shown to give various combinations of gearing. Determine all of the train values possible between the pulley and gear 12.

7.9. Figure 7.27 shows part of a gear train for a vertical milling machine. Compound gears 1 and 2 can slide so that either gear 1 meshes with gear 5 or gear 2 meshes with gear 3. In the same manner, gear 13 meshes with gear 15 or gear 14 meshes with gear 16. (a) With gear 2 meshing with gear 3, determine the two possible spindle speeds for a motor speed of 1800 rpm. Will the spindle rotate in the same or opposite direction to the

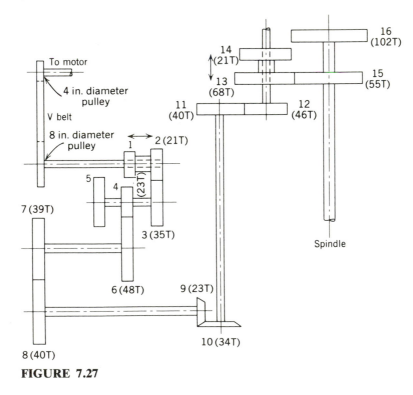

FIGURE 7.27

motor? (*b*) With gear 13 meshing with gear 15 and a spindle speed of 130 rpm, determine the number of teeth for gears 1 and 5 if gears 1, 2, 3, and 5 are standard and have the same diametral pitch.

7.10. A conventional automotive transmission is shown diagrammatically in Fig. 7.28. The transmission of power is as follows: Low gear: gear 3 shifted to mesh with gear 6. Transmission of power through gears 1, 4, 6, 3. Second gear: gear 2 shifted to mesh with gear 5. Transmission through gears 1, 4, 5, 2. High gear: gear 2 shifted so that clutch teeth on end of gear 2 mesh with clutch teeth on end of gear 1. Direct drive results. Reverse gear: gear 3 shifted to mesh with gear 8. Transmission through gears 1, 4, 7, 8, 3. A car equipped with this transmission has a differential ratio of 2.9:1 and a tire outside diameter of 26 in. Determine the engine speed of the car under the following conditions: (*a*) low gear and car traveling 20 mph; (*b*) high gear and car traveling 60 mph; (*c*) reverse gear and car traveling 4 mph.

7.11. In the planetary clutch shown in Fig. 7.29, the stop 6 may be engaged or disengaged. When engaged, a planetary gear train results, and, when disengaged, an ordinary gear train results because arm 5 will remain stationary. If gear 2 rotates in the direction shown in 300 rpm, determine (*a*) the speed of the ring gear 4 when the stop 6 is disengaged as shown, and (*b*) the speed of arm 5 when the stop 6 is engaged with the ring gear 4.

7.12. Considering a bevel gear differential as used in automotive drives, prove that, when one of the rear wheels on a car is jacked up, it will turn twice as fast as the differential carrier.

7.13. If a truck is rounding a right-hand curve at 15 mph, determine the speed in rpm of the differential carrier. The radius of curvature of the curve is 100 ft to the center of the truck, and the truck tread is 6 ft. The outside diameter of the tires is 36 in.

7.14. For the bevel gear planetary drive shown in Fig. 7.30, determine the ratio ω_4/ω_3 when gear 1 is stationary.

7.15. For the ball bearing shown in Fig. 7.31, the inner race 1 is stationary and the outer race 2 rotates with a tubular shaft at 1600 rpm. Assuming pure rolling between the balls and races, determine the speed of the ball retainer 4.

7.16. A mechanism known as Ferguson's paradox is shown in Fig. 7.32. For one revolution

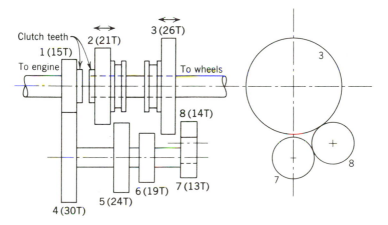

FIGURE 7.28

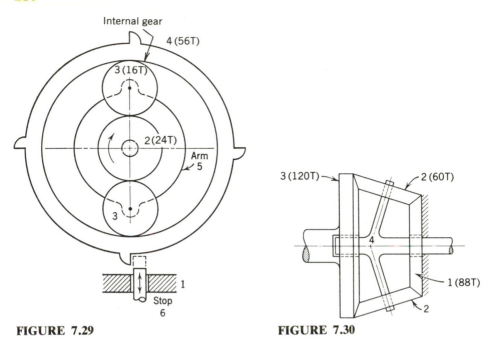

FIGURE 7.29   **FIGURE 7.30**

of the arm in the direction shown, find the number of revolutions of gears 3, 4, and 5 and their directions of rotation. The gears are nonstandard.

7.17. Shaft A rotates in the direction shown in Fig. 7.33 at 640 rpm. If shaft B is to rotate at 8 rpm in the direction shown, calculate the angular-velocity ratio ω_2/ω_4. What would the ratio ω_2/ω_4 have to be for shaft B to rotate at 8 rpm in the opposite direction?

7.18. In the mechanism in Fig. 7.34, gear 2 rotates at 60 rpm in the direction shown. Determine the speed and direction of rotation of gear 12.

7.19. A mechanism known as Humpage's gear is shown in Fig. 7.35. Find the angular velocity ratio ω_A/ω_B.

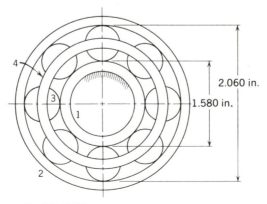

FIGURE 7.31

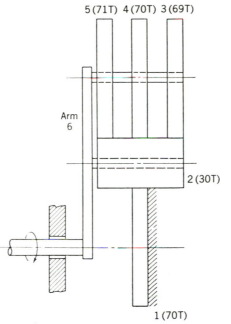

5 (71T) 4 (70T) 3 (69T)

Arm
6

2 (30T)

1 (70T) **FIGURE 7.32**

7.20. In the planetary gear train shown in Fig. 7.36, determine the angular-velocity ratio ω_2/ω_7. Compare this ratio with that obtained if the arm 4 is connected directly to the output shaft and gears 5, 6, and 7 are omitted.

7.21. In the gear train for Problem 7.20, gear 2 rotates at 600 rpm in the direction shown and gear 1 (and gear 6) rotates at 300 rpm in the opposite direction. Calculate the speed and direction of rotation of gear 7.

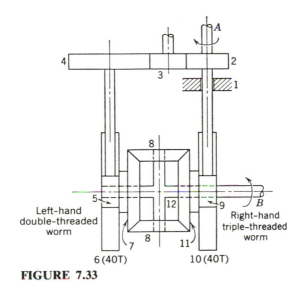

4
3
2
1
A
8
5
12
9
B
Left-hand
double-threaded
worm
Right-hand
triple-threaded
worm
8
7
11
6 (40T) 10 (40T)

FIGURE 7.33

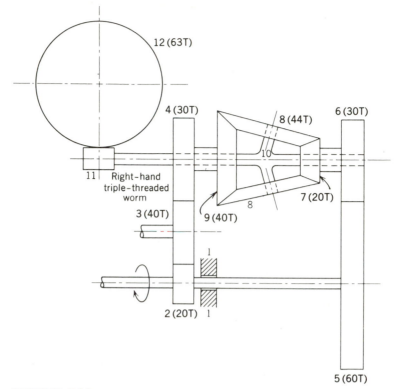

FIGURE 7.34

7.22. A planetary gear train for a two-speed aircraft supercharger drive is shown in Fig. 7.37. Gear 2 is driven by a 63-tooth gear (not shown) which operates at 2400 rpm. At high speed, gear 2 connects to the supercharger shaft through additional gearing. At low speed, gear 7 is held stationary and shaft B is connected to the supercharger shaft with the same gear ratio as was used between gear 2 and the supercharger shaft. If the supercharger operates at 24,000 rpm at high speed, calculate the low-speed value.

7.23. Figure 7.38 shows the planetary gear and power shaft assembly for an aircraft servo. If shaft A connects to the motor, determine the angular-velocity ratio ω_A/ω_B.

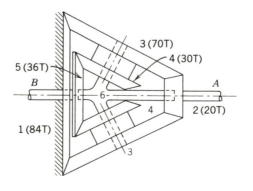

FIGURE 7.35

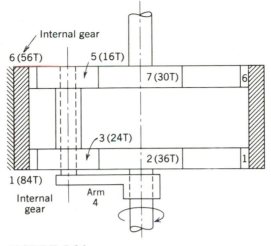

FIGURE 7.36

7.24. Figure 7.39 shows a planetary gear train for a large reduction. (*a*) If shaft *A* connects to the motor, determine the angular-velocity ratio ω_A/ω_B. (*b*) Will gears 2, 3, and 4 and gears 5, 6, and 7 be standard or nonstandard? Why? (*c*) If the number of teeth in gear 3 is changed from 51 teeth to 52 teeth, calculate the angular-velocity ratio ω_A/ω_B.

7.25. An aircraft propeller reduction drive is shown diagrammatically in Fig. 7.40. Determine the propeller speed in magnitude and direction if the engine turns at 2450 rpm in the direction indicated.

7.26. In the planetary reduction unit shown in Fig. 7.41, gear 2 turns at 300 rpm in the direction indicated. Determine the speed and direction of rotation of gear 5.

7.27. In the gear train for Problem 7.26, gear 2 turns at 300 rpm in the direction shown, and gear 1 rotates at 50 rpm in the opposite direction. Calculate the speed and direction of rotation of gear 5.

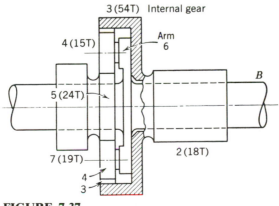

FIGURE 7.37

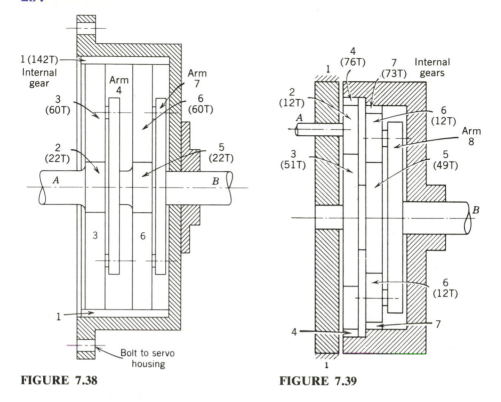

FIGURE 7.38

FIGURE 7.39

7.28. In the planetary gear train shown in Fig. 7.42, gear 2 turns at 600 rpm in the direction indicated. Determine the speed and direction of rotation of arm 6 if gear 5 rotates at 350 rpm in the same direction as gear 2.

7.29. If in the gear train for Problem 7.28, gear 2 rotates at 1000 rpm in the direction shown and gear 5 is held stationary, arm 6 will rotate at 625 rpm in the same direction

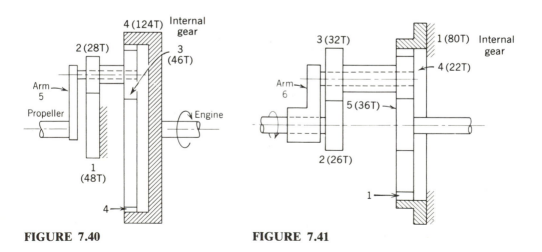

FIGURE 7.40

FIGURE 7.41

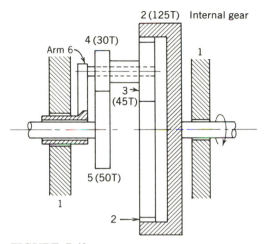

FIGURE 7.42

as gear 2. Determine the speed and direction of rotation which must be given to gear 5 to make arm 6 stand still if gear 2 continues to rotate at 1000 rpm.

7.30. For the gear train of Fig. 7.43, shaft *A* rotates at 300 rpm and shaft *B* at 600 rpm in the directions shown. Determine the speed and direction of rotation of shaft *C*.

7.31. In Fig. 7.44, shaft *A* turns at 100 rpm in the direction shown. Calculate the speed of shaft *B* and give its direction of rotation.

7.32. In the planetary gear train shown in Fig. 7.45, shaft *A* rotates at 450 rpm and shaft *B* at 600 rpm in the directions shown. Calculate the speed of shaft *C* and give its direction of rotation.

7.33. Shaft *A* rotates in Fig. 7.46 at 350 rpm and shaft *B* at 400 rpm in the directions shown. Determine the speed and direction of rotation of shaft *C*.

7.34. In the bevel gear planetary train shown in Fig. 7.47, shaft *A* rotates in the direction shown at 1250 rpm and shaft *B* in the direction shown at 600 rpm. Determine the speed of shaft *C* in magnitude and direction.

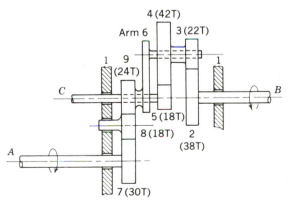

FIGURE 7.43

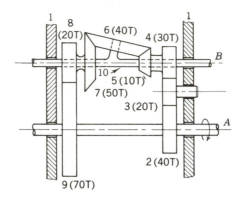

FIGURE 7.44

7.35. For the planetary gear train of Fig. 7.37, calculate the maximum number of planets possible without overlapping and the numbers of equally spaced planets that can be used in the train.

7.36. In a planetary train similar to that of Fig. 7.15, gear 1 has 41 teeth, gear 2 has 18 teeth, and gear 3 has 78 teeth. Gears 1 and 3 are standard and gear 2 is nonstandard. Determine the maximum number of equally spaced planets that can be used.

7.37. Calculate the maximum number of equally spaced compound planets that can be used in the gear train of Fig. 7.36.

7.38. For the planetary gear train shown in Fig. 7.41, calculate the maximum number of compound planets that can be used.

7.39. In the planetary gear train shown in Fig. 7.48, the carrier (link 4) is the driving member and the sun gear (link 3) is the driven member. The internal gear is held stationary. The sun gear is to rotate 2.5 times the speed of the carrier. The pitch diameter of the internal gear is to be approximately 280 mm. (*a*) Design the gear train by determining the numbers of teeth for the internal gear, the sun gear, and the planets using 2.5-module, 20° standard spur gear teeth. Hold the 280-mm pitch diameter as closely as possible. (*b*) Determine whether three equally spaced planets can be used.

7.40. In the planetary gear train shown in Fig. 7.48, the carrier (link 4) is the driving member and the sun gear (link 3) is the driven member. The internal gear is held stationary.

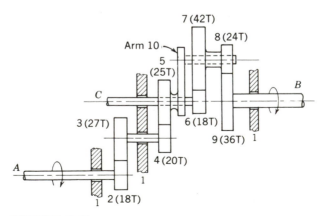

FIGURE 7.45

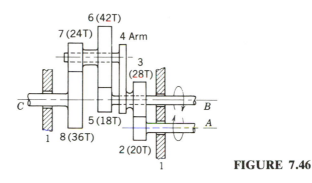

FIGURE 7.46

The sun gear is to rotate 2.5 times the speed of the carrier. The pitch diameter of the internal gear is to be approximately 11.0 in. (*a*) Design the gear train by determining the numbers of teeth for the internal gear, the sun gear, and the planets using 10 diametral-pitch 20° full-depth spur gear teeth. Hold the 11.0-in. pitch diameter as closely as possible. (*b*) Determine whether three equally spaced planets can be used in this drive.

7.41. In the planetary gear train shown in Fig. 7.48, the carrier (link 4) is the driving member and the sun gear (link 3) is the driven member. The internal gear is held stationary. The sun gear is to rotate 2.5 times the speed of the carrier. The pitch diameter of the internal gear is to be approximately 12.5 in. (*a*) Design the gear train by determining the numbers of teeth for the internal gear, the sun gear, and the planets using 8 diametral-pitch, 20° full-depth spur gear teeth. Hold the 12.5-in. pitch diameter as closely as possible. (*b*) Determine whether three equally spaced planets can be used in this drive.

7.42. Design a three-gear planetary train having an output to input speed ratio of 1:8, with the output shaft turning in the same direction as the input. Use the configuration of Fig. 7.49 and indicate which shaft is the input. Select the smallest gears possible from the following available stock sizes: even tooth numbers from 12 to 100 and every fourth number from 100 to 160. Also find the maximum number of planets (gear 2) that could be used.

7.43. Refer to the three-gear planetary train with bevel gears of Fig. 7.50 and find the most conservative design to reduce an input of 125 rpm to 75 rpm. Use the same range of tooth sizes available in Problem 7.42.

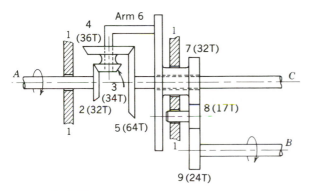

FIGURE 7.47

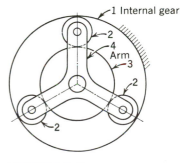

FIGURE 7.48

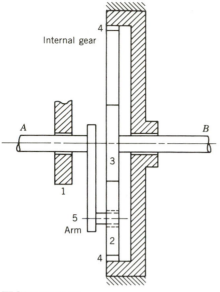

FIGURE 7.49

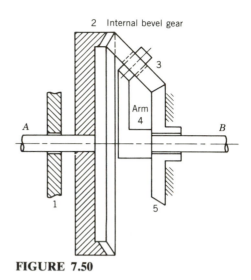

FIGURE 7.50

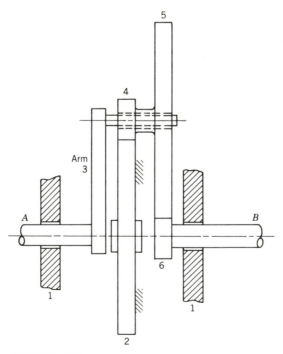

FIGURE 7.51

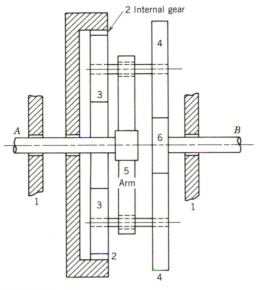

FIGURE 7.52

7.44. Design a planetary gear train with an output to input speed ratio of 1:142, with the output shaft turning opposite the input. Use the three-gear configurations of Fig. 7.49 and Fig. 7.50 or the basic four-gear planetary train shown in Fig. 7.51. Available stock sizes for bevel and spur gears are as follows: all tooth numbers from 12 to 40 and even tooth numbers from 40 to 180. Sketch the selected gear train and indicate the input shaft.

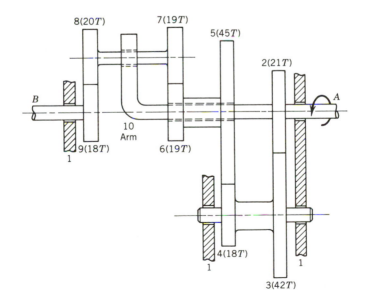

FIGURE 7.53

7.45. Design the smallest possible four-gear planetary train with a fixed annular gear as in Fig. 7.52 to reduce an input of 265 rpm to 15 rpm. Indicate which shaft is chosen as the input. Specifications required gear 2 to have 150 teeth. Available stock sizes are as follows: even tooth numbers from 12 to 40, every four from 40 to 100, and every five from 100 to 150. Also determine if two planet gears are possible as shown.

7.46. In Fig. 7.33, shaft A rotates at 640 rpm in the direction shown and transmits 10 hp to gear 2. Calculate the power circulating in the branch control circuit and make a schematic power flow diagram. Shaft B connected to arm 12 is the output shaft. Gear 2 has 20 teeth and gear 4 has 40 teeth.

7.47. In Fig. 7.34, 5 hp is transmitted to gear 2, which turns at 60 rpm in the direction shown. Determine the power circulating in the branch control circuit and make a schematic power flow diagram. Arm 10 is the output shaft.

7.48. Figure 7.44 shows a planetary train in which shaft A turns at 100 rpm in the direction shown and transmits 20 hp to gear 2. Determine the power circulating in the branch control circuit and make a schematic power flow diagram. Arm 10 is the output shaft.

7.49. In the spur gear differential shown in Fig. 7.53, shaft A turns at 250 rpm in the direction shown and transmits 30 hp. Shaft B is the output shaft. Calculate the power circulating in the branch control circuit and make a schematic power flow diagram.

Chapter Eight

Velocity and Acceleration Analysis

8.1 INTRODUCTION

Because motion is inherent in machinery, kinematic quantities such as velocity and acceleration are of engineering importance in the analysis and design of machine components. Kinematic values in machines have reached extraordinary magnitudes. Rotative speeds, once considered high at 10,000 rpm, are approaching 100,000 rpm. Large rotors of jet engines operate at 10,000 to 15,000 rpm, and small turbine wheels rotate at 30,000 to 60,000 rpm.

Size and rotative speed in rotors are related such that the smaller the size, the greater the allowable rotative speed. A more basic quantity in rotors is peripheral speed, which depends on rotative speed and size ($V = \omega R$). Peripheral speeds in turbomachinery are reaching 50,000 to 100,000 ft/min. Peripheral speeds in electric armatures (10,000 ft/min) and automotive crankshafts (3000 ft/min) are lower than in aeronautical rotors. Although the rotor, or crank, speeds of linkage mechanisms are low, the trend is toward higher speeds because of the demand for higher rates of productivity from the machines used in printing, paper making, thread spinning, automatic computing, packaging, bottling, automatic machining, and numerous other applications.

The centripetal acceleration at a rotor periphery depends on the square of the rotative speed and size ($A^n = \omega^2 R$). In turbines, such accelerations are approaching values of 1 to 3 million ft/s^2, or about $30,000g$ to $100,000g$, values that may be compared with the acceleration of $10g$ withstandable by airplane pilots or the $1000g$ of automotive pistons.

Acceleration is related to force (MA), by Newton's principle, and, in turn, related to stress and deformation, which may or may not be critical in a machine

part, depending on the materials used. The speed of a machine is limited ulti-
mately by the properties of the materials of which it consists and the conditions
that influence the properties of the materials used. High temperature arising from
the compression of gases and the combustion of fuels, together with that arising
from friction, is a condition in high-speed power machines that influences the
strength of the materials. The degree to which the temperature rises also depends
on the provisions made for the transmission of heat by coolants such as air, oil,
water, or Freon.

The successful design of a machine depends on the exploitation of knowledge in the fields of dynamics, stress analysis, thermodynamics, heat transmission,
and properties of materials. However, it is the purpose of this chapter to deal
solely with kinematic relationships in machines. In subsequent chapters, acceleration and force are discussed in connection with the determination of forces
acting on individual links of a mechanism and in connection with machine balance.

For bodies rotating about a fixed axis, such as rotors, kinematic values are
quickly determined from well-known elementary formulas ($V = \omega R$, $A^n = \omega^2 R$,
$A^t = \alpha R$). However, mechanisms such as the slider crank and its inversions are
combinations of links consisting not only of a rotor but of oscillating and reciprocating members as well. Because of the relative velocities and relative accelerations among the several members, together with the many geometrical relative
positions possible, the kinematic analysis of a linkage is relatively complex compared to that of a rotor. The principles and methods illustrated in this chapter
are primarily those for the analysis of linkages consisting of combinations of
rotors, bars, sliders, cams, gears, and rolling elements.

In the following discussions, the individual links of a mechanism are assumed
to be rigid bodies in which the distance between two given particles of a moving
link remains fixed. Links which undergo large deformations during motion, such
as springs, fall in another category and are analyzed as vibrating members. A
current research topic of considerable importance that should also be mentioned
is the study of mechanisms having links which undergo small elastic deformations.[1]

Most elementary mechanisms are in plane motion or may be analyzed as
such. Mechanisms in which all of the particles move in parallel planes are said
to be in *plane motion*. An illustration is a four-bar linkage (Fig. 8.1) consisting
of two rockers and a connecting rod. This arrangement is often referred to as a
double-rocker mechanism.

The motion of a link is expressed in terms of the linear displacements, linear
velocities, and linear accelerations of the individual particles that constitute the
link. However, the motion of a link may also be expressed in terms of angular
displacements, angular velocities, and angular accelerations of lines moving with
the rigid link.

In Fig. 8.1, the linear velocity $\mathbf{V}_A$ and the linear acceleration $\mathbf{A}_A$ of particle
A are shown by the fixed vectors at A. Because of the connecting pin at A,
particle A_2 on link 2 and particle A_3 on link 3 have the same motion, and the

[1]A. Midha, A. G. Erdman, and D. A. Frohrib, "Finite Element Approach to Mathematical Modeling
of High-Speed Elastic Linkages," *Mechanism and Machine Theory*, **13**, 1978, pp. 603–618.

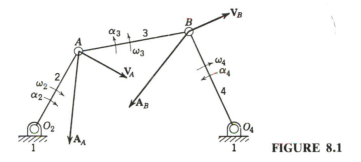

FIGURE 8.1

vectors shown at *A* represent the motions of both particles. The angular motions of links 2 and 3 are different as given by the angular velocities ω_2, ω_3 and the angular accelerations α_2, α_3. Usually, the angular motion of a driving link is known, or assumed, such as ω_2 and α_2 of Fig. 8.1, and the motions of the connecting and driven links are to be determined.

8.2 LINEAR MOTION OF A PARTICLE

In useful mechanisms, the particles of the links are constrained to move on given paths, many of which, such as circles and straight lines, are obvious. In Fig. 8.1, the particles of links 2 and 4 are constrained to move on circular paths. The particles of link 3, however, are in motion along generally curvilinear paths less simple than circles or straight lines.

A particle in motion on a curvilinear path is said to be in *curvilinear translation*. The basic kinematic relationships for a particle translating in a plane are well known from the study of mechanics. These are reviewed in the following paragraphs with reference to Fig. 8.2 and were contributed by Professor J. Y. Harrison, University of New South Wales, Australia.

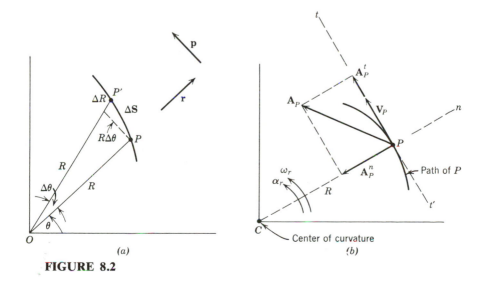

(a)

(b)

FIGURE 8.2

The linear velocity $\mathbf{V}_P$ of a particle P is the instantaneous rate of change of the position of the particle, or displacement, with respect to time. Referring to Fig. 8.2a, in a small interval of time Δt, the particle is displaced $\Delta \mathbf{S}$ along the curved path from position P to position P'. At the same time, the radius vector of the particle changes from $\mathbf{R}$ to $\mathbf{R} + \Delta \mathbf{R}$ and undergoes an angular displacement $\Delta\theta$. Therefore, the displacement $\Delta \mathbf{S}$ is made up of two components: one due to the angular displacement $\Delta\theta$ of radius R and the other due to the change in length ΔR.

From Fig. 8.2a,

$$\Delta \mathbf{S} = R\, \Delta\theta\, \mathbf{p} + \Delta R \mathbf{r}$$

where $\mathbf{p}$ and $\mathbf{r}$ are unit vectors perpendicular and parallel to R, respectively. The equation for the velocity of P can be determined as follows:

$$\mathbf{V}_P = \lim_{\Delta t \to 0} \left(\frac{\Delta S}{\Delta t} \right) = \lim_{\Delta t \to 0} \left(\frac{R\, \Delta\theta}{\Delta t}\, \mathbf{p} + \frac{\Delta R}{\Delta t}\, \mathbf{r} \right)$$

Therefore,

$$\mathbf{V}_P = R\omega_r \mathbf{p} + \frac{dR}{dt}\, \mathbf{r} \qquad \text{where} \qquad \omega_r = \frac{d\theta}{dt}$$

or, by using the vector cross product

$$\mathbf{V}_P = \omega_r \times \mathbf{R} + \frac{dR}{dt}\, \mathbf{r} \tag{8.1}$$

The acceleration of P is given by

$$\mathbf{A}_P = \frac{d}{dt} \left(R\omega_r \mathbf{p} + \frac{dR}{dt}\, \mathbf{r} \right)$$

$$= \frac{dR}{dt}\, \omega_r \mathbf{p} + R\dot{\omega}_r \mathbf{p} - R\omega_r^2 \mathbf{r} + \frac{d^2R}{dt^2}\, \mathbf{r} + \frac{dR}{dt}\, \omega_r \mathbf{p}$$

$$= 2\omega_r \frac{dR}{dt}\, \mathbf{p} + R\dot{\omega}_r \mathbf{p} - R\omega_r^2 \mathbf{r} + \frac{d^2R}{dt^2}\, \mathbf{r} \tag{8.2}$$

The acceleration of P therefore consists of two components, one of magnitude $2\omega_r(dR/dt) + R\dot{\omega}_r$ in the direction of the unit vector $\mathbf{p}$, the other of magnitude $(d^2R/dt^2) - R\omega_r^2$ in the direction of the unit vector $\mathbf{r}$. The equation for $\mathbf{A}_P$ can also be expressed by using the vector cross product as

$$\mathbf{A}_P = 2\omega_r \frac{dR}{dt}\, \mathbf{p} + \alpha_r \times \mathbf{R} + \omega_r \times (\omega_r \times \mathbf{R}) + \frac{d^2R}{dt^2}\, \mathbf{r} \tag{8.2a}$$

where $\alpha_r = \dot{\omega}_r \mathbf{p}$.

When the origin of the coordinate system coincides with the center of curvature, $dR/d\theta$ and $d^2R/d\theta^2$ are zero so that dR/dt and d^2R/dt^2 are also zero. Under this condition, Eq. 8.1 can be simplified to give

$$\mathbf{V}_P = \boldsymbol{\omega}_r \times \mathbf{R} \qquad (8.3)$$

and

$$|\mathbf{V}_P| = R\omega_r$$

From Eq. 8.2a,

$$\mathbf{A}_P = \boldsymbol{\omega}_r \times (\boldsymbol{\omega}_r \times \mathbf{R}) + \boldsymbol{\alpha}_r \times \mathbf{R} \qquad (8.4)$$

The term $\boldsymbol{\omega}_r \times (\boldsymbol{\omega}_r \times \mathbf{R})$ is the normal component of the acceleration with direction from point P toward the center of curvature, and $\boldsymbol{\alpha}_r \times \mathbf{R}$ is the tangential component with direction tangent to the curve at point P. Equation 8.4 may therefore be written as

$$\mathbf{A}_P = \mathbf{A}_P^n + \mathbf{A}_P^t$$

where

$$\mathbf{A}_P^n = \boldsymbol{\omega}_r \times (\boldsymbol{\omega}_r \times \mathbf{R}) \qquad (8.4a)$$

$$|\mathbf{A}_P^n| = R\omega_r^2 = V_P\omega_r = \frac{V_P^2}{R}$$

and

$$\mathbf{A}_P^t = \boldsymbol{\alpha}_r \times \mathbf{R}$$
$$|\mathbf{A}_P^t| = R\alpha_r \qquad (8.4b)$$

If the condition should arise where the origin of the coordinate system is on the normal to the curve through the point, $dR/d\theta$ and therefore dR/dt will be zero, and Eqs. 8.1 and 8.2a can be modified accordingly.

Figure 8.2b shows the instantaneous directional orientation with respect to the tangent and normal to the path for the velocity vector $\mathbf{V}_P$ and the component vectors of acceleration $\mathbf{A}_P^n$ and $\mathbf{A}_P^t$; the radius of curvature is assumed to be constant. It is important to note that the direction of $\mathbf{A}_P^n$ is normal to the path and that its sense is toward the center of curvature C of path. The direction of $\mathbf{A}_P^t$ is tangent to the path and its sense is for increasing velocity. The resultant acceleration $\mathbf{A}_P$ is the vector sum of $\mathbf{A}_P^n$ and $\mathbf{A}_P^t$ as shown.

From Eqs. 8.3, 8.4a, and 8.4b, the magnitudes of the vectors describing the linear motion of a particle can be calculated, and they appear repeatedly in the development of kinematic relationships of particles in mechanisms where the origin of the coordinate system coincides with the center of curvature.

8.3 ANGULAR MOTION

Angular velocity and angular acceleration are the first and second derivatives, respectively, of the angular displacement θ of a line with respect to time t. In machine analysis, the angular motion of a link is expressed by the angular motion of any line visualized fixed to the link. In Fig. 8.3, line AB is in angular motion because of its angular displacement with respect to time. Lines BC and AC undergo the same angular displacements with respect to time as line AB because triangle ABC is fixed in position with link 3 as a rigid body. Since all lines of link 3 have the same angular motion, the angular velocity and angular acceleration of these lines are ω_3 and α_3 of the link, with the subscript denoting the link number.

Angular motion of a link may be the same or different from the angular motions of the radii of curvature of the paths of the individual particles of the link. In Fig. 8.3, since all particles of link 2 are moving on circular paths having a common center of curvature at the fixed center O_2, it is obvious that ω_r and α_r of the radii of curvature of the paths of all particles are equal to the respective angular velocity and angular acceleration ω_2 and α_2 of the link. In the case of the connecting link 3 in Fig. 8.3, which is not rotating about a fixed center, ω_r and α_r of the radius of curvature of the path of any given particle are not the same as ω_3 and α_3 of link 3.

It is an important concept in mechanics that a particle, which has the infinitely small size of a point, can only translate; it cannot rotate. *Angular* motion is motion of a line, and since a particle is a point, not a line, it is not considered to be in angular motion. This concept must be fully understood to understand the *relative* motion among particles. For example, the velocity of the particle on link 2 at O_2 in Fig. 8.3 relative to the velocity of any particle on the fixed link 1 is zero. *Linear* velocity is implied, and it is incorrect to hold that, by virtue of the angular motion of link 2, the particle O_2 has the angular velocity of the link.

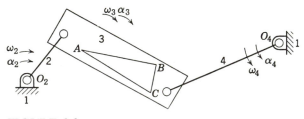

FIGURE 8.3

8.4 RELATIVE MOTION

As will be shown in a later section, the relative motion between particles is very important in the kinematic analysis of mechanisms. In Fig. 8.4a, P and Q are particles moving relative to a fixed reference plane at the respective velocities of $\mathbf{V}_P$ and $\mathbf{V}_Q$, and it is necessary to determine the relative velocity $\mathbf{V}_{PQ}$ between the two particles. In determining $\mathbf{V}_{PQ}$, use will be made of the fact that the addition of equal velocities to each particle does not change the relative velocity of the two particles. Therefore, if P and Q are each given a velocity equal and opposite to $\mathbf{V}_Q$, the particle Q becomes stationary in the fixed plane, and P acquires an additional velocity component $-\mathbf{V}_Q$ relative to the fixed plane. The new absolute velocity of $P(\mathbf{V}_P - \mathbf{V}_Q)$ therefore becomes the relative velocity $\mathbf{V}_{PQ}$ because Q is now fixed relative to the reference plane. This is shown by the vector diagram of Fig. 8.4b from which the equation for $\mathbf{V}_{PQ}$ becomes

$$\mathbf{V}_{PQ} = \mathbf{V}_P - \mathbf{V}_Q \tag{8.5}$$

In a similar manner, $\mathbf{V}_{QP}$ can be obtained by the addition of $-\mathbf{V}_P$ to each particle. This is shown in Fig. 8.4c, and $\mathbf{V}_{QP}$ is given by the equation

$$\mathbf{V}_{QP} = \mathbf{V}_Q - \mathbf{V}_P$$

The vector equation for the acceleration of particle P relative to particle Q is similar in form to Eq. 8.5:

$$\mathbf{A}_{PQ} = \mathbf{A}_P - \mathbf{A}_Q \tag{8.6}$$

The angular motion of a line may be given relative to another line in motion. In Fig. 8.5, the angular velocities ω_2 and ω_3 of the lines on links 2 and 3, respectively, are taken relative to line a–a on the fixed link. If $-\omega_3$ is added to links 2 and 3, link 3 becomes stationary and the new absolute velocity of link 2 ($\omega_2 - \omega_3$) therefore becomes the relative angular velocity ω_{23} because link 3 is now fixed.

Therefore,

$$\omega_{23} = \omega_2 - \omega_3 \tag{8.7}$$

In a similar manner,

$$\alpha_{23} = \alpha_2 - \alpha_3 \tag{8.8}$$

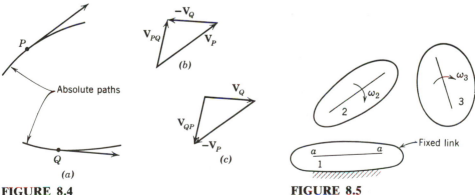

FIGURE 8.4   FIGURE 8.5

8.5 METHODS OF VELOCITY AND ACCELERATION ANALYSIS

Of the many methods of determining velocities and accelerations in mechanisms, three find wide usage. These, which are presented in the following sections, are (*a*) analysis using vector mathematics to express the velocity and acceleration of a point with respect to a moving and a fixed coordinate system; (*b*) analysis using equations of relative motion which are solved either analytically or graphically by velocity and acceleration polygons; and (*c*) analysis using vector loop closure equations written in complex form. In addition, velocities by instant centers will be considered as well as graphical or computer differentiation of displacement–time and velocity–time curves to yield velocities and accelerations, respectively.

Of the methods of velocity and acceleration analysis listed above, the use of either of the first two maintains the physical concept of the problem. However, the third method, using vectors in complex form, tends to become too mechanical in its operation so that the physical aspects of the problem are soon lost. It should also be mentioned that the first and the third methods lend themselves to computer solutions, which is a decided advantage if a mechanism is to be analyzed for a complete cycle.

8.6 VELOCITY AND ACCELERATION ANALYSIS BY VECTOR MATHEMATICS

In Fig. 8.6, the motion of point P is known with respect to the xyz-coordinate system which, in turn, is moving relative to the fixed coordinate system XYZ. The position of point P relative to the XYZ-system can be expressed as

$$\mathbf{R}_P = \mathbf{R}_O + \mathbf{R} \tag{8.9}$$

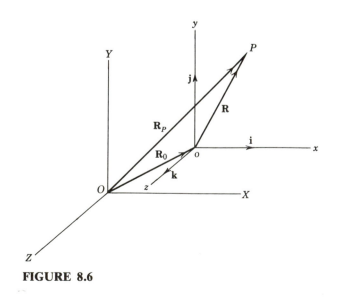

FIGURE 8.6

If unit vectors $\mathbf{i}$, $\mathbf{j}$, and $\mathbf{k}$ are fixed to the x-, y-, and z-axes, respectively,

$$\mathbf{R} = x\mathbf{i} + y\mathbf{j} + z\mathbf{k} \tag{8.10}$$

The absolute velocity of point P relative to the XYZ-system, $\mathbf{V}_P$, may be obtained by differentiating Eq. 8.9 with respect to time to give

$$\mathbf{V}_P = \dot{\mathbf{R}}_P = \dot{\mathbf{R}}_O + \dot{\mathbf{R}} \tag{8.11}$$

Differentiating Eq. 8.10 with respect to time gives

$$\dot{\mathbf{R}} = (\dot{x}\mathbf{i} + \dot{y}\mathbf{j} + \dot{z}\mathbf{k}) + (x\dot{\mathbf{i}} + y\dot{\mathbf{j}} + z\dot{\mathbf{k}}) \tag{8.12}$$

The term $(\dot{x}\mathbf{i} + \dot{y}\mathbf{j} + \dot{z}\mathbf{k})$ is the velocity of point P relative to the moving coordinate system xyz. For convenience, let

$$\dot{x}\mathbf{i} + \dot{y}\mathbf{j} + \dot{z}\mathbf{k} = \mathbf{V} \tag{8.13}$$

Consider next the terms in the second set of parentheses of Eq. 8.12. From the fact that the velocity of the tip of a vector $\mathbf{R}$, which passes through a fixed base point and rotates about the base point with an angular velocity ω, can be shown to be $\omega \times \mathbf{R}$, the velocities of the tips of the unit vectors $\mathbf{i}$, $\mathbf{j}$, and $\mathbf{k}$ can be expressed as

$$\dot{\mathbf{i}} = \omega \times \mathbf{i}$$
$$\dot{\mathbf{j}} = \omega \times \mathbf{j}$$
$$\dot{\mathbf{k}} = \omega \times \mathbf{k}$$

where ω is the angular velocity of the moving coordinate system xyz relative to the fixed system XYZ. Making the above substitutions,

$$x\dot{\mathbf{i}} + y\dot{\mathbf{j}} + z\dot{\mathbf{k}} = x(\omega \times \mathbf{i}) + y(\omega \times \mathbf{j}) + z(\omega \times \mathbf{k}) = \omega \times (x\mathbf{i} + y\mathbf{j} + z\mathbf{k})$$

and using the relation expressed in Eq. 8.10,

$$x\dot{\mathbf{i}} + y\dot{\mathbf{j}} + z\dot{\mathbf{k}} = \omega \times \mathbf{R} \tag{8.14}$$

Equation 8.12 then becomes

$$\dot{\mathbf{R}} = \mathbf{V} + \omega \times \mathbf{R} \tag{8.15}$$

Equation 8.11 can now be rewritten as follows by letting $\mathbf{V}_O = \dot{\mathbf{R}}_O$ and substituting for $\dot{\mathbf{R}}$ from Eq. 8.15:

$$\mathbf{V}_P = \mathbf{V}_O + \mathbf{V} + \omega \times \mathbf{R} \tag{8.16}$$

where

$\mathbf{V}_P$ = velocity of point P in the XYZ-system

$\mathbf{V}_O$ = velocity of origin of xyz-system relative to XYZ-system

$\mathbf{V}$ = velocity of point P relative to xyz-system

$\boldsymbol{\omega}$ = angular velocity of xyz-system relative to XYZ-system

$\mathbf{R}$ = distance from origin of xyz-system to point P

The acceleration of point P relative to the XYZ-system may now be obtained by differentiating Eq. 8.16:

$$\mathbf{A}_P = \dot{\mathbf{V}}_P = \dot{\mathbf{V}}_O + \dot{\mathbf{V}} + \dot{\boldsymbol{\omega}} \times \mathbf{R} + \boldsymbol{\omega} \times \dot{\mathbf{R}} \qquad (8.17)$$

To evaluate $\dot{\mathbf{V}}$, it is necessary to differentiate Eq. 8.13:

$$\dot{\mathbf{V}} = (\ddot{x}\mathbf{i} + \ddot{y}\mathbf{j} + \ddot{z}\mathbf{k}) + (\dot{x}\dot{\mathbf{i}} + \dot{y}\dot{\mathbf{j}} + \dot{z}\dot{\mathbf{k}}) \qquad (8.18)$$

The term $(\ddot{x}\mathbf{i} + \ddot{y}\mathbf{j} + \ddot{z}\mathbf{k})$ is the acceleration of point P relative to the moving coordinate system xyz. Let

$$\ddot{x}\mathbf{i} + \ddot{y}\mathbf{j} + \ddot{z}\mathbf{k} = \mathbf{A} \qquad (8.19)$$

Considering the terms in the second set of parentheses of Eq. 8.18,

$$\dot{x}\dot{\mathbf{i}} + \dot{y}\dot{\mathbf{j}} + \dot{z}\dot{\mathbf{k}} = \dot{x}(\boldsymbol{\omega} \times \mathbf{i}) + \dot{y}(\boldsymbol{\omega} \times \mathbf{j}) + \dot{z}(\boldsymbol{\omega} \times \mathbf{k}) = \boldsymbol{\omega} \times (\dot{x}\mathbf{i} + \dot{y}\mathbf{j} + \dot{z}\mathbf{k})$$

But from Eq. 8.13,

$$\dot{x}\mathbf{i} + \dot{y}\mathbf{j} + \dot{z}\mathbf{k} = \mathbf{V}$$

Therefore,

$$\dot{x}\dot{\mathbf{i}} + \dot{y}\dot{\mathbf{j}} + \dot{z}\dot{\mathbf{k}} = \boldsymbol{\omega} \times \mathbf{V} \qquad (8.20)$$

Equation 8.18 then becomes

$$\dot{\mathbf{V}} = \mathbf{A} + \boldsymbol{\omega} \times \mathbf{V} \qquad (8.21)$$

Also from Eq. 8.15,

$$\boldsymbol{\omega} \times \dot{\mathbf{R}} = \boldsymbol{\omega} \times \mathbf{V} + \boldsymbol{\omega} \times (\boldsymbol{\omega} \times \mathbf{R}) \qquad (8.22)$$

Substituting $\dot{\mathbf{V}}$ from Eq. 8.21 and $\boldsymbol{\omega} \times \dot{\mathbf{R}}$ from Eq. 8.22 into Eq. 8.17 and letting $\mathbf{A}_O = \dot{\mathbf{V}}_O$, the equation for the acceleration of point P relative to the XYZ-system becomes

$$\mathbf{A}_P = \mathbf{A}_O + \mathbf{A} + 2\boldsymbol{\omega} \times \mathbf{V} + \dot{\boldsymbol{\omega}} \times \mathbf{R} + \boldsymbol{\omega} \times (\boldsymbol{\omega} \times \mathbf{R}) \qquad (8.23)$$

where

$2\boldsymbol{\omega} \times \mathbf{V}$ = the Coriolis component of acceleration

$\mathbf{A}_P$ = acceleration of point P in the XYZ-system

$\mathbf{A}_O$ = acceleration of xyz-system origin relative to the XYZ-system

$\mathbf{A}$ = acceleration of point P relative to xyz-system[2]

$\boldsymbol{\omega}$ = angular velocity of xyz-system relative to XYZ-system

$\mathbf{V}$ = velocity of point P relative to xyz-system

$\mathbf{R}$ = distance from xyz-system origin to point P

Example 8.1. Consider the mechanism shown in Fig. 8.7. Link 2 rotates in the direction shown at a constant angular velocity. The velocity and acceleration of point A are therefore known, and it is necessary to find the velocity and acceleration of point B. Select coordinate axes as shown with point O_2 as the origin of the XY-system and point A as the origin of the xy-system.

The equation for the velocity of point B can be written from Eq. 8.16 as follows:

$$\mathbf{V}_B = \mathbf{V}_O + \mathbf{V} + \boldsymbol{\omega} \times \mathbf{R}$$

where

$\mathbf{V}_B$ = direction perpendicular to O_4B, magnitude unknown

$\mathbf{V}_O = |\mathbf{V}_A| = (O_2A)\omega_2 = \frac{3}{12} \times 24 = 6.0$ ft/s, direction perpendicular to O_2A

$\mathbf{V}$ = 0 because B is a fixed point in the xy-system

$\boldsymbol{\omega} \times \mathbf{R}$ = direction perpendicular to $AB(\omega = \omega_3, R = AB)$, magnitude unknown

The direction of $\boldsymbol{\omega} \times \mathbf{R}$ can be determined by knowing that the vector representing $\boldsymbol{\omega}$ will be perpendicular to the xy-plane. When $\boldsymbol{\omega}$ is crossed into $\mathbf{R}$, the product $\boldsymbol{\omega} \times \mathbf{R}$ will

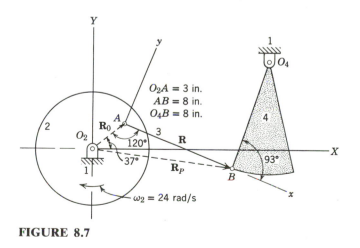

FIGURE 8.7

---

[2]It should be noted that to specify the normal and tangential components of $\mathbf{A}$, the path of point P relative to the xyz-system must be known.

be in the *xy*-plane and perpendicular to **R** by the right-hand rule. This can be shown in Fig. 8.8, where the direction of $\boldsymbol{\omega} \times \mathbf{R}$ is the same regardless of whether $\boldsymbol{\omega}$ is clockwise or counterclockwise. However, the sense of the vector is opposite for the two cases.

The equation for $\mathbf{V}_B$ can be solved graphically by a polygon or analytically by unit vectors. Solution by the latter method follows. *Note that all components have been taken relative to the* xy-*axes.* This problem could also be worked by taking components relative to the *XY*-axes.

$$\mathbf{V}_B = \mathbf{V}_O + \mathbf{V} + \boldsymbol{\omega} \times \mathbf{R}$$

$$\mathbf{V}_B = V_B(\cos 3°\mathbf{i} + \sin 3°\mathbf{j}) = 0.9986V_B\mathbf{i} + 0.0523V_B\mathbf{j}$$

$$\mathbf{V}_O = \mathbf{V}_A = V_A(\cos 30°\mathbf{i} - \sin 30°\mathbf{j}) = 6(0.8660\mathbf{i} - 0.5000\mathbf{j})$$

$$= 5.2\mathbf{i} - 3.0\mathbf{j}$$

$$\mathbf{V} = 0$$

$$\boldsymbol{\omega} \times \mathbf{R} = (\omega \times R)\mathbf{j}$$

Substituting the above relations in the equation for $\mathbf{V}_B$,

$$0.9986V_B\mathbf{i} + 0.0523V_B\mathbf{j} = 5.2\mathbf{i} - 3.0\mathbf{j} + (\omega \times R)\mathbf{j}$$

Summing **i** *components,*

$$0.9986V_B\mathbf{i} = 5.2\mathbf{i}$$

$$V_B = 5.21 \text{ ft/s}$$

Therefore,

$$\mathbf{V}_B = (0.9986)(5.21)\mathbf{i} + (0.0523)(5.21)\mathbf{j} = 5.2\mathbf{i} + 0.271\mathbf{j}$$

Summing **j** *components,*

$$0.0523V_B\mathbf{j} = -3.0\mathbf{j} + (\omega \times R)\mathbf{j}$$

$$(0.0523)(5.21)\mathbf{j} = -3.0\mathbf{j} + (\omega \times R)\mathbf{j}$$

$$(\omega \times R) = 3.271 \text{ ft/s}$$

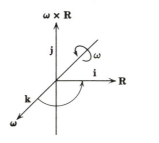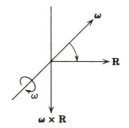

FIGURE 8.8

Therefore,

$$\boldsymbol{\omega} \times \mathbf{R} = 3.271\mathbf{j} \text{ ft/s}$$

and

$$\omega = \omega_3 = \frac{3.271}{R} = \frac{3.271}{\frac{8}{12}} = 4.91 \text{ rad/s} \qquad \text{(ccw)}$$

$$\omega_4 = \frac{V_B}{O_4 B} = \frac{5.21}{\frac{8}{12}} = 7.82 \text{ rad/s} \qquad \text{(ccw)}$$

The equation for the acceleration of point B can be written from Eq. 8.23 as follows:

$$\mathbf{A}_B = \mathbf{A}_O + \mathbf{A} + 2\boldsymbol{\omega} \times \mathbf{V} + \dot{\boldsymbol{\omega}} \times \mathbf{R} + \boldsymbol{\omega} \times (\boldsymbol{\omega} \times \mathbf{R})$$

where

$$\mathbf{A}_B^n = \frac{V_B^2}{O_4 B} = \frac{5.21^2}{\frac{8}{12}} = 40.4 \text{ ft/s}^2, \text{ direction from } B \text{ toward } O_4$$

$\mathbf{A}_B^t =$ direction perpendicular to O_4B, magnitude unknown

$\mathbf{A}_O = |\mathbf{A}_A| = |\mathbf{A}_A^n| = (O_2A)\omega_2^2 = \frac{3}{12} \times 24^2 = 144 \text{ ft/s}^2$, direction from A toward O_2 $\qquad (\mathbf{A}_A^t = 0)$

$\mathbf{A} = 0$ because B is a fixed point in the xy-system

$2\boldsymbol{\omega} \times \mathbf{V} = 0$ because $V = 0$

$\dot{\boldsymbol{\omega}} \times \mathbf{R} =$ direction perpendicular to AB, magnitude unknown

$\boldsymbol{\omega} \times (\boldsymbol{\omega} \times \mathbf{R}) = -\omega^2 R$, direction from B toward A

$\boldsymbol{\omega} = 4.91\mathbf{k} \text{ rad/s}$ from velocity solution

$\omega^2 R = (4.91)^2 \times \frac{8}{12} = 16.1 \text{ ft/s}^2$

The direction of $\dot{\boldsymbol{\omega}} \times \mathbf{R}$ can be determined from the fact that the direction of the vector representing $\dot{\boldsymbol{\omega}}$ will be perpendicular to the xy-plane. When $\dot{\boldsymbol{\omega}}$ is crossed into $\mathbf{R}$, the product $\dot{\boldsymbol{\omega}} \times \mathbf{R}$ will be in the xy-plane and perpendicular to $\mathbf{R}$. The direction of $\boldsymbol{\omega} \times (\boldsymbol{\omega} \times \mathbf{R})$ can be determined from Fig. 8.9, where ω is counterclockwise as determined from the velocity solution.

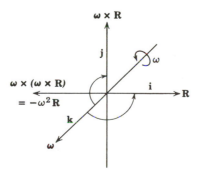

FIGURE 8.9

The equation for A_B is solved by unit vectors in the following manner, again using the xy-coordinate system.

$$\mathbf{A}_B = \mathbf{A}_O + \mathbf{A} + 2\boldsymbol{\omega} \times \mathbf{V} + \dot{\boldsymbol{\omega}} \times \mathbf{R} + \boldsymbol{\omega} \times (\boldsymbol{\omega} \times \mathbf{R})$$

where

$$\mathbf{A}_B^n = A_B^n(-\sin 3°\mathbf{i} + \cos 3°\mathbf{j}) = 40.4(-0.0523\mathbf{i} + 0.9986\mathbf{j}) = -2.1\mathbf{i} + 40.3\mathbf{j}$$

$$\mathbf{A}_B^t = A_B^t(\cos 3°\mathbf{i} + \sin 3°\mathbf{j}) = 0.9986A_B^t\mathbf{i} + 0.0523A_B^t\mathbf{j}$$

$$\mathbf{A}_O = \mathbf{A}_A = A_A^n(-\cos 60°\mathbf{i} - \sin 60°\mathbf{j})$$

$$= 144(-0.500\mathbf{i} - 0.8660\mathbf{j}) = -72\mathbf{i} - 124.8\mathbf{j}$$

$$\mathbf{A} = 0$$

$$2\boldsymbol{\omega} \times \mathbf{V} = 0$$

$$\dot{\boldsymbol{\omega}} \times \mathbf{R} = (\dot{\omega} \times R)\mathbf{j}$$

$$\boldsymbol{\omega} \times (\boldsymbol{\omega} \times \mathbf{R}) = -16.1\mathbf{i} \text{ ft/s}^2$$

Substituting the above relations into the equation for $\mathbf{A}_B$,

$$-2.1\mathbf{i} + 40.3\mathbf{j} + 0.9986A_B^t\mathbf{i} + 0.0523A_B^t\mathbf{j} = -72\mathbf{i} - 124.8\mathbf{j}$$
$$+ (\dot{\omega} \times R)\mathbf{j} - 16.1\mathbf{i}$$

*Summing **i** components,*

$$-2.1\mathbf{i} + 0.9986A_B^t\mathbf{i} = -72\mathbf{i} - 16.1\mathbf{i}$$

$$0.9986A_B^t\mathbf{i} = -86.0\mathbf{i}$$

$$A_B^t = -86.1 \text{ ft/s}^2$$

Therefore,

$$\mathbf{A}_B^t = (0.9986)(-86.1)\mathbf{i} + (0.0523)(-86.1)\mathbf{j} = -86.0\mathbf{i} - 4.5\mathbf{j}$$

*Summing **j** components,*

$$40.3\mathbf{j} + 0.0523A_B^t\mathbf{j} = -124.8\mathbf{j} + (\dot{\omega} \times R)\mathbf{j}$$

$$40.3\mathbf{j} - 4.5\mathbf{j} = -124.8\mathbf{j} + (\dot{\omega} \times R)\mathbf{j}$$

$$(\dot{\omega} \times R) = 160.5 \text{ ft/s}^2$$

Therefore,

$$\dot{\boldsymbol{\omega}} \times \mathbf{R} = 160.5\mathbf{j} \text{ ft/s}^2$$

and

$$\dot{\omega} = \alpha_3 = \frac{160.5}{R} = \frac{160.5}{\frac{8}{12}} = 241 \ \text{rad/s}^2 \qquad (\text{ccw})$$

$$\alpha_4 = \frac{A_B^t}{O_4B} = \frac{86.1}{\frac{8}{12}} = 129 \ \text{rad/s}^2 \qquad (\text{cw})$$

$$\mathbf{A}_B = \mathbf{A}_B^n + \mathbf{A}_B^t = -2.1\mathbf{i} + 40.3\mathbf{j} - 86.0\mathbf{i} - 4.5\mathbf{j} = -88.1\mathbf{i} + 35.8\mathbf{j}$$

$$A_B = |\mathbf{A}_B| = \sqrt{88.1^2 + 35.8^2} = 95.1 \ \text{ft/s}^2$$

To obtain a better understanding of the vectors involved in the velocity and acceleration analysis of the linkage of Example 8.1, a graphical solution of the vector equations will be given. It is often helpful to make a rough sketch of the vector polygons as an aid in visualizing and checking the analytical solution. The linkage of Fig. 8.7 is therefore redrawn in Fig. 8.10a, and polygons are shown which give the magnitudes and directions of the vectors which were previously determined analytically. Figure 8.10b shows the graphical representation of the velocity equation

$$\mathbf{V}_B = \mathbf{V}_O + \mathbf{V} + \boldsymbol{\omega} \times \mathbf{R}$$

where
$$\mathbf{V}_O = \mathbf{V}_A$$
$$\mathbf{V} = 0$$

Therefore,

$$\mathbf{V}_B = \mathbf{V}_A + \boldsymbol{\omega} \times \mathbf{R}$$

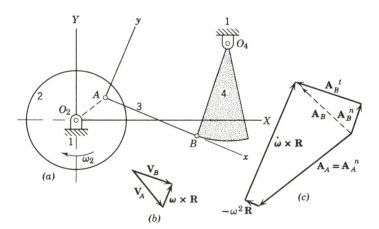

(a)

(b)

(c)

FIGURE 8.10

The addition of the vectors $\mathbf{V}_A$ and $\boldsymbol{\omega} \times \mathbf{R}$ to give $\mathbf{V}_B$ can easily be seen in the polygon of Fig. 8.10b.

Figure 8.10c shows the graphical representation of the acceleration equation

$$\mathbf{A}_B = \mathbf{A}_O + \mathbf{A} + 2\boldsymbol{\omega} \times \mathbf{V} + \dot{\boldsymbol{\omega}} \times \mathbf{R} + \boldsymbol{\omega} \times (\boldsymbol{\omega} \times \mathbf{R})$$

where

$$\mathbf{A}_B = \mathbf{A}_B^n + \mathbf{A}_B^t$$
$$\mathbf{A}_O = \mathbf{A}_A = \mathbf{A}_A^n \qquad (\mathbf{A}_A^t = 0)$$
$$\mathbf{A} = 0$$
$$2\boldsymbol{\omega} \times \mathbf{V} = 0$$
$$\boldsymbol{\omega} \times (\boldsymbol{\omega} \times \mathbf{R}) = -\omega^2 \mathbf{R}$$

which results in

$$\mathbf{A}_B^n + \mathbf{A}_B^t = \mathbf{A}_A + \dot{\boldsymbol{\omega}} \times \mathbf{R} - \omega^2 \mathbf{R}$$

The addition of these vectors can easily be seen in the polygon of Fig. 8.10c.

In comparing the analytical solution with the graphical solution, it is obvious that the graphical is much quicker but less accurate. If the analysis of only one position is required, one would undoubtedly choose the graphical solution. If, however, the analysis of several positions or of a complete cycle is necessary, the analytical solution would be preferred possibly with the aid of a computer.

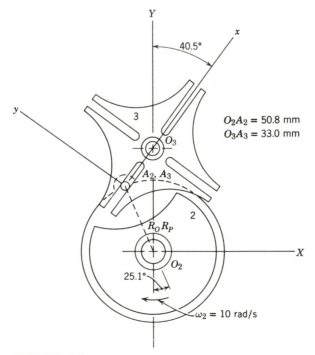

$O_2A_2 = 50.8$ mm
$O_3A_3 = 33.0$ mm

$\omega_2 = 10$ rad/s

FIGURE 8.11

Example 8.2. As a second example, consider the mechanism shown in Fig. 8.11, where the angular velocity of link 2 is constant, and it is required to find the angular velocity and angular acceleration of link 3. The coordinate system xy is fixed in link 3 as shown with its origin at point A_3. The system XY has its origin at point O_2.

The equation for the velocity of point A_3 cannot be evaluated directly from Eq. 8.16 because, by placing the origin of the xy-system at point A_3, $\mathbf{V}_O$ equals $\mathbf{V}_{A_3}$, and an identity results. It is therefore necessary to write Eq. 8.16 for $\mathbf{V}_{A_2}$ as follows:

$$\mathbf{V}_{A_2} = \mathbf{V}_O + \mathbf{V} + \omega_3 \times \mathbf{R}$$

where

$$\mathbf{V}_{A_2} = (O_2A_2)\omega_2 = 50.8 \times 10 = 508 \text{ mm/s, direction perpendicular to } O_2A$$

$$\mathbf{V}_O = \mathbf{V}_{A_3} = \text{direction perpendicular to } O_3A_3, \text{ magnitude unknown}$$

$$\mathbf{V} = \text{direction parallel to } O_3A_3, \text{ magnitude unknown}$$

$$\omega_3 \times \mathbf{R} = 0 \text{ because } R = 0$$

The equation for $\mathbf{V}_{A_2}$ is solved by unit vectors, with all components taken relative to the xy-axes; ω_3 is calculated from $|\mathbf{V}_{A_3}|$.

$$\mathbf{V}_{A_2} = \mathbf{V}_O + \mathbf{V} + \omega \times \mathbf{R}$$

$$\mathbf{V}_{A_2} = V_{A_2}(\cos 24.4\mathbf{i} - \sin 24.4\mathbf{j}) = 508(0.9107\mathbf{i} - 0.4131\mathbf{j})$$

$$= 462.6\mathbf{i} - 209.9\mathbf{j}$$

$$\mathbf{V}_O = \mathbf{V}_{A_3} = -V_{A_3}\mathbf{j}$$

$$\mathbf{V} = V\mathbf{i}$$

$$\omega \times \mathbf{R} = 0$$

Substituting the above relations in the equation for $\mathbf{V}_{A_2}$,

$$462.6\mathbf{i} - 209.9\mathbf{j} = -V_{A_3}\mathbf{j} + V\mathbf{i}$$

Summing **i** *components,*

$$462.6\mathbf{i} = V\mathbf{i}$$

$$V = 462.6 \text{ mm/s}$$

Therefore,

$$\mathbf{V} = 462.6\mathbf{i} \text{ mm/s}$$

Summing **j** *components,*

$$-209.9\mathbf{j} = -V_{A_3}\mathbf{j}$$

$$V_{A_3} = 209.9 \text{ mm/s}$$

Therefore,

$$\mathbf{V}_{A_3} = -209.9\mathbf{j} \text{ mm/s}$$

$$\omega_3 = \frac{V_{A_3}}{O_3A_3} = \frac{209.9}{33.0} = 6.36 \text{ rad/s} \qquad \text{(ccw)}$$

The equation for the acceleration $\mathbf{A}_{A_2}$ can be written from Eq. 8.23 as follows:

$$\mathbf{A}_{A_2} = \mathbf{A}_O + \mathbf{A} + 2\boldsymbol{\omega} \times \mathbf{V} + \dot{\boldsymbol{\omega}} \times \mathbf{R} + \boldsymbol{\omega} \times (\boldsymbol{\omega} \times \mathbf{R})$$

where

$$\mathbf{A}_{A_2}^n = (O_2A_2)\omega_2^2 = 50.8 \times 10^2 = 5080 \text{ mm/s}^2, \text{ direction from } A_2 \text{ toward } O_2$$

$$\mathbf{A}_{A_2}^t = 0$$

$$\mathbf{A}_O = \mathbf{A}_{A_3}$$

$$\mathbf{A}_{A_3}^n = \frac{V_{A_3}^2}{O_3A_3} = \frac{209.9^2}{33.0} = 1335 \text{ mm/s}^2, \text{ direction from } A_3 \text{ toward } O_3$$

$$\mathbf{A}_{A_3}^t = \text{ direction perpendicular to } O_3A_3, \text{ magnitude unknown}$$

$$\mathbf{A}^n = 0 \text{ because radius of curvature is infinite (the path of point } A_2 \text{ relative to the } xy\text{-system is a straight line along the centerline of the slot)}$$

$$\mathbf{A}^t = \text{ direction parallel to } O_3A_3, \text{ magnitude unknown}$$

$$2\boldsymbol{\omega} \times \mathbf{V} = 2 \times 6.36 \times 462.6 = 5884 \text{ mm/s}^2 \text{ direction along positive } y\text{-axis (see Fig. 8.12)}$$

$$\dot{\boldsymbol{\omega}} \times \mathbf{R} = 0 \text{ because } R = 0$$

$$\boldsymbol{\omega} \times (\boldsymbol{\omega} \times \mathbf{R}) = 0 \text{ because } R = 0$$

The equation for $\mathbf{A}_{A_2}$ is solved by unit vectors and α_3 is calculated from $|\mathbf{A}_{A_3}^t|$ in the following manner.

$$\mathbf{A}_{A_2} = \mathbf{A}_O + \mathbf{A} + 2\boldsymbol{\omega} \times \mathbf{V} + \dot{\boldsymbol{\omega}} \times \mathbf{R} + \boldsymbol{\omega} \times (\boldsymbol{\omega} \times \mathbf{R})$$

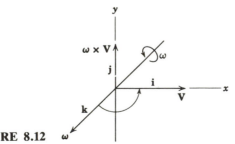

FIGURE 8.12

where

$$\mathbf{A}^n_{A_2} = A^n_{A_2}(-\sin 24.4\mathbf{i} - \cos 24.4\mathbf{j}) = 5080(-0.4131\mathbf{i} - 0.9107\mathbf{j}) = -2099\mathbf{i} - 4626\mathbf{j}$$

$$\mathbf{A}^t_{A_2} = 0$$

$$\mathbf{A}_O = \mathbf{A}_{A_3}$$

$$\mathbf{A}^n_{A_3} = 1335\mathbf{i} \text{ mm/s}^2$$

$$\mathbf{A}^t_{A_3} = A^t_{A_3}\mathbf{j} \text{ (assume as positive)}$$

$$\mathbf{A}^n = 0$$

$$\mathbf{A}^t = A^t\mathbf{i}$$

$$2\boldsymbol{\omega} \times \mathbf{V} = 5884\mathbf{j} \text{ mm/s}^2$$

$$\dot{\boldsymbol{\omega}} \times \mathbf{R} = 0$$

$$\boldsymbol{\omega} \times (\boldsymbol{\omega} \times \mathbf{R}) = 0$$

Substituting the above relations in the equation for $\mathbf{A}_{A_2}$,

$$-2099\mathbf{i} - 4626\mathbf{j} = 1335\mathbf{i} + A^t_{A_3}\mathbf{j} + A^t\mathbf{i} + 5884\mathbf{j}$$

Summing **i** *components,*

$$-2099\mathbf{i} = 1335\mathbf{i} + A^t\mathbf{i}$$

$$A^t = -3434 \text{ mm/s}^2$$

Therefore,

$$\mathbf{A}^t = -3434\mathbf{i} \text{ mm/s}^2$$

Summing **j** *components,*

$$-4626\mathbf{j} = A^t_{A_3}\mathbf{j} + 5884\mathbf{j}$$

$$A^t_{A_3} = -10{,}510\mathbf{j} \text{ mm/s}^2$$

Therefore,

$$\mathbf{A}^t_{A_3} = -10{,}510\mathbf{j} \text{ mm/s}^2$$

$$\mathbf{A}_{A_3} = \mathbf{A}^n_{A_3} + \mathbf{A}^t_{A_3} = 1335\mathbf{i} - 10{,}510\mathbf{j}$$

$$|\mathbf{A}_{A_3}| = \sqrt{1335^2 + 10{,}510^2} = 10{,}590 \text{ mm/s}^2$$

$$\alpha_3 = \frac{A^t_{A_3}}{O_3 A_3} = \frac{10{,}510}{33.0} = 318 \text{ rad/s}^2 \quad \text{(ccw)}$$

It should be mentioned that the origin of the xy-system was taken at point A_3 with point A_2 as P because the path of point A_2 relative to point A_3 (and hence the xy-system) is a straight line. If the origin of the xy-system had been taken at point A_2 with point A_3 as P, the solution would have been more difficult because the path of A_3 relative to A_2 is not readily known.

To present the graphical solution of Example 8.2, the linkage of Fig. 8.11 is redrawn in Fig. 8.13a. Figure 8.13b shows the graphical representation of the velocity equation

$$\mathbf{V}_{A_2} = \mathbf{V}_O + \mathbf{V} + \boldsymbol{\omega} \times \mathbf{R}$$

where

$$\mathbf{V}_O = \mathbf{V}_{A_3}$$

$$\boldsymbol{\omega} \times \mathbf{R} = 0$$

Therefore,

$$\mathbf{V}_{A_2} = \mathbf{V}_{A_3} + \mathbf{V}$$

The addition of the vectors $\mathbf{V}_{A_3}$ and $\mathbf{V}$ to give $\mathbf{V}_{A_2}$ can easily be seen in the polygon of Fig. 8.13b. The value of $\boldsymbol{\omega}_3$ is calculated in the same manner as in the analytical solution, namely,

$$\omega_3 = \frac{V_{A_3}}{O_3 A_3} = 6.35 \text{ rad/s} \qquad \text{(ccw)}$$

Figure 8.13c shows the graphical representation of the acceleration equation

$$\mathbf{A}_{A_2} = \mathbf{A}_O + \mathbf{A} + 2\boldsymbol{\omega} \times \mathbf{V} + \dot{\boldsymbol{\omega}} \times \mathbf{R} + \boldsymbol{\omega} \times (\boldsymbol{\omega} \times \mathbf{R})$$

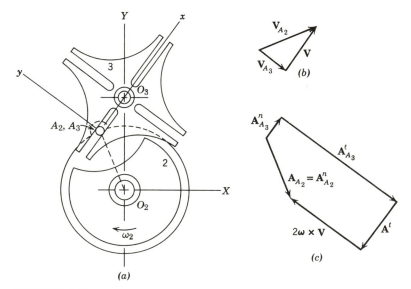

(a)

(b)

(c)

FIGURE 8.13

where

$$\mathbf{A}_{A_2} = \mathbf{A}_{A_2}^n \quad (\mathbf{A}_{A_2}^t = 0)$$

$$\mathbf{A}_O = \mathbf{A}_{A_3} = \mathbf{A}_{A_3}^n + \mathbf{A}_{A_3}^t$$

$$\mathbf{A} = \mathbf{A}^t \quad (\mathbf{A}^n = 0)$$

$$\dot{\boldsymbol{\omega}} \times \mathbf{R} = 0$$

$$\boldsymbol{\omega} \times (\boldsymbol{\omega} \times \mathbf{R}) = 0$$

Therefore,

$$\mathbf{A}_{A_2}^n = \mathbf{A}_{A_3}^n + \mathbf{A}_{A_3}^t + \mathbf{A}^t + 2\boldsymbol{\omega} \times \mathbf{V}$$

The addition of the four vectors to give $\mathbf{A}_{A_2}$ can easily be seen in the polygon of Fig. 8.13c. The value of α_3 is calculated in the same manner as in the analytical solution, namely,

$$\alpha_3 = \frac{A_{A_3}^t}{O_3 A_3} = 318 \text{ rad/s}^2 \quad \text{(ccw)}$$

8.7 DETERMINATION OF VELOCITY IN MECHANISMS BY VECTOR POLYGONS

Vector polygons are a convenient tool for determining velocity in mechanisms. These polygons may be solved graphically, analytically, or by some combination of the two. Graphical methods may be used to determine the linear velocities of all particles of a mechanism quickly with relatively little calculation, as illustrated in several examples that follow. However, a fundamental insight of the relative motion of the particles in the mechanism is needed.

In Fig. 8.14 are shown three types of linkages in which the driving link (link 2) is the same but in which the motion transmitted to the driven link depends on a different type of constraint. In Fig. 8.14a, motion constraint is achieved through pin connections; in 8.14b, by sliding in a guide; and in 8.14c, by rolling contact. The absolute velocity of any particle on link 2 is quickly determined if

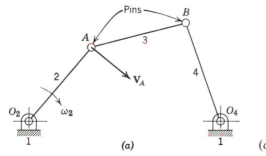

FIGURE 8.14
(*continued next page*)

(*a*)

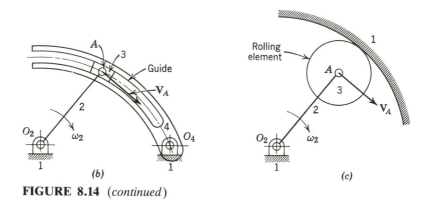

FIGURE 8.14 (*continued*)

the driving angular velocity ω_2 is known. The magnitude of $\mathbf{V}_A$, for example, may be calculated from Eq. 8.1a:

$$|\mathbf{V}_A| = R\omega_r$$
$$= (O_2A)\omega_2$$

The direction of $\mathbf{V}_A$ is known to be tangent to the circular path of A, and the sense of $\mathbf{V}_A$ is known from the sense of ω_2. However, to determine the linear velocity of any particle on the driven links or followers, a knowledge of the relative motion of pairs of particles is required.

8.8 RELATIVE VELOCITY OF PARTICLES IN MECHANISMS

Referring to Eq. 8.5 and Fig. 8.4, the relative velocity $\mathbf{V}_{PQ}$ of one particle relative to another may be determined from the vector difference of the absolute velocities $\mathbf{V}_P$ and $\mathbf{V}_Q$ provided the absolute velocities are known. However, in a linkage analysis, only one of the absolute velocities is usually known and the other is to be determined. The unknown absolute velocity $\mathbf{V}_P$, for example, may be determined from Eq. 8.5 in the following form:

$$\mathbf{V}_P = \mathbf{V}_Q + \mathbf{V}_{PQ} \tag{8.24}$$

Although $\mathbf{V}_Q$ may be known, it is necessary that the relative velocity $\mathbf{V}_{PQ}$ also be known. In linkages, the motions of particles P and Q are not independent as in Fig. 8.4 but are constrained relative to each other so that their relative motion is controlled. In the following section, the basic types of motion constraint are discussed to show the determination of the magnitude, direction, and sense of $\mathbf{V}_{PQ}$.

312

8.9 RELATIVE VELOCITY OF PARTICLES IN A COMMON LINK

Considering the rigid body (link 3) in Fig. 8.15a, any particle such as Q may be at the absolute velocity $\mathbf{V}_Q$ and the link at an absolute angular velocity ω_3. If observations are made relative to Q, then Q is at rest as shown in Fig. 8.15b. However, since particle Q has no angular motion, the angular velocity ω_3 of the link relative to Q is unchanged. Therefore, as in Fig. 8.15b, relative to Q, the link rotates at the absolute angular velocity ω_3 about Q as if Q were a fixed center.

Relative to Q, any other particle in the link such as P is constrained to move on a circular path as shown in Fig. 8.15c because the link is a rigid body and the distance PQ is fixed. The relative velocity $\mathbf{V}_{PQ}$ of P relative to Q is tangent to the relative path as shown. Since the radius of curvature R of the relative path is equal to PQ and the angular velocity of the radius of curvature ω_r is equal to ω_3, the magnitude of $\mathbf{V}_{PQ}$ may be determined from Eq. 8.1a as follows:

$$|\mathbf{V}_{PQ}| = (PQ)\omega_3 \qquad \textbf{(8.25)}$$

Because the relative path is circular, dR/dt is zero.

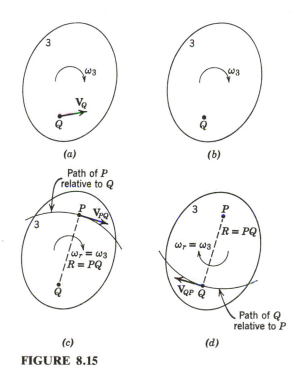

(a) (b) (c) (d)

FIGURE 8.15

It is to be observed from Eq. 8.25 that if the link has no absolute angular velocity, the relative velocity $\mathbf{V}_{PQ}$ of any two particles of the link is zero. The link is then in pure translation, and the absolute velocities of all particles of the link are identical.

In Fig. 8.15c, the direction of $\mathbf{V}_{PQ}$ is tangent to the relative circular path and is shown as a fixed vector at P. The sense of $\mathbf{V}_{PQ}$ is determined by making its turning sense about Q the same as the sense of ω_3. In Fig. 8.15d is shown the vector $\mathbf{V}_{QP}$ denoting the velocity of Q relative to P. It may be seen that relative to P, the angular velocity ω_3 of link 3 is the same in magnitude and sense as relative to Q. Therefore, the magnitudes of $\mathbf{V}_{QP}$ and $\mathbf{V}_{PQ}$ are the same. Their directions are also the same since both are normal to the line PQ. However, the sense of $\mathbf{V}_{QP}$ is opposite to that of $\mathbf{V}_{PQ}$.

As illustrated in the following example, Eqs. 8.24 and 8.25 and the knowledge of the direction and sense of the relative velocity of two particles in a given link are necessary in the kinematic analysis of mechanisms.

Example 8.3. Link 2 of the four-bar linkage of Fig. 8.16a is the driving link having a uniform angular velocity ω_2 of 30 rad/s. For the phase shown, draw the velocity polygon

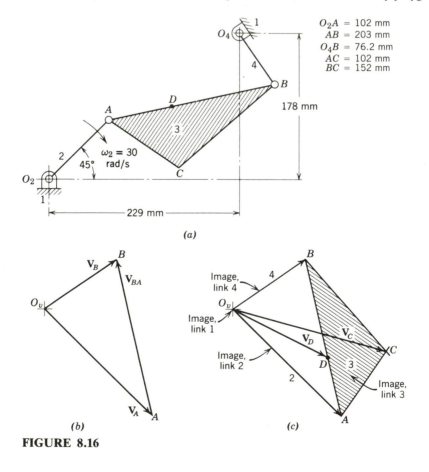

O_2A	= 102 mm
AB	= 203 mm
O_4B	= 76.2 mm
AC	= 102 mm
BC	= 152 mm

FIGURE 8.16

and determine the velocity $\mathbf{V}_B$ of point B, the angular velocities ω_3 and ω_4, and the relative angular velocities ω_{32} and ω_{43}. Also determine the velocity images of all links to show how the linear velocity of any point in the linkage may be determined. Since each vector has a magnitude m and a direction d, we can conveniently tabulate the knowns and unknowns in the vector equation. One vector equation can be solved for two scalar unknowns. Velocity equations can be written as follows.

I. $\mathbf{V}_B = \mathbf{V}_A + \mathbf{V}_{BA}$

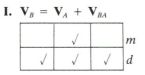

where

$\quad\quad \mathbf{V}_B$ = direction perpendicular to O_4B, magnitude unknown

$\quad\quad \mathbf{V}_A = (O_2A)\omega_2 = (102)30 = 3060$ mm/s, direction perpendicular to O_2A

$\quad\quad \mathbf{V}_{BA}$ = direction perpendicular to BA, magnitude unknown

Measured from the polygon, $V_B = 1800$ mm/s and $V_{BA} = 3180$ mm/s.

$$\omega_3 = \frac{V_{BA}}{BA} = \frac{3180}{203} = 15.7 \text{ rad/s} \quad\quad (\text{ccw})$$

$$\omega_4 = \frac{V_B}{O_4B} = \frac{1800}{76.2} = 23.6 \text{ rad/s} \quad\quad (\text{ccw})$$

$$\omega_{32} = \omega_3 - \omega_2 = 15.7 - (-30) = 45.7 \text{ rad/s} \quad\quad (\text{ccw})$$

$$\omega_{43} = \omega_4 - \omega_3 = 23.6 - 15.7 = 7.9 \text{ rad/s} \quad\quad (\text{ccw})$$

II. $\mathbf{V}_C = \mathbf{V}_A + \mathbf{V}_{CA}$

III. $\mathbf{V}_C = \mathbf{V}_B + \mathbf{V}_{CB}$

where

$\quad\quad \mathbf{V}_C$ = direction unknown, magnitude unknown

$\quad\quad \mathbf{V}_{CA}$ = direction perpendicular to CA, magnitude unknown

$\quad\quad \mathbf{V}_{CB}$ = direction perpendicular to CB, magnitude unknown

Measured from the polygon, $V_C = 3050$ mm/s, $V_{CA} = 1600$ mm/s, and $V_{CB} = 2390$ mm/s.

Equation I expresses $\mathbf{V}_B$ in terms of $\mathbf{V}_A$ and $\mathbf{V}_{BA}$. As indicated, the components $\mathbf{V}_B$ and $\mathbf{V}_{BA}$ are known only in direction, while $\mathbf{V}_A$ is known in magnitude, sense, and direction. In constructing the velocity polygon Fig. 8.16*b* starting with the right side of Eq. I, the vector $\mathbf{V}_A$ is drawn from pole O_v, and its tip is labeled "A." Next, add the direction of $\mathbf{V}_{BA}$ starting at point A. As can be seen, it is impossible to complete the solution using only these two components. Therefore, consider the left side of the equation and draw the direction of $\mathbf{V}_B$ from O_v. The intersection of the direction of $\mathbf{V}_B$ and the direction of $\mathbf{V}_{BA}$ completes the polygon. Arrowheads are now added to the vectors $\mathbf{V}_B$ and $\mathbf{V}_{BA}$ so that the addition of the vectors of the polygon agrees with the addition of the terms of Eq. I. The tip of the vector $\mathbf{V}_B$ is labeled "B."

The magnitudes and senses of ω_3 and ω_4 can now be determined from V_{BA} and V_B, respectively, as shown. The values of ω_{32} and ω_{43} can also be determined as indicated.

To determine $\mathbf{V}_C$, it is necessary to use Eqs. II and III, which give the relations between $\mathbf{V}_C$ and $\mathbf{V}_A$ and $\mathbf{V}_B$. The directions of $\mathbf{V}_{CA}$ and $\mathbf{V}_{CB}$ are known as indicated. The velocity vectors $\mathbf{V}_A$ and $\mathbf{V}_B$ are redrawn in Fig. 8.16c to give a clearer diagram. Use Eq. II and draw the direction of the vector $\mathbf{V}_{CA}$ from point A in Fig. 8.16c. Next, consider Eq. III and draw the direction of the vector $\mathbf{V}_{CB}$ from point B. The intersection of the direction of $\mathbf{V}_{CA}$ and the direction of $\mathbf{V}_{CB}$ completes the polygon. This intersection is point C, which gives $\mathbf{V}_C$. The vector addition in the polygon is checked to see that it agrees with that of Eqs. II and III.

The shaded triangle ABC of Fig. 8.16c is known as the velocity image of link 3, and as such has the same shape as link 3. The velocity of any point D as shown on link 3 can be determined by locating its corresponding position on the velocity image of link 3. The vector from O_v to D is $\mathbf{V}_D$ (2080 mm/s) as shown in Fig. 8.16c. The velocity image of link 1 is at the pole O_v because link 1 is fixed and has zero velocity. The images of links 2 and 4 are lines O_vA and O_vB, respectively, which correspond to O_2A and O_4B, respectively, in the configuration diagram.

In the analysis above, the angular velocity ω_3 was determined from the relation

$$\omega_3 = \frac{V_{BA}}{BA}$$

It should also be mentioned that after the velocity image of link 3 has been completed, ω_3 can also be found from

$$\omega_3 = \frac{V_{CA}}{CA} = \frac{V_{CB}}{CB} = \frac{V_{DA}}{DA}$$

In other words, all relative velocities of points on a link are proportional to the distances between these points.

8.10 RELATIVE VELOCITY OF COINCIDENT PARTICLES ON SEPARATE LINKS

In many mechanisms such as in Fig. 8.14b, constraint of relative motion is achieved by guiding a particle A on one link along a prescribed path relative to another link by a guiding surface. Such constraint is to be found in cams and the inversions of the slider crank, where a surface on one link controls the motion of a particle on another link by relative sliding or rolling.

In Fig. 8.17, particle P_3 on link 3 is in motion along a curvilinear path traced on link 2 because of the guiding slot in link 2. The path of P_3 relative to link 2 is shown with tangent t–t and normal n–n constructed at P_3. Consider a particle Q_2 on link 2 which is coincident in position with particle P_3 on link 3. It may be seen that regardless of the absolute angular velocities ω_2 and ω_3 of links 2 and 3, the guide constrains the motion of P_3 so that it cannot displace relative to Q_2 in the normal direction n–n, and therefore there cannot be a relative velocity of the two particles in this direction. However, the guide permits freedom for particle

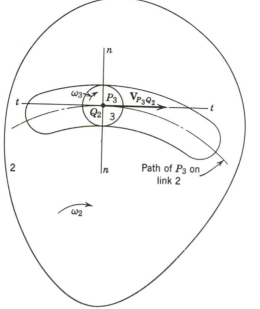

FIGURE 8.17

P_3 to displace relative to Q_2 in the tangential direction t–t, and therefore the relative velocity $\mathbf{V}_{P_3Q_2}$ can be only in the tangential direction of the guide.

In mechanisms where guiding constraint is utilized, the knowledge that the relative velocity of coincident particles can be only in the tangential direction of the guide is sufficient to solve velocity problems as illustrated in the following example.

Example 8.4. The disk cam of Fig. 8.18*a* drives an oscillating roller follower and a radial point follower simultaneously. The cam rotates counterclockwise at a constant angular velocity ω_2 of 10 rad/s. Springs (not shown) are used to maintain contact of the

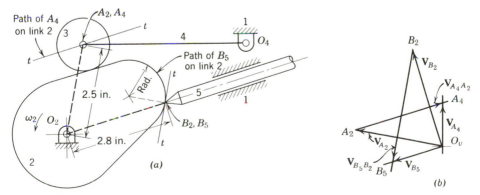

FIGURE 8.18

followers with the cam. For the phase shown, determine the velocity $\mathbf{V}_{A_4}$ of point A_4 on the oscillating follower and the velocity $\mathbf{V}_{B_5}$ of point B_5 on the point follower. Velocity equations can be written as follows:

I. $\mathbf{V}_{A_4} = \mathbf{V}_{A_2} + \mathbf{V}_{A_4A_2}$

where

$\quad\quad \mathbf{V}_{A_4}$ = direction perpendicular to O_4A_4, magnitude unknown

$\quad\quad \mathbf{V}_{A_2} = (O_2A_2)\omega_2 = (2.5)10 = 25$ in./s, direction perpendicular to O_2A_2

$\quad\quad \mathbf{V}_{A_4A_2}$ = direction parallel to straight side of cam, magnitude unknown

Measured on the polygon of Fig. 8.18b, $V_{A_4} = 12.3$ in./s and $V_{A_4A_2} = 26.3$ in./s.

II. $\mathbf{V}_{B_5} = \mathbf{V}_{B_2} + \mathbf{V}_{B_5B_2}$

where

$\quad\quad \mathbf{V}_{B_5}$ = direction along centerline of follower 5, magnitude unknown

$\quad\quad \mathbf{V}_{B_2} = (O_2B_2)\omega_2 = (2.8)10 = 28$ in./s, direction perpendicular to O_2B_2

$\quad\quad \mathbf{V}_{B_5B_2}$ = direction along tangent to cam contour at point B_2, magnitude unknown

Measured on the polygon of Fig. 8.18b, $V_{B_5} = 14.7$ in./s and $V_{B_5B_2} = 31.6$ in./s.

Considering first the oscillating follower link 4, it can be seen that the straight side of the cam 2 is a guiding surface which constrains point A_4 on link 4 to follow a straight-line path relative to link 2. Point A_2 on link 2 and point A_4 on link 4 are coincident, and Eq. I shows the relation of their velocities. As indicated, the components $\mathbf{V}_{A_4}$ and $\mathbf{V}_{A_4A_2}$ are known in direction while $\mathbf{V}_{A_2}$ is known in magnitude, sense, and direction.

The construction of the velocity polygon of Fig. 8.18b is started with the right side of Eq. I, and the vector $\mathbf{V}_{A_2}$ is drawn from pole O_v with its tip labeled "A_2." Next, add the direction of $\mathbf{V}_{A_4A_2}$ starting at point A_2. Because it is impossible to complete the solution using only these two components, consider the left side of the equation and draw the direction of $\mathbf{V}_{A_4}$ from O_v. The intersection of the direction of $\mathbf{V}_{A_4}$ and the direction of $\mathbf{V}_{A_4A_2}$ completes the polygon. Arrowheads are now added to the vectors $\mathbf{V}_{A_4}$ and $\mathbf{V}_{A_4A_2}$ so that the addition of the vectors of the polygon agrees with the addition of the terms of Eq. I. The tip of vector $\mathbf{V}_{A_4}$ is labeled "A_4."

The velocity polygon for the determination of the velocity $\mathbf{V}_{B_5}$ of the point follower can be drawn in a similar manner from Eq. II. Points B_2 and B_5 are coincident, and, as indicated, the components $\mathbf{V}_{B_5}$ and $\mathbf{V}_{B_5B_2}$ are known in direction while $\mathbf{V}_{B_2}$ is known in magnitude, sense, and direction. Figure 8.18b shows the polygon drawn from the same pole point O_v as was the polygon for the determination of $\mathbf{V}_{A_4}$.

8.11 RELATIVE VELOCITY OF COINCIDENT PARTICLES AT THE POINT OF CONTACT OF ROLLING ELEMENTS

A third type of constraint in mechanisms is that which occurs because one link is constrained to roll on another link without slipping at the point of contact. In Fig. 8.19 are shown the rolling pitch circles of a pair of gears in mesh with the

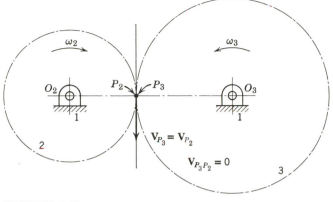

FIGURE 8.19

coincident particles at the point of contact, P_3 on link 3 and P_2 on link 2. Because the circles have pure rolling contact, these particles have identical velocities so that $\mathbf{V}_{P_3} = \mathbf{V}_{P_2}$, and the relative velocity between the two particles will be zero. Example 8.5 illustrates the use of this principle.

Example 8.5. In Fig. 8.20a is shown a mechanism consisting of three bars, two gears, and a rack. The velocity $\mathbf{V}_A$ of point A is 122 m/s in the direction shown. Determine the angular velocities ω_4 and ω_5 of the two gears and show the velocity images of the two gears. Determine also the velocity $\mathbf{V}_D$ of point D on gear 5. Velocity equations can be written as follows:

I. $\mathbf{V}_B = \mathbf{V}_A + \mathbf{V}_{BA}$

where

$\mathbf{V}_B$ = direction parallel to pitch line of rack, magnitude unknown

$\mathbf{V}_A$ = 122 m/s (given), direction perpendicular to O_2A

$\mathbf{V}_{BA}$ = direction perpendicular to BA, magnitude unknown

Measured on the polygon of Fig. 8.20b, V_B = 104 m/s and V_{BA} = 116 m/s.

II. $\mathbf{V}_C = \mathbf{V}_B + \mathbf{V}_{CB}$

where

$\mathbf{V}_C$ = direction perpendicular to O_6C, magnitude unknown

$\mathbf{V}_{CB}$ = direction perpendicular to CB, magnitude unknown

Measured on the polygon of Fig. 8.20b, V_C = 36.6 m/s and V_{CB} = 112 m/s. Measured on the polygon of Fig. 8.20c, $V_{BP_4} = V_B$ = 104 m/s, V_{CM_5} = 207 m/s, $V_{M_5} = V_{M_4}$ = 206 m/s, V_D = 215 m/s, and

$$\omega_4 = \frac{V_{BP_4}}{BP} = \frac{104}{\frac{102}{1000}} = 1020 \text{ rad/s} \qquad \text{(cw)}$$

$$\omega_5 = \frac{V_{CM_5}}{CM} = \frac{207}{\frac{51}{1000}} = 4060 \text{ rad/s} \qquad \text{(ccw)}$$

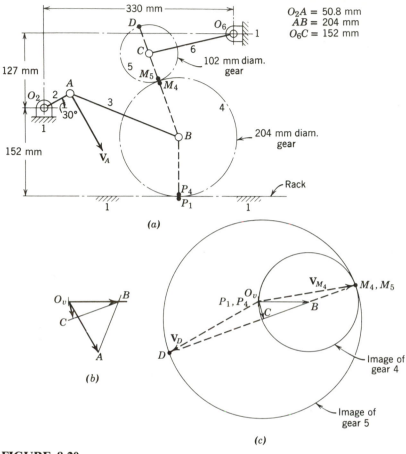

$O_2A = 50.8$ mm
$AB = 204$ mm
$O_6C = 152$ mm

FIGURE 8.20

Since the distance BC between the centers of the two gears is constant in all phases of the mechanism, an equivalent link joining the two centers may be visualized. Therefore, a five-bar linkage is first analyzed to determine the velocities $\mathbf{V}_B$ and $\mathbf{V}_C$ of the centers of the gears. The velocity polygon of Fig. 8.20b shows the determination of $\mathbf{V}_B$ and $\mathbf{V}_{BA}$ from Eq. I. In a similar manner, $\mathbf{V}_C$ and $\mathbf{V}_{CB}$ are determined from Eq. II.

In Fig. 8.20c, the velocity vectors $\mathbf{V}_B$ and $\mathbf{V}_C$ of Fig. 10.20b are redrawn for the construction of the velocity images of gears 4 and 5. Because the velocity $\mathbf{V}_{P_1}$ of point P_1 is zero and $\mathbf{V}_{P_4} = \mathbf{V}_{P_1}(\mathbf{V}_{P_4 P_1} = 0)$, the image of both points P_1 and P_4 is at the pole point O_v as shown. With point P_4 located on the polygon, the velocity image of gear 4 is drawn with B as a center and radius BP_4. The image of point M_4 on the circle is determined by drawing a line through B on the polygon perpendicular to the line M_4B on the configuration diagram. The image of point M_5 is the same as that of point M_4 because $\mathbf{V}_{M_5} = \mathbf{V}_{M_4}$. The image of gear 5 is, therefore, drawn with C as a center and radius CM_5. The image of point D is located on a diameter of the circle opposite point M_5.

The magnitudes and senses of ω_4 and ω_5 can now be determined from $\mathbf{V}_{BP_4}$ and $\mathbf{V}_{CM_5}$, respectively, as shown.

8.12 INSTANTANEOUS CENTERS OF VELOCITY

In the foregoing paragraphs and examples, the velocity analyses of linkages were made from an understanding of relative velocity and the influence of motion constraint on relative velocity. In the following, another concept is utilized to determine the linear velocity of particles in mechanisms, namely, the concept of the instantaneous center of velocity. This concept is based on the fact that at a given instant a pair of coincident points on two links in motion will have identical velocities relative to a fixed link and, therefore, will have zero velocity relative to each other. At this instant either link will have pure rotation relative to the other link about the coincident points. A special case of this is where one link is moving and the other is fixed. A pair of coincident points on these two links will then have zero absolute velocity, and the moving link at this instant will be rotating relative to the fixed link about the coincident points. In both cases the coincident set of points is referred to as an *instantaneous center of velocity* (sometimes referred to as *instant center,* or *centro*). From the foregoing, it can be seen that an instantaneous center is (*a*) a point in both bodies, (*b*) a point at which the two bodies have no relative velocity, and (*c*) a point about which one body may be considered to rotate relative to the other body at a given instant. It is easily seen that when two links, either both moving or one moving and one fixed, are directly connected together, the center of the connecting joint is an instantaneous center for the two links. When two links, either both moving or one moving and one fixed, are not directly connected, however, an instant center for the two links will also exist for a given phase of the linkage as will be shown in the following section.

In the four-bar linkage of Fig. 8.21, it is obvious that relative to the fixed link, points O_2 and O_4 are locations of particles on links 2 and 4, respectively, which are at zero velocity. It is less obvious that on link 3, which has both translating and angular motion, a particle is also at zero velocity relative to the fixed link. Referring to the velocity polygon shown in Fig. 8.21, the velocity image of link 3 appears as the line AB and none of the particles on this line is

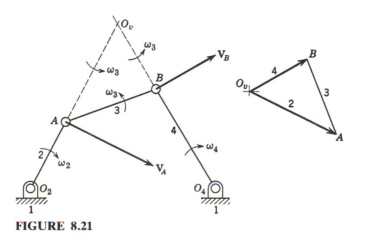

FIGURE 8.21

at zero velocity. However, if link 3 is visualized large enough in extent as a rigid body to include O_v of the polygon, a particle of zero velocity is then included in the image. To determine the location of O_v, the instantaneous center of link 3 relative to link 1, on the mechanism, a triangle similar to O_vBA of the polygon is constructed on the mechanism so that the sides of the two similar triangles are mutually perpendicular. It is important to note that for the particles on link 3 at A and at B, the fixed vectors $\mathbf{V}_A$ and $\mathbf{V}_B$ on link 3 are normal to the lines drawn from the instantaneous center O_v to A and B.

Since A and the instantaneous center O_v are particles on a common rigid link, the magnitude of $\mathbf{V}_A$ may be determined from $V_A = \omega_3(O_vA)$. Similarly, $V_B = \omega_3(O_vB)$. The magnitude of the velocity of any particle on link 3 may be determined from the product of ω_3 and the radial distance from the instantaneous center to the particle, and the direction of the velocity vector is normal to the radial line.

It may also be seen that the instantaneous center of link 3 relative to link 1 changes position with respect to time because of the changes in the shape of the velocity polygon as the mechanism passes through a cycle of phases. However, for links in pure rotation, the instantaneous centers are fixed centers, such as O_2 and O_4 of links 2 and 4, respectively, of Fig. 8.21.

The determination of velocities by instantaneous centers does not require the velocity polygon of free vectors and is judged by many to be the quicker method. By the method of instantaneous centers, the velocity vectors are shown directly as fixed vectors.

In the solution of a problem, such as in Fig. 8.22, the locations of the instantaneous centers of the moving links relative to the fixed link are generally determined first. For links 2 and 4, O_2 and O_4 are obviously points of zero velocity. For links such as link 3, only the *directions* of the velocities of two particles on the link need to be known since the intersection of the normals to the velocity direction lines determines the instantaneous center.

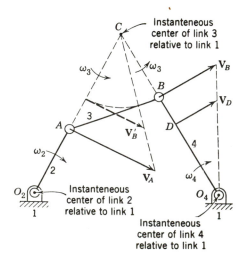

FIGURE 8.22

Fixed velocity vectors may be determined almost entirely by graphical construction. In Fig. 8.22, assuming ω_2 is the only information given, V_A may be computed from $\omega_2(O_2A)$ and $\mathbf{V}_A$ drawn normal to O_2A using the instantaneous center of link 2 relative to link 1. Considering particles A and B as part of link 3, the magnitude of $\mathbf{V}_B$ may be determined from similar triangles, as shown by the graphical construction, since V_A and V_B are proportional to the distance of A and B from the instantaneous center of link 3 relative to link 1. The equation which justifies the use of similar triangles in determining $\mathbf{V}_B$ may be written as $\omega_3 = V_A/(CA) = V_B/(CB)$. The velocity of any particle on link 4 such as D may be determined graphically from similar triangles as shown using the instantaneous center of link 4 relative to link 1.

For links that are in pure translation, such as the slider in a slider-crank mechanism, the direction lines of the velocities of all of its particles are parallel, and the normals, also being parallel, intersect at infinity. Thus, the instantaneous center of a link in translation is at an infinite distance from the link, in a direction normal to the path of translation.

8.13 INSTANTANEOUS CENTER NOTATION

In the foregoing, instantaneous centers of velocity were determined for each of the moving links relative to the fixed link. The system of labeling these points is shown in Fig. 8.23, where the instantaneous center of link 3 relative to the fixed link is labeled 31 to indicate the motion of "3 relative to 1." Link 1 has the same instantaneous center relative to link 3 when link 3 is considered the fixed link, in which case link 1 appears to be rotating in the opposite sense ($\omega_{13} = -\omega_{31}$) relative to link 3. Since points 31 and 13 are the same point, either designation is acceptable although the simpler notation 13 is preferred. The instantaneous center of link 2 relative to link 1 is labeled 21 or 12, and that of link 4 relative to link 1 is labeled 41 or 14 as shown.

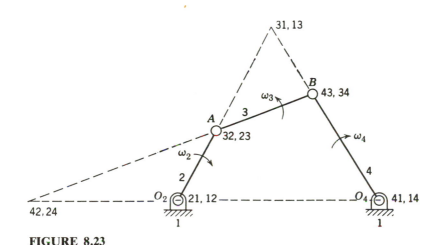

FIGURE 8.23

Also of interest is the instantaneous center of one link relative to another where both links are moving relative to the fixed link. Such a center is shown at point A in Fig. 8.23, where both A_2 and A_3 have a common absolute velocity $\mathbf{V}_A$ because of the pinned joint so that the relative velocities $\mathbf{V}_{A_3A_2}$ and $\mathbf{V}_{A_2A_3}$ are zero. It is obvious that point A is the instantaneous center 32 about which link 3 is rotating relative to link 2 at an angular velocity ω_{32}. Point A is also the instantaneous center 23. In a similar manner point B is the instantaneous center 43 or 34. The instantaneous center 42 or 24 is also shown in Fig. 8.23. However, the method of determining its location will not be presented until the next section.

8.14 KENNEDY'S THEOREM

For three independent bodies in general plane motion, Kennedy's theorem states that the three instantaneous centers lie on a common straight line. In Fig. 8.24, three independent links (1, 2, and 3) are shown in motion relative to each other. There are three instantaneous centers (12, 13, and 23), whose instantaneous locations are to be determined.

If link 1 is regarded as a fixed link, or datum link, the velocities of particles A_2 and B_2 on link 2 and the velocities of D_3 and E_3 on link 3 may be regarded as absolute velocities relative to link 1. The instantaneous center 12 may be located from the intersection of the normals to the velocity direction lines drawn from A_2 and B_2. Similarly, the center 13 is located from normals drawn from particles D_3 and E_3. The instantaneous centers 12 and 13 are relative to link 1.

The third instantaneous center 23 remains to be determined. On a line drawn through the centers 12 and 13, there exists a particle C_2 on link 2 at an absolute velocity $\mathbf{V}_{C_2}$ having the same direction as the absolute velocity $\mathbf{V}_{C_3}$ of a particle C_3 on link 3. Since $\mathbf{V}_{C_2}$ is proportional to the distance of C_2 from 12, the magnitude of $\mathbf{V}_{C_2}$ is determined from the graphical construction shown, and

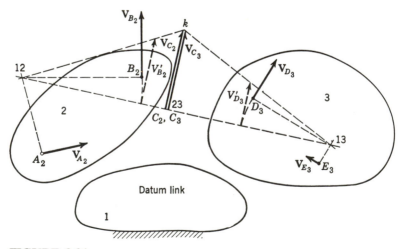

FIGURE 8.24

V_{C_3} is determined in a similar manner. From the intersection of the construction lines at k, a common location of C_2 and C_3 is determined such that the absolute velocities V_{C_2} and V_{C_3} are identical. This location is the instantaneous center 23, since the absolute velocities of the coincident particles are common and the relative velocities $V_{C_2C_3}$ and $V_{C_3C_2}$ are zero. It should be obvious that 23 is on a straight line with 12 and 13 in order for the directions of V_{C_2} and V_{C_3} to be common.

Kennedy's theorem is extemely useful in determining the locations of instantaneous centers in mechanisms having a large number of links, many of which are in general plane motion.

8.15 DETERMINATION OF INSTANTANEOUS CENTERS BY KENNEDY'S THEOREM

In a mechanism consisting of n links, there are $n - 1$ instantaneous centers relative to any given link. For n number of links, there is a total of $n(n - 1)$ instantaneous centers. However, since for each location of instantaneous centers there are two centers, the total number N of locations is given by

$$N = \frac{n(n - 1)}{2}$$

The number of locations of centers increases rapidly with numbers of links as shown below.

n LINKS	N CENTERS
4	6
5	10
6	15
7	21

Example 8.6. For the Whitworth mechanism shown in Fig. 8.25, determine the 15 locations of instantaneous centers of zero velocity.

Solution. Because of the large number of locations to be determined, it is desirable to use a system of accounting for the centers as they are determined. The circle diagram shown in Fig. 8.25 is one of the simplest means of accounting. The numbers of the links are designated on the periphery of the circle, and the chord linking any two numbers represents an instantaneous center. In the upper circle are shown eight centers which may be determined by inspection. Five of the centers (12, 14, 23, 45, and 56) are at pin-jointed connections as shown. Two centers (16 and 34) are at infinity, since link 6 is in translation relative to link 1, and link 3 is in translation relative to link 4. Because the absolute velocity directions of points B and C of link 5 are known, the intersection of the normals locates 15. Thus, eight centers are located by inspection, as shown by the solid lines on the circle diagram.

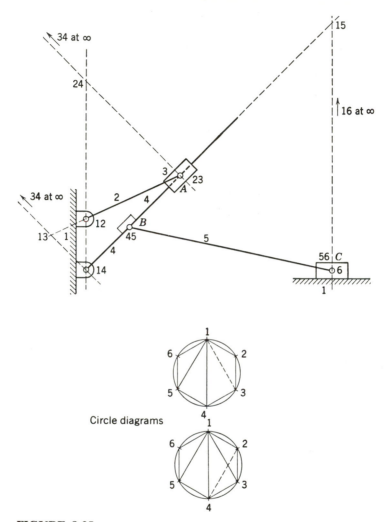

FIGURE 8.25

For centers less obviously determined, Kennedy's theorem may be used. In the upper circle, to locate center 13, a dashed line is drawn such that it closes two triangles. The triangle 1–2–3 represents the three centers (12, 23, and 13) of links 1, 2, and 3, which according to Kennedy's theorem lie on a straight line. Similarly, triangle 1–3–4 represents the centers 13, 34, and 14, which also lie on a straight line. The intersection of the two lines on the mechanism locates the center 13, which must lie on both lines. The dashed line may be made solid to indicate that the unknown center has been located. The lower circle shows the next step in which the center 24 is located using triangles 2–3–4 and 1–2–4. It may be seen that 24 is the logical center to determine rather than 25 or 26, which cannot be drawn as common to two triangles until other centers have been determined.

In Fig. 8.25, 10 of the 15 centers are shown. Figure 8.26 shows the same mechanism with all 15 centers located.

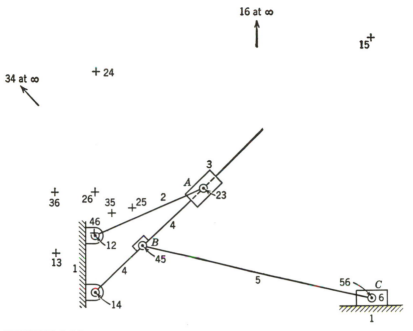

FIGURE 8.26

8.16 DETERMINATION OF VELOCITY BY INSTANTANEOUS CENTERS

Kennedy's theorem may be used to great advantage in determining directly the absolute velocity of any given particle of a mechanism without necessarily determining the velocities of intermediate particles as required by the vector polygon method. In connection with the Whitworth mechanism of Fig. 8.25, for example, the velocity of the tool support (link 6) may be determined from the known speed of the driving link 2 without first determining the velocities of points on the connecting links 3, 4, and 5.

Example 8.7. For the Whitworth mechanism shown in Fig. 8.27, determine the absolute velocity $\mathbf{V}_C$ of the tool support when the driving link 2 rotates at a speed such that $\mathbf{V}_A = 30$ ft/s as shown.

Solution. Two solutions for $\mathbf{V}_C$ are shown in Fig. 8.27. In the first of these (Fig. 8.27a), links 1, 3, and 5 are involved such that instantaneous centers 13, 15, and 35 are used. $\mathbf{V}_A$ is the known absolute velocity of a particle on link 3 relative to link 1; thus, links 3 and 1 are involved. The absolute velocity $\mathbf{V}_C$ is to be determined for a particle on link 5 also relative to link 1, thus involving links 5 and 1. According to Kennedy's theorem, the instantaneous centers 13, 15, and 35 are on a common straight line as shown in Fig. 8.27a. Using center 13, the absolute velocity $\mathbf{V}_{P_3}$ for a particle P_3 located at 35 on link 3 may be determined graphically from similar triangles by swinging $\mathbf{V}_A$ to position $\mathbf{V}'_A$ using center 13 as a pivot point. Point 35 represents the location of coincident particles P_3 on link 3 and P_5 on link 5, for which the absolute velocities are common (see Fig. 8.24). Thus,

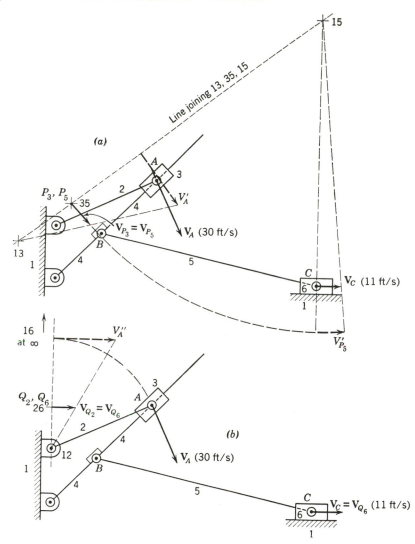

FIGURE 8.27

$\mathbf{V}_{P_3}$ is also the absolute velocity $\mathbf{V}_{P_5}$ of a particle on link 5. Since both P_5 and C are points on link 5, the absolute velocity $\mathbf{V}_C$ may be determined from similar triangles by swinging $\mathbf{V}_{P_5}$ to position $\mathbf{V}'_{P_5}$ using center 15 as a pivot point. The length of $\mathbf{V}_C$ is measured to determine magnitude of velocity.

In the above solution, the centers 13 and 15 relative to the fixed link are *pivot points,* and the center 35 of the moving links is the *transfer point.* By properly identifying these points, the determination of velocities becomes systematic.

The second solution (Fig. 8.27b) for $\mathbf{V}_C$ is similar to the first, in which pivot points 12 and 16 are used because $\mathbf{V}_A$ represents the absolute velocity of a particle on link 2 and $\mathbf{V}_C$ is the absolute velocity of a particle on link 6. Center 26 is the transfer point representing the location of coincident particles Q_2 and Q_6 on links 2 and 6, for which the absolute velocities $\mathbf{V}_{Q_2}$ and $\mathbf{V}_{Q_6}$ are common. $\mathbf{V}_{Q_2}$ is determined graphically from $\mathbf{V}_A$ using center 12 as a pivot point. Since pivot point 16 is at infinity, link 6 is in pure translation relative to link 1 so that $\mathbf{V}_C$ is the same in magnitude and direction as $\mathbf{V}_{Q_2}$ and $\mathbf{V}_{Q_6}$, as shown.

8.17 ROLLING ELEMENTS

The method of instantaneous centers is frequently applied to mechanisms consisting of rolling elements as in epicyclic gear trains (Fig. 8.28). As shown previously, the relative velocity of the coincident particles at the point of contact of two rolling links is zero. Thus, an instantaneous center exists at the point of contact.

For the reduction drive shown in Fig. 8.28, the instantaneous centers are as shown. The speed reduction ratio ω_{31}/ω_{41} (the internal gear speed to carrier speed when the sun gear is fixed) may be determined from linear velocities of particles as shown. Assuming that the absolute angular velocity ω_{41} of the carrier is known, $\mathbf{V}_A$ may be determined considering A as a particle on link 4. $\mathbf{V}_A$ is also the absolute velocity of a particle on link 2; therefore, using the center 12, the absolute velocity $\mathbf{V}_{P_2}$ of P_2 on link 2 may be determined graphically from similar triangles. Since center 23 is the location of coincident particles on links 2 and 3 having a common absolute velocity, ω_{31} may be calculated from $\mathbf{V}_{P_3}$.

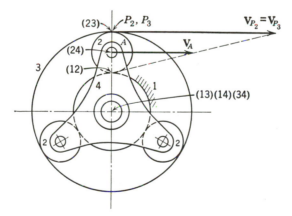

FIGURE 8.28

8.18 GRAPHICAL DETERMINATION OF ACCELERATION IN MECHANISMS BY VECTOR POLYGONS

As in the determination of velocities of particles in a mechanism, the linear accelerations of particles may also be determined by graphical construction of acceleration polygons and acceleration images. It is important that the relative acceleration of pairs of particles be understood.

8.19   RELATIVE ACCELERATION OF PARTICLES IN MECHANISMS

If the acceleration $\mathbf{A}_Q$ of a particle Q is known, the acceleration of another particle $\mathbf{A}_P$ may be determined by adding the relative acceleration vector $\mathbf{A}_{PQ}$ as shown in the following vector equation:

$$\mathbf{A}_P = \mathbf{A}_Q + \mathbf{A}_{PQ} \tag{8.26}$$

As discussed in the sections on relative velocity, it is shown that the relative velocity of a pair of particles depends on the type of constraint used in a given mechanism. Similarly, the relative acceleration $\mathbf{A}_{PQ}$ in mechanisms depends on the type of built-in constraint.

8.20   RELATIVE ACCELERATION OF PARTICLES IN A COMMON LINK

As shown in Fig. 8.29a, when two particles P and Q in the same rigid link are considered, the fixed distance PQ constrains particle P to move on a circular arc relative to Q regardless of the absolute linear motion of Q. Therefore, since the path of P relative to Q is circular, the acceleration vector $\mathbf{A}_{PQ}$ may be represented by the perpendicular components of acceleration $\mathbf{A}_{PQ}^n$ and $\mathbf{A}_{PQ}^t$ normal and tangent, respectively, to the relative path at P. Regardless of the linear absolute acceleration of Q, the angular motions of the link relative to Q are the same as relative to the fixed link because a particle such as Q has no angular motion. For the circular path of P relative to Q, the angular velocity ω_r of the radius of curvature PQ is the same as the absolute angular velocity ω_3 of the link. Also, the angular acceleration α_r of the radius of curvature is the same as the absolute angular acceleration α_3 of the link.

The magnitude of the normal relative acceleration $\mathbf{A}_{PQ}^n$ may be determined from Eq. 8.4a:

$$|\mathbf{A}_{PQ}^n| = (PQ)\omega_3^2 = \frac{V_{PQ}^2}{PQ} \tag{8.27}$$

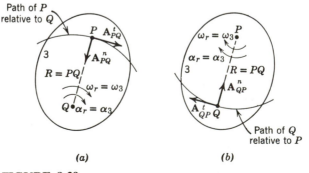

(a)   (b)

FIGURE 8.29

The magnitude of the tangential relative acceleration $\mathbf{A}_{PQ}^t$ may be determined from Eq. 8.4b:

$$|\mathbf{A}_{PQ}^t| = (PQ)\alpha_3 \qquad (8.28)$$

Because the relative path is circular, dR/dt is zero.

Observe that the direction of $\mathbf{A}_{PQ}^n$ is normal to the relative path and that its sense is toward the center of curvature Q so that the vector is directed from P toward Q as shown in Fig. 8.29a. The direction of $\mathbf{A}_{PQ}^t$ is tangent to the relative path (normal to line PQ), and the sense of the vector depends on the sense of α_r. In Fig. 8.29b, the relative acceleration vectors $\mathbf{A}_{QP}^n$ and $\mathbf{A}_{QP}^t$ of Q relative to P are shown where the magnitudes and senses of ω_3 and α_3 are the same as in Fig. 8.29a. The relative path shown is that of Q observed at P. It is to be noted that $\mathbf{A}_{QP}^n = -\mathbf{A}_{PQ}^n$ and $\mathbf{A}_{QP}^t = -\mathbf{A}_{PQ}^t$, where the minus signs indicate "opposite in sense."

Example 8.8. When the mechanism is in the phase shown in Fig. 8.30a, link 2 rotates with the angular velocity ω_2 of 30 rad/s and an angular acceleration α_2 of 240 rad/s² in the directions given. Determine the acceleration $\mathbf{A}_B$ of point B, the acceleration $\mathbf{A}_C$ of point C, the angular acceleration α_3 of link 3, the angular acceleration α_4 of link 4, and

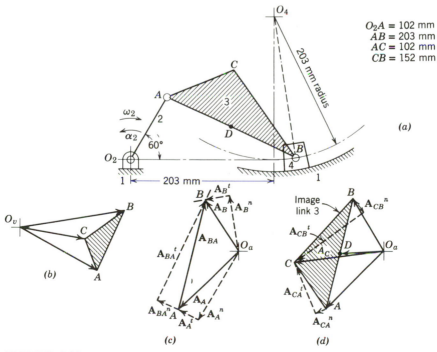

O_2A = 102 mm
AB = 203 mm
AC = 102 mm
CB = 152 mm

FIGURE 8.30

the relative acceleration α_{34}. Velocity and acceleration equations can be written as follows:

I. $\mathbf{V}_B = \mathbf{V}_A + \mathbf{V}_{BA}$

II. $\mathbf{V}_C = \mathbf{V}_A + \mathbf{V}_{CA}$

III. $\mathbf{V}_C = \mathbf{V}_B + \mathbf{V}_{CB}$

where

$\mathbf{V}_B$ = direction perpendicular to O_4B, magnitude unknown

$\mathbf{V}_A = (O_2A)\omega_2 = (102)30 = 3060$ mm/s, direction perpendicular to O_2A

$\mathbf{V}_{BA}$ = direction perpendicular to BA, magnitude unknown

$\mathbf{V}_C$ = direction unknown, magnitude unknown

$\mathbf{V}_{CA}$ = direction perpendicular to CA, magnitude unknown

$\mathbf{V}_{CB}$ = direction perpendicular to CB, magnitude unknown

Measured on the polygon of Fig. 8.30b, $V_B = 3660$ mm/s, $V_{BA} = 2300$ mm/s, $V_{CA} = 1130$ mm/s, and $V_{CB} = 1750$ mm/s.

IV. $\mathbf{A}_B = \mathbf{A}_A + \mathbf{A}_{BA}$

$\mathbf{A}_B^n + \mathbf{A}_B^t = \mathbf{A}_A^n + \mathbf{A}_A^t + \mathbf{A}_{BA}^n + \mathbf{A}_{BA}^t$

where

$\mathbf{A}_B^n = \dfrac{V_B^2}{O_4B} = \dfrac{3660^2}{203} = 66{,}000$ mm/s^2, direction from B toward O_4

$\mathbf{A}_B^t$ = direction perpendicular to $\mathbf{A}_B^n$, magnitude unknown

$\mathbf{A}_A^n = \dfrac{V_A^2}{O_2A} = \dfrac{3060^2}{102} = 91{,}800$ mm/s^2, direction from A toward O_2

$\mathbf{A}_A^t = (O_2A)\alpha_2 = (102)240 = 24{,}500$ mm/s, direction perpendicular to $\mathbf{A}_A^n$

$\mathbf{A}_{BA}^n = \dfrac{V_{BA}^2}{BA} = \dfrac{2300^2}{203} = 26{,}100$ mm/s^2, direction from B toward A

$\mathbf{A}_{BA}^t$ = direction perpendicular to $\mathbf{A}_{BA}^n$, magnitude unknown

Measured on the polygon of Fig. 8.30c, $A_B = 70{,}400$ mm/s^2, $A_B^t = 24{,}700$ mm/s^2, $A_{BA}^t = 129{,}000$ mm/s^2, and

$$\alpha_3 = \frac{A_{BA}^t}{BA} = \frac{129{,}000}{203} = 635 \text{ rad/s}^2 \qquad (ccw)$$

$$\alpha_4 = \frac{A_B^t}{O_4B} = \frac{24{,}700}{203} = 122 \text{ rad/s}^2 \qquad (cw)$$

$$\alpha_{34} = \alpha_3 - \alpha_4 = 635 - (-122) = 757 \text{ rad/s}^2 \qquad (ccw)$$

V. $\mathbf{A}_C = \mathbf{A}_A + \mathbf{A}_{CA}^n + \mathbf{A}_{CA}^t$

VI. $\mathbf{A}_C = \mathbf{A}_B + \mathbf{A}_{CB}^n + \mathbf{A}_{CB}^t$

where

$\mathbf{A}_C$ = direction unknown, magnitude unknown

$\mathbf{A}_{CA}^n = \dfrac{V_{CA}^2}{CA} = \dfrac{1130^2}{102} = 12,500 \text{ mm/s}^2$, direction from C toward A

$\mathbf{A}_{CA}^t$ = direction perpendicular to $\mathbf{A}_{CA}^n$, magnitude unknown

$\mathbf{A}_{CB}^n = \dfrac{V_{CB}^2}{CB} = \dfrac{1750^2}{152} = 20,100 \text{ mm/s}^2$, direction from C toward B

$\mathbf{A}_{CB}^t$ = direction perpendicular to $\mathbf{A}_{CB}^n$, magnitude unknown

Measured on the polygon of Fig. 8.30d, $A_C = 104,000 \text{ mm/s}^2$.

The velocity polygon of Fig. 8.30b shows the determination of $\mathbf{V}_B$ and $\mathbf{V}_{BA}$ from Eq. I. In a similar manner $\mathbf{V}_C$, $\mathbf{V}_{CA}$, and $\mathbf{V}_{CB}$ are determined from Eqs. II and III. The shaded triangle ABC of the velocity polygon is the velocity image of link 3.

Equation IV expresses $\mathbf{A}_B$ in terms of $\mathbf{A}_A$ and $\mathbf{A}_{BA}$, and all of the components of this equation are known as indicated in magnitude, sense, and direction or in direction. In constructing the acceleration polygon Fig. 8.30c starting with the right side of Eq. IV, the vector $\mathbf{A}_A^n$ is drawn from pole O_a to which is added $\mathbf{A}_A^t$. This gives the vector $\mathbf{A}_A$ whose tip is labeled "A." Next, add the vector $\mathbf{A}_{BA}^n$ starting at point A, and to it add the direction of $\mathbf{A}_{BA}^t$. As can be seen, it is impossible to complete the solution using only the components on the right side of Eq. IV. Therefore, consider the left side of the equation and draw vector $\mathbf{A}_B^n$ from O_a and to it add the direction of $\mathbf{A}_B^t$. The intersection of the direction of $\mathbf{A}_{BA}^t$ and the direction of $\mathbf{A}_B^t$ completes the polygon. Arrowheads are now added to the vectors $\mathbf{A}_{BA}^t$ and $\mathbf{A}_B^t$ so that the addition of the vectors of the polygon agrees with the addition of the terms of Eq. IV. The resultant of the vectors $\mathbf{A}_B^n$ and $\mathbf{A}_B^t$ gives $\mathbf{A}_B$ whose tip is labeled "B." The resultant of $\mathbf{A}_{BA}^n$ and $\mathbf{A}_{BA}^t$ is also shown on the polygon.

The magnitudes and senses of α_3 and α_4 can now be determined from $\mathbf{A}_{BA}^t$ and $\mathbf{A}_B^t$, respectively, as shown.

To determine $\mathbf{A}_C$, it is necessary to use Eqs. V and VI, which give the relations between $\mathbf{A}_C$ and $\mathbf{A}_A$ and $\mathbf{A}_B$. The components of these equations are known as indicated. For clarity, the acceleration vectors $\mathbf{A}_A$ and $\mathbf{A}_B$ are redrawn in Fig. 8.30d from Fig. 8.30c without their normal and tangential components. Use Eq. V, and draw the vector $\mathbf{A}_{CA}^n$ from point A in Fig. 8.30d and to it add the direction of $\mathbf{A}_{CA}^t$. Consider next Eq. VI and draw the vector $\mathbf{A}_{CB}^n$ from point B and to it add the direction of $\mathbf{A}_{CB}^t$. The intersection of the direction of $\mathbf{A}_{CA}^t$ and the direction of $\mathbf{A}_{CB}^t$ completes the polygon. This intersection is point C, which gives $\mathbf{A}_C$. Arrowheads are now added to the vectors $\mathbf{A}_{CA}^t$ and $\mathbf{A}_{CB}^t$ so that the vector addition checks with Eqs. V and VI. The shaded triangle ABC of Fig. 8.30d is the acceleration image of link 3.

The acceleration of any point D as shown on link 3 can be determined by locating its corresponding position on the acceleration image of link 3. The vector from O_a to D is $\mathbf{A}_D$ as shown in Fig. 8.30d.

8.21 RELATIVE ACCELERATION OF COINCIDENT PARTICLES ON SEPARATE LINKS. CORIOLIS COMPONENT OF ACCELERATION

The next mechanism to be considered is one in which there is relative sliding between two links, as between links 3 and 4 as shown in Fig. 8.31, and it is required to determine ω_4 and α_4 given ω_2 and α_2. In this mechanism, points A_2

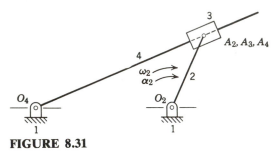

FIGURE 8.31

and A_3 are the same point, and point A_4 is their projection on link 4. To find ω_4 and α_4, the velocity and acceleration of the two coincident points A_2 and A_4, each on separate links, must be analyzed.[3]

The equation for the velocity of point A_4 can be written as follows:

$$\mathbf{V}_{A_4} = \mathbf{V}_{A_2} + \mathbf{V}_{A_4A_2} \tag{8.29}$$

In this equation, $\mathbf{V}_{A_2}$ is known in magnitude, sense, and direction and $\mathbf{V}_{A_4}$ and $\mathbf{V}_{A_4A_2}$ are known in direction. The velocity polygon can easily be drawn, and $\mathbf{V}_{A_4}$ determined, from which ω_4 can be calculated.

The acceleration of point A_4 can be determined from the following equation:

$$\mathbf{A}_{A_4} = \mathbf{A}_{A_2} + \mathbf{A}_{A_4A_2} \tag{8.30}$$

which can be expanded as follows:

$$\mathbf{A}_{A_4}^n + \mathbf{A}_{A_4}^t = \mathbf{A}_{A_2}^n + \mathbf{A}_{A_2}^t + \mathbf{A}_{A_4A_2}^n + \mathbf{A}_{A_4A_2}^t + 2\omega_2 \times \mathbf{V}_{A_4A_2} \tag{8.31}$$

In going from Eq. 8.30 to Eq. 8.31, the following substitution was made:

$$\mathbf{A}_{A_4A_2} = \mathbf{A}_{A_4A_2}^n + \mathbf{A}_{A_4A_2}^t + 2\omega_2 \times \mathbf{V}_{A_4A_2}$$

To determine the relative acceleration between two moving coincident points, it is necessary to add a third component as shown. This component is known as *Coriolis component*, which was developed using vector mathematics in section 8.6. Also, because points A_4 and A_2 are coincident, the terms $\mathbf{A}_{A_4A_2}^n$ and $\mathbf{A}_{A_4A_2}^t$ do not represent the usual normal and tangential components of two points on the same rigid body as previously considered. For this reason, they often appear in the literature written with a capital script $\mathcal{A}$. The magnitude of $\mathbf{A}_{A_4A_2}^n$ can be calculated from the relation

$$|\mathbf{A}_{A_4A_2}^n| = \frac{V_{A_4A_2}^2}{R} \tag{8.32}$$

---

[3]Point A_3 could have been used instead of A_2 as the point coincident with A_4. However, point A_2 is generally preferred because it is on a link directly connected to the ground and its motion can be easily visualized.

where R is the radius of curvature of the path of point A_4 relative to point A_2. This component is directed from the coincident points along the radius toward the center of curvature. The tangential component $\mathbf{A}^t_{A_4A_2}$ is known in direction and is tangent to the path of $\mathbf{A}_4$ relative to $\mathbf{A}_2$ at the coincident points. The magnitude of the Coriolis component $2\boldsymbol{\omega}_2 \times \mathbf{V}_{A_4A_2}$ is easily calculated because $\boldsymbol{\omega}_2$ is given data and $\mathbf{V}_{A_4A_2}$ can be determined from the velocity polygon. The direction of this component is normal to the path of A_4 relative to A_2, and its sense is the same as that of $\mathbf{V}_{A_4A_2}$ rotated about its origin 90° in the direction of ω_2. An example of this method of determining the direction will be given in a later section.

In Eq. 8.31, all of the components can easily be determined in magnitude, sense, and direction or in direction except $\mathbf{A}^n_{A_4A_2}$. This component calculated from $V^2_{A_4A_2}/R$ can only be determined if the instantaneous radius of curvature R of the path of A_4 relative to A_2 is known. Unfortunately, because this path is not easily determined for the mechanism shown in Fig. 8.31, it is necessary to rewrite Eq. 8.31 in the following form:

$$\mathbf{A}^n_{A_2} + \mathbf{A}^t_{A_2} = \mathbf{A}^n_{A_4} + \mathbf{A}^t_{A_4} + \mathbf{A}^n_{A_2A_4} + \mathbf{A}^t_{A_2A_4} + 2\boldsymbol{\omega}_4 \times \mathbf{V}_{A_2A_4} \quad (8.33)$$

With Eq. 8.31 written in this form, $\mathbf{A}^n_{A_2A_4}$ can easily be evaluated as zero because the path of A_2 relative to link 4 (which contains point A_4) is a straight line and R is infinite. The acceleration polygon can now be drawn and $\mathbf{A}^t_{A_4}$ determined, from which α_4 is calculated.

While it is easy to see in Fig. 8.31 that the path of point A_2 relative to point A_4 is a straight line by inverting the mechanism and letting link 4 be the fixed link, it is very difficult to visualize the path of A_4 relative to A_2. As a means of determining this path, consider Fig. 8.32, where link 2 is now the fixed link. In this figure, link 1 is placed in a number of angular positions relative to link 2, and the relative position of A_4 is determined for each position of link 1. It may be seen that the position of link 4 is always in a direction from O_4 through A_2 and that A_4 is a fixed distance from O_4. As shown, the path of A_4 on link 2 is curvilinear and tangent to link 4 at point A_2. Unfortunately, the path is not circular so that the radius of curvature is difficult to determine.

Consider next the case where link 4 of Fig. 8.31 has been replaced by a curved link of circular form as shown in Fig. 8.33. In this linkage, the path of A_2 relative to A_4 is a circular arc of known radius and center of curvature. The magnitude of $\mathbf{A}^n_{A_2A_4}$ is therefore not zero, and the vector representing this component will be directed from point A toward the center of curvature C.

The Coriolis component is always in the same direction as the $\mathbf{A}^n_{A_2A_4}$ component, if one exists, but its sense may or may not be the same. Considering the Coriolis term $2\boldsymbol{\omega}_4 \times \mathbf{V}_{A_2A_4}$ for the linkage of Fig. 8.33, its direction and sense can easily be determined as follows. Draw the vector representing the relative velocity $\mathbf{V}_{A_2A_4}$ in its correct direction and sense. Rotate this vector 90° about its origin in the same sense as ω_4. This will give the direction and sense of the Coriolis component as shown in Fig. 8.34. As can be seen, the terms $\mathbf{A}^n_{A_2A_4}$ and $2\boldsymbol{\omega}_4 \times \mathbf{V}_{A_2A_4}$ have the same sense for this case and will therefore add together. Obviously, this method of determining the direction and sense of Coriolis applies even if the $\mathbf{A}^n_{A_2A_4}$ component is zero.

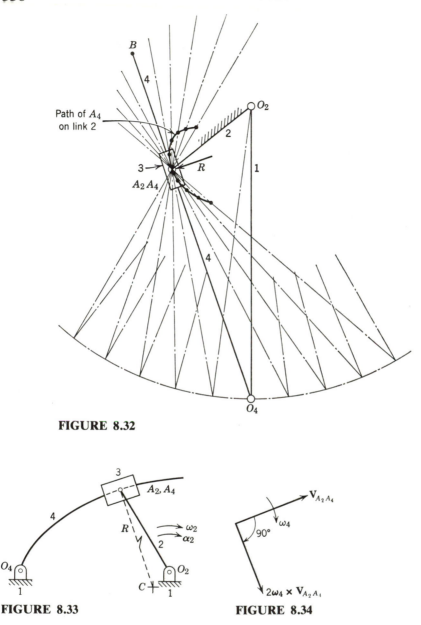

FIGURE 8.32

FIGURE 8.33

FIGURE 8.34

Example 8.9. In the crank-shaper mechanism shown in Fig. 8.35a link 2 rotates at a constant angular velocity ω_2 of 10 rad/s. Determine the acceleration $\mathbf{A}_{A_4}$ of point A_4 on link 4 and the angular acceleration α_4 when the mechanism is in the phase shown. Velocity and acceleration equations can be written as follows:

I. $\mathbf{V}_{A_4} = \mathbf{V}_{A_2} + \mathbf{V}_{A_4 A_2}$

where

$\mathbf{V}_{A_4}$ = direction perpendicular to O_4A_4, magnitude unknown

$\mathbf{V}_{A_2}$ = $(O_2A_2)\omega_2$ = (4)10 = 40 in./s, direction perpendicular to O_2A_2

$\mathbf{V}_{A_4A_2}$ = direction parallel to O_4A_4, magnitude unknown

Measured on the polygon of Fig. 8.35b, V_{A_4} = 13 in./s, $V_{A_4A_2}$ = 38 in./s, and

$$\omega_4 = \frac{V_{A_4}}{O_4A_4} = \frac{13}{10} = 1.3 \text{ rad/s} \qquad (\text{ccw})$$

II. $\mathbf{A}_{A_4} = \mathbf{A}_{A_2} + \mathbf{A}_{A_4A_2}$

III. $\mathbf{A}_{A_2} = \mathbf{A}_{A_4} + \mathbf{A}_{A_2A_4}$

$\mathbf{A}_{A_2}^n + \mathbf{A}_{A_2}^t = \mathbf{A}_{A_4}^n + \mathbf{A}_{A_4}^t + \mathbf{A}_{A_2A_4}^n + \mathbf{A}_{A_2A_4}^t + 2\omega_4 \times \mathbf{V}_{A_2A_4}$

where

$$\mathbf{A}_{A_2}^n = \frac{V_{A_2}^2}{O_2A_2} = \frac{40^2}{4} = 400 \text{ in./s}^2, \text{ direction from } A_2 \text{ toward } O_2$$

$$\mathbf{A}_{A_2}^t = 0 \qquad (\alpha_2 = 0)$$

$$\mathbf{A}_{A_4}^n = \frac{V_{A_4}^2}{O_4A_4} = \frac{13^2}{10} = 16.9 \text{ in./s}^2, \text{ direction from } A_4 \text{ toward } O_4$$

$\mathbf{A}_{A_4}^t$ = direction perpendicular to $\mathbf{A}_{A_4}^n$, magnitude unknown

$$\mathbf{A}_{A_2A_4}^n = \frac{V_{A_2A_4}^2}{R} = 0 \qquad (R = \infty)$$

$2\omega_4 \times \mathbf{V}_{A_2A_4}$ = 2(1.3)38 = 98.8 in./s^2, direction perpendicular to $\mathbf{V}_{A_2A_4}$

$\mathbf{A}_{A_2A_4}^t$ = direction perpendicular to $2\omega_4 \times \mathbf{V}_{A_2A_4}$, magnitude unknown

Measured on the polygon of Fig. 8.35c, A_{A_4} = 475 in./s^2, $A_{A_4}^t$ = 474 in./s^2, and

$$\alpha_4 = \frac{A_{A_4}^t}{O_4A_4} = \frac{474}{10} = 47.4 \text{ rad/s}^2 \qquad (\text{cw})$$

Link 4 is a guide link which constrains points A_2 and A_3 to follow a straight-line path on link 4. Two pairs of coincident points may be considered, either A_2 and A_4 or A_3 and A_4. For this illustration, A_2 and A_4 are chosen, and the straight guide path is the relative path of A_2 on link 4. Thus, the vectors $\mathbf{V}_{A_2A_4}$ and $\mathbf{A}_{A_2A_4}$ are involved, and the $\mathbf{A}_{A_2A_4}^n$ component of $\mathbf{A}_{A_2A_4}$ can easily be determined because $R = \infty$.

The velocity polygon of Fig. 8.35b shows the determination of $\mathbf{V}_{A_4}$ and $\mathbf{V}_{A_4A_2}$ from Eq. I. The calculation for ω_4 is also shown.

Equation II expresses $\mathbf{A}_{A_4}$ in terms of $\mathbf{A}_{A_2}$ and $\mathbf{A}_{A_4A_2}$. However, because the path of point A_4 relative to point A_2 is not easily determined, Eq. II is rewritten in the form of Eq. III so as to use the component $\mathbf{A}_{A_2A_4}$ as discussed above.

All of the components of Eq. III are known as indicated in magnitude, sense, and direction or in direction only. In constructing the acceleration polygon of Fig. 8.35c starting with the right side of Eq. III, the vector $\mathbf{A}_{A_4}^n$ is drawn first, followed by the direction of

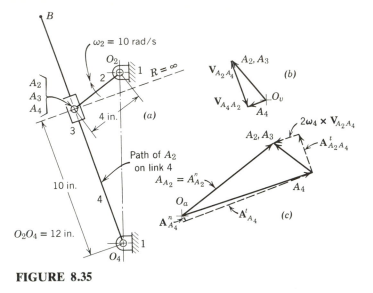

FIGURE 8.35

A'_{A_4}. This is all that can be laid off from the right side of Eq. III at present. Therefore, consider the left side of Eq. III and draw the vector A_{A_2}. Next, draw the vector $2\omega_4 \times V_{A_2A_4}$ so that its tip meets the tip of vector A_{A_2}. Draw $A'_{A_2A_4}$ perpendicular to the Coriolis component until it intersects the direction of the vector representing A'_{A_4}; this completes the polygon. Arrowheads are now added to the vectors A'_{A_4} and $A'_{A_2A_4}$ so that the addition of the vectors of the polygon agrees with the addition of the terms of Eq. III. The magnitude and sense of α_4 can now be determined from A'_{A_4} as shown.

Example 8.10. In the mechanism shown in Fig. 8.36a, link 2 drives link 3 through a pin at point B. Link 2 rotates at a uniform angular velocity ω_2 of 50 rad/s, and the radius of curvature R of the slot in link 3 is 305 mm. Determine the acceleration A_{B_3} of point B_3 on link 3 and the angular acceleration α_3 for the position shown. Velocity and acceleration equations can be written as follows:

I. $V_{B_3} = V_{B_2} + V_{B_3B_2}$

where

V_{B_3} = direction perpendicular to O_3B_3, magnitude unknown

$V_{B_2} = (O_2B_2)\omega_2 = (50.8)50 = 2540$ mm/s, direction perpendicular to O_2B_2

$V_{B_3B_2}$ = direction perpendicular to **R**, magnitude unknown

Measured on the polygon of Fig. 8.36b, $V_{B_3} = 1650$ mm/s, $V_{B_3B_2} = 2540$ mm/s, and

$$\omega_3 = \frac{V_{B_3}}{O_3B_3} = \frac{1650}{208} = 7.93 \text{ rad/s} \quad \text{(ccw)}$$

II. $A_{B_3} = A_{B_2} + A_{B_3B_2}$

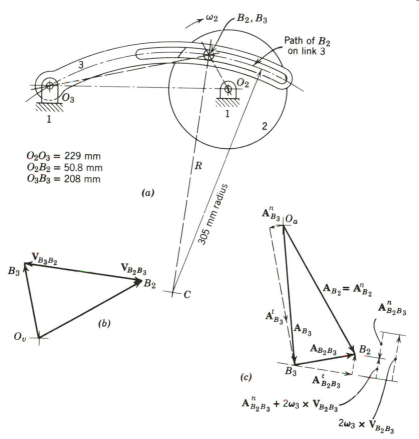

$O_2O_3 = 229$ mm
$O_2B_2 = 50.8$ mm
$O_3B_3 = 208$ mm

(a)

(b)

(c)

FIGURE 8.36

III. $\mathbf{A}_{B_2} = \mathbf{A}_{B_3} + \mathbf{A}_{B_2B_3}$

$\mathbf{A}_{B_2}^n + \mathbf{A}_{B_2}^t = \mathbf{A}_{B_3}^n + \mathbf{A}_{B_3}^t + \mathbf{A}_{B_2B_3}^n + \mathbf{A}_{B_2B_3}^t + 2\boldsymbol{\omega}_3 \times \mathbf{V}_{B_2B_3}$

where

$$\mathbf{A}_{B_2}^n = \frac{V_{B_2}^2}{O_2B_2} = \frac{2540^2}{50.8} = 127{,}000 \text{ mm/s}^2, \text{ direction from } B_2 \text{ toward } O_2$$

$$\mathbf{A}_{B_2}^t = 0 \quad (\alpha_2 = 0)$$

$$\mathbf{A}_{B_3}^n = \frac{V_{B_3}^2}{O_3B_3} = \frac{1650^2}{208} = 13{,}100 \text{ mm/s}^2, \text{ direction from } B_3 \text{ toward } O_3$$

$$\mathbf{A}_{B_3}^t = \text{direction perpendicular to } \mathbf{A}_{B_3}^n, \text{ magnitude unknown}$$

$$\mathbf{A}_{B_2B_3}^n = \frac{V_{B_2B_3}^2}{R} = \frac{2540^2}{305} = 21{,}200 \text{ mm/s}^2, \text{ direction from } B_2 \text{ toward } C$$

$$2\boldsymbol{\omega}_3 \times \mathbf{V}_{B_2B_3} = 2(7.93)2540 = 40{,}300 \text{ mm/s}^2, \text{ direction perpendicular to } \mathbf{V}_{B_2B_3}$$

$$\mathbf{A}_{B_2B_3}^t = \text{direction perpendicular to } 2\boldsymbol{\omega}_3 \times \mathbf{V}_{B_2B_3}, \text{ magnitude unknown}$$

Measure on the polygon of Fig. 8.36c, $\mathbf{A}_{B_3} = 122{,}000$ mm/s^2, $\mathbf{A}_{B_3}^t = 120{,}000$ mm/s^2, and

$$\alpha_3 = \frac{A_{B_3}^t}{O_3 B_3} = \frac{120{,}000}{208} = 577 \text{ rad/s}^2 \quad \text{(cw)}$$

Link 3 is a guide link which constrains point B_2 on link 2 to follow a circular path on link 3. Points B_2 and B_3 on link 3 are coincident, and the circular guide path is the relative path of B_2 on link 3. Therefore, the vectors $\mathbf{V}_{B_2 B_3}$ and $\mathbf{A}_{B_2 B_3}$ are involved in the analysis.

The velocity polygon of Fig. 8.36b shows the determination of $\mathbf{V}_{B_3}$ and $\mathbf{V}_{B_3 B_2}$ from Eq. I. The calculation for ω_3 is also shown.

Equation II gives $\mathbf{A}_{B_3}$ in terms of $\mathbf{A}_{B_2}$ and $\mathbf{A}_{B_3 B_2}$. Because the path of B_2 relative to B_3 is known to be a circular arc and the path of B_3 relative to B_2 is not easily determined, Eq. II is rewritten in the form of Eq. III so as to use the component $\mathbf{A}_{B_2 B_3}$.

All of the components of Eq. III are known as indicated in magnitude, sense, and direction, or in direction only. The acceleration polygon of Fig. 8.36c is started with the right side of Eq. III by drawing the vector $\mathbf{A}_{B_3}^n$ followed by the direction of $\mathbf{A}_{B_3}^t$. This is all that can be laid off on the right side of Eq. III at the moment, so consider the left side of the equation and draw the vector $\mathbf{A}_{B_2}$. The vectors $\mathbf{A}_{B_2 B_3}^n$ and $2\boldsymbol{\omega}_3 \times \mathbf{V}_{B_2 B_3}$ have opposite sense. Determine the resultant of these two vectors, and add it to the polygon so that its tip meets the tip of vector $\mathbf{A}_{B_2}$. Draw $\mathbf{A}_{B_2 B_3}^t$ perpendicular to $\mathbf{A}_{B_2 B_3}^n$ until it intersects the direction of the vector representing $\mathbf{A}_{B_3}^t$; this completes the polygon. Arrowheads are now added to the vectors $\mathbf{A}_{B_3}^t$ and $\mathbf{A}_{B_2 B_3}^t$ so that the addition of the vectors of the polygon agrees with the addition of the terms of Eq. III. The magnitude and sense of α_3 can now be determined from $\mathbf{A}_{B_3}^t$ as shown.

8.22 RELATIVE ACCELERATION OF COINCIDENT PARTICLES AT THE POINT OF CONTACT OF ROLLING ELEMENTS

An important type of constraint in mechanisms is that which occurs because one link is constrained to roll on another link without relative sliding of the two surfaces at the point of contact. In Fig. 8.37 are shown the rolling pitch circles of a pair of gears in mesh with particles P_3 on link 3 and P_2 on link 2 coincident in position at the point of contact of the rolling circles. As concluded in an earlier paragraph, the relative velocity $\mathbf{V}_{P_3 P_2}$ of the coincident particles is zero, and the absolute velocities $\mathbf{V}_{P_3}$ and $\mathbf{V}_{P_2}$ are identical.

The relative acceleration $\mathbf{A}_{P_3 P_2}$ of the coincident particles may be represented by component accelerations, a component $\mathbf{A}_{P_3 P_2}^t$ in the t–t direction of the common tangent to the surfaces at the point of contact, and a component $\mathbf{A}_{P_3 P_2}^n$ in a direction normal to the surfaces at the point of contact. The tangential component of relative acceleration $\mathbf{A}_{P_3 P_2}^t$ is the vector difference of the absolute tangential accelerations $\mathbf{A}_{P_3}^t$ and $\mathbf{A}_{P_2}^t$ shown in Fig. 8.37. Like the tangential velocities $\mathbf{V}_{P_3}$ and $\mathbf{V}_{P_2}$, the tangential accelerations $\mathbf{A}_{P_3}^t$ and $\mathbf{A}_{P_2}^t$ are identical because of the condition of no slipping of the surfaces at the point of contact. No slipping requires that there be no relative motion of the two particles in the direction of possible sliding, which is the tangent direction. Thus, because $\mathbf{A}_{P_3}^t$ and $\mathbf{A}_{P_2}^t$ are identical, the tangential component of acceleration of P_3 relative to P_2 is zero.

The normal component of relative acceleration $\mathbf{A}_{P_3 P_2}^n$ is the vector difference

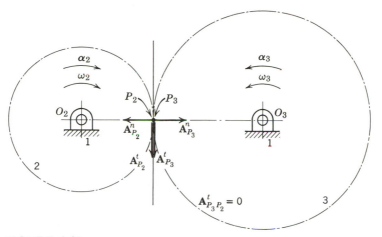

FIGURE 8.37

of the absolute accelerations $\mathbf{A}_{P_3}^n$ and $\mathbf{A}_{P_2}^n$ shown in Fig. 8.37 in the normal direction. It may be seen that the absolute normal acceleration of P_3 is toward O_3 and that of P_2 is toward O_2. These are parallel vectors, but the senses of the vectors are opposite so that the magnitude of $\mathbf{A}_{P_3P_2}^n$ is the sum of the magnitudes of $\mathbf{A}_{P_3}^n$ and $\mathbf{A}_{P_2}^n$. Thus, it is important to observe that a normal relative acceleration $\mathbf{A}_{P_3P_2}^n$ exists although the tangential relative acceleration is zero.

In a mechanism such as is shown in Fig. 8.37 where the centers of the gears are fixed, it is not necessary to draw an acceleration polygon to determine $\mathbf{A}_{P_3}$ and α_3. The angular acceleration α_3 can easily be determined from α_2 and from the ratio of the gear radii using the fact that $\mathbf{A}_{P_3}^t = \mathbf{A}_{P_2}^t$. After α_3 and ω_3 have been found, the components $\mathbf{A}_{P_3}^n$ and $\mathbf{A}_{P_3}^t$ can be calculated and combined to give $\mathbf{A}_{P_3}$. In more complex cases where gear centers are in motion, as in the following example, it is recommended that solutions be undertaken using polygon construction.

Example 8.11. In the mechanism shown in Fig. 8.38a, gear 2 rotates about O_2 with a constant angular velocity ω_2 of 10 rad/s, and gear 3 rolls on gear 2. Determine the acceleration $\mathbf{A}_{P_3}$ of point P_3 on gear 3 and the velocity and acceleration images of gears 2 and 3. Velocity and acceleration equations can be written as follows:

I. $\mathbf{V}_B = \mathbf{V}_A + \mathbf{V}_{BA}$

II. $\mathbf{V}_{P_2} = \mathbf{V}_A + \mathbf{V}_{P_2A}$

where

$\mathbf{V}_B$ = direction perpendicular to O_4B, magnitude unknown

$\mathbf{V}_A = (O_2A)\omega_2 = (2)10 = 20$ in./s, direction perpendicular to O_2A

$\mathbf{V}_{BA}$ = direction perpendicular to line joining points B and A, magnitude unknown

$\mathbf{V}_{P_2}$ = direction perpendicular to O_2P_2, magnitude unknown

$\mathbf{V}_{P_2A}$ = direction perpendicular to P_2A, magnitude unknown

Measured on the polygon, $V_B = 16$ in./s, $V_{BA} = 16$ in./s, $V_{P_2} = 41$ in./s, $V_{P_2A} = 25$ in./s, and $V_{P_3B} = 41$ in./s.

III. $\mathbf{A}_B = \mathbf{A}_A + \mathbf{A}_{BA}$

$\qquad \mathbf{A}_B^n + \mathbf{A}_B^t = \mathbf{A}_A^n + \mathbf{A}_A^t + \mathbf{A}_{BA}^n + \mathbf{A}_{BA}^t$

where

$$\mathbf{A}_B^n = \frac{V_B^2}{O_4 B} = \frac{16^2}{8} = 32 \text{ in./s}^2, \text{ direction from } B \text{ toward } O_4$$

$\mathbf{A}_B^t = $ direction perpendicular to $\mathbf{A}_B^n$, magnitude unknown

$$\mathbf{A}_A^n = \frac{V_A^2}{O_2 A} = \frac{20^2}{2} = 200 \text{ in./s}^2, \text{ direction from } A \text{ toward } O_2$$

$\mathbf{A}_A^t = 0 \qquad (\alpha_2 = 0)$

$$\mathbf{A}_{BA}^n = \frac{V_{BA}^2}{BA} = \frac{16^2}{4} = 64 \text{ in./s}^2, \text{ direction from } B \text{ toward } A$$

$\mathbf{A}_{BA}^t = $ direction perpendicular to $\mathbf{A}_{BA}^n$, magnitude unknown

IV. $\mathbf{A}_{P_2} = \mathbf{A}_A + \mathbf{A}_{P_2A}^n + \mathbf{A}_{P_2A}^t$

where

$\mathbf{A}_{P_2} = $ direction unknown, magnitude unknown

$$\mathbf{A}_{P_2A}^n = \frac{V_{P_2A}^2}{P_2 A} = \frac{25^2}{2.5} = 250 \text{ in./s}^2, \text{ direction from } P_2 \text{ toward } A$$

$\mathbf{A}_{P_2A}^t = 0 \qquad (\alpha_2 = 0)$

V. $\mathbf{A}_{P_3} = \mathbf{A}_{P_2} + \mathbf{A}_{P_3P_2}^n + \mathbf{A}_{P_3P_2}^t$

where

$\qquad \mathbf{A}_{P_3} = $ direction unknown, magnitude unknown

$\qquad \mathbf{A}_{P_3P_2}^n = $ direction parallel to line AB, magnitude unknown

$\qquad \mathbf{A}_{P_3P_2}^t = 0$

VI. $\mathbf{A}_{P_3} = \mathbf{A}_B + \mathbf{A}_{P_3B}^n + \mathbf{A}_{P_3B}^t$

where

$\qquad \mathbf{A}_{P_3} = $ direction unknown, magnitude unknown

$$\mathbf{A}_{P_3B}^n = \frac{V_{P_3B}^2}{P_3 B} = \frac{41^2}{1.5} = 1120 \text{ in./s}^2, \text{ direction from } P_3 \text{ toward } B$$

$\mathbf{A}_{P_3B}^t = $ direction perpendicular to $\mathbf{A}_{P_3B}^n$, magnitude unknown

Measured on the polygon, $A_{P_3} = 965 \text{ in./s}^2$.

It may be seen that the motions of the centers of the gears at A and B are the same as the pins of an equivalent four-bar linkage connecting points O_2, A, B, and O_4. The velocity

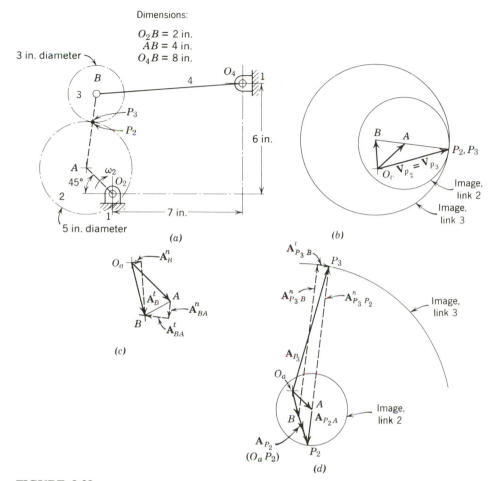

FIGURE 8.38

polygon in Fig. 8.38*b* shows the determination of $\mathbf{V}_B$ and $\mathbf{V}_{BA}$ from Eq. I. In a similar manner, $\mathbf{V}_{P_2}$ and $\mathbf{V}_{P_2A}$ are determined from Eq. II. The point P_3 is also known because $\mathbf{V}_{P_2} = \mathbf{V}_{P_3}$. The velocity image of link 2 is a circle with point A as the center and a radius AP_2. The velocity image of link 3 is a circle with point B as the center and a radius BP_3.

The acceleration polygon of Fig. 8.38*c* shows the determination of $\mathbf{A}_B$ and $\mathbf{A}_{BA}$ from Eq. III whose components are known as indicated.

To have a clearer diagram for determining $\mathbf{A}_{P_2}$ and $\mathbf{A}_{P_3}$, the acceleration vectors $\mathbf{A}_A$ and $\mathbf{A}_B$ from Fig. 8.38*c* are redrawn to a different scale in Fig. 8.38*d*. The acceleration polygon shows the determination of $\mathbf{A}_{P_2}$ from Eq. IV knowing $\mathbf{A}_A$ and $\mathbf{A}_{P_2A}^n(\mathbf{A}_{P_2A}^t = 0)$. The acceleration image of link 2 is a circle with point A as the center and a radius AP_2.

To determine $\mathbf{A}_{P_3}$, it is necessary to use Eqs. V and VI. The polygon shows the vector $\mathbf{A}_{P_2}$ to which is added the direction of $\mathbf{A}_{P_3P_2}^n$ from Eq. V ($\mathbf{A}_{P_3P_2}^t = 0$). Next, from Eq. VI, the vector $\mathbf{A}_{P_3B}^n$ and the direction of $\mathbf{A}_{P_3B}^t$ are added to the vector $\mathbf{A}_B$. The intersection of the direction of $\mathbf{A}_{P_3P_2}^n$ and the direction of $\mathbf{A}_{P_3B}^t$ closes the polygon and determines point P_3. The image of link 3 is a circle with B as a center and a radius BP_3.

8.23 ANALYTICAL VECTOR SOLUTION OF RELATIVE VELOCITY AND ACCELERATION EQUATIONS

Another method of velocity and acceleration analysis is to use the equations of relative motion but to express the components of these equations in unit vector form. By doing this, an analytical solution can be developed in place of a graphical one which uses velocity and acceleration polygons.

Consider the four-bar linkage shown in Fig. 8.39. The velocity and acceleration equation for points A and B can be expressed as follows:

$$\mathbf{V}_B = \mathbf{V}_A + \mathbf{V}_{BA}$$

$$\mathbf{A}_B = \mathbf{A}_A + \mathbf{A}_{BA}$$

$$\mathbf{A}_B^n + \mathbf{A}_B^t = \mathbf{A}_A^n + \mathbf{A}_A^t + \mathbf{A}_{BA}^n + \mathbf{A}_{BA}^t$$

The basic equations from which the magnitudes of the above components can readily be calculated are

$$V = r\omega \tag{8.34}$$

$$A^n = V\omega \tag{8.35}$$

$$A^t = r\alpha \tag{8.36}$$

It is obvious that these equations cannot give direction or sense. By writing them as cross products, however, directions as well as magnitudes can easily be determined. Equations 8.34, 8.35, and 8.36 may be rewritten as cross products as follows:

$$\mathbf{V} = \boldsymbol{\omega} \times \mathbf{r} \tag{8.37}$$

$$\mathbf{A}^n = \boldsymbol{\omega} \times \mathbf{V} = \boldsymbol{\omega} \times (\boldsymbol{\omega} \times \mathbf{r}) \tag{8.38}$$

$$\mathbf{A}^t = \boldsymbol{\alpha} \times \mathbf{r} = \dot{\boldsymbol{\omega}} \times \mathbf{r} \tag{8.39}$$

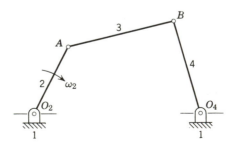

FIGURE 8.39

To illustrate this method in general terms, consider Eq. 8.37 and let

$$\boldsymbol{\omega} = \omega\mathbf{k}$$

and

$$\mathbf{r} = x_A\mathbf{i} + y_A\mathbf{j}$$

By keeping in mind that

$$\mathbf{i} \times \mathbf{i} = \mathbf{j} \times \mathbf{j} = \mathbf{k} \times \mathbf{k} = 0$$
$$\mathbf{i} \times \mathbf{j} = -\mathbf{j} \times \mathbf{i} = \mathbf{k}$$
$$\mathbf{j} \times \mathbf{k} = -\mathbf{k} \times \mathbf{j} = \mathbf{i}$$
$$\mathbf{k} \times \mathbf{i} = -\mathbf{i} \times \mathbf{k} = \mathbf{j}$$

Eq. 8.37 can be expanded to give

$$\mathbf{V} = \boldsymbol{\omega} \times \mathbf{r} = -\omega y_A\mathbf{i} + \omega x_A\mathbf{j}$$

from which the magnitude, direction, and sense of the velocity of the point in question can easily be determined. It is easier, however, to solve Eq. 8.37 if it is written as a determinant; thus,

$$\mathbf{V} = \boldsymbol{\omega} \times \mathbf{r} = \begin{vmatrix} \mathbf{i} & \mathbf{j} & \mathbf{k} \\ 0 & 0 & \omega \\ x_A & y_A & 0 \end{vmatrix} = -\omega y_A\mathbf{i} + \omega x_A\mathbf{j}$$

Consider next Eq. 8.38 written as a determinant:

$$A^n = \boldsymbol{\omega} \times (\boldsymbol{\omega} \times \mathbf{r}) = \omega\mathbf{k} \times \begin{vmatrix} \mathbf{i} & \mathbf{j} & \mathbf{k} \\ 0 & 0 & \omega \\ x_A & y_A & 0 \end{vmatrix}$$

$$= \begin{vmatrix} \mathbf{i} & \mathbf{j} & \mathbf{k} \\ 0 & 0 & \omega \\ -\omega y_A & \omega x_A & 0 \end{vmatrix} = -\omega^2 x_A\mathbf{i} - \omega^2 y_A\mathbf{j}$$

From the above discussion, illustrated with Eqs. 8.37 and 8.38, it can be seen that it is an easy matter to express the components of the relative motion equations in unit vector form. A complete solution is then obtained by substituting in the equations of relative motion and summing the $\mathbf{i}$ and $\mathbf{j}$ components. A complete solution is illustrated in the following example, which analyzes the same mechanism as Example 8.2.

Example 8.12. Consider the Geneva mechanism as analyzed in Example 8.2, and let it now be required to determine ω_3 and α_3 using the equations of relative motion with components expressed in unit vector form. A skeleton diagram of the mechanism is shown in Fig. 8.40.

The velocity and acceleration equations are written as follows considering the velocity and acceleration of point A_2 relative to A_3 because the path of point A_2 is known to be a straight line relative to A_3.

I. $\mathbf{V}_{A_2} = \mathbf{V}_{A_3} + \mathbf{V}_{A_2 A_3}$

where

$$\mathbf{V}_{A_2} = \boldsymbol{\omega}_2 \times \mathbf{r}_2$$

$$\mathbf{V}_{A_3} = \boldsymbol{\omega}_3 \times \mathbf{r}_3$$

$$\mathbf{V}_{A_2 A_3} = V_{A_2 A_3}(\sin 40.5\mathbf{i} + \cos 40.5\mathbf{j}) = V_{A_2 A_3}(0.6494\mathbf{i} + 0.7604\mathbf{j}) = 0.6494 V_{A_2 A_3}\mathbf{i} + 0.7604 V_{A_2 A_3}\mathbf{j}$$

$$\boldsymbol{\omega}_2 = -10\mathbf{k} \text{ rad/s}$$

$$\boldsymbol{\omega}_3 = \omega_3\mathbf{k}$$

$$\mathbf{r}_2 = r_2(-\sin 25.1\mathbf{i} + \cos 25.1\mathbf{j}) = 50.8(-0.4242\mathbf{i} + 0.9056\mathbf{j}) = -21.55\mathbf{i} + 46.00\mathbf{j}$$

$$\mathbf{r}_3 = r_3(-\sin 40.5\mathbf{i} - \cos 40.5\mathbf{j}) = 33.0(-0.6494\mathbf{i} - 0.7604\mathbf{j}) = -21.43\mathbf{i} - 25.09\mathbf{j}$$

Substituting the values of $\boldsymbol{\omega}_2$, $\mathbf{r}_2$ and $\boldsymbol{\omega}_3$, $\mathbf{r}_3$ into the equations for $\mathbf{V}_{A_2}$ and $\mathbf{V}_{A_3}$, respectively, gives

$$\mathbf{V}_{A_2} = \boldsymbol{\omega}_2 \times \mathbf{r}_2 = \begin{vmatrix} \mathbf{i} & \mathbf{j} & \mathbf{k} \\ 0 & 0 & -10 \\ -21.55 & 46.00 & 0 \end{vmatrix} = 460.0\mathbf{i} + 215.5\mathbf{j}$$

$$\mathbf{V}_{A_3} = \boldsymbol{\omega}_3 \times \mathbf{r}_3 = \begin{vmatrix} \mathbf{i} & \mathbf{j} & \mathbf{k} \\ 0 & 0 & \omega_3 \\ -21.43 & -25.09 & 0 \end{vmatrix} = 25.09\omega_3\mathbf{i} - 21.43\omega_3\mathbf{j}$$

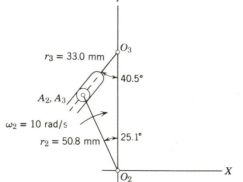

FIGURE 8.40

Substituting the above values for $\mathbf{V}_{A_2}$, $\mathbf{V}_{A_3}$, and $\mathbf{V}_{A_2A_3}$ into Eq. I, gives the following:

$$460.0\mathbf{i} + 215.5\mathbf{j} = 25.09\omega_3\mathbf{i} - 21.43\omega_3\mathbf{j} + 0.6494\,V_{A_2A_3}\mathbf{i} + 0.7604\,V_{A_2A_3}\mathbf{j}$$

By summing **i** *components,*

$$460.0\mathbf{i} = 25.09\omega_3\mathbf{i} + 0.6494\,V_{A_2A_3}\mathbf{i}$$

By summing **j** *components,*

$$215.5\mathbf{j} = -21.43\omega_3\mathbf{j} + 0.7604\,V_{A_2A_3}\mathbf{j}$$

Therefore,

$$25.09\omega_3 + 0.6494\,V_{A_2A_3} = 460.0$$

$$-21.43\omega_3 + 0.7604\,V_{A_2A_3} = 215.5$$

Multiplying the second equation by 25.09/21.43 and adding the two equations gives

$$1.540\,V_{A_2A_3} = 712.3$$

$$V_{A_2A_3} = 462.5 \text{ mm/s}$$

Therefore,

$$\mathbf{V}_{A_2A_3} = 462.5(0.6494\mathbf{i} + 0.7604\mathbf{j}) = 300.3\mathbf{i} + 351.7\mathbf{j}$$

and

$$\omega_3 = \frac{460.0 - (0.6494)(462.5)}{25.09} = 6.36 \text{ rad/s} \qquad \text{(ccw)}$$

II. $\mathbf{A}_{A_2} = \mathbf{A}_{A_3} + \mathbf{A}_{A_2A_3}$

$$\mathbf{A}_{A_2}^n + \mathbf{A}_{A_2}^t = \mathbf{A}_{A_3}^n + \mathbf{A}_{A_3}^t + \mathbf{A}_{A_2A_3}^n + 2\omega_3 \times \mathbf{V}_{A_2A_3} + \mathbf{A}_{A_2A_3}^t$$

where

$$\mathbf{A}_{A_2}^n = \omega_2 \times \mathbf{V}_{A_2} = \omega_2 \times (\omega_2 \times \mathbf{r}_2)$$

$$= -10\mathbf{k} \times \begin{vmatrix} \mathbf{i} & \mathbf{j} & \mathbf{k} \\ 0 & 0 & -10 \\ -21.55 & 46.00 & 0 \end{vmatrix}$$

$$= \begin{vmatrix} \mathbf{i} & \mathbf{j} & \mathbf{k} \\ 0 & 0 & -10 \\ 460.0 & 215.5 & 0 \end{vmatrix}$$

$$\mathbf{A}_{A_2}^n = 2155\mathbf{i} - 4600\mathbf{j}$$

$$\mathbf{A}_{A_2}^t = 0 \qquad (\alpha_2 = 0)$$

$$\mathbf{A}_{A_3}^n = \boldsymbol{\omega}_3 \times \mathbf{V}_{A_3} = \boldsymbol{\omega}_3 \times (\boldsymbol{\omega}_3 \times \mathbf{r}_3)$$

$$= 6.36\mathbf{k} \times \begin{vmatrix} \mathbf{i} & \mathbf{j} & \mathbf{k} \\ 0 & 0 & 6.36 \\ -21.43 & -25.09 & 0 \end{vmatrix}$$

$$= \begin{vmatrix} \mathbf{i} & \mathbf{j} & \mathbf{k} \\ 0 & 0 & 6.36 \\ 159.6 & -136.3 & 0 \end{vmatrix}$$

$$\mathbf{A}_{A_3}^n = 866.9\mathbf{i} + 1015\mathbf{j}$$

$$\mathbf{A}_{A_3}^t = \dot{\boldsymbol{\omega}}_3 \times \mathbf{r}_3 = \begin{vmatrix} \mathbf{i} & \mathbf{j} & \mathbf{k} \\ 0 & 0 & \dot{\omega}_3 \\ -21.43 & -25.09 & 0 \end{vmatrix} = 25.09\dot{\omega}_3\mathbf{i} - 21.43\dot{\omega}_3\mathbf{j}$$

$$\mathbf{A}_{A_2A_3}^n = 0 \quad (\mathbf{R} = \infty)$$

$$2\boldsymbol{\omega}_3 \times \mathbf{V}_{A_2A_3} = \begin{vmatrix} \mathbf{i} & \mathbf{j} & \mathbf{k} \\ 0 & 0 & 12.70 \\ 300.3 & 351.7 & 0 \end{vmatrix} = -4474\mathbf{i} + 3820\mathbf{j}$$

$$\mathbf{A}_{A_2A_3}^t = A_{A_2A_3}^t(0.6494\mathbf{i} + 0.7604\mathbf{j})$$

$$= 0.6494A_{A_2A_3}^t\mathbf{i} + 0.7604A_{A_2A_3}^t\mathbf{j}$$

Substituting the above values into the component form of Eq. II gives

$$2155\mathbf{i} - 4600\mathbf{j} = 866.9\mathbf{i} + 1015\mathbf{j} + 25.09\dot{\omega}_3\mathbf{i} - 21.43\dot{\omega}_3\mathbf{j}$$
$$- 4474\mathbf{i} + 3820\mathbf{j} + 0.6494A_{A_2A_3}^t\mathbf{i} + 0.7604A_{A_2A_3}^t\mathbf{j}$$

By summing **i** *components,*

$$2155\mathbf{i} = 866.9\mathbf{i} + 25.09\dot{\omega}_3\mathbf{i} - 4474\mathbf{i} + 0.6494A_{A_2A_3}^t\mathbf{i}$$

By summing **j** *components,*

$$-4600\mathbf{j} = 1015\mathbf{j} - 21.43\dot{\omega}_3\mathbf{j} + 3820\mathbf{j} + 0.7604A_{A_2A_3}^t\mathbf{j}$$

Therefore,

$$25.09\dot{\omega}_3 + 0.6494A_{A_2A_3}^t = 5762$$

$$-21.43\dot{\omega}_3 + 0.7604A_{A_2A_3}^t = -9435$$

Multiplying the second equation by 25.09/21.43 and adding the two equations gives

$$1.540A_{A_2A_3}^t = -5288$$

$$A_{A_2A_3}^t = -3434 \text{ mm/s}^2$$

Therefore,

$$\mathbf{A}_{A_2A_3}^t = -3434(0.6494\mathbf{i} + 0.7604\mathbf{j}) = -2230\mathbf{i} - 2611\mathbf{j}$$

and

$$\dot{\omega}_3 = \frac{5762 - (0.6494)(-3434)}{25.09} = 319 \text{ rad/s}^2 \qquad \text{(ccw)}$$

$$\mathbf{A}_{A_3}^t = 25.09\dot{\omega}_3\mathbf{i} - 21.43\dot{\omega}_3\mathbf{j}$$

$$= 319(25.09\mathbf{i} - 21.43\mathbf{j})$$

$$\mathbf{A}_{A_3}^t = 8004\mathbf{i} - 6836\mathbf{j}$$

$$|\mathbf{A}_{A_3}^t| = \sqrt{8004^2 + 6836^2} = 10{,}530 \text{ mm/s}^2$$

$$\mathbf{A}_{A_3}^n = 866.9\mathbf{i} + 1015\mathbf{j}$$

$$|\mathbf{A}_{A_3}^n| = \sqrt{866.9^2 + 1015^2} = 1335 \text{ mm/s}^2$$

$$\mathbf{A}_{A_3} = \mathbf{A}_{A_3}^n + \mathbf{A}_{A_3}^t$$

$$|\mathbf{A}_{A_3}| = \sqrt{1335^2 + 10{,}530^2}$$

$$= 10{,}610 \text{ mm/s}^2$$

The results of this example, worked using the equations of relative motion, are seen to agree with the results of Example 8.2, which was worked using vector methods with fixed and moving coordinate systems.

8.24 VELOCITY AND ACCELERATION ANALYSIS BY NUMERICAL OR GRAPHICAL DIFFERENTIATION

A kinematic method that should not be overlooked is numerical or graphical differentiation. The displacement curve is differentiated to obtain the velocity curve, and the velocity curve is then differentiated to obtain the acceleration curve. Once the displacement curve is known, either of these methods may be used regardless of the type or complexity of the mechanism being analyzed. Graphical techniques, such as the one to be demonstrated in this section, are often inaccurate, especially in obtaining the acceleration curves. Fortunately, numerical differentiation based on finite difference methods are ideally suited for programming on a digital computer, and it is possible to obtain very accurate results. All the graphical and numerical methods use one or more points on a curve to approximate the derivative. In the graphical approach, the slope is found by drawing the tangent to the curve at the specified point. The method is illustrated graphically in Fig. 8.41 for a linkage in which the driving link 2 rotates at constant angular velocity and the driven link 4 oscillates as shown. Twelve phases of the mechanism are shown to scale K_s for equal increments of time as given by the equal angular displacements of link 2. The velocity and acceleration of point B are desired. Curves are shown for the coordinate displacements X and Y of the point B as it traverses its curvilinear path. See Fig. 8.41b.

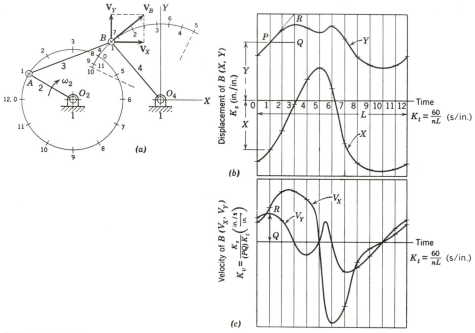

FIGURE 8.41

The abscissa of the displacement curve is a line of arbitrary length L divided into 12 equal parts to represent equal time intervals in one revolution of link 2. Since the time for one revolution of link 2 is $1/n$ min, or $60/n$ sec (n = rpm), the time scale for the abscissa is $K_t = 60/nL$ s/in. The displacements X, Y of point B are shown on the ordinate of the displacement curve to the same scale K_s as used in the layout of the mechanism.

Graphical differentiation is accomplished by drawing a tangent to the displacement curve at some point (such as for position 1 in Fig. 8.41b) and determining the slope of the curve from the triangle PRQ. The slope represents velocity or the derivative of displacement with respect to time:

$$V = \frac{(QR)K_s}{(PQ)K_t} \qquad (8.40)$$

In Eq. 8.40, the lengths QR and PQ are inches on the paper, and the scales K_s and K_t are required to convert slope to units of velocity. If K_s is in inches per inch and K_t in seconds per inch, velocity is then in inches per second.

To plot a curve of velocity against time as shown in Fig. 8.41c, slopes at the incremental points on the displacement curve are evaluated graphically. If PQ is taken as the same length for all triangles drawn to determine slope, then the distance QR is the variable showing the variations in velocity. QR may be transferred from the triangle of the displacement curve to the velocity curve as the ordinate. As shown in Fig. 8.41c, the velocity curves for the coordinate velocities $\mathbf{V}_X$ and $\mathbf{V}_Y$ of point B are plotted. However, since QR is in inches, the velocity scale K_v for the curves must be determined:

$$V = (QR)K_v$$

$$K_v = \frac{V}{QR}$$

By substituting for V from Eq. 8.40,

$$K_v = \frac{(QR)K_s}{(QR)(PQ)K_t}$$

$$K_v = \frac{K_s}{(PQ)K_t} \tag{8.41}$$

Thus, Eq. 8.41 gives the velocity scale in terms of the other scales and the length PQ, which, although an arbitrarily chosen length in inches, is the same for all triangles.

The velocity $\mathbf{V}_B$ is the vector resultant of the component coordinate velocities $\mathbf{V}_X$ and $\mathbf{V}_Y$. As shown in Fig. 8.41a, for position 1 of the mechanism $\mathbf{V}_B$ is the resultant of its components and should be normal to line O_4B. As shown by inspection of the velocity curves, the maximum velocity of point B is near positions 6 and 7. Also, the curves show that at the extreme positions of link 4, namely, positions 5 and 10, the velocity of B is zero.

To determine the coordinate accelerations $\mathbf{A}_X$ and $\mathbf{A}_Y$ of B, the velocity curves may be differentiated graphically in a similar manner and curves may be shown of acceleration against time. The acceleration scale may be calculated from the following expression:

$$K_a = \frac{K_v}{(P'Q')K_t} \tag{8.42}$$

where K_a is the acceleration scale and $P'Q'$ is an arbitrary length similar to PQ.

The accuracy of differentiating graphically depends on the care taken in drawing tangents and on the number of increments into which the abscissa of the displacement curve is divided. Accuracy increases as the number of increments is increased and the individual increments are made smaller.

As has been shown above, graphical differentiation is a very simple method of determining velocity and acceleration curves from a displacement–time curve when a complete cycle of a mechanism is to be analyzed. The method is rapid in plotting one curve from another but, unfortunately, the accuracy is limited. It is obvious that in a case where the equations for displacement, velocity, and acceleration are readily available, as in the slider-crank mechanism, it is easier to calculate the values and plot the curves if desired than to resort to graphical differentiation. In other mechanisms, however, such as that shown in Fig. 8.41, graphical differentiation is much quicker than analytical methods provided that sufficient accuracy can be obtained.

The accuracy of this method can be greatly improved by using a digital computer to perform the differentiation instead of doing it graphically. This can easily be done if the displacement–time values, or the equation from which they can be calculated, are available. The example that follows shows a comparison of the values of velocity for the piston in a slider-crank mechanism found by computer differentiation and by formula.

Example 8.13. A slider-crank mechanism with a crank of 2 in. and a connecting rod of 8 in. operates at a crank speed of 3300 rpm. Determine the piston velocity (ft/s) for 90° of crank rotation starting from top dead center in increments of 1° by the following methods:

1. Numerical differentiation of the displacement–time values calculated from the equation

$$x = R(1 - \cos \theta) + \frac{R^2}{2L} \sin^2 \theta$$

   from Chapter 2 using finite difference methods.
2. Direct calculation of velocity from the equation

$$V = R\omega\left(\sin \theta + \frac{R}{2L} \sin 2\theta\right)$$

Show the improvement in accuracy for numerical differentiation by also taking increments of 0.1° and 0.01°.

Solution

$$\omega = \frac{2\pi n}{60} = 345.40 \text{ rad/s}$$

Time for stroke (180°) = 0.00909 s

Time for 1° of crank rotation = 0.0000505 s

$$K_v = \frac{K_s}{(PQ)K_t} = \frac{\frac{1}{12}}{0.0000505} = 1650.1 \text{ ft/s/in.}$$

The space scale is taken full size and converted to ft/in. The term $(PQ)K_t$ is the value of the increment in seconds and changes if the increment is changed.

After the value of piston displacement x has been calculated for each angular increment, the change in displacement Δx between increments is determined. The value of Δx is proportional to the velocity for the particular point under consideration, and the product of Δx and K_v gives the velocity in ft/s, assuming a constant value of Δt. An illustration of this is shown in Fig. 8.42.

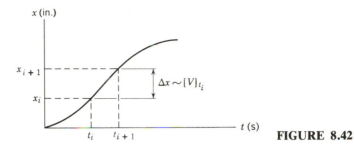

FIGURE 8.42

Values of velocities determined by formula and by numerical differentiation for increments of 1°, 0.1°, and 0.01° are shown in Table 8.1. It is interesting to note how closely the velocities by differentiation with 0.01° increments match the velocities calculated by formula.

Although not included in this example, piston accelerations can be determined in a similar manner from velocities.

It should be mentioned that the values of V with 1° increments will more nearly match those of V by formula if the latter are calculated at the midpoints of the intervals, that is, at 0.5°, 1.5°, 2.5°, and so on. This can be seen in the tabulation for the first 10 single-degree increments shown on the following page.

TABLE 8.1 Velocity Analysis by Numerical Differentiation

θ, deg	V, ft/s, Formula	V, ft/s, 1° Increment	V, ft/s, 0.1° Increment	V, ft/s, 0.01° Increment
1	1.26	0.63	1.19	1.25
2	2.51	1.88	2.45	2.51
3	3.77	3.14	3.70	3.76
4	5.02	4.39	4.96	5.01
5	6.27	5.65	6.21	6.26
6	7.52	6.89	7.45	7.49
7	8.76	8.14	8.70	8.76
8	10.00	9.38	9.94	10.01
9	11.23	10.62	11.17	11.23
10	12.46	11.85	12.40	12.45
20	24.33	23.76	24.27	24.31
30	35.03	34.53	34.98	35.04
40	44.11	43.70	44.07	44.11
50	51.21	50.91	51.18	51.22
60	56.11	55.92	56.09	56.02
70	58.75	58.67	58.74	58.69
80	59.18	59.21	59.19	59.17
90	57.59	57.72	57.61	57.59

θ, DEG	V, FT/S, FORMULA
0.5	0.628
1.5	1.883
2.5	3.138
3.5	4.391
4.5	5.642
5.5	6.890
6.5	8.135
7.5	9.376
8.5	10.613
9.5	11.844

The reason for this is that V by numerical differentiation more nearly represents the velocity at the midpoint instead of at the end of the interval. As the increments become smaller, this difference will decrease until it becomes negligible as in the case of the 0.01° increments.

8.25   KINEMATIC ANALYSIS BY COMPLEX NUMBERS

In addition to the methods of velocity and acceleration analysis already presented, analytical solutions with vectors expressed in complex form are often used.

A simple kinematic case is shown in Fig. 8.43a in which link 2 rotates about a fixed axis O_2. It is desired to determine the velocity and acceleration vectors $\mathbf{V}_P$ and $\mathbf{A}_P$ of particle P when the link is in the phase given by θ_2 and the known instantaneous angular velocity and angular acceleration are ω_2 and α_2.

The position of particle P may be represented by the vector $\mathbf{r}_P$ shown in Fig. 8.43b. By establishing real and imaginary axes as shown, $\mathbf{r}_P$ may be expressed by a complex number in any of the following equivalent forms:

$$\mathbf{r}_P = a + ib$$
$$\mathbf{r}_P = r_P(\cos \theta_2 + i \sin \theta_2) \tag{8.43}$$
$$\mathbf{r}_P = r_P e^{i\theta_2}$$

Although all forms of the complex number are useful ones, the simplest form for differentiation is the exponential form in which r_P is the magnitude of the position vector and $e^{i\theta_2}$ represents a vector of unit length at a counterclockwise angular position θ_2. Differentiation of Eq. 8.43 yields the velocity vector $\mathbf{V}_P$.

$$\mathbf{V}_P = \dot{\mathbf{r}}_P = r_P \dot{\theta}_2 (ie^{i\theta_2})$$
$$\mathbf{V}_P = r_P \omega_2 (ie^{i\theta_2}) \tag{8.44}$$

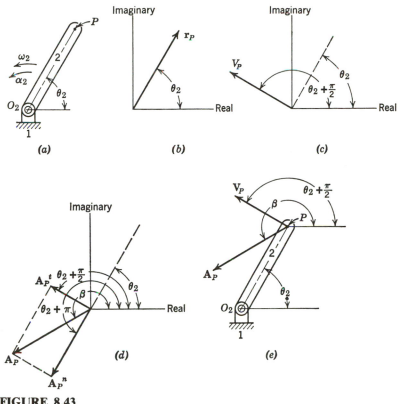

FIGURE 8.43

where $\dot{\mathbf{r}}_P = d\mathbf{r}_P/dt$ and $\dot{\theta}_2 = d\theta_2/dt = \omega_2$. The term in parenthesis in Eq. 8.44 is the unit vector multiplied by i and is equivalent to $i(\cos\theta_2 + i\sin\theta_2)$. By using trigonometric relationships, it may be shown that $i(\cos\theta_2 + i\sin\theta_2)$ is equal to $\cos(\theta_2 + \pi/2) + i\sin(\theta_2 + \pi/2)$ so that $ie^{i\theta_2} = e^{i(\theta_2 + \pi/2)}$. Thus,

$$\mathbf{V}_P = r_P\omega_2 e^{i(\theta_2 + \pi/2)} \tag{8.45}$$

As shown in Fig. 8.43c, the direction of the velocity vector $\mathbf{V}_P$ is given by the angle $(\theta_2 + \pi/2)$ and is shown to be at an angle 90° greater than the angle of $\mathbf{r}_P$. Thus, multiplication of the unit vector by i rotates the vector 90° in the counterclockwise sense. Also, each subsequent multiplication of the unit vector by i rotates the vector an additional 90° increment in the counterclockwise sense.

Differentiation of the velocity Eq. 8.44 gives the acceleration vector $\mathbf{A}_P$ as follows:

$$
\begin{aligned}
\mathbf{A}_P = \ddot{\mathbf{r}}_P &= r_P\omega_2^2(i^2e^{i\theta_2}) + r_P\dot{\omega}_2(ie^{i\theta_2}) \\
&= r_P\omega_2^2(i^2e^{i\theta_2}) + r_P\alpha_2(ie^{i\theta_2})
\end{aligned}
\tag{8.46}
$$

where $\alpha_2 = d\omega_2/dt = \dot{\omega}_2$. The first right-hand term of Eq. 8.46 represents the normal component of acceleration A_P^n, in which $r_P\omega_2^2$ is the magnitude and i^2 indicates that the direction is 180° greater than θ_2 as shown in Fig. 8.43d. The second term is the tangential component of acceleration A_P^t of magnitude $r_P\alpha_2$ and direction 90° greater than θ_2 as indicated by i. To designate the directions of the component accelerations, Eq. 8.46 may be rewritten as follows:

$$\mathbf{A}_P = r_P\omega_2^2 e^{i(\theta_2 + \pi)} + r_P\alpha_2 e^{i(\theta_2 + \pi/2)} \tag{8.47}$$

Equations 8.46 and 8.47 show that the acceleration vector $\mathbf{A}_P$ is the resultant of two perpendicular vectors. To determine the magnitude of the resultant vector and its angular position, the following algebraic steps may be made beginning with Eq. 8.46:

$$\mathbf{A}_P = -r_P\omega_2^2(\cos\theta_2 + i\sin\theta_2) + r_P\alpha_2(i\cos\theta_2 - \sin\theta_2)$$
$$= -(r_P\omega_2^2\cos\theta_2 + r_P\alpha_2\sin\theta_2) + i(-r_P\omega_2^2\sin\theta_2 + r_P\alpha_2\cos\theta_2)$$
$$= a + ib \tag{8.48}$$

As Eq. 8.48 shows, the acceleration $\mathbf{A}_P$ may also be expressed as the resultant of two component vectors in which "a" is the real component and "b" is the perpendicular imaginary component. The magnitude of $\mathbf{A}_P$ may be determined as follows:

$$\mathbf{A}_P = \sqrt{a^2 + b^2}$$
$$= \sqrt{(r_P\omega_2^2\cos\theta_2 + r_P\alpha_2\sin\theta_2)^2 + (-r_P\omega_2^2\sin\theta_2 + r_P\alpha_2\cos\theta_2)^2}$$
$$= \sqrt{(r_P\omega_2^2)^2 + (r_P\alpha_2)^2} \tag{8.49}$$

The direction of $\mathbf{A}_P$ is given by the angle β in Fig. 8.43d, and this angle may be determined as follows:

$$\tan\beta = \frac{b}{a} = \frac{(-\omega_2^2\sin\theta_2 + \alpha_2\cos\theta_2)}{-(\omega_2^2\cos\theta_2 + \alpha_2\sin\theta_2)} \tag{8.50}$$

Using the angle β, the acceleration vector $\mathbf{A}_P$ may be expressed as a single vector instead of two vectors as follows:

$$\mathbf{A}_P = A_P e^{i\beta} \tag{8.51}$$

In Fig. 8.43e, the velocity and acceleration vectors $\mathbf{V}_P$ and $\mathbf{A}_P$ are shown as fixed vectors at the particle P on the link.

It is important to note that the preceding relationships are based on the assumption that ω_2 and α_2 are known quantities for all phases θ_2 of the link. In many problems related to machinery, the link may rotate at constant angular velocity so that ω_2 is constant and α_2 is zero. If, for example, α_2 is not zero but is a constant, then ω_2 is a function of time or θ_2. Considering the case where

α_2 = constant = k, and ω_2 is zero at the initial condition θ_2 = 0, the dependence of ω_2 on θ_2 may be determined as follows:

$$\frac{d\omega_2}{dt} = \alpha_2 = k$$

By expressing the derivative $d\omega_2/dt$ as $(d\omega_2/d\theta_2)(d\theta_2/dt) = \omega_2(d\omega_2/d\theta_2)$,

$$\omega_2\left(\frac{d\omega_2}{d\theta_2}\right) = k$$

$$\int \omega_2 \, d\omega_2 = k \int d\theta_2$$

$$\omega_2^2 = 2k\theta_2 + C_1 \qquad\qquad \textbf{(8.52)}$$

C_1 is the constant of integration and is equal to zero for ω_2 = 0 at θ_2 = 0.

8.26 ANALYSIS OF THE SLIDER CRANK BY LOOP CLOSURE EQUATIONS AND COMPLEX NUMBERS

In the slider-crank mechanism of Fig. 8.44a, the crank rotates at constant angular velocity ω_2, and the velocity $\mathbf{V}_B$ and acceleration $\mathbf{A}_B$ of the slider are to be determined. As shown, the position of particle B relative to the fixed point O_2 is given by the vector $\mathbf{r}_B$. Referring to Fig. 8.44b, it may be seen that two independent vector equations may be written for $\mathbf{r}_B$, namely, $\mathbf{r}_B = \mathbf{r}_1$ and $\mathbf{r}_B = \mathbf{r}_2 + \mathbf{r}_3$.

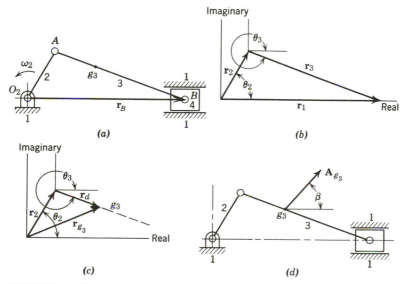

(a)

(b)

(c)

(d)

FIGURE 8.44

The obvious result of combining the equations for $\mathbf{r}_B$ is the following vector equation:

$$\mathbf{r}_1 = \mathbf{r}_2 + \mathbf{r}_3 \tag{8.53}$$

If the vectors are to be represented by complex numbers, real and imaginary axes may be shown as in Fig. 8.44b, and Eq. 8.53 may be written as follows:

$$r_1 e^{i\theta_1} = r_2 e^{i\theta_2} + r_3 e^{i\theta_3} \tag{8.54}$$

where r_2 and r_3 are the fixed lengths of links 2 and 3, and r_1 is a variable length giving the position of the slider. The angle θ_1 of r_1 is fixed at $\theta_1 = 0$ so that $e^{i\theta_1} = 1$. Thus,

$$\mathbf{r}_1 = r_2 e^{i\theta_2} + r_3 e^{i\theta_3} \tag{8.55}$$

Two successive differentiations of Eq. 8.55 yield expressions giving the velocity $\mathbf{V}_B$ and acceleration $\mathbf{A}_B$ as follows:

$$\mathbf{V}_B = \dot{\mathbf{r}}_1 = r_2 \omega_2 (ie^{i\theta_2}) + r_3 \omega_3 (ie^{i\theta_3}) \tag{8.56}$$

$$\mathbf{A}_B = \ddot{\mathbf{r}}_1 = r_2 (i\alpha_2 - \omega_2^2) e^{i\theta_2} + r_3 (i\alpha_3 - \omega_3^2) e^{i\theta_3} \tag{8.57}$$

It may be seen from an inspection of Eqs. 8.55, 8.56, and 8.57 that, although the differentiations are made to determine the kinematic values of particle B, the equations also involve the angular velocities and accelerations of links 2 and 3 as well as their angular positions. In these equations, r_2, r_3, θ_2, ω_2, and α_2 are the known quantities, and the unknown quantities to be determined are six in number, namely, r_1, $\dot{r}_1$, $\ddot{r}_1$, θ_3, ω_3, and α_3.

Two of the unknowns, r_1 and θ_3, may be determined from Eq. 8.55 by separately equating the real and imaginary parts of the equation as follows:

$$\mathbf{r}_1 = r_2(\cos \theta_2 + i \sin \theta_2) + r_3(\cos \theta_3 + i \sin \theta_3)$$

$$r_1 = r_2 \cos \theta_2 + r_3 \cos \theta_3 \quad \text{(real)} \tag{8.58}$$

$$0 = r_2 \sin \theta_2 + r_3 \sin \theta_3 \quad \text{(imaginary)} \tag{8.59}$$

Equation 8.59 may be solved to determine θ_3.

$$\theta_3 = \sin^{-1} \left(\frac{-r_2}{r_3} \sin \theta_2 \right) \tag{8.60}$$

Equation 8.58 may then be used to determine r_1.

In a similar manner, the unknowns $\dot{r}_1$ and ω_3 may be obtained from Eq.

8.56 by separately equating the real and imaginary parts of the equation:

$$\dot{r}_1 = r_2\omega_2(i \cos \theta_2 - \sin \theta_2) + r_3\omega_3(i \cos \theta_3 - \sin \theta_3)$$

$$\dot{r}_1 = -r_2\omega_2 \sin \theta_2 - r_3\omega_3 \sin \theta_3 \qquad \text{(real)} \qquad \textbf{(8.61)}$$

$$0 = r_2\omega_2 \cos \theta_2 + r_3\omega_3 \cos \theta_3 \qquad \text{(imaginary)} \qquad \textbf{(8.62)}$$

Equation 8.62 allows the determination of ω_3.

$$\omega_3 = -\omega_2\left(\frac{r_2 \cos \theta_2}{r_3 \cos \theta_3}\right) \qquad \textbf{(8.63)}$$

Equation 8.61 may then be used to determine $\dot{r}_1 = V_B$.

The remaining unknowns, $\ddot{r}_1$ and α_3, are determined from the real and imaginary parts of Eq. 8.57:

$$\ddot{r}_1 = -r_2(\omega_2^2 \cos \theta_2 + \alpha_2 \sin \theta_2)$$
$$\quad - r_3(\omega_3^2 \cos \theta_3 + \alpha_3 \sin \theta_3) \qquad \text{(real)} \qquad \textbf{(8.64)}$$

$$0 = r_2(\alpha_2 \cos \theta_2 - \omega_2^2 \sin \theta_2)$$
$$\quad + r_3(\alpha_3 \cos \theta_3 - \omega_3^2 \sin \theta_3) \qquad \text{(imaginary)} \qquad \textbf{(8.65)}$$

From Eq. 8.65, the unknown α_3 may be determined.

$$\alpha_3 = \frac{r_2(\omega_2^2 \sin \theta_2 - \alpha_2 \cos \theta_2)}{r_3 \cos \theta_3} + \frac{\omega_3^2 \sin \theta_3}{\cos \theta_3} \qquad \textbf{(8.66)}$$

From Eq. 8.64, $\ddot{r}_1 = A_B$ may then be determined. For constant angular velocity of the crank, the angular acceleration α_2 is zero, so that Eqs. 8.64 and 8.66 giving A_B and α_3 are somewhat simplified.

Those kinematic quantities of engineering interest, such as the velocity $\mathbf{V}_B$ and acceleration $\mathbf{A}_B$ of the slider and ω_3 and α_3 of the connecting rod, may be determined numerically from the preceding equations for all phases θ_2 of the crank and for arbitrary values of the crank speed ω_2 and L/R ratio (r_3/r_2). Although the calculations to be undertaken involve voluminous arithmetical operations, such operations may be assigned to the digital computer with the advantage that a great number of variations of the problem may be solved to optimize a design.

The velocity and acceleration of other particles of the mechanism may also be of engineering interest. For example, as discussed in Chapter 9, the accelerations of the mass centers of the individual links are important because they are related to the forces acting on the links. In considering the acceleration $\mathbf{A}_{g_3}$ of the mass center g_3 of link 3 in Fig. 8.44a, the following equations result from the vector addition shown in Fig. 8.44c:

$$\mathbf{r}_{g_3} = \mathbf{r}_2 + \mathbf{r}_d = r_2 e^{i\theta_2} + r_d e^{i\theta_3} \tag{8.67}$$

$$\mathbf{V}_{g_3} = \dot{\mathbf{r}}_{g_3} = r_2 \omega_2 (i e^{i\theta_2}) + r_d \omega_3 (i e^{i\theta_3}) \tag{8.68}$$

$$\mathbf{A}_{g_3} = \ddot{\mathbf{r}}_{g_3} = r_2 (i\alpha_2 - \omega_2^2) e^{i\theta_2} + r_d (i\alpha_3 - \omega_3^2) e^{i\theta_3} \tag{8.69}$$

For constant angular speed of the crank, $\alpha_2 = 0$, so that

$$\mathbf{A}_{g_3} = -r_2 \omega_2^2 e^{i\theta_2} + r_d (i\alpha_3 - \omega_3^2) e^{i\theta_3}$$

$$= (-r_2 \omega_2^2 \cos \theta_2 - r_d \alpha_3 \sin \theta_3 - r_d \omega_3^2 \cos \theta_3)$$
$$+ i(-r_2 \omega_2^2 \sin \theta_2 + r_d \alpha_3 \cos \theta_3 - r_d \omega_3^2 \sin \theta_3)$$

$$= a_{g_3} + ib_{g_3} \tag{8.70}$$

The magnitude of A_{g_3} may be determined from $A_{g_3} = \sqrt{a_{g_3}^2 + b_{g_3}^2}$, and the angle β which $\mathbf{A}_{g_3}$ makes with the real axis may be determined from $\tan \beta = b_{g_3}/a_{g_3}$. The vector $\mathbf{A}_{g_3}$ is shown as a fixed vector in Fig. 8.44d.

Example 8.14. The slider crank of an internal-combustion engine (Fig. 8.44a) includes a crank of 2.0 in. length and a connecting rod of 8.0 in. length. The crank speed of the engine is constant at 3000 rpm (314 rad/s). Determine the acceleration of the mass center $\mathbf{A}_{g_3}$ of the connecting rod when the crank angle is $\theta_2 = 30°$. The mass center g_3 is located 2.0 in. from the crank pin at A. In addition, determine curves showing (1) the magnitude of $\mathbf{A}_{g_3}$ versus θ_2 and (2) the angle β which $\mathbf{A}_{g_3}$ makes with the real axis versus θ_2.

Solution. The calculation of the acceleration A_{g_3} may be made using Eq. 8.70 and the following given data: $r_2 = 2.0$ in., $r_3 = 8.0$ in., $r_d = 2.0$ in., $\omega_2 = 314$ rad/s, and $\theta_2 = 30°$. However, before the calculation can be undertaken, the unknowns θ_3, ω_3, and α_3 must first be determined.

The connecting-rod angle θ_3 may be determined from Eq. 8.60 as follows:

$$\sin \theta_3 = -\frac{r_2}{r_3} \sin \theta_2 = -\frac{2.0}{8.0} \sin 30°$$

$$= -0.125$$

$$\theta_3 = -7.18° \text{ or } 352.82°$$

$$\cos \theta_3 = 0.992$$

It may be seen that for $\sin \theta_3 = -0.125$, there are two positions of the connecting rod, either $\theta_3 = 352.82°$ or $187.18°$, depending on whether the slider is to the right or to the left of the crank center O_2.

The angular velocity ω_3 and the angular acceleration α_3 of the connecting rod may be determined from Eqs. 8.63 and 8.66, respectively:

$$\omega_3 = -\omega_2 \frac{r_2 \cos \theta_2}{r_3 \cos \theta_3} = -314 \left(\frac{2.0}{8.0}\right) \frac{\cos 30°}{\cos 352.82°}$$

$$= -68.56 \text{ rad/s}$$

$$\alpha_3 = \left(\frac{r_2}{r_3}\right) \frac{\omega_2^2 \sin \theta_2 - \alpha_2 \cos \theta_2}{\cos \theta_3} + \frac{\omega_3^2 \sin \theta_3}{\cos \theta_3}$$

$$= \left(\frac{2.0}{8.0}\right) \frac{314^2 \sin 30° - 0}{\cos 352.82°} + \frac{(-68.56)^2 \sin 352.82°}{\cos 352.82°}$$

$$= 11{,}840 \text{ rad/s}^2$$

With the preceding quantities determined, the real and imaginary components of the acceleration $\mathbf{A}_{g_3}$ may be determined by evaluating a_{g_3} and b_{g_3} in Eq. 8.70 as follows:

$$a_{g_3} = -r_2\omega_2^2 \cos \theta_2 - r_d\alpha_3 \sin \theta_3 - r_d\omega_3^2 \cos \theta_3$$

$$= -2.0(314)^2(0.866) - 2.0(11{,}840)(-0.125)$$

$$-2.0(-68.56)^2(0.992)$$

$$= -177{,}300 \text{ in./s}^2$$

$$b_{g_3} = -r_2\omega_2^2 \sin \theta_2 + r_d\alpha_3 \cos \theta_3 - r_d\omega_3^2 \sin \theta_3$$

$$= -2.0(314)^2(0.500) + 2.0(11{,}840)(0.992)$$

$$-2.0(-68.56)^2(-0.125)$$

$$= -74{,}020 \text{ in./s}^2$$

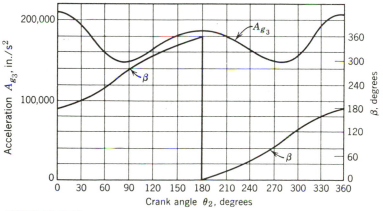

FIGURE 8.45

The magnitude of $\mathbf{A}_{g_3}$ is the vector sum of the components:

$$A_{g_3} = \sqrt{a_{g_3}^2 + b_{g_3}^2} = \sqrt{(-177,300)^2 + (-74,020)^2}$$

$$= 192,200 \text{ in./s}^2$$

and the angle of $\mathbf{A}_{g_3}$ with the real or horizontal axis is given by β:

$$\tan \beta = \frac{b_{g_3}}{a_{g_3}} = \frac{-74,020}{-177,300} = 0.417$$

$$\beta = 202.67°$$

Calculations similar to those illustrated for increasing crank angle θ_2 in 10° increments make possible the plotting of the curves of A_{g_3} and β for one cycle of the crank as shown in Fig. 8.45.

8.27 ANALYSIS OF THE INVERTED SLIDER CRANK BY LOOP-CLOSURE EQUATIONS AND COMPLEX NUMBERS

Of the inversions of the slider crank, the crank shaper (Fig. 8.46a) is interesting to analyze by complex numbers because the Coriolis component of acceleration is involved. In Fig. 8.46b are shown the vectors giving the position $\mathbf{r}_{B_2}$ of particle B_2 on the crank at the pin connection to the slider. Two independent vector equations for the position of B_2, namely, $\mathbf{r}_{B_2} = \mathbf{r}_4$ and $\mathbf{r}_{B_2} = \mathbf{r}_1 + \mathbf{r}_2$, may be combined to give the following vector equation:

$$\mathbf{r}_4 = \mathbf{r}_1 + \mathbf{r}_2$$

$$r_4 e^{i\theta_4} = r_1 + r_2 e^{i\theta_2} \tag{8.71}$$

Differentiating Eq. 8.71 yields the following velocity equation:

$$r_4\omega_4(ie^{i\theta_4}) + \dot{r}_4 e^{i\theta_4} = r_2\omega_2(ie^{i\theta_2}) \tag{8.72}$$

From an inspection of Eq. 8.72 term by term, it may be seen that the equation is another form of the equation $\mathbf{V}_{B_4} + \mathbf{V}_{B_2B_4} = \mathbf{V}_{B_2}$ for the coincident particles B_4 and B_2. Differentiating Eq. 8.72 yields the following acceleration equation:

$$r_4\omega_4^2(i^2 e^{i\theta_4}) + r_4\alpha_4(ie^{i\theta_4}) + 2\dot{r}_4\omega_4(ie^{i\theta_4}) + \ddot{r}_4 e^{i\theta_4}$$
$$= r_2\omega_2^2(i^2 e^{i\theta_2}) + r_2\alpha_2(ie^{i\theta_2}) \tag{8.73}$$

Inspection of Eq. 8.73 term by term shows that the equation is an alternate form of the equation $\mathbf{A}_{B_4}^n + \mathbf{A}_{B_4}^t + 2\boldsymbol{\omega}_4 \times \mathbf{V}_{B_2B_4} + \mathbf{A}_{B_2B_4}^t = \mathbf{A}_{B_2}^n + \mathbf{A}_{B_2}^t$.

In the crank-shaper mechanism, link 2 is the driving link usually rotating at a known constant angular velocity ω_2, with α_2 equal to zero. Thus, referring

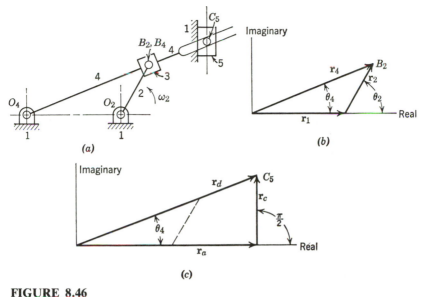

FIGURE 8.46

to Eqs. 8.71, 8.72, and 8.73, it may be seen that r_1, r_2, θ_2, ω_2, and α_2 are the known quantities, and the six unknowns to be determined are θ_4, ω_4, α_4, r_4, $\dot{r}_4$, and $\ddot{r}_4$. By equating the real and imaginary parts of each of the Eqs. 8.71, 8.72, and 8.73, six equations are obtained which make possible the determination of the six unknowns.

After θ_4, ω_4, and α_4 have been determined for a known value of ω_2 and an arbitrary value of θ_2, it becomes possible to determine numerically the velocity and acceleration of other particles of the mechanism. For example, since the crank shaper is a quick-return mechanism, it is of interest to determine the velocity $\mathbf{V}_{C_5}$ of the tool-holding slider (link 5) of Fig. 8.46a for comparison of the magnitudes of the slider velocity during the working and return strokes of the mechanism. In Fig. 8.46c is shown the vector polygon which includes the position vector $\mathbf{r}_c$ of particle C_5. From the polygon,

$$\mathbf{r}_d = \mathbf{r}_a + \mathbf{r}_c$$

$$\mathbf{r}_c = \mathbf{r}_d - \mathbf{r}_a$$

$$r_c e^{i\pi/2} = r_d e^{i\theta_4} - r_a \tag{8.74}$$

Differentiating Eq. 8.74 gives the following velocity expression:

$$\mathbf{V}_{C_5} = \dot{r}_c e^{i\pi/2} = r_d \omega_4 (i e^{i\theta_4}) + \dot{r}_d e^{i\theta_4} \tag{8.75}$$

In Eqs. 8.74 and 8.75 r_a is a known fixed length, and θ_4 and ω_4 are known from previously developed equations. By equating the real and imaginary parts of each of Eqs. 8.74 and 8.75, four equations become available for the determination of the four unknowns r_c, $\dot{r}_c$, r_d, and $\dot{r}_d$, of which $\dot{r}_c$ is the velocity magnitude V_{C_5} of the slider.

8.28 ANALYSIS OF THE FOUR-BAR LINKAGE BY LOOP CLOSURE EQUATIONS AND COMPLEX NUMBERS

Although it may at first seem surprising, velocity and acceleration analysis of the four-bar linkage is generally a much simpler task than position analysis. To see the reason for this, consider once again the planar four-bar linkage of Fig. 2.6, which for convenience has been repeated here as Fig. 8.47. The loop closure equation for this mechanism written in complex number form is

$$r_2 e^{i\theta_2} + r_3 e^{i\theta_3} - r_4 e^{i\theta_4} - r_1 = 0 \tag{8.76}$$

This equation may be expanded into real and imaginary parts and written in the form used in Chapter 2:

$$r_2 \cos \theta_2 + r_3 \cos \theta_3 - r_4 \cos \theta_4 - r_1 = 0 \tag{8.77}$$

$$r_2 \sin \theta_2 + r_3 \sin \theta_3 - r_4 \sin \theta_4 = 0 \tag{8.78}$$

In position analysis, the values of r_1, r_2, r_3, r_4, and the input angle θ_2 are given, and the problem is to find the angles θ_3 and θ_4. Since these unknowns are embedded within the sine and cosine terms, the equations are said to be *transcendental*. A relatively simple method for solving the position analysis problem using the

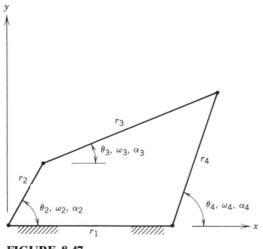

FIGURE 8.47

law of cosines was presented in Chapter 2, section 2.1. Methods for analytically solving Eqs. 8.77 and 8.78 in complex number form are somewhat more involved. For this reason, they have been presented in Appendix 1. It should also be noted that numerical techniques such as the Newton–Raphson method presented in Chapter 2 are easy to implement, provide relatively quick solutions, and are easily extended to handle the position analysis problem for more complex mechanisms.

Velocity loop closure equations for the four-bar linkage are obtained by differentiating the position loop closure equations with respect to time as follows:

$$- r_2(\sin \theta_2)\omega_2 - r_3(\sin \theta_3)\omega_3 + r_4(\sin \theta_4)\omega_4 = 0 \tag{8.79}$$

$$r_2(\cos \theta_2)\omega_2 + r_3(\cos \theta_3)\omega_3 - r_4(\cos \theta_4)\omega_4 = 0 \tag{8.80}$$

In velocity analysis, it is assumed that the values of r_1, r_2, r_3, r_4, and θ_2 are given, and that the values of θ_3 and θ_4 have already been determined from a position analysis. In addition, the angular velocity of the input link ω_2 must also be given. The only unknowns in the above equations are the angular velocities of links 3 and 4. These equations may therefore be written in the following form:

$$A\omega_3 + B\omega_4 = C \tag{8.81}$$

$$D\omega_3 + E\omega_4 = F \tag{8.82}$$

where the values of A through F are calculated from

$$
\begin{aligned}
A &= -r_3 \sin \theta_3 \\
B &= r_4 \sin \theta_4 \\
C &= r_2(\sin \theta_2)\omega_2 \\
D &= r_3 \cos \theta_3 \\
E &= -r_4 \cos \theta_4 \\
F &= -r_2(\cos \theta_2)\omega_2
\end{aligned}
\tag{8.83}
$$

This form clearly shows Eqs. 8.81 and 8.82 to be linear in the two unknowns ω_3 and ω_4. Solving this pair of equations gives

$$\omega_3 = \frac{FB - EC}{DB - EA} \tag{8.84}$$

$$\omega_4 = \frac{DC - FA}{DB - EA} \tag{8.85}$$

Differentiation of the velocity equations (Eqs. 8.79 and 8.80) with respect to time yields the acceleration equations:

$$-r_2(\cos\theta_2)\omega_2^2 - r_2(\sin\theta_2)\alpha_2 - r_3(\cos\theta_3)\omega_3^2 - r_3(\sin\theta_3)\alpha_3$$
$$+ r_4(\cos\theta_4)\omega_4^2 + r_4(\sin\theta_4)\alpha_4 = 0 \quad (8.86)$$

$$-r_2(\sin\theta_2)\omega_2^2 + r_2(\cos\theta_2)\alpha_2 - r_3(\sin\theta_3)\omega_3^2 + r_3(\cos\theta_3)\alpha_3$$
$$+ r_4(\sin\theta_4)\omega_4^2 - r_4(\cos\theta_4)\alpha_4 = 0 \quad (8.87)$$

In acceleration analysis, it is assumed that the values of r_1, r_2, r_3, r_4, θ_2, ω_2, and α_2 are given and that the values of θ_3, θ_4, ω_3, and ω_4 have already been determined from position and velocity analysis. Thus, the only unknowns in the above equations are the angular accelerations of links 3 and 4, α_3 and α_4, and Eqs. 8.86 and 8.87 may be rewritten as follows:

$$A\alpha_3 + B\alpha_4 = C' \quad (8.88)$$

$$D\alpha_3 + E\alpha_4 = F' \quad (8.89)$$

where A through F can be calculated as follows:

$$A = -r_3\sin\theta_3$$
$$B = r_4\sin\theta_4$$
$$C' = r_2(\cos\theta_2)\omega_2^2 + r_2(\sin\theta_2)\alpha_2 + r_3(\cos\theta_3)\omega_3^2 - r_4(\cos\theta_4)\omega_4^2 \quad (8.90)$$
$$D = r_3\cos\theta_3$$
$$E = -r_4\cos\theta_4$$
$$F' = r_2(\sin\theta_2)\omega_2^2 - r_2(\cos\theta_2)\alpha_2 + r_3(\sin\theta_3)\omega_3^2 - r_4(\sin\theta_4)\omega_4^2$$

It is interesting to note that the values of A, B, D, and E are the same as those used in velocity analysis, and hence these values do not need to be recalculated. The form of Eqs. 8.88 and 8.89 clearly shows them to be linear in the two unknowns α_3 and α_4. Solving this pair of equations gives

$$\alpha_3 = \frac{F'B - EC'}{DB - EA} \quad (8.91)$$

$$\alpha_4 = \frac{DC' - F'A}{DB - EA} \quad (8.92)$$

Example 8.15. For the four-bar linkage of Fig. 8.47, find the angular velocities and the angular accelerations of links 3 and 4. The link lengths are $r_1 = 7$ in., $r_2 = 3$ in., $r_3 = 8$ in., $r_4 = 6$ in., the input link position is $\theta_2 = 60°$, the input link velocity is $\omega_2 = 1$ rad/s, and the input link acceleration is $\alpha_2 = 1$ rad/s². A position analysis of this mechanism has been performed in Chapter 2, Example 2.2. The results of this analysis are $\theta_3 = 22.812°$, $\theta_4 = 71.798°$. The values of variables A through F may now be calculated from Eqs. 8.83:

$$A = -8 \sin 22.812 = -3.102$$

$$B = 6 \sin 71.798 = 5.700$$

$$C = 3(\sin 60)(1) = 2.598$$

$$D = 8 \cos 22.812 = 7.374$$

$$E = -6 \cos 71.798 = -1.874$$

$$F = -3(\cos 60)(1) = -1.500$$

```
10  '***************************************************************
20  '* MECHANISM DESIGN - DISPLACEMENT, VELOCITY & ACCELERATION ANALYSIS
30  '* -  Uses Newton-Raphson root finding method to determine unknown
40  '*      angles of links 3 and 4 of a four bar linkage.  Also calc.s
50  '*      angular velocities & accelerations of links 3 and 4.
60  '* -  Mabie and Reinholtz, 4th Ed.
70  '* -  Program revised by - Steve Wampler ( 7/ 9/85)
80  '***************************************************************
90  CLS:MAX.PASS%=10:D2R=3.14159/180 ' constant to convert deg.s to rad.s
100 INPUT "Enter ang. pos. of links 2,3,4 (deg.s)";THETA2,THETA3,THETA4
110 THETA2=THETA2*D2R:THETA3=THETA3*D2R:THETA4=THETA4*D2R
120 INPUT "Enter ang. vel. of link 2 (rad.s/sec)";OMEGA2
130 INPUT "Enter ang. acc. of link 2 (rad.s/sec^2)";ALPHA2
140 INPUT "Enter list of link lengths r1,r2,r3,r4";R1,R2,R3,R4
150 INPUT "Enter ang. step size for input link rotation (deg.s)";ANG.INC
160 PRINT " THETA2    THETA3    THETA4    OMEGA3    OMEGA4    ALPHA3    ALPHA4"
170 PRINT " (deg.)    (deg.)    (deg.)    (rad/s)   (rad/s)  (rad/s^2) (rad/s^2)
180 WHILE INKEY$=""
190    FUNC.1=1 'force next WHILE statement to be true
200    WHILE (ABS(FUNC.1)>.001 OR ABS(FUNC.2)>.001) AND PASS%<MAX.PASS%
210       FUNC.1=R1+(R4*COS(THETA4))-(R2*COS(THETA2))-(R3*COS(THETA3))
220       FUNC.2=(R4*SIN(THETA4))-(R2*SIN(THETA2))-(R3*SIN(THETA3))
230       DF1DT3=R3*SIN(THETA3):DF1DT4=-R4*SIN(THETA4)   ' take partials
240       DF2DT3=-R3*COS(THETA3):DF2DT4=R4*COS(THETA4)
250       DF2DT4=R4*COS(THETA4)   'Partial of func. 2 w/respect to theta4
260       DEL=DF1DT3*DF2DT4-DF1DT4*DF2DT3   'calc. del function
270       DELTA.THETA4=(DF2DT3*FUNC.1-DF1DT3*FUNC.2)/DEL
280       DELTA.THETA3=-(DF2DT4*FUNC.1-DF1DT4*FUNC.2)/DEL
290       THETA3=THETA3+DELTA.THETA3:THETA4=THETA4+DELTA.THETA4
300       PASS%=PASS%+1
310    WEND ' go back if func.s<>0 and number of passes not to high
320    IF PASS%<MAX.PASS% THEN GOSUB 350 ELSE GOSUB 480
330    PASS%=0:THETA2=THETA2+ANG.INC*D2R ' increment input link
340 WEND:END
350 ' calc. vel. and acc. of links 3 and 4 and print results
360 A=-R3*SIN(THETA2):B=R4*SIN(THETA4):C=R2*SIN(THETA2)*OMEGA2
370 D=R3*COS(THETA3):E=-R4*COS(THETA4):F=-R2*COS(THETA2)*OMEGA2
380 OMEGA3=(F*B-E*C)/(D*B-E*A):OMEGA4=(D*C-F*A)/(D*B-E*A) ' velocities
390 C1=R2*COS(THETA2)*OMEGA2^2+R2*SIN(THETA2)*ALPHA2 'part of C.PRIME
400 C.PRIME=C1+R3*COS(THETA3)*OMEGA3^2-R4*COS(THETA4)*OMEGA4^2
410 F1=R2*SIN(THETA2)*OMEGA2^2-R2*COS(THETA2)*ALPHA2 'part of F.PRIME
420 F.PRIME=F1+R3*SIN(THETA3)*OMEGA3^2-R4*SIN(THETA4)*OMEGA4^2
430 ALPHA3=(F.PRIME*B-E*C.PRIME)/(D*B-E*A) ' angular accelerations
440 ALPHA4=(D*C.PRIME-F.PRIME*A)/(D*B-E*A)
450 PRINT USING"####.##   ";THETA2/D2R;THETA3/D2R;THETA4/D2R;
460 PRINT USING"####.##   ";OMEGA3;OMEGA4;ALPHA3;ALPHA4
470 RETURN
480 ' mech. does not assemble message
490 PRINT "Mechanism does not assemble at THETA2 =";THETA2/D2R" deg."
500 RETURN
```

FIGURE 8.48

THETA2 (deg.)	THETA3 (deg.)	THETA4 (deg.)	OMEGA3 (rad/s)	OMEGA4 (rad/s)	ALPHA3 (rad/s^2)	ALPHA4 (rad/s^2)
0.00	46.57	75.52	-0.75	-0.75	-0.41	0.49
10.00	39.51	69.24	-0.65	-0.50	0.10	1.08
20.00	33.71	65.65	-0.51	-0.22	0.33	1.26
30.00	29.35	64.63	-0.37	0.01	0.36	1.19
40.00	26.25	65.66	-0.26	0.19	0.31	1.05
50.00	24.15	68.20	-0.17	0.31	0.27	0.91
60.00	22.81	71.80	-0.10	0.40	0.23	0.81
70.00	22.07	76.11	-0.05	0.46	0.22	0.72
80.00	21.79	80.90	-0.01	0.50	0.22	0.66
90.00	21.91	85.96	0.03	0.52	0.22	0.60
100.00	22.37	91.17	0.06	0.52	0.24	0.54
110.00	23.14	96.41	0.09	0.52	0.26	0.48
120.00	24.20	101.57	0.12	0.51	0.29	0.42
130.00	25.57	106.58	0.15	0.49	0.32	0.36
140.00	27.22	111.35	0.18	0.46	0.35	0.29
150.00	29.18	115.82	0.21	0.43	0.39	0.22
160.00	31.45	119.93	0.24	0.39	0.42	0.15
170.00	34.01	123.63	0.27	0.35	0.44	0.09
180.00	36.87	126.87	0.30	0.30	0.46	0.02
190.00	40.00	129.62	0.33	0.25	0.47	-0.04
200.00	43.38	131.87	0.35	0.20	0.46	-0.10
210.00	46.95	133.59	0.37	0.15	0.45	-0.16
220.00	50.66	134.78	0.38	0.09	0.42	-0.21
230.00	54.44	135.44	0.38	0.04	0.37	-0.27
240.00	58.20	135.56	0.37	-0.02	0.31	-0.34
250.00	61.84	135.11	0.36	-0.07	0.22	-0.42
260.00	65.26	134.07	0.33	-0.14	0.12	-0.51
270.00	68.31	132.36	0.28	-0.21	-0.02	-0.63
280.00	70.82	129.92	0.22	-0.28	-0.20	-0.77
290.00	72.59	126.64	0.13	-0.38	-0.43	-0.94
300.00	73.38	122.37	0.02	-0.48	-0.71	-1.14
310.00	72.90	116.95	-0.12	-0.60	-1.03	-1.34
320.00	70.85	110.26	-0.29	-0.74	-1.34	-1.50
330.00	66.98	102.26	-0.48	-0.86	-1.55	-1.48
340.00	61.29	93.23	-0.65	-0.93	-1.49	-1.12
350.00	54.18	83.92	-0.75	-0.91	-1.06	-0.38
360.00	46.57	75.52	-0.75	-0.75	-0.41	0.49

FIGURE 8.49 The output from the BASIC computer program of Fig. 8.48. The position analysis is performed using the Newton-Raphson method (Chapter 2, Section 2.3): velocities and accelerations are calculated in closed form from Eqs. 8.84, 8.85, 8.91, and 8.92.

Therefore,

$$\omega_3 = \frac{(-1.500)(5.700) - (-1.874)(2.598)}{(7.374)(5.700) - (-1.874)(-3.102)} = -0.102 \text{ rad/s}$$

$$\omega_4 = \frac{(7.374)(2.598) - (-1.500)(-3.102)}{(7.374)(5.700) - (-1.874)(-3.102)} = 0.400 \text{ rad/s}$$

Substituting the known quantities into the expressions for C' and F' in Eq. 8.90 gives

$$C' = 3(\cos 60)(1)^2 + 3(\sin 60)(1) + 8(\cos 22.812)(-0.102)^2$$
$$- 6(\cos 71.798)(0.400)^2 = 3.875$$

$$F' = 3(\sin 60)(1)^2 + 3(\cos 60)(1) + 8(\sin 22.812)(-0.102)^2$$
$$- 6(\sin 71.798)(0.400)^2 = 0.218$$

Therefore,

$$\alpha_3 = \frac{(0.218)(5.700) - (-1.874)(3.875)}{(7.374)(5.700) - (-1.874)(-3.102)} = 0.235 \text{ rad/s}^2$$

$$\alpha_4 = \frac{(7.374)(3.875) - (0.218)(-3.102)}{(7.374)(5.700) - (-1.874)(-3.102)} = 0.808 \text{ rad/s}^2$$

A BASIC-language computer program for iterative position analysis of the four-bar linkage was described in section 2.2. The velocity and acceleration analysis methods discussed in this section have been added to that program, and the new listing is shown in Fig. 8.48. The corresponding output for 10° increments of the input angle is shown in Fig. 8.49.

8.29 COMPLEX MECHANISMS

The addition of extra links to the basic four-bar mechanism increases the complexity of the kinematic analysis of the mechanism. In Fig. 8.50 is shown a six-bar linkage in which the addition of links 5 and 6 to the basic four-bar linkage (1, 2, 3, 4) forms a second four-bar linkage consisting of links 3, 4, 5, and 6. By the addition of the two links, the number of kinematic unknowns is increased by six (θ_5, θ_6, ω_5, ω_6, α_5, and α_6) so that the total number of unknowns is 12. Thus, 12 independent equations are required to determine the unknowns.

By combining the independent vector equations for the position of point B, the following equation is obtained to include the kinematic quantities of the links in the loop formed by the lower four-bar linkage:

$$\mathbf{r}_2 + \mathbf{r}_3 = \mathbf{r}_1 + \mathbf{r}_4 \qquad\qquad (8.93)$$

Similarly, a second combined vector equation may be written for the position of

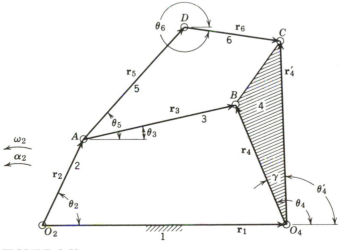

FIGURE 8.50

point C such as to involve a loop of links including links 5 and 6:

$$\mathbf{r}_2 + \mathbf{r}_5 + \mathbf{r}_6 = \mathbf{r}_1 + \mathbf{r}_4' \tag{8.94}$$

The independent equations 8.93 and 8.94 may be written in complex number form:

$$r_2 e^{i\theta_2} + r_3 e^{i\theta_3} = r_1 + r_4 e^{i\theta_4} \tag{8.95}$$

$$r_2 e^{i\theta_2} + r_5 e^{i\theta_5} + r_6 e^{i\theta_6} = r_1 + r_4' e^{i\theta_4'} \tag{8.96}$$

By equating the real and imaginary components of the preceding equations, four independent equations are obtained from which the four unknown angles θ_3, θ_4, θ_5, and θ_6 may be determined as functions of θ_2. The angle θ_4' may be given as θ_4 minus the fixed angle γ of link 4 shown in Fig. 8.50. The determination of the unknown angles is complicated trigonometrically and requires the determination of auxiliary lengths and angles as illustrated in the discussion of the four-bar linkage.

Differentiation of Eqs. 8.95 and 8.96 and equating the real and imaginary components result in four additional independent equations which may be used in the determination of the four angular velocities ω_3, ω_4, ω_5, and ω_6 as functions of ω_2. The solution of the numerous simultaneous equations is best accomplished by the use of determinants. Further differentiation and equating of real and imaginary parts yield four equations for the determination of the four unknown angular accelerations.

The preceding method of analysis may be applied to plane mechanisms of higher order of complexity. If two more links are added to the linkage of Fig. 8.50, making an eight-link mechanism, a third independent vector equation enclosing another independent loop of links makes available the additional equations required for solution.

8.30 VELOCITY AND ACCELERATION ANALYSIS USING THE INTEGRATED MECHANISMS PROGRAM (IMP)

In Chapter 2 (section 2.4), the use of the Integrated Mechanisms Program (IMP) was introduced as a convenient method of displacement analysis of four-bar linkages. At that time, it was stated that IMP could also be used for velocity and acceleration analyses. The use of IMP for the determination of velocities and accelerations is presented below.

Figures 2.9b and 2.9c from section 2.4 are repeated as Figs. 8.51a and 8.51b here.

The following revolute statements, revolute data, and point data are also

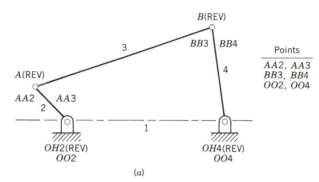

Points

AA2, AA3
BB3, BB4
OO2, OO4

(a)

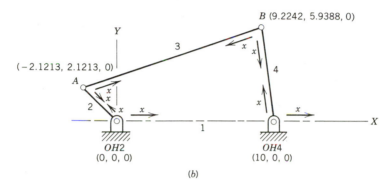

(b)

FIGURE 8.51

reproduced from section 2.4 as follows:

```
GROUND = FRAME
REVOLUTE ( FRAME , LNK2 ) = OH2
REVOLUTE ( LNK2 , LNK3 ) = A
REVOLUTE ( LNK3 , LNK4 ) = B
REVOLUTE ( LNK4 , FRAME ) = OH4
DATA : REVOLUTE ( OH2 ) = Ø , Ø , Ø / Ø , Ø , 1 / 1 , Ø , Ø / − 2 . 1213 , 2 . 1213 , Ø
DATA : REVOLUTE ( A ) = − 2 . 1213 , 2 . 1213 , Ø / − 2 . 1213 , 2 . 1213 , 1 / $
               Ø , Ø , Ø / 9 . 2242 , 5 . 9388 , Ø
DATA : REVOLUTE ( B ) = 9 . 2242 , 5 . 9388 , Ø / 9 . 2242 , 5 . 9388 , 1 / $
               − 2 . 1213 , 2 . 1213 , Ø / 1Ø , Ø , Ø
DATA : REVOLUTE ( OH4 ) = 1Ø , Ø , Ø / 1Ø , Ø , 1 / 9 . 2242 , 5 . 9388 , Ø / $
               12 , Ø , Ø
POINT ( LNK2 ) = OO2 , AA2
DATA : POINT ( OO2 , OH2 ) = Ø , Ø , Ø
DATA : POINT ( AA2 , A ) = Ø , Ø , Ø
```

```
POINT(LNK3)=AA3,  BB3
DATA:POINT(AA3,A)=Ø,Ø,Ø
DATA:POINT(BB3,B)=Ø,Ø,Ø
POINT(LNK4)=BB4,OO4
DATA:POINT(BB4,B)=Ø,Ø,Ø
DATA:POINT(OO4,OH4)=Ø,Ø,Ø
ZOOM(7)=5,1.5,Ø
RETURN
```

It is now necessary to enter the angular position of link 2 (135°) for which the analysis is required and to enter the angular velocity of link 2 (500 rad/s, cw) which is constant. These will be specified for revolute OH2. The velocity and acceleration of point B are to be determined as well as the angular velocity and acceleration of link 4. These requirements will be entered for point BB4 and revolute OH4. The velocity of point AA2 will also be determined as a check on the computer analysis.

```
DATA:POSI(OH2)=135
DATA:VELO(OH2)=-500
PRINT:VELO(AA2,BB4,OH4)
PRINT:ACCEL(BB4,OH4)
EXECUTE
```

Table 8.2 gives the velocity of point A and the velocity and acceleration of point B. Also included are the angular velocity and acceleration of link 4. For comparison, values found by unit vectors are included.

TABLE 8.2

	Velocity			**Acceleration**	
	AA2	*BB4*	*OH4(ω_4)[a]*	*BB4*	*OH4(α_4)[a]*
IMP	1499.99 in./s	1369.39 in./s	−228.64 rad/s	453242 in./s^2	−54717.5 rad/s^2
Unit vectors		1371.1 in./s	−228.64 rad/s	453903 in./s^2	−54740.3 rad/s^2

[a]Values of ω_4 and α_4 are given relative to the frame.

Problems

8.1. A turbine operates at 15,000 rpm. Calculate the velocity and acceleration of the tip of the rotor blade which is 10 in. from the center of rotation.

8.2. The tip of a turbine blade has a linear velocity of 600 m/s. Calculate the angular velocity in revolutions per minute for the following wheel diameters: 70, 400, 750, and 900 mm.

8.3. In an automotive engine, the maximum piston acceleration is $1000g$ ($g = 32.2$ ft/s^2) at a given constant crank speed. The crank radius is $2\frac{1}{2}$ in., and the connecting rod length is 10 in. Determine the crank speed in rpm and the linear speed of the crank-pin center in feet per second and feet per minute. Determine also the maximum piston velocity in ft/s and ft/min at this crank speed. See Eqs. 2.18 and 2.19 of Chapter 2.

8.4. The particle Q of the body shown in Fig. 8.52 is in motion along a curvilinear path. The radius of curvature of the path, its angular velocity and acceleration, and its rate of change with time are as shown. Determine the magnitudes of $\mathbf{A}_Q^n$, $\mathbf{A}_Q^t$, and $\mathbf{A}_Q$ and show as vectors on a sketch of the figure.

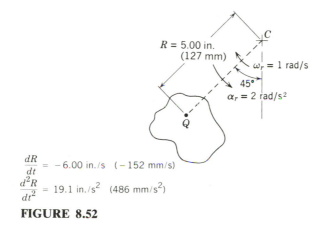

$$\frac{dR}{dt} = -6.00 \text{ in./s} \quad (-152 \text{ mm/s})$$

$$\frac{d^2R}{dt^2} = 19.1 \text{ in./s}^2 \quad (486 \text{ mm/s}^2)$$

FIGURE 8.52

8.5. A point mass P travels a curvilinear path about point A as shown in Fig. 8.53. If $A_P = 640$ mm/s^2, determine ω_R.

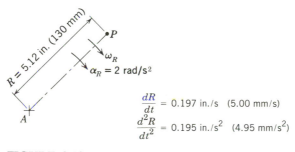

$$\frac{dR}{dt} = 0.197 \text{ in./s} \quad (5.00 \text{ mm/s})$$

$$\frac{d^2R}{dt^2} = 0.195 \text{ in./s}^2 \quad (4.95 \text{ mm/s}^2)$$

FIGURE 8.53

8.6. In Fig. 8.54, link 2 and disk 3 rotate about the same fixed axis O; $\omega_2 = 15$ rad/s (cw) and $\alpha_2 = 0$, $\omega_3 = 10$ rad/s (cw) and $\alpha_3 = 30$ rad/s^2 (ccw). Point P_2 in link 2 is coincident with point P_3 in disk 3. Determine $\mathbf{V}_{P_2P_3}$, $\mathbf{A}_{P_2P_3}$, ω_{23}, and α_{23}.

8.7. Given points A and B in a common link as shown in Fig. 8.55 calculate the velocity and acceleration of point B relative to point A.

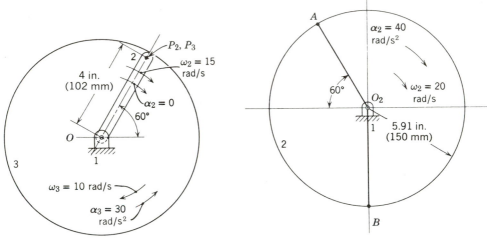

FIGURE 8.54   **FIGURE 8.55**

8.8. The wheel in Fig. 8.56 rolls without slipping. The velocity and acceleration of P are as shown. For each phase, make a sketch of the wheel and graphically determine the values of $\mathbf{A}_P^n$ and $\mathbf{A}_P^t$ using convenient scales. Calculate R and ω_r of the path of P and locate the center of curvature C. What information is needed to determine α_r?

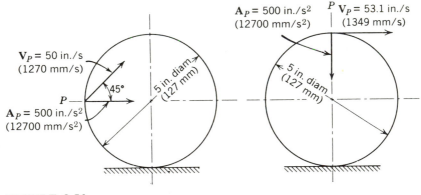

FIGURE 8.56

8.9. The centrifugal blower shown in Fig. 8.57 rotates at an angular velocity of 900 rad/s. The velocity of a particle of gas P relative to the blade is $\mathbf{V}_{PB} = 82.3$ m/s. Determine the velocity of the blade tip and the absolute velocity of particle P.

8.10. In Fig. 8.58 is shown a gas particle P leaving the passage of a turbine wheel at a velocity $\mathbf{V}_{PB}$ relative to the blades. The angle α is the departure angle of the particle relative to the passage and is measured from the plane of rotation. If the wheel speed is 10,000 rpm, determine angle α so that the absolute velocity of P is 100 m/s in a direction parallel to the exit duct axis. What is the sense of rotation of the wheel when viewed from the right? The radius from the axis of rotation to particle P is 300 mm.

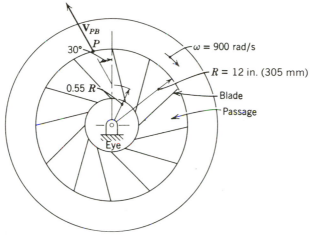

FIGURE 8.57

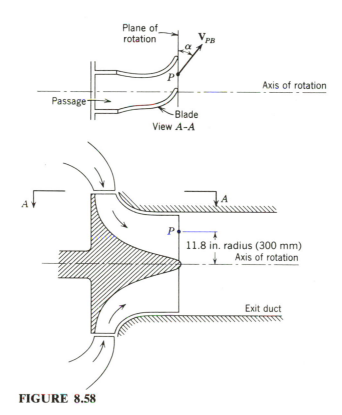

FIGURE 8.58

8.11. For the linkage shown in Fig. 8.59, link 2 rotates at a constant angular velocity. (*a*) Determine ω_4 and α_4 using unit vectors. (*b*) Draw the velocity polygon and determine V_B and V_D. (*c*) Determine the above values using computer analysis.

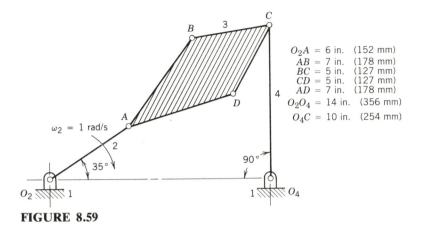

O_2A	= 6 in.	(152 mm)
AB	= 7 in.	(178 mm)
BC	= 5 in.	(127 mm)
CD	= 5 in.	(127 mm)
AD	= 7 in.	(178 mm)
O_2O_4	= 14 in.	(356 mm)
O_4C	= 10 in.	(254 mm)

FIGURE 8.59

8.12. For the linkage of Fig. 8.60, link 2 rotates at a constant angular velocity. (*a*) Determine V_B and A_B using unit vectors. (*b*) Draw the velocity polygon and determine V_C, V_D, ω_3, and ω_4. Indicate the velocity image of link 3. (*c*) Solve for the above values using computer analysis.

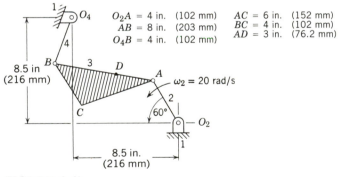

O_2A	= 4 in.	(102 mm)	AC	= 6 in.	(152 mm)
AB	= 8 in.	(203 mm)	BC	= 4 in.	(102 mm)
O_4B	= 4 in.	(102 mm)	AD	= 3 in.	(76.2 mm)

FIGURE 8.60

8.13. For the linkage of Fig. 8.61, link 2 rotates at a constant angular velocity. (*a*) Determine V_C and A_C using unit vectors. (*b*) Draw the velocity polygon and determine V_C, ω_3, and ω_4. Indicate the velocity image of each link. (*c*) Use computer methods to determine the above values.

8.14. For the crank-shaper mechanism shown in Fig. 8.62, link 2 rotates at a constant angular velocity. (*a*) Calculate V_{A_4}, ω_4, A_{A_4}, and α_4 using unit vectors. (*b*) Draw the velocity polygon and determine the relative velocities $V_{A_2A_3}$ and $V_{A_3A_4}$. Also calculate ω_{24}, ω_{34}, and ω_{32}. (*c*) Determine the above values using computer analysis.

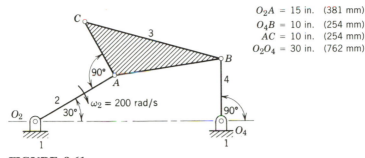

$$O_2A = 15 \text{ in.} \quad (381 \text{ mm})$$
$$O_4B = 10 \text{ in.} \quad (254 \text{ mm})$$
$$AC = 10 \text{ in.} \quad (254 \text{ mm})$$
$$O_2O_4 = 30 \text{ in.} \quad (762 \text{ mm})$$

FIGURE 8.61

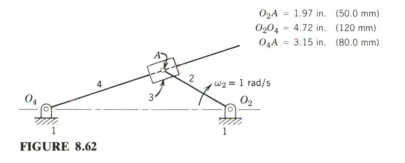

$$O_2A = 1.97 \text{ in.} \quad (50.0 \text{ mm})$$
$$O_2O_4 = 4.72 \text{ in.} \quad (120 \text{ mm})$$
$$O_4A = 3.15 \text{ in.} \quad (80.0 \text{ mm})$$

FIGURE 8.62

8.15. For the linkage shown in Fig. 8.63, link 2 rotates at a constant angular velocity. Determine V_B, ω_4, A_B, and α_4 by (*a*) unit vectors; (*b*) velocity and acceleration polygons; (*c*) proportion from results of part *b* for 3000 rpm.

8.16. For the linkage shown in Fig. 8.64, link 2 rotates at a constant angular velocity of 160 rad/s. Determine V_B, ω_4, A_B, and α_4 by (*a*) unit vectors; (*b*) velocity and acceleration polygons; (*c*) computer analysis.

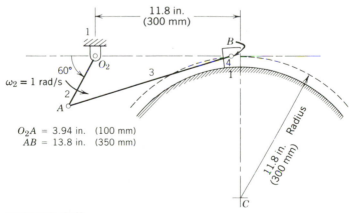

$$O_2A = 3.94 \text{ in.} \quad (100 \text{ mm})$$
$$AB = 13.8 \text{ in.} \quad (350 \text{ mm})$$

FIGURE 8.63

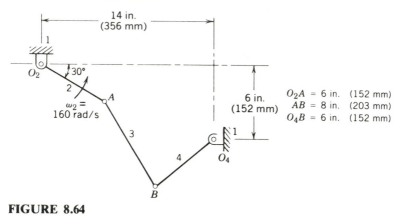

FIGURE 8.64

8.17. In the mechanism shown in Fig. 8.65, link 4 rotates at a constant angular velocity and $V_B = 24.4$ m/s. Determine α_2 by (*a*) unit vectors; (*b*) velocity and acceleration polygons; (*c*) computer analysis.

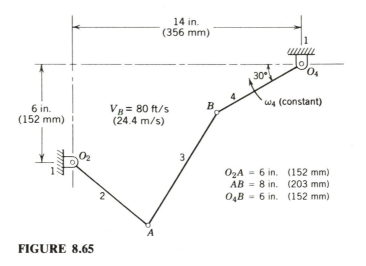

FIGURE 8.65

8.18. For the linkage shown in Fig. 8.66, link 4 rotates at a constant angular velocity. Determine α_2 by (*a*) unit vectors; (*b*) velocity and acceleration polygons; (*c*) computer analysis.

8.19. In Fig. 8.67, a cam rotates at a constant angular velocity driving a radial roller follower. Determine the acceleration of the follower by (*a*) unit vectors; (*b*) velocity and acceleration polygons; (*c*) velocity and acceleration polygons to determine the follower acceleration when the cam is rotated 45° from the position shown. For convenience in drawing, rotate the follower relative to the cam.

8.20. For the crank-shaper mechanism shown in Fig. 8.68, link 2 rotates at a constant angular velocity. Determine ω_4, A_{A_4}, and α_4 by (*a*) unit vectors; (*b*) velocity and acceleration polygons; (*c*) computer analysis.

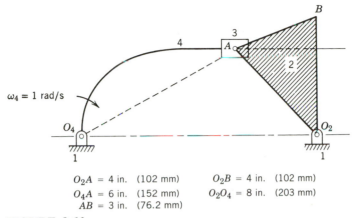

O_2A = 4 in. (102 mm)	O_2B = 4 in. (102 mm)
O_4A = 6 in. (152 mm)	O_2O_4 = 8 in. (203 mm)
AB = 3 in. (76.2 mm)	

FIGURE 8.66

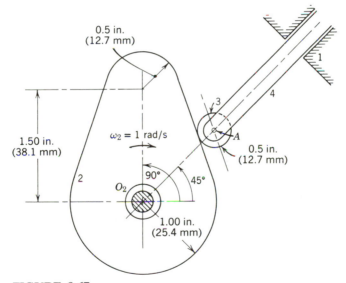

FIGURE 8.67

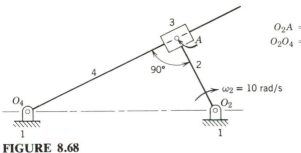

O_2A = 3.94 in. (100 mm)
O_2O_4 = 8.86 in. (225 mm)

FIGURE 8.68

8.21. The driving link 2 of the Whitworth quick-return mechanism shown in Fig. 8.69 rotates at a constant angular velocity. Determine the velocity and acceleration of the tool holder (link 6) by (*a*) unit vectors; (*b*) velocity and acceleration polygons; (*c*) computer analysis.

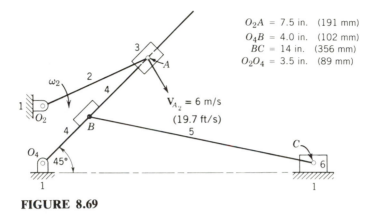

O_2A = 7.5 in.   (191 mm)
O_4B = 4.0 in.   (102 mm)
BC = 14 in.   (356 mm)
O_2O_4 = 3.5 in.   (89 mm)

V_{A_2} = 6 m/s
(19.7 ft/s)

FIGURE 8.69

8.22. Link 2 of the Geneva mechanism shown in Fig. 8.70 rotates at a constant angular velocity. Determine ω_3 and α_3 by (*a*) unit vectors; (*b*) velocity and acceleration polygons.

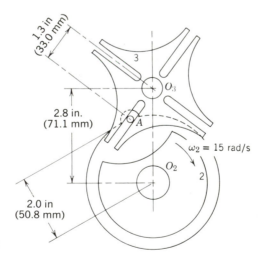

1.3 in
(33.0 mm)

2.8 in.
(71.1 mm)

ω_2 = 15 rad/s

2.0 in
(50.8 mm)

FIGURE 8.70

8.23. A cam and follower are shown in Fig. 8.71, with point P on body 3 and point Q on body 2. Determine V_Q, ω_2, A_Q, and α_2 by (*a*) unit vectors for the curved follower; (*b*) velocity and acceleration polygons for both the flat-faced follower and the curved follower.

8.24. In Fig. 8.72, a cam rotating at a constant angular velocity drives an oscillating roller follower. Determine V_{B_4}, ω_4, A_{B_4}, and α_4 by (*a*) unit vectors; (*b*) velocity and acceleration polygons.

8.25. An offset slider-crank mechanism is shown in Fig. 8.73. If ω_2 is constant, determine V_B, ω_3, A_B, and α_3 by (*a*) unit vectors; (*b*) velocity and acceleration polygons; (*c*) computer analysis.

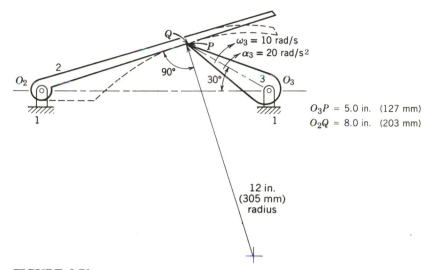

FIGURE 8.71

$O_3P = 5.0$ in.   (127 mm)
$O_2Q = 8.0$ in.   (203 mm)

8.26. In the slider-crank mechanism of Fig. 8.74a, the velocity and acceleration of the slider are given and ω_2 is constant. Determine V_A, ω_2, ω_3, and α_3 by (a) unit vectors; (b) velocity and acceleration polygons; (c) computer analysis. (d) With the value of V_A found for Fig. 8.74a, determine the acceleration of the slider in Fig. 8.74b and α_3 using an acceleration polygon.

8.27. For the linkage shown in Fig. 8.75, link 2 rotates at a constant angular velocity. Determine ω_4 and α_4 by (a) unit vectors; (b) velocity and acceleration polygons; (c) computer analysis.

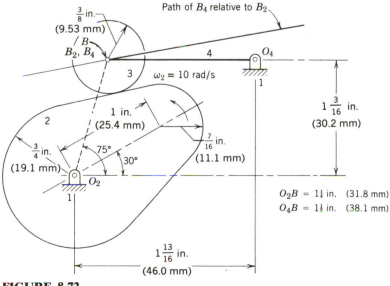

$O_2B = 1\frac{1}{4}$ in.   (31.8 mm)
$O_4B = 1\frac{1}{2}$ in.   (38.1 mm)

FIGURE 8.72

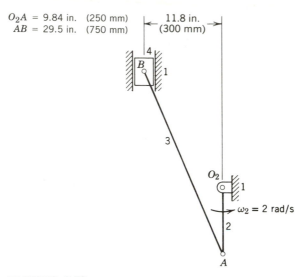

$O_2A = 9.84$ in. (250 mm)
$AB = 29.5$ in. (750 mm)

11.8 in. (300 mm)

$\omega_2 = 2$ rad/s

FIGURE 8.73

8.28. For the mechanism shown in Fig. 8.76, link 2 rotates at constant angular velocity. Determine the velocity and acceleration of point D by (a) unit vectors; (b) velocity and acceleration polygons. By proportion, calculate the velocity and acceleration of point D if the angular velocity of the driving link is increased to 1200 rpm. (c) Computer analysis for $\omega_2 = 1$ rad/s.

8.29. For the linkage shown in Fig. 8.77, link 2 rotates at a constant angular velocity of 3600 rpm. Determine the velocity and acceleration of point F by (a) unit vectors; (b)

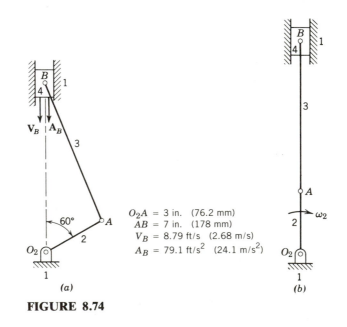

$O_2A = 3$ in. (76.2 mm)
$AB = 7$ in. (178 mm)
$V_B = 8.79$ ft/s (2.68 m/s)
$A_B = 79.1$ ft/s^2 (24.1 m/s^2)

(a)

(b)

FIGURE 8.74

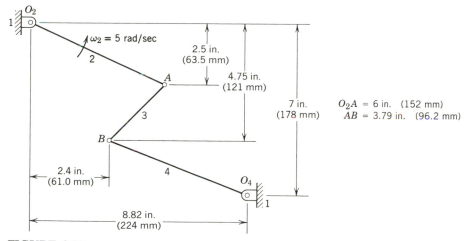

$O_2A = 6$ in. (152 mm)
$AB = 3.79$ in. (96.2 mm)

FIGURE 8.75

velocity and acceleration polygons constructed for a unit speed of $\omega_2 = 1$ rad/s and using proportion to increase to 3600 rpm; (c) computer analysis.

8.30. The wheel of the mechanism of Fig. 8.78 rolls without slipping. (a) Construct the velocity image of the wheel and determine V_C and ω_4. (b) Construct the acceleration image of the wheel and determine A_C and α_4.

8.31. A linkage that was used on steam locomotives is shown in Fig. 8.79. For a locomotive speed of 96.6 km/h, determine the velocities and accelerations of points C, R, and S and ω_2, ω_3, α_2, and α_3 by (a) unit vectors; (b) velocity and acceleration polygons.

8.32. In the epicyclic gear train shown in Fig. 8.80a, the carrier (link 2) rotates such that $\omega_2 = 12$ rad/s and $\alpha_2 = 48$ rad/s^2 at the instant shown. (a) Construct the velocity polygon and show the velocity image of the planet (link 3). Determine ω_3. (b) Construct the acceleration polygon and show the acceleration image of the planet. Determine α_3.

8.33. The carrier (link 2) of the epicyclic gear train shown in Fig. 8.80b rotates clockwise at a uniform speed of 10 rad/s. (a) Draw the velocity polygon and show the velocity

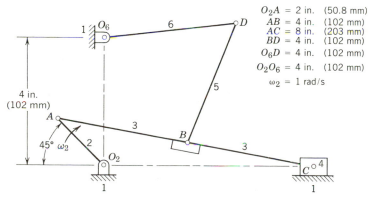

$O_2A = 2$ in. (50.8 mm)
$AB = 4$ in. (102 mm)
$AC = 8$ in. (203 mm)
$BD = 4$ in. (102 mm)
$O_6D = 4$ in. (102 mm)
$O_2O_6 = 4$ in. (102 mm)
$\omega_2 = 1$ rad/s

FIGURE 8.76

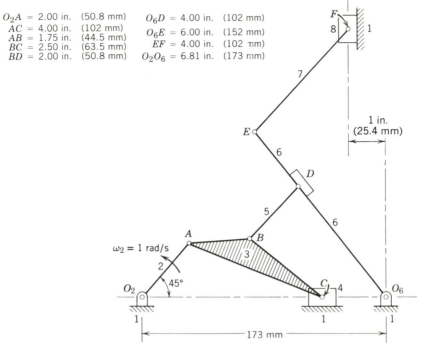

O₂A = 2.00 in. (50.8 mm) O_6D = 4.00 in. (102 mm)
AC = 4.00 in. (102 mm) O_6E = 6.00 in. (152 mm)
AB = 1.75 in. (44.5 mm)   EF = 4.00 in. (102 mm)
BC = 2.50 in. (63.5 mm) O_2O_6 = 6.81 in. (173 mm)
BD = 2.00 in. (50.8 mm)

ω_2 = 1 rad/s

45°

173 mm

1 in.
(25.4 mm)

FIGURE 8.77

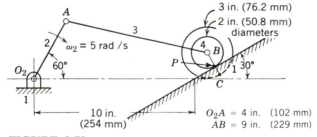

3 in. (76.2 mm)
2 in. (50.8 mm)
diameters

ω_2 = 5 rad /s

60°

30°

10 in.
(254 mm)

O_2A = 4 in. (102 mm)
AB = 9 in. (229 mm)

FIGURE 8.78

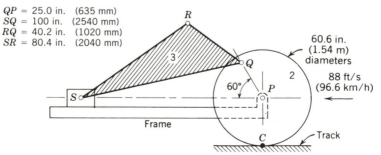

QP = 25.0 in. (635 mm)
SQ = 100 in. (2540 mm)
RQ = 40.2 in. (1020 mm)
SR = 80.4 in. (2040 mm)

60.6 in.
(1.54 m)
diameters

88 ft/s
(96.6 km/h)

60°

Frame

Track

FIGURE 8.79

384

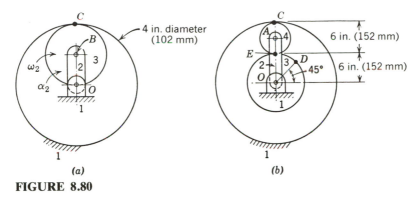

FIGURE 8.80

images of gears 3 and 4. Determine the velocity of point D_3. (*b*) Draw the acceleration polygon and show the acceleration images of gears 3 and 4. Determine the acceleration of point D_3.

8.34. For the mechanism shown in Fig. 8.81, link 2 rotates at a constant angular velocity. Determine the velocity and acceleration of the slider and ω_{32} and α_{32} by (*a*) unit vectors; (*b*) velocity and acceleration polygons.

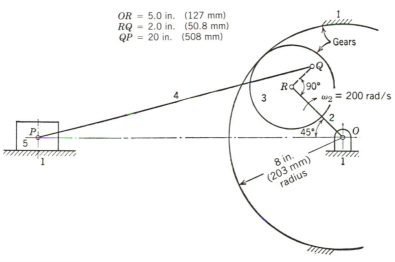

FIGURE 8.81

8.35. In the mechanism shown in Fig. 8.82, members 4 and 5 are gears in mesh. Link 2 rotates at constant angular velocity. (*a*) Construct the velocity polygons and show the images of gears 4 and 5 when $\omega_5 = 0$ and when $\omega_5 = 5$ rad/s. (*b*) Construct the acceleration polygons and show the images of gears 4 and 5 when $\omega_5 = 0$ and when $\omega_5 = 5$ rad/s.

8.36. For the mechanism shown in Fig. 8.83, link 2 rotates at constant angular velocity and slider 5 moves with constant linear velocity. Determine the velocity and acceleration

O_2B = 1.5 in.   (38.1 mm)
CB = 3.0 in.   (76.2 mm)

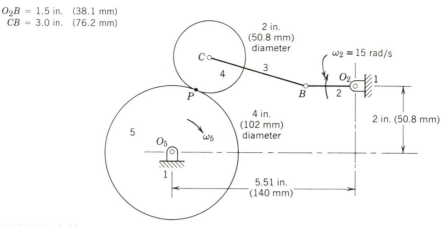

FIGURE 8.82

of point C by (*a*) unit vectors; (*b*) velocity and acceleration polygons; (*c*) computer analysis.

8.37. In Fig. 8.84 is shown a double slider-crank mechanism with the cranks rotating at constant angular velocities. Determine the velocity and acceleration of point D relative to point B by (*a*) unit vectors; (*b*) velocity and acceleration polygons; (*c*) computer analysis.

8.38. A double slider-crank mechanism is shown in Fig. 8.85. Crank 2 rotates at constant angular velocity. Determine the velocity and acceleration of each slider by (*a*) unit vectors; (*b*) velocity and acceleration polygons; (*c*) computer analysis.

8.39. A diagram of a three-cylinder radial engine is shown in Fig. 8.86. If the crank rotates at constant angular velocity, determine the velocity and acceleration of each piston by (*a*) unit vectors; (*b*) velocity and acceleration polygons; (*c*) computer analysis.

8.40. A toggle mechanism is shown in Fig. 8.87 with link 2 rotating at a constant angular velocity. Determine the velocity and acceleration of the slider by (*a*) unit vectors; (*b*) velocity and acceleration polygons; (*c*) computer analysis.

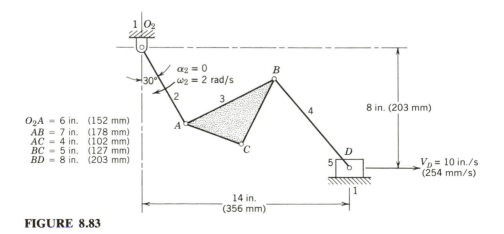

O_2A = 6 in.   (152 mm)
AB = 7 in.   (178 mm)
AC = 4 in.   (102 mm)
BC = 5 in.   (127 mm)
BD = 8 in.   (203 mm)

FIGURE 8.83

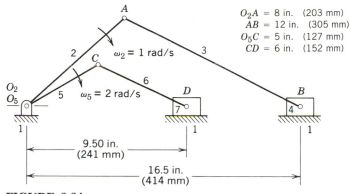

O_2A = 8 in. (203 mm)
AB = 12 in. (305 mm)
O_5C = 5 in. (127 mm)
CD = 6 in. (152 mm)

$\omega_2 = 1$ rad/s

$\omega_5 = 2$ rad/s

9.50 in.
(241 mm)

16.5 in.
(414 mm)

FIGURE 8.84

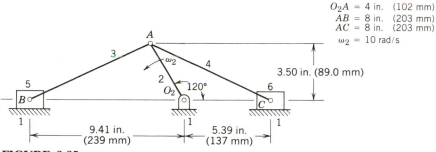

O_2A = 4 in. (102 mm)
AB = 8 in. (203 mm)
AC = 8 in. (203 mm)
$\omega_2 = 10$ rad/s

3.50 in. (89.0 mm)

ω_2

120°

9.41 in.
(239 mm)

5.39 in.
(137 mm)

FIGURE 8.85

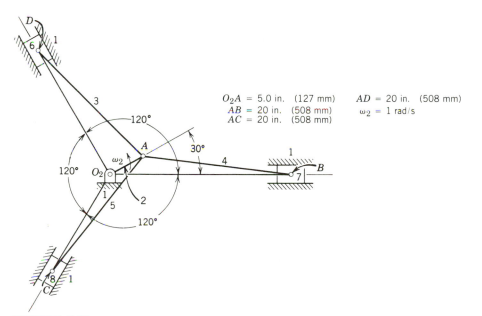

120°

120°

120°

ω_2

30°

O_2A = 5.0 in. (127 mm) AD = 20 in. (508 mm)
AB = 20 in. (508 mm) $\omega_2 = 1$ rad/s
AC = 20 in. (508 mm)

FIGURE 8.86

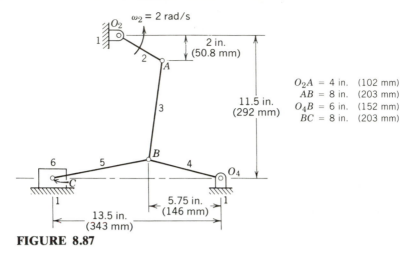

$$O_2A = 4 \text{ in.} \quad (102 \text{ mm})$$
$$AB = 8 \text{ in.} \quad (203 \text{ mm})$$
$$O_4B = 6 \text{ in.} \quad (152 \text{ mm})$$
$$BC = 8 \text{ in.} \quad (203 \text{ mm})$$

FIGURE 8.87

8.41. For the linkage shown in Fig. 8.88, slider 5 moves with constant linear velocity. Determine V_{AC}, ω_{34}, A_{AC}, and α_{34} by (*a*) unit vectors; (*b*) velocity and acceleration polygons; (*c*) computer analysis.

8.42. For the mechanism shown in Fig. 8.89, determine V_{g_3} and A_{g_3} of the center of gravity of the connecting rod. Also, determine the velocity of sliding between links 5 and 4, $V_{C_5C_4}$. Solve for these quantities by (*a*) unit vectors; (*b*) velocity and acceleration polygons; (*c*) computer analysis.

8.43. For the slider-crank mechanism shown in Fig. 8.90, (*a*) determine all instantaneous

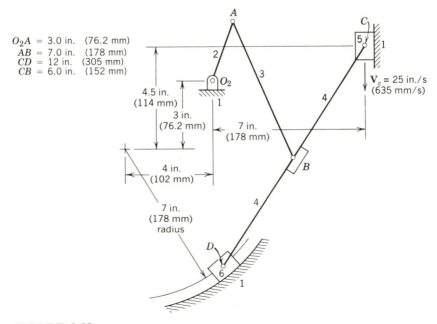

$$O_2A = 3.0 \text{ in.} \quad (76.2 \text{ mm})$$
$$AB = 7.0 \text{ in.} \quad (178 \text{ mm})$$
$$CD = 12 \text{ in.} \quad (305 \text{ mm})$$
$$CB = 6.0 \text{ in.} \quad (152 \text{ mm})$$

FIGURE 8.88

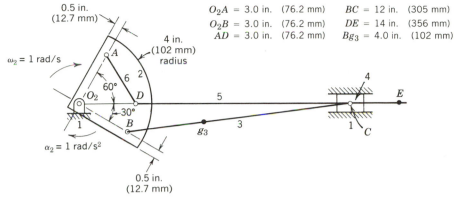

O_2A = 3.0 in. (76.2 mm) BC = 12 in. (305 mm)
O_2B = 3.0 in. (76.2 mm) DE = 14 in. (356 mm)
AD = 3.0 in. (76.2 mm) Bg_3 = 4.0 in. (102 mm)

FIGURE 8.89

centers; (*b*) determine velocity of point *B* by instantaneous center methods; (*c*) check $\mathbf{V}_B$ found in part *b* by drawing a velocity polygon.

8.44. For the offset slider-crank shown in Fig. 8.91, determine the velocity of the slider in mm/s using instantaneous centers if ω_2 = 2 rad/s.

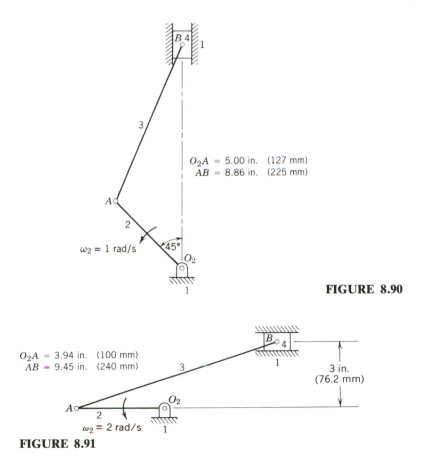

O_2A = 5.00 in. (127 mm)
AB = 8.86 in. (225 mm)

FIGURE 8.90

O_2A = 3.94 in. (100 mm)
AB = 9.45 in. (240 mm)

FIGURE 8.91

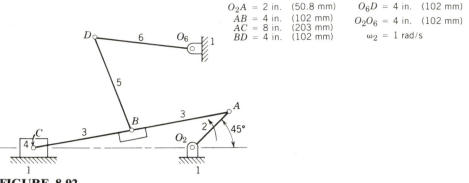

O_2A = 2 in. (50.8 mm) O_6D = 4 in. (102 mm)
AB = 4 in. (102 mm) O_2O_6 = 4 in. (102 mm)
AC = 8 in. (203 mm) ω_2 = 1 rad/s
BD = 4 in. (102 mm)

FIGURE 8.92

8.45. Given the mechanism shown in Fig. 8.92, locate all instantaneous centers.

8.46. Without determining other instantaneous centers, locate the instantaneous centers 13 and 17 in Fig. 8.77. On what line is 37 located?

8.47. Locate the six instantaneous centers of Fig. 8.78.

8.48. Determine all instantaneous center locations for the mechanism of Fig. 8.81; V_R is 83.3 ft/s and V_P is to be determined. Determine V_P using instantaneous centers 12, 15, 25. Check your answer by determining V_P using instantaneous centers 13, 14, 34 and also 13, 15, 35.

O_2A = 4 in. (102 mm)

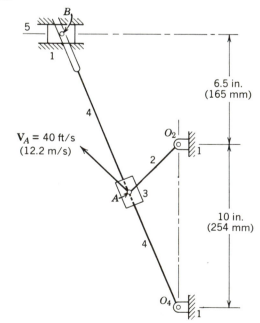

FIGURE 8.93

8.49. In Fig. 8.82, $\omega_2 = 15$ rad/s and $\omega_5 = 5$ rad/s. Determine ω_4 using the instantaneous centers 15, 14, 45.

8.50. For the crank-shaper mechanism shown in Fig. 8.93, determine (*a*) all instantaneous center locations, and (*b*) the velocity of the tool holder (link 5) using the known velocity $\mathbf{V}_{A_2} = 12.2$ m/s.

8.51. (*a*) Given the mechanism shown in Fig. 8.94, locate all instantaneous centers. (*b*) Determine $\mathbf{V}_D$ by instantaneous center methods if $\mathbf{V}_A = 25$ in./s with ω_2 turning counterclockwise.

$$O_2A = 3.0 \text{ in.} \quad (76.2 \text{ mm}) \qquad CD = 12.0 \text{ in.} \quad (305 \text{ mm})$$
$$AB = 7.0 \text{ in.} \quad (178 \text{ mm}) \qquad CB = 6.0 \text{ in.} \quad (152 \text{ mm})$$

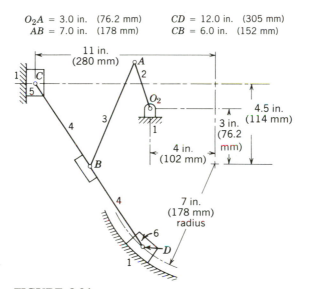

FIGURE 8.94

8.52. The claw mechanism shown in Fig. 8.95 is used to shift items to the left with intermittent motion. Gears 2 and 3 are in mesh at P, and the velocity of P is given by a vector 1 in. long. Using instantaneous centers, determine the vector representing the velocity of C of the claw.

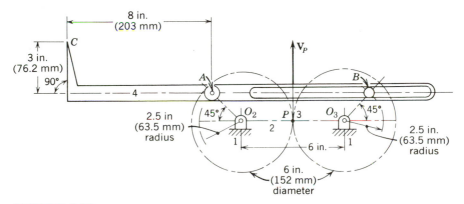

FIGURE 8.95

8.53. In the mechanism shown in Fig. 8.96, gears 2 and 3 are in mesh. The velocity of point A is given by a vector 1 in. long. By using instantaneous centers, determine the velocity of the slider (link 5). Also determine ω_4.

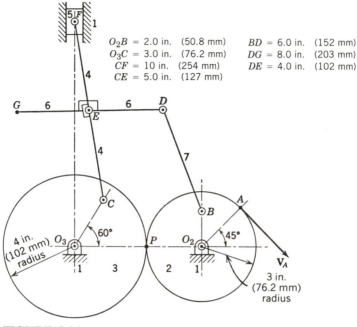

O_2B =	2.0 in.	(50.8 mm)	BD = 6.0 in.	(152 mm)
O_3C =	3.0 in.	(76.2 mm)	DG = 8.0 in.	(203 mm)
CF =	10 in.	(254 mm)	DE = 4.0 in.	(102 mm)
CE =	5.0 in.	(127 mm)		

FIGURE 8.96

8.54. In the epicyclic gear train of Fig. 8.97, the sun gear (link 3) and the internal gear (link 5) have the same sense of rotation; $\mathbf{V}_{R_5}$ is one-half the length of $\mathbf{V}_{P_3}$. Determine the location of instantaneous center 14. Determine the angular-velocity ratios ω_4/ω_3 and ω_2/ω_3.

8.55. Using real and imaginary axes, sketch to scale the vectors given by the following complex numbers: $\mathbf{r} = 8e^{i\pi/3}$, $\mathbf{r} = -4e^{i\pi}$, $\mathbf{r} = 8(\cos 60° + i \sin 60°)$, $\mathbf{r} = 10 - 40i$, $\mathbf{r} = -4 - 8i$, $\dot{\mathbf{r}} = i(\cos 120° + i \sin 120°)$, $\ddot{\mathbf{r}} = i^2(\cos 120° + i \sin 120°)$.

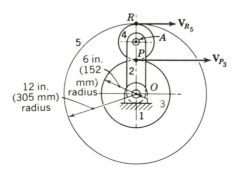

FIGURE 8.97

8.56. The link shown in Fig. 8.43a is made to rotate counterclockwise about O_2 at a uniform angular acceleration α_2 beginning from rest at $\theta_2 = 0$. From the equations of section 8.25, determine expressions for the magnitude $\mathbf{A}_P$ and angle β in terms of α_2 rather than ω_2. Numerically, evaluate $\mathbf{A}_P$ and β for $\alpha_2 = 10$ rad/s^2, $O_2P = 4$ in., and $\theta_2 = 120°$.

8.57. In Example 8.14 are shown the calculations of $\mathbf{A}_{g_3}$ and β of the acceleration $\mathbf{A}_{g_3}$ of the connecting rod in the slider crank in Fig. 8.44. Determine $\mathbf{A}_{g_3}$ and β when the crank angle θ_2 is 120° instead of 30°.

8.58. Using the equations in section 8.26, continue the calculations of Example 8.14 to determine $\mathbf{V}_B$ and $\mathbf{A}_B$ of the slider for $\theta_2 = 30°$.

8.59. The crank of the Scotch yoke in Fig. 8.98 rotates at constant ω_2. Beginning with the vector equation $\mathbf{r}_{B_4} = \mathbf{r}_2 + \mathbf{r}_a$, derive expressions for the magnitudes $\mathbf{V}_{B_4}$ and $\mathbf{A}_{B_4}$ of the yoke using the equations of complex numbers.

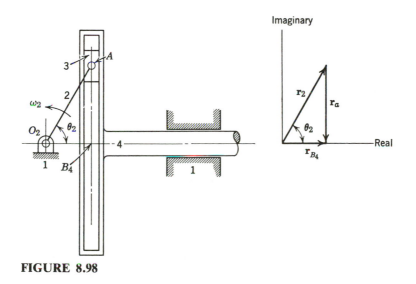

FIGURE 8.98

8.60. In the mechanism of Fig. 8.99, the crank drives the linkage at constant ω_2. Using complex number equations, determine expressions giving θ_4, ω_4, and α_4 in terms of ω_2 and known lengths.

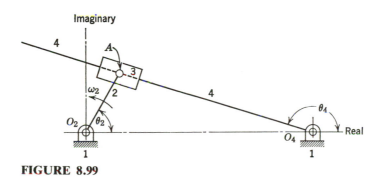

FIGURE 8.99

8.61. In Fig. 8.100 is shown an offset slider crank for which the instantaneous values of ω_2 and α_2 are known at the phase θ_2. Beginning with the vector equation $\mathbf{r}_a + \mathbf{r}_d = \mathbf{r}_2 + \mathbf{r}_3$ and using the equations of complex numbers, determine expressions for θ_3, ω_3, α_3, r_a, $\dot{r}_a$, and $\ddot{r}_a$.

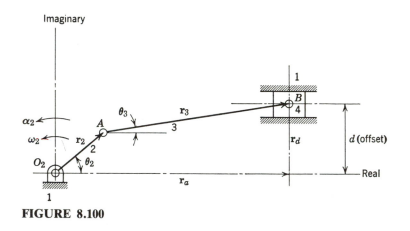

FIGURE 8.100

8.62. With reference to the four-bar linkage of Fig. 8.47, the angular positions of the links are numerically determined in section 2.1 for the data given in the figure and for $\theta_2 = 60°$. Evaluate numerically the values of ω_3, ω_4, α_3, and α_4 when $\theta_2 = 60°$, $\omega_2 = 1$ rad/s, and $\alpha_2 = 0$. Evaluate also the magnitude and angle of the acceleration vector of point B.

8.63. Equations 8.84 and 8.85 for the four-bar linkage show that ω_3 and ω_4 may be zero when $(\theta_4 - \theta_2)$ or $(\theta_3 - \theta_2)$ are zero, respectively. By sketches, show possible positions of the linkage for (a) $\omega_3 = 0$ and (b) $\omega_4 = 0$. The same equations show that ω_3 and ω_4 may be infinite when link angles are such as to make the denominators of the equations zero. Sketch possible positions of the linkage for (a) $\omega_3 = \infty$ and (b) $\omega_4 = \infty$. Are these positions practical?

Chapter Nine

Force Analysis
of Machinery

9.1 INTRODUCTION

To design the parts of a machine or mechanism for strength, it is necessary to determine the forces and torques acting on the individual links. Each component of a complete machine, however small, should be carefully analyzed for its role in transmitting force. A four-bar mechanism, for example, in reality consists of eight links if the pins or bearings connecting the primary members are included. Bearings, pins, screws, and other fasteners are often critical elements in machinery because of the concentration of force at these elements. Mechanisms that transmit force by direct surface contact on small areas of contact, such as cams, gears, and Geneva wheel pins, are also important in this respect.

In machines doing useful work, the forces associated with the principal function of the machine are usually known or assumed. For example, in a piston-type engine or piston-type compressor, the gas force on the piston is known or assumed; in a quick-return mechanism such as the crank shaper or the Whitworth machine, the resistance of the cutting tool is assumed. Such forces are termed *static* forces since in a machine analysis they are classed differently from inertia forces, which are expressed in terms of the accelerated motion of the individual links.

In mechanisms operating at high speeds, the forces on an individual link that produce the accelerated motion of the link are often greater than the static forces related to the primary function of the machine. In many rotary machines, such as bladed compressor and turbine wheels, precautions are necessary to avoid runaway conditions in which speeds may exceed structurally safe design speeds.

9.2 CENTRIFUGAL FORCE IN ROTOR BLADES

In rotors, inertia force, the product of mass and acceleration, is known as *centrifugal force*. In high-speed rotors with blades (such as compressor and turbine wheels, supercharger wheels, fans, and propellers), centrifugal forces tend to separate the blades from the rotor. Figure 9.1 shows a simple type of bladed rotor. To determine the centrifugal force producing a resisting centripetal force at the base (section *a–a*) of any given blade, an integration is required since the acceleration is a function of R. Assuming that the rotor is at constant angular velocity ω, the inertia force dF acting on the element of the blade shown is the product of the mass of the element dM and the centripetal acceleration $A^n = \omega^2 R$ from Eq. 8.4a. Therefore,

$$dF = (dM)A^n = \omega^2 R \, dM \qquad (9.1)$$

It is recalled from the study of mechanics that inertia force is opposite in sense to the centripetal acceleration, hence the term *centrifugal* force. The mass of the element is the product of mass density w/g (w is weight density in pounds per cubic inch and g is 386 in./s^2) and the volume of the element $bt(dR)$; b, t, and R are in inches.

$$dF = \omega^2 R \, \frac{w}{g} \, bt \, dR$$

$$F = bt \, \frac{w}{g} \, \omega^2 \int_{R=R_i}^{R=R_o} R \, dR \qquad (9.2)$$

The average tensile stress s_b at the base of the blade due to the inertia force

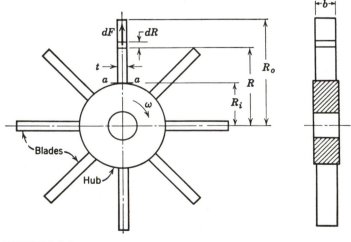

FIGURE 9.1

is P/A, in which $P = F$ and $A = bt$:

$$s_b = \frac{F}{bt} = \frac{w}{g} \omega^2 \int_{R=R_i}^{R=R_o} R \, dR \qquad (9.3)$$

Equation 9.3 shows that the stress at the base of the blade is independent of the cross-sectional area $A = bt$ but depends on rotor speed, mass density, and the inner and outer radii of the blades.

Rotor blades such as the wide blades of a fan are idealized in Fig. 9.2. The fan has the form of a disk with slots between the blades. The element of inertia force dF is the same as given by Eq. 9.1, in which the mass of the element is

$$dM = \frac{w}{g} tR \, d\phi \, dR$$

and

$$dF = \frac{w}{g} t\omega^2 R^2 \, dR \, d\phi$$

Therefore,

$$F = \frac{w}{g} t\omega^2 \int_{\phi=0}^{\phi=2\pi/N} \int_{R=R_i}^{R=R_o} R^2 \, dR \, d\phi$$

$$F = \frac{2\pi}{N} \frac{w}{g} t\omega^2 \int_{R=R_i}^{R=R_o} R^2 \, dR \qquad (9.4)$$

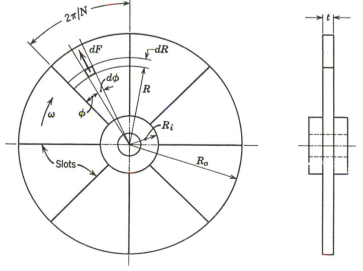

FIGURE 9.2

where N is the number of blades and F is the force tending to separate the blade from the hub. The average stress at the base of the blade with cross-sectional area $A = 2\pi R_i t/N$ is

$$s_b = \frac{w}{g}\frac{\omega^2}{R_i}\int_{R=R_i}^{R=R_o} R^2\, dR \tag{9.5}$$

In aircraft propellers, the blades are set at a blade angle as shown in Fig. 9.3. In such cases, inertia forces produce a twisting moment on the blade. With reference to Fig. 9.3, the inertia force dF on an element $t\, dx\, dR$ is

$$dF = \omega^2 a\, dM$$

in which $dM = (w/g)t\, dx\, dR$. The inertia force dF, which is due to A^n of the element of mass, may be shown as the components dF_n and dF_t, in which dF_n

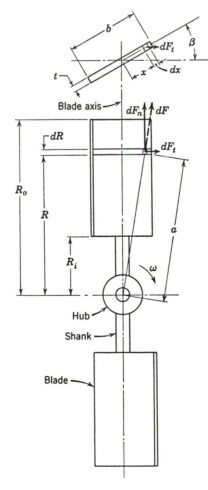

FIGURE 9.3

produces a tensile force on the blade parallel to the blade axis and dF_t produces a twisting moment dM_t about the axis of the blade because of the moment arm $x \sin \beta$.

$$dF_n = \frac{R}{a} dF = \omega^2 R \, dM$$

$$dF_n = \frac{w}{g} t\omega^2 R \, dx \, dR \qquad (9.6)$$

$$dF_t = \frac{x \cos \beta}{a} dF = \omega^2 x \cos \beta \, dM$$

$$dF_t = \frac{w}{g} t\omega^2 x \cos \beta \, dx \, dR \qquad (9.7)$$

$$dM_t = x \sin \beta \, dF_t$$

$$dM_t = \frac{w}{g} t \cos \beta \sin \beta \omega^2 x^2 \, dx \, dR \qquad (9.8)$$

The total inertia force of the blade producing tension in the shank parallel to the axis of the blade is

$$F_n = \frac{w}{g} t\omega^2 2 \int_{x=0}^{x=b/2} \int_{R=R_i}^{R=R_o} R \, dx \, dR \qquad (9.9)$$

and the total twisting moment on the shank is

$$M_t = \frac{w}{g} t\omega^2 \cos \beta \sin \beta 2 \int_{x=0}^{x=b/2} \int_{R=R_i}^{R=R_o} x^2 \, dx \, dR \qquad (9.10)$$

9.3 INERTIA FORCE, INERTIA TORQUE

From the study of mechanics, it is known that the following equations of motion apply for a rigid body in plane motion.

$$\Sigma \mathbf{F} = M\mathbf{A}_g \qquad (9.11)$$

$$\Sigma T = I\alpha \qquad (9.12)$$

in which $\Sigma \mathbf{F}$ is the vector sum, or the resultant $\mathbf{R}$, of a system of forces acting on the body in the plane of motion; M is the mass of the body; and $\mathbf{A}_g$ is the acceleration of the mass center g (center of gravity) of the body. ΣT is the sum of the moments of the forces and torques about an axis through the mass center normal to the plane of motion; I is the moment of inertia of the body about the same axis through the mass center; and α is the angular acceleration of the body in the plane of motion. The unit of mass M commonly used is the slug (lb · s²/ ft), and the unit of moment of inertia I is slug · ft² (lb · s² · ft). For the International System of Units, the unit of mass is kilogram (kg), and the unit of moment of inertia is kg · m².

Figure 9.4 shows a rigid body in plane motion acted upon by forces for which the resultant **R** is determined from the polygon of free force vectors shown. Since **R** represents Σ **F**, Eq. 9.11 may be rewritten

$$\mathbf{R} = M\mathbf{A}_g \qquad (9.13)$$

For the case in which the forces are known, the acceleration $\mathbf{A}_g$ of the body may be calculated from Eq. 9.13 provided the mass is also known. The direction of $\mathbf{A}_g$ is parallel to **R** and in the same sense as **R**.

The line of action of **R** is determined as shown in Fig. 9.4, and, from the principle of moments, Re is equal to Σ T. Equation 9.12 may be rewritten

$$Re = I\alpha \qquad (9.14)$$

The angular acceleration α of the body may be determined from Eq. 9.14 if the forces and the moment of inertia I of the body are known; α is in the same angular sense as the moment Re.

Equations of motion in the forms of Eqs. 9.11 through 9.14 are useful when accelerations are to be determined including magnitude, direction, and sense. However, for mechanisms with constrained motion, accelerations are usually known from a kinematic analysis as discussed in Chapter 8, and the forces and moments which produce the accelerations are to be determined.

When $\mathbf{A}_g$ of a given link is known and $M\mathbf{A}_g$ is calculable, a simplification in concept results if $M\mathbf{A}_g$, expressed in units of force, is regarded as a force vector $\mathbf{F}_o$ and is shown as the equilibrant of **R** on the free-body diagram of the link. In Fig. 9.5, the body of Fig. 9.4 is shown with $\mathbf{F}_o$ as an equilibrant. As a vector, $\mathbf{F}_o$

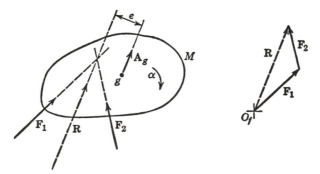

FIGURE 9.4

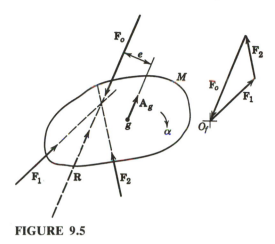

FIGURE 9.5

is shown parallel in direction to $\mathbf{A}_g$, which is also parallel to $\mathbf{R}$ and is equal in magnitude to $\mathbf{R}$ from Eq. 9.13. To be the equilibrant of $\mathbf{R}$, however, $\mathbf{F}_o$ must be shown in a sense opposite to $\mathbf{A}_g$. Also, the line of action of $\mathbf{F}_o$ must be such that its moment about the mass center is equal and *opposite* to the moment of R. Equation 9.14 may be used to determine the distance e of the line of action of $\mathbf{F}_o$:

$$e = \frac{I\alpha}{R}$$

$$e = \frac{I\alpha}{F_o} = \frac{I\alpha}{MA_g} \tag{9.15}$$

It should be noted that the moment of $\mathbf{F}_o$ about the mass center is opposite in sense to α. By showing $\mathbf{F}_o$ opposite in sense to $\mathbf{A}_g$ and the moment of $\mathbf{F}_o$ opposite in sense to α, it appears to represent a resistance to the accelerated motion of the link and in a sense is a measure of the inertia of the link. Thus, $\mathbf{F}_o$ is termed an *inertia force*.

In Fig. 9.5, $\mathbf{F}_o$ is shown as the equilibrant of $\mathbf{R}$. An alternative representation as in Fig. 9.6 is to show $\mathbf{F}_o$ at the mass center g and to add an *inertia torque* or *inertia couple* T_o in a sense opposite to α. The magnitudes of $\mathbf{F}_o$ and T_o are given by the following equations:

$$F_o = MA_g \tag{9.16}$$

$$T_o = I\alpha \tag{9.17}$$

When $\mathbf{A}_g$ is zero and α is of a value other than zero, only the inertia couple T_o remains.

It may be seen from Fig. 9.5 that, by showing known mass acceleration

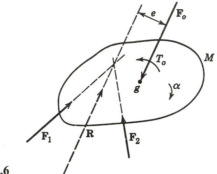

FIGURE 9.6

effects of a link as an inertia force, the equations of motion (9.11 and 9.12) may be interpreted as equations of static equilibrium and may be written as:

$$\Sigma \,\mathbf{F} = 0$$

$$\Sigma \, T = 0$$

in which $\Sigma \, \mathbf{F}$ includes $\mathbf{F}_o$, and $\Sigma \, T$ includes T_o. This is sometimes referred to as the concept of *dynamic equilibrium*. In Fig. 9.5, the polygon of free force vectors, including $\mathbf{F}_o$, closes as required for static equilibrium.

The inertia force method is a simple and useful method since kinetic problems involving rigid-body linkages in plane motion are reduced to problems of static equilibrium. Because of the constrained motion of linkages, accelerations and inertia forces and couples of the individual links may be determined first, and the forces producing accelerated motion may then be determined from the laws of static equilibrium.

9.4   FORCE DETERMINATION

In the force analysis of a complete mechanism, a free-body diagram of each link should usually be made to indicate the forces acting on the link. In determining the directions of these forces, the following laws from the study of statics will be recalled.

1. A rigid body acted on by two forces is in static equilibrium only if the two forces are collinear and equal in magnitude but opposite in sense. If only the points of application of the two forces are known, such as A and B of Fig. 9.7, the directions of the two forces may be determined from the direction of the line joining A and B.

2. For a rigid body acted upon by three forces in static equilibrium, the lines of action of the three forces are concurrent at some point such as k in Fig. 9.8. Thus, if the lines of action of two of the forces are known, the line of action of the third force must pass through its point of application and the point of con-

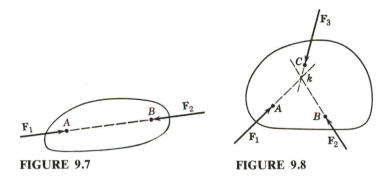

FIGURE 9.7   **FIGURE 9.8**

currency k. In some cases, a greater number than three forces on a body may be reduced to three by determining the resultant of the known forces.

3. A rigid body acted upon by a couple is in static equilibrium only if acted upon by another coplanar couple equal in magnitude and opposite in sense as shown in Fig. 9.9.

In the case of a static force analysis, the vector sum of the forces on each link must equal zero for equilibrium. This must also be true for a dynamic analysis when inertia forces are used. To use the concept of inertia forces is therefore an advantage because both static and dynamic cases can be treated in the same manner. In both types of analyses, the vector equations can be solved analytically or graphically to determine the unknown forces.

Factors which determine whether an analytical or a graphical solution should be made are the type of mechanism and the number of positions to be analyzed. For relatively simple mechanisms such as cams and gears, an analytical solution is generally used. For the analysis of a linkage at only one position, a graphical solution is much quicker than an analytical one. However, if several positions or a complete cycle are to be studied, analytical methods should be selected. This is especially true if computer facilities are available or pocket or desk calculators with vector resolver capabilities. It should be mentioned, however, that even when an analytical solution is used, it is often desirable to check the results at one position by graphical means.

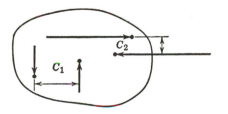

FIGURE 9.9

9.5 METHODS OF LINKAGE FORCE ANALYSIS

Two methods for analyzing the forces in mechanisms are currently used: (*a*) the superposition method and (*b*) the matrix method. Both of these methods will be discussed in this text. The superposition method is best suited for hand calculation or graphical layout, whereas the matrix method is best suited for computer solution. In the superposition method, a separate analysis of the mechanism is made for each moving link considering the inertial and external forces and torques acting on that link alone. A mechanism having *n* moving links therefore requires *n* separate analyses. The results of these analyses are then summed together to determine the total forces and torques in the mechanism. In the matrix method, the equations of motion are written for each moving link taken as a free body. This results in a system of $3n$ linear equations in $3n$ unknowns, which must be solved simultaneously.

Two variations of the superposition method find wide usage. The first method uses inertia force and inertia torque directly and is best suited for analytical work. The second method eliminates the need to consider the inertia torque by offsetting the inertia force by an amount *e*. This method is preferrable in graphical solutions. Both these methods are illustrated in the following section.

9.6 LINKAGE FORCE ANALYSIS BY SUPERPOSITION

The principle of superposition may be used in the force analysis of a rigid body in static equilibrium. This principle states that a resultant effect may be determined from the summation of several effects which are equivalent to the total effect. By this method, a linkage with several forces acting on it can easily be analyzed by determining the effect of these forces taken one at a time. The results of the several single-force analyses are then added together to give the total forces acting on each joint in the linkage. Superposition can also be used advantageously to combine the results of a static and an inertia force analysis that have been made independently.

Although this method is easy to use, it has the disadvantage that the linkage must be analyzed several times, which often becomes tedious. Another disadvantage is that an accurate analysis cannot be made if friction forces are to be considered. In linkages with turning pairs, this problem does not usually arise because friction forces are small enough to be neglected. However, with sliding pairs, such as with the piston and cylinder in the slider-crank mechanism, superposition would not be a suitable method of analysis if friction is to be considered between the piston and cylinder. Errors would occur because of a change in direction of the force between the piston and cylinder in the separate solutions required for superposition.

An analytical force analysis by superposition is given in Example 9.1 using inertia forces and inertia torques directly in the equations of dynamic equilibrium.

Example 9.2 gives a similar solution done graphically using only the inertia forces and offsetting them a distance e to produce a couple equivalent to the inertia torque.

Example 9.1. In Fig. 9.10a is shown the linkage from Fig. 8.7 of Example 8.1 for which a velocity and acceleration analysis has been made. It is required to determine the bearing forces on each link and the shaft torque T_s at O_2 by superposition using unit vectors.

Solution. From the solution of Example 8.1,

$$\omega_3 = 4.91 \text{ rad/s} \quad (\text{ccw})$$

$$\omega_4 = 7.82 \text{ rad/s} \quad (\text{ccw})$$

$$\alpha_3 = 241 \text{ rad/s}^2 \quad (\text{ccw})$$

$$\alpha_4 = -129 \text{ rad/s}^2 \quad (\text{cw})$$

$$\mathbf{A}_A = -72\mathbf{i} - 124.8\mathbf{j}$$

$$|\mathbf{A}_A| = 144 \text{ ft/s}^2$$

$$\mathbf{A}_B = -88.1\mathbf{i} + 35.8\mathbf{j}$$

$$|\mathbf{A}_B| = 95.1 \text{ ft/s}^2$$

By using the above values, the following accelerations were calculated and their directions determined:

$$|\mathbf{A}_{g_3}| = 91.6 \text{ ft/s}^2$$

$$|\mathbf{A}_{g_4}| = 62.7 \text{ ft/s}^2$$

The angles at which the vector $\mathbf{A}_{g_3}$ and $\mathbf{A}_{g_4}$ act are shown in Fig. 9.10a.
    The magnitudes of the inertia forces and torques can be calculated as follows:

$$F_{o_2} = 0 \quad (A_{g_2} = 0)$$

$$F_{o_3} = M_3 A_{g_3} = \frac{4 \times 91.6}{32.2} = 11.4 \text{ lb}$$

$$F_{o_4} = M_4 A_{g_4} = \frac{8 \times 62.7}{32.2} = 15.6 \text{ lb}$$

$$T_{o_3} = -I_3 \alpha_3 = -0.006 \times 241 = -1.446 \text{ lb} \cdot \text{ft} = -17.35 \text{ lb} \cdot \text{in.}$$

$$T_{o_4} = -I_4 \alpha_4 = -0.026 \times -129 = 3.351 \text{ lb} \cdot \text{ft} = 40.21 \text{ lb} \cdot \text{in.}$$

The vectors $\mathbf{F}_{o_3}$, $\mathbf{F}_{o_4}$, $\mathbf{T}_{o_3}$ and $\mathbf{T}_{o_4}$ are shown in Fig. 9.10b and c in their correct positions and with the correct orientations.

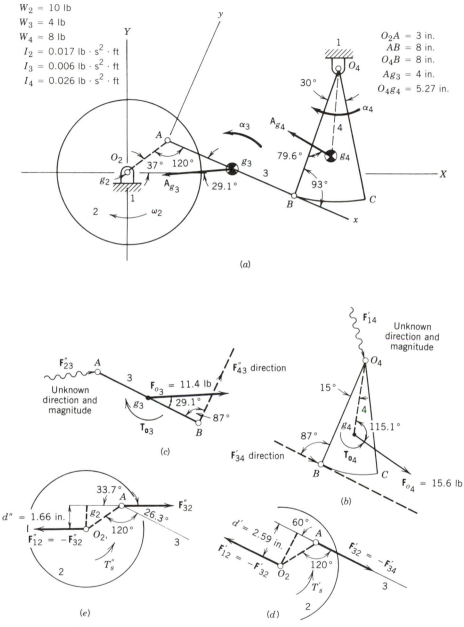

$W_2 = 10$ lb
$W_3 = 4$ lb
$W_4 = 8$ lb
$I_2 = 0.017$ lb · s^2 · ft
$I_3 = 0.006$ lb · s^2 · ft
$I_4 = 0.026$ lb · s^2 · ft

$O_2A = 3$ in.
$AB = 8$ in.
$O_4B = 8$ in.
$Ag_3 = 4$ in.
$O_4g_4 = 5.27$ in.

FIGURE 9.10

The solution of this problem will be carried out using superposition considering (*a*) only $\mathbf{F}_{o_4}$ and $\mathbf{T}_{o_4}$ acting, (*b*) only $\mathbf{F}_{o_3}$ and $\mathbf{T}_{o_3}$ acting, and (*c*) addition of the two analyses. All components have been taken relative to the *xy*-axes fixed in link 3.

FORCE ANALYSIS WITH ONLY $\mathbf{F}_{o_4}$ AND $\mathbf{T}_{o_4}$ ACTING.   Figure 9.10*b* shows a free-body diagram of link 4 upon which act forces $\mathbf{F}_{o_4}$, $\mathbf{F}'_{34}$, and $\mathbf{F}'_{14}$ and torque $\mathbf{T}_{o_4}$, where $\mathbf{F}'_{34}$ is the force link 3 exerts on link 4 and $\mathbf{F}'_{14}$ is the force link 1 exerts on link 4. The single prime is used to indicate that these are only that part of the actual forces which act between the links due to $\mathbf{F}_{o_4}$ and $\mathbf{T}_{o_4}$. Note that the direction of $\mathbf{F}'_{34}$ is known, because link 3 becomes a two-force member in this portion of the superposition process. Both the direction and the magnitude of $\mathbf{F}'_{14}$ are unknown, however. Since link 4 is in equilibrium under the action of forces $\mathbf{F}_{o_4}$, $\mathbf{F}'_{34}$, and $\mathbf{F}'_{14}$ and torque $\mathbf{T}_{o_4}$, moments may be summed about any convenient point and set equal to zero.

By summing moments about point O_4,

$$F_{o_4}(\overline{O_4 g_4}) \sin 115.1° + F'_{34}(\overline{O_4 B}) \sin 87° + T_{o_4} = 0$$

$$(15.6)(5.27) \sin 115.1° + F'_{34}(8) \sin 87° + 40.21 = 0$$

$$F'_{34} = -14.35 \text{ lb}$$

Link 4 must also be in translational equilibrium under the action of the given forces, so that

$$\mathbf{F}_{o_4} + \mathbf{F}'_{14} + \mathbf{F}'_{34} = 0$$

Expressing $\mathbf{F}_{o_4}$ and $\mathbf{F}'_{34}$ in the *xy*-coordinate system gives

$$\mathbf{F}_{o_4} = 15.6 \,(\cos 7.4° \,\mathbf{i} - \sin 7.4° \,\mathbf{j})$$

$$= 15.5\mathbf{i} - 2.01\mathbf{j}$$

$$\mathbf{F}'_{34} = F'_{34}\mathbf{i} = -14.35\mathbf{i}$$

And the translational equilibrium equation for link 4 becomes

$$15.5\mathbf{i} - 2.01\mathbf{j} - 14.35\mathbf{i} + F'_{14x}\mathbf{i} + F'_{14y}\mathbf{j} = 0$$

where $\mathbf{F}'_{14x}$ and $\mathbf{F}'_{14y}$ are the *x*- and *y*-components of $\mathbf{F}'_{14}$, respectively.

By summing **i** *components,*

$$15.5\mathbf{i} - 14.35\mathbf{i} + F'_{14x}\mathbf{i} = 0$$

$$F'_{14x} = -1.15 \text{ lb}$$

By summing **j** *components,*

$$-2.01\mathbf{j} + F'_{14y}\mathbf{j} = 0$$

$$F'_{14y} = 2.01 \text{ lb}$$

To calculate the shaft torque T'_s necessary to hold link 2 in equilibrium under the action of the couple produced by F'_{32} and F'_{12}, refer to Fig. 9.10d, where

$$F'_{32} = F'_{43} = 14.3 \text{ lb}$$

$$d' = 2.59 \text{ in.}$$

Therefore,

$$T'_s = F'_{32}d' = (14.3)(2.59)$$

$$= 37.2 \text{ lb} \cdot \text{in.} \quad (\text{ccw})$$

Torque $\mathbf{T}'_s$ could also have been easily determined using vector equations for $\mathbf{F}'_{32}$ and d' and the relation

$$\mathbf{T}'_s = -(\mathbf{F}_{32} \times d')$$

FORCE ANALYSIS WITH ONLY $\mathbf{F}_{o_3}$ AND $\mathbf{T}_{o_3}$ ACTING.   Figure 9.10c shows a free-body diagram of link 3 under the action of three forces $\mathbf{F}_{o_3}$, $\mathbf{F}''_{23}$, and $\mathbf{F}''_{43}$ and torque $\mathbf{T}_{o_3}$. Here, the double prime denotes part b of the superposition problem. The direction of $\mathbf{F}_{o_3}$ is known, and that of $\mathbf{F}''_{43}$ is along line O_4B because link 4 becomes a two-force member when $\mathbf{F}_{o_4}$ and $\mathbf{T}_{o_4}$ are omitted. Link 3 is in equilibrium under the action of forces $\mathbf{F}_{o_3}$, $\mathbf{F}''_{23}$, and $\mathbf{F}''_{43}$ and torque $\mathbf{T}_{o_4}$. Once again, moments may be summed about any convenient point and set equal to zero.

By summing moments about point A,

$$F_{o_3}(\overline{A_{g3}}) \sin 29.1° + F''_{43}(\overline{AB}) \cos 3° + T_{o_3} = 0$$

$$(11.4)(4) \sin 29.1° + F''_{43}(8) \cos 3° - 17.35 = 0$$

$$F''_{43} = -0.604 \text{ lb}$$

Link 3 must be in translational equilibrium under the action of the given forces, so that

$$\mathbf{F}_{o_3} + \mathbf{F}''_{43} + \mathbf{F}''_{23} = 0$$

Expressing $\mathbf{F}_{o_3}$ and $\mathbf{F}''_{43}$ in the xy-coordinate system gives

$$\mathbf{F}_{o_3} = 11.4(\cos 29.1° \, \mathbf{i} + \sin 29.1° \, \mathbf{j}) = 9.94\mathbf{i} + 5.53\mathbf{j}$$

$$\mathbf{F}''_{43} = 0.604(\cos 87° \, \mathbf{i} - \sin 87° \, \mathbf{j}) = 0.04\mathbf{i} - 0.60\mathbf{j}$$

and the translational equilibrium equation for link 3 becomes

$$9.94\mathbf{i} + 5.53\mathbf{j} + 0.04\mathbf{i} - 0.60\mathbf{j} + F''_{23x}\mathbf{i} + F''_{23y}\mathbf{j} = 0$$

By summing $\overline{\mathbf{i}}$ components,

$$9.94\mathbf{i} + 0.04\mathbf{i} + F''_{23x}\mathbf{i} = 0$$

$$F''_{23x} = -9.98 \text{ lb}$$

By summing $\bar{\mathbf{j}}$ components,

$$5.53\mathbf{j} - 0.60\mathbf{j} + F''_{23y}\mathbf{j} = 0$$

$$F''_{23y} = -4.93 \text{ lb}$$

$$F''_{23} = (F''^2_{23x} + F''^2_{23y})^{1/2} = 11.1 \text{ lb}$$

The shaft torque T''_s necessary to hold link 2 in equilibrium under the action of the couple produced by $\mathbf{F}''_{32}$ and $\mathbf{F}''_{12}$ can be calculated from Fig. 9.10e, where

$$F''_{32} = 11.1 \text{ lb}$$

and

$$d'' = 1.66 \text{ in.}$$

Therefore,

$$T''_s = F''_{32} d'' = 11.1 \times 1.66$$

$$= 18.5 \text{ lb} \cdot \text{in.} \quad \text{(ccw)}$$

If desirable, T''_s can be calculated using the following vector equations:

$$\mathbf{F}''_{32} = -\mathbf{F}''_{23} = 9.98\mathbf{i} + 4.93\mathbf{j}$$

$$\mathbf{d}'' = 0.736\mathbf{i} - 1.49\mathbf{j}$$

and

$$\mathbf{T}''_s = -(\mathbf{F}''_{32} \times \mathbf{D}'')$$

$$= 18.5 \text{ k lb} \cdot \text{in.} \quad \text{(ccw)}$$

TOTAL FORCES:

$$\mathbf{F}_{32} = \mathbf{F}'_{32} + \mathbf{F}''_{32} = \mathbf{F}'_{43} + \mathbf{F}''_{32}$$

$$= 14.3\mathbf{i} + 9.98\mathbf{i} + 4.93\mathbf{j}$$

$$= 24.3\mathbf{i} + 4.93\mathbf{j}$$

and

$$|\mathbf{F}_{32}| = 24.8 \text{ lb}$$

$$\mathbf{F}_{43} = \mathbf{F}'_{43} + \mathbf{F}''_{43}$$

$$= 14.3\mathbf{i} + 0.032\mathbf{i} - 0.604\mathbf{j}$$

$$= 14.3\mathbf{i} - 0.604\mathbf{j}$$

and

$$|\mathbf{F}_{43}| = 14.4 \text{ lb}$$

$$\mathbf{F}_{14} = \mathbf{F}'_{14} + \mathbf{F}''_{14} = \mathbf{F}'_{14} + \mathbf{F}''_{43}$$

$$= -1.13\mathbf{i} + 2.01\mathbf{j} + 0.032\mathbf{i} - 0.604\mathbf{j}$$

$$= -1.10\mathbf{i} - 1.41\mathbf{j}$$

and

$$|\mathbf{F}_{14}| = 1.78 \text{ lb}$$

$$T_s = T'_s + T''_s$$

$$= 37.2 + 18.5$$

$$= 55.7 \text{ lb} \cdot \text{in.} \quad \text{(ccw)}$$

Therefore,

$$F_{14} = 1.78 \text{ lb}$$

$$F_{43} = 14.4 \text{ lb}$$

$$F_{32} = 24.8 \text{ lb}$$

$$F_{12} = 24.8 \text{ lb}$$

$$T_s = 55.7 \text{ lb} \cdot \text{in.} \quad \text{(ccw)}$$

Example 9.2. A graphical analysis of the linkage of Example 9.1 will now be made using superposition and inertia forces only and using S.I. units. The inertia torques are eliminated from consideration by offsetting the inertia forces from the mass centers of the links to produce an equivalent couple.

Solution. The following values were determined from an acceleration polygon (not shown) for the linkage.

$$A_{g_2} = 0$$

$$A_{g_3} = 27.92 \text{ m/s}^2$$

$$A_{g_4} = 19.11 \text{ m/s}^2$$

$$\alpha_3 = 241 \text{ rad/s}^2 \quad \text{(ccw)}$$

$$\alpha_4 = 129 \text{ rad/s}^2 \quad \text{(cw)}$$

Vectors representing the above values are shown on the configuration diagram (Fig. 9.11a). The magnitudes of the inertia forces and offset distances can be calculated as follows:

$$F_{o_2} = 0 \quad (A_{g_2} = 0)$$

$$F_{o_3} = M_3 A_{g_3} = 1.81 \times 27.92 = 50.6 \text{ N}$$

$F_{o_4} = M_4 A_{g_4} = 3.63 \times 19.11 = 69.4 \text{ N}$

$e_3 = \dfrac{I_3 \alpha_3}{F_{o_3}} = \dfrac{0.008 \times 241}{50.6} = 0.0381 \text{ m} = 38.11 \text{ mm}$

$e_4 = \dfrac{I_4 \alpha_4}{F_{o_4}} = \dfrac{0.035 \times 129}{69.4} = 0.0651 \text{ m} = 65.1 \text{ mm}$

The vectors $\mathbf{F}_{o_3}$ and $\mathbf{F}_{o_4}$ are shown on the configuration diagram (Fig. 9.11a) in their correct positions relative to their respective acceleration vectors, that is, parallel to acceleration vector, opposite in sense, and offset a distance e so that $F_o \times e$ gives a torque whose sense is opposite to that of α.

The solution of this problem will be carried out using superposition considering (a) only $\mathbf{F}_{o_4}$ acting, (b) only $\mathbf{F}_{o_3}$ acting, and (c) addition of the two analyses.

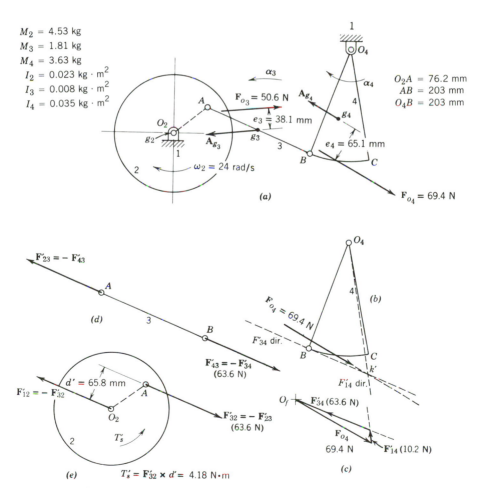

FIGURE 9.11

FORCE ANALYSIS WITH ONLY $\mathbf{F}_{o_4}$ ACTING.   Figure 9.11*b* shows a free-body diagram of link 4 upon which act $\mathbf{F}_{o_4}$, $\mathbf{F}'_{34}$, and $\mathbf{F}'_{14}$. The direction of $\mathbf{F}_{o_4}$ is known and that of $\mathbf{F}'_{34}$ is along link 3 because link 3 becomes a two-force member when $\mathbf{F}_{o_3}$ is omitted from the analysis. The direction lines of these two vectors intersect at point k'. Link 4 is in equilibrium under the action of these three forces with no couple acting upon it; therefore, the direction of vector $\mathbf{F}'_{14}$ must pass through points k' and O_4.

Figure 9.11*c* shows the force polygon for the determination of the magntidues and senses of the vectors $\mathbf{F}'_{34}$ and $\mathbf{F}'_{14}$. Because link 4 is in equilibrium, $\mathbf{F}_{o_4} + \mathbf{F}'_{14} + \mathbf{F}'_{34} = 0$ and the polygon closes as shown.

Figure 9.11*d* shows the free-body diagram of link 3 with force $\mathbf{F}'_{23}$ acting at point A and $\mathbf{F}'_{43}$ acting at point B. Link 3 is in equilibrium under the action of these two forces. The directions of these forces are along the link, and their magnitudes are equal to the magnitude of $\mathbf{F}'_{34}$ as shown.

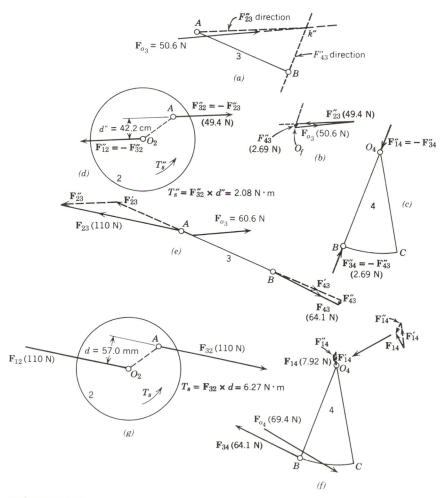

FIGURE 9.12

Figure 9.11e shows the free-body diagram of link 2. Force $\mathbf{F}'_{32}$ acts at point A as shown and is balanced by the equal and opposite force $\mathbf{F}'_{12}$ applied at point O_2. Because these forces are not collinear and form a couple $\mathbf{F}'_{32} \times d'$, it is necessary to apply a torque T'_s with a sense opposite to that of the couple to hold link 2 in equilibrium. Torque T'_s is applied to the shaft upon which link 2 is mounted.

FORCE ANALYSIS WITH ONLY $\mathbf{F}_{o_3}$ ACTING.   Figure 9.12a shows a free-body diagram of link 3 under the action of three forces $\mathbf{F}_{o_3}$, $\mathbf{F}''_{23}$, and $\mathbf{F}''_{43}$. The direction of $\mathbf{F}_{o_3}$ is known and that of $\mathbf{F}''_{43}$ is along line O_4B because link 4 becomes a two-force member when $\mathbf{F}_{o_4}$ is omitted from the analysis. The intersection of the known directions of $\mathbf{F}_{o_3}$ and $\mathbf{F}''_{43}$ gives point k''. The direction of $\mathbf{F}''_{23}$ must pass through points k'' and A because link 3 is in equilibrium under the action of these three forces with no couple acting upon it.

Figure 9.12b shows the force polygon for the determination of vectors $\mathbf{F}''_{23}$ and $\mathbf{F}''_{43}$. Because link 3 is in equilibrium, $\mathbf{F}_{o_3} + \mathbf{F}''_{23} + \mathbf{F}''_{43} = 0$ and the polygon closes as shown.

Figure 9.12c shows the free-body diagram of link 4 with force $\mathbf{F}''_{34}$ acting at point B and $\mathbf{F}''_{14}$ acting at point O_4. Link 4 is in equilibrium under the action of these two forces so that they act along the line O_4B and their magnitudes are equal to that of $\mathbf{F}''_{43}$ as shown.

Figure 9.12d shows the free-body diagram of link 2. Forces $\mathbf{F}''_{32}$ and $\mathbf{F}''_{12}$ act at points A and O_2, respectively. Because a couple is formed by $\mathbf{F}''_{32} \times d''$, it is necessary to apply torque T''_s with a sense opposite to that of the couple to hold link 2 in equilibrium.

TOTAL FORCES.   Figure 9.12e shows the free-body diagram of link 3 in equilibrium with forces $\mathbf{F}_{o_3}$, $\mathbf{F}_{23}$, and $\mathbf{F}_{43}$ where by superposition $\mathbf{F}_{23} = \mathbf{F}'_{23} + \mathbf{F}''_{23}$ and $\mathbf{F}_{43} = \mathbf{F}'_{43} + \mathbf{F}''_{43}$. Figure 9.12f shows the free-body diagram of link 4 in equilibrium under the action of forces $\mathbf{F}_{o_4}$, $\mathbf{F}_{34}$, and $\mathbf{F}_{14}$. Force $\mathbf{F}_{14} = \mathbf{F}'_{14} + \mathbf{F}''_{14}$ as shown. Figure 9.12g shows link 2 in equilibrium with forces $\mathbf{F}_{32}$ and $\mathbf{F}_{12}$ and torque $T_s = T'_s + T''_s$.

9.7   LINKAGE FORCE ANALYSIS BY MATRIX METHODS

Although the superposition method is computationally easy to use, it becomes tedious because the linkage must be analyzed several times. The matrix method, on the other hand, requires only a single analysis but results in a set of linear equations which must be solved simultaneously for all the unknown forces and torques. If a computer or calculator program for solving sets of linear equations is available, the matrix method will require less effort on the part of the designer. For hand calculation, however, the superposition method will be easier to use. It should also be pointed out that the superposition method can be checked graphically at each step and gives directly the separate force effects due to the mass and inertia of each link.

As an example of force analysis using matrix methods, consider the four-bar linkage of Fig. 9.13. Note that the mass centers of the moving links g_2, g_3, and g_4 need not be along the lines connecting the joints. As with the superposition method, the position and linear acceleration of the mass center of each moving link and the angular acceleration of each moving link must be known from previous analysis. In the matrix method, each link must be shown separately in

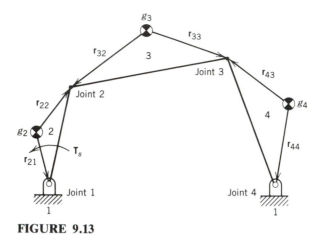

FIGURE 9.13

a free-body diagram; this has been done in Fig. 9.14. From the free-body diagrams, the equations of motion for each moving link may be written in vector form as follows:

Link 2:

$$\mathbf{F}_{32} - \mathbf{F}_{21} = M_2 \mathbf{A}_{g_2} \tag{9.18}$$

$$\mathbf{r}_{22} \times \mathbf{F}_{32} - \mathbf{r}_{21} \times \mathbf{F}_{21} + \mathbf{T}_s = I_2 \boldsymbol{\alpha}_2 \tag{9.19}$$

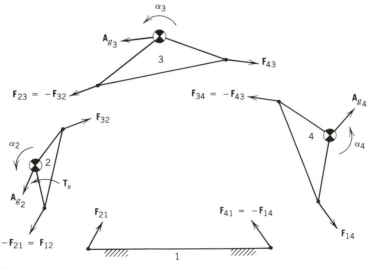

FIGURE 9.14

Link 3:

$$\mathbf{F}_{43} - \mathbf{F}_{32} = M_3\mathbf{A}_{g_3} \tag{9.20}$$

$$\mathbf{r}_{33} \times \mathbf{F}_{43} - \mathbf{r}_{32} \times \mathbf{F}_{32} = I_3\boldsymbol{\alpha}_3 \tag{9.21}$$

Link 4:

$$\mathbf{F}_{14} - \mathbf{F}_{43} = M_4\mathbf{A}_{g_4} \tag{9.22}$$

$$\mathbf{r}_{44} \times \mathbf{F}_{14} - \mathbf{r}_{43} \times \mathbf{F}_{43} = I_4\boldsymbol{\alpha}_4 \tag{9.23}$$

where

$\mathbf{r}_{ij}$ = the vector from the center of gravity of link i to joint j

$\mathbf{F}_{ik}$ = the force link i exerts on link k (note that $\mathbf{F}_{ik} = -\mathbf{F}_{ki}$ as shown in the figure)

g_i = the center of gravity of link i

$\mathbf{A}_{g_i}$ = the acceleration of the center of gravity g_i

$\boldsymbol{\alpha}_i$ = the angular acceleration of link i

M_i = the mass of link i

I_i = the mass moment of inertia of link i about its center of gravity

$\mathbf{T}_s$ = the driving torque applied to the input link

The vector force equations (Eqs. 9.18, 9.20, and 9.22) are separated into *x*- and *y*-components as follows:

$$F_{32x} - F_{21x} = M_2A_{g_2x} \tag{9.24}$$

$$F_{32y} - F_{21y} = M_2A_{g_2y} \tag{9.25}$$

$$F_{43x} - F_{32x} = M_3A_{g_3x} \tag{9.26}$$

$$F_{43y} - F_{32y} = M_3A_{g_3y} \tag{9.27}$$

$$F_{14x} - F_{43x} = M_4A_{g_4x} \tag{9.28}$$

$$F_{14y} - F_{43y} = M_4A_{g_4y} \tag{9.29}$$

Expanding the vector cross products in Eqs. 9.19, 9.21, and 9.23 by using the relation $\mathbf{r} \times \mathbf{F} = r_xF_y - r_yF_x$ gives

$$r_{22x}F_{32y} - r_{22y}F_{32x} - r_{21x}F_{21y} + r_{21y}F_{21x} = I_2\alpha_2 - T_s \tag{9.30}$$

$$r_{33x}F_{43y} - r_{33y}F_{43x} - r_{32x}F_{32y} + r_{32y}F_{32x} = I_3\alpha_3 \tag{9.31}$$

$$r_{44x}F_{14y} - r_{44y}F_{14x} - r_{43x}F_{43y} + r_{43y}F_{43x} = I_4\alpha_4 \tag{9.32}$$

$$
\begin{bmatrix}
-1 & 0 & 1 & 0 & 0 & 0 & 0 & 0 & 0 \\
0 & -1 & 0 & 1 & 0 & 0 & 0 & 0 & 0 \\
r_{21y} & -r_{21x} & -r_{22y} & r_{22x} & 0 & 0 & 0 & 0 & 1 \\
0 & 0 & -1 & 0 & 1 & 0 & 0 & 0 & 0 \\
0 & 0 & 0 & -1 & 0 & 1 & 0 & 0 & 0 \\
0 & 0 & r_{32y} & -r_{32x} & -r_{33y} & r_{33x} & 0 & 0 & 0 \\
0 & 0 & 0 & 0 & -1 & 0 & 1 & 0 & 0 \\
0 & 0 & 0 & 0 & 0 & -1 & 0 & 1 & 0 \\
0 & 0 & 0 & 0 & r_{43y} & -r_{43x} & -r_{44y} & r_{44x} & 0
\end{bmatrix}
\begin{bmatrix}
F_{21x} \\ F_{21y} \\ F_{32x} \\ F_{32y} \\ F_{43x} \\ F_{43y} \\ F_{14x} \\ F_{14y} \\ T_s
\end{bmatrix}
=
\begin{bmatrix}
M_2A_{g2x} \\ M_2A_{g2y} \\ I_2\alpha_2 \\ M_3A_{g3x} \\ M_3A_{g3y} \\ I_3\alpha_3 \\ M_4A_{g4x} \\ M_4A_{g4y} \\ I_4\alpha_4
\end{bmatrix}
$$

FIGURE 9.15

Equations 9.24 through 9.32 form a set of nine linear equations in the nine unknowns F_{21x}, F_{21y}, F_{32x}, F_{32y}, F_{43x}, F_{43y}, F_{14x}, F_{14y}, and T_s. With some minor rearrangements, these equations may be cast in the matrix form shown in Fig. 9.15. Most computers and many programmable calculators have programs available for solving such systems of linear equations. Note that it would be an easy matter to add the effects of other known external forces to the right-hand side of this matrix equation.

Example 9.3. Figure 9.16 shows the linkage from Fig. 8.7 of Example 8.1 for which a velocity and acceleration analysis has been made. It is required to determine the bearing forces on each link and the shaft torque T_s at O_2 using the matrix method.

The first step in this analysis is to determine vectors A_{gi} and r_{ij} as follows:

$$A_{g2} = 0i + 0j \quad (\text{ft/s}^2)$$

$$r_{21} = 0i + 0j \quad (\text{in.})$$

$$r_{22} = 3 \angle 37° = 2.40i + 1.81j \quad (\text{in.})$$

$$A_{g3} = 91.6 \angle 186.1° = -91.08i - 9.73j \quad (\text{ft/s}^2)$$

$$r_{32} = 4 \angle 157° = -3.68i + 1.56j \quad (\text{in.})$$

$$r_{33} = 4 \angle -23° = 3.68i - 1.56j \quad (\text{in.})$$

$$A_{g4} = 62.7 \angle 149.6° = -54.08i + 31.73j \quad (\text{ft/s}^2)$$

$$r_{44} = 5.27 \angle 85° = 0.46i + 5.25j \quad (\text{in.})$$

$$r_{43} = r_{44} + \overline{O_4B} = r_{44} + 8 \angle 250° = -2.28i - 2.27j \quad (\text{in.})$$

Next, the inertia forces and torques are calculated.

$$M_2A_{g2x} = (10 \text{ lb} \cdot \text{s}^2/32.2 \text{ ft})(0 \text{ ft/s}^2) = 0 \text{ lb}$$

$$M_2A_{g2y} = (10 \text{ lb} \cdot \text{s}^2/32.2 \text{ ft})(0 \text{ ft/s}^2) = 0 \text{ lb}$$

$$I_2\alpha_2 = (0.017 \text{ lb} \cdot \text{s}^2 \cdot \text{ft})(12 \text{ in./ft})(0 \text{ rad/s}^2) = 0 \text{ lb} \cdot \text{in.}$$

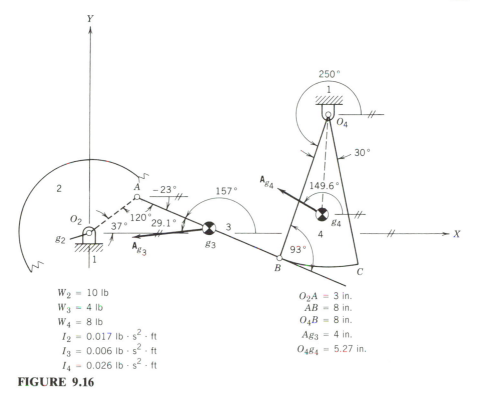

W_2 = 10 lb
W_3 = 4 lb
W_4 = 8 lb
I_2 = 0.017 lb · s^2 · ft
I_3 = 0.006 lb · s^2 · ft
I_4 = 0.026 lb · s^2 · ft

O_2A = 3 in.
AB = 8 in.
O_4B = 8 in.
Ag_3 = 4 in.
O_4g_4 = 5.27 in.

FIGURE 9.16

$$M_3 A_{g_{3x}} = (4 \text{ lb} \cdot \text{s}^2/32.2 \text{ ft})(-91.08 \text{ ft/s}^2) = -11.31 \text{ lb}$$

$$M_3 A_{g_{3y}} = (4 \text{ lb} \cdot \text{s}^2/32.2 \text{ ft})(-9.73 \text{ ft/s}^2) = -1.21 \text{ lb}$$

$$I_3 \alpha_3 = (0.006 \text{ lb} \cdot \text{s}^2 \cdot \text{ft})(12 \text{ in.}/\text{ft})(241 \text{ rad/s}^2) = 17.35 \text{ lb} \cdot \text{in.}$$

$$M_4 A_{g_{4x}} = (8 \text{ lb} \cdot \text{s}^2/32.2 \text{ ft})(-54.08 \text{ ft/s}^2) = -13.44 \text{ lb}$$

$$M_4 A_{g_{4y}} = (8 \text{ lb} \cdot \text{s}^2/32.2 \text{ ft})(31.73 \text{ ft/s}^2) = 7.88 \text{ lb}$$

$$I_4 \alpha_4 = (0.026 \text{ lb} \cdot \text{s}^2 \cdot \text{ft})(12 \text{ in.}/\text{ft})(-129 \text{ rad/s}^2) = -40.25 \text{ lb} \cdot \text{in.}$$

The above values are substituted into the matrix analysis equation of Fig. 9.15. The resulting system of analysis equations may be solved using any one of a wide variety of methods. In this case, the equations have been solved by matrix inversion. A BASIC computer program for force analysis of four-bar linkages using matrix methods and inversion is listed in Appendix 3. The results of this program are given below.

F_{21x} = 24.29 lb
F_{21y} = −4.95 lb F_{21} = 24.80 lb
F_{32x} = 24.29 lb
F_{32y} = −4.95 lb F_{32} = 24.80 lb
F_{43x} = 12.98 lb

$$F_{43y} = -6.16 \text{ lb} \qquad\qquad F_{43} = 14.36 \text{ lb}$$
$$F_{14x} = -0.46 \text{ lb}$$
$$F_{14y} = 1.73 \text{ lb} \qquad\qquad F_{14} = 1.79 \text{ lb}$$
$$T_s = 55.70 \text{ lb} \cdot \text{in.} \qquad \text{(ccw)}$$

These results check with those obtained in Example 9.1 in which the superposition method was used to solve this same problem. The results of that example, however, are expressed in the coordinate system attached to the coupler link, whereas the results of this example are expressed in the fixed coordinate system.

9.8 FORCE ANALYSIS USING THE INTEGRATED MECHANISMS PROGRAM (IMP)

In Chapter 2 (section 2.4), the use of the Integrated Mechanisms Program (IMP) was introduced as a convenient method of displacement analysis of four-bar linkages. A later presentation showed how IMP could also be used for velocity and acceleration analysis. The use of IMP for force analysis is presented below.

Figure 9.17 is reproduced from section 2.4; the positions of the centers of gravity of links 2, 3, and 4 have been added to the figure. Listed below the figure are the specifications of the link lengths, weights of links, and the moments of inertia of the links.

The statements for the revolutes to be listed in the input to the IMP program are the same as for the Example 2.3 and are reproduced below. In addition, a statement for gravity must be added to the revolute statements as shown. This value is 386 in./s² for link lengths expressed in inches.

```
GROUND=FRAME
REVOLUTE(FRAME,LNK2)=OH2
REVOLUTE(LNK2,LNK3)=A
REVOLUTE(LNK3,LNK4)=B
REVOLUTE(LNK4,FRAME)=OH4
ZERO(GRAVITY)=386
```

Next, the coordinates of the revolutes are listed from Example 2.3.

```
DATA:REVOLUTE(OH2)=0,0,0/0,0,1/1,0,0/-2.1213,2.1213,0
DATA:REVOLUTE(A)=-2.1213,2.1213,0/-2.1213,2.1213 ,1/$
                 0,0,0/9.2242,5.9388,0
DATA:REVOLUTE(B)=9.2242,5.9388,0/9.2242,5.9388,1/$
                 -2.1213,2.1213,0/10,0,0
DATA:REVOLUTE(OH4)=10,0,0/10,0,1/9.2242,5.9388,0/12,0,0
```

The data for the points defining the links and their centers of gravity are listed next as OO2, G2, and AA2 for link 2; AA3, G3, and BB3 for link 3; and BB4, G4, and OO4 for link 4. These are given below with the distance of G2 taken relative to revolute OH2, distance G3 taken relative to revolute A, and distance G4 taken relative to revolute OH4. It should be mentioned that these

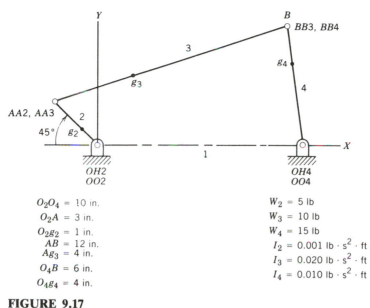

$O_2O_4 = 10$ in.
$O_2A = 3$ in.
$O_2g_2 = 1$ in.
$AB = 12$ in.
$Ag_3 = 4$ in.
$O_4B = 6$ in.
$O_4g_4 = 4$ in.

$W_2 = 5$ lb
$W_3 = 10$ lb
$W_4 = 15$ lb
$I_2 = 0.001$ lb $\cdot$ s$^2 \cdot$ ft
$I_3 = 0.020$ lb $\cdot$ s$^2 \cdot$ ft
$I_4 = 0.010$ lb $\cdot$ s$^2 \cdot$ ft

FIGURE 9.17

centers of gravity could have been taken as $G2$ relative to A, $G3$ relative to B, and $G4$ relative to B.

```
POINT(LNK2)=OO2,G2,AA2
DATA:POINT(OO2,OH2)=0,0,0
DATA:POINT(G2,OH2)=1,0,0
DATA:POINT(AA2,A)=0,0,0
POINT(LNK3)=AA3,G3,BB3
DATA:POINT(AA3,A)=0,0,0
DATA:POINT(G3,A)=4,0,0
DATA:POINT(BB3,B)=0,0,0
POINT(LNK4)=BB4,G4,OO4
DATA:POINT(BB4,B)=0,0,0
DATA:POINT(G4,OH4)=4,0,0
DATA:POINT(OO4,OH4)=0,0,0
```

The data for the weights of the links in force units and the coordinates of the centers of gravity relative to the local coordinate systems attached to the links at the revolutes $OH2$, A, and $OH4$ as specified above are as follows:

```
DATA:WEIGHT(LNK2,OH2)=5;1,0,0
DATA:WEIGHT(LNK3,A)=10;4,0,0
DATA:WEIGHT(LNK4,OH4)=15;4,0,0
```

The data for the moments of inertia of the links are listed relative to the local coordinate systems at the revolutes $OH2$, A, and $OH4$. The moments of inertia are *weight* moments of inertia and are transferred from the centers of gravity of the links to the revolutes specified using the parallel-axis theorem.

A sample calculation follows for the weight moment of inertia of link 2 taken relative to revolute *OH2*:

$$I_2 = 0.001 \text{ lb} \cdot \text{s}^2 \cdot \text{ft} = 0.012 \text{ lb} \cdot \text{s}^2 \cdot \text{in.}$$

$$W_2 = 5 \text{ lb}$$

$$d_2 = 1.0 \text{ in. (distance from } OH2 \text{ to } g_2)$$

Therefore,

$$I_{OH2} = I_2 + \frac{W_2}{g}(d^2) = 0.012 + \frac{5}{386}(1^2)$$
$$= 0.02495 \text{ lb} \cdot \text{s}^2 \cdot \text{in.} \quad \text{(mass units)}$$

The moment of inertia in weight units (I_w) *is equal to the moment of inertia in mass units* (I) *multiplied by g.*[1] Therefore,

$$I_{OH2(w)} = (0.02495)(386) = 9.631 \text{ lb} \cdot \text{in.}^2 \quad \text{(weight units)}$$

In a similar manner, for Link 3, $I_{A(w)} = 252.6$ lb $\cdot$ in.^{2}; for Link 4, $I_{OH4(w)} = 286.3$ lb $\cdot$ in.^{2}

The data for the inertias are listed as follows:

```
DATA: INERTIA(LNK2,OH2)=0,0,9.631,0,0,0
DATA: INERTIA(LNK3,A)=0,0,252.6,0,0,0
DATA: INERTIA(LNK4,OH4)=0,0,286.3,0,0,0
ZOOM(7)=5,1.5,0
ZERO(POSI)=0.001
RETURN
```

It is now necessary to enter the angular position of link 2 (135°) for which the analysis is required and to enter the angular velocity of link 2 (500 rad/s, cw) which is constant. These will be specified for revolute *OH2*. It is required to determine the forces F_{12}, F_{14}, F_{23}, and F_{34} and the shaft torque T_s that must be applied to link 2 to hold the linkage in equilibrium.

```
DATA:POSI(OH2)=135
DATA:VELO(OH2)=-500
PRINT:FORCE(OH2,OH4,A,B)
EXECUTE
```

Table 9.1 gives the forces F_{12}, F_{14}, F_{23}, and F_{34} and the torque on link 2, T_s, as calculated by IMP. For comparison, values found by unit vectors are included.

---

[1]*Marks' Standard Handbook for Mechanical Engineers,* 8th edition, McGraw-Hill Book Company, New York, 1978, pp. 3–10.

TABLE 9.1

	IMP	Unit Vectors
F_{12}	21,742 lb	21,746 lb
F_{14}	14,093 lb	14,138 lb
F_{23}	18,886 lb	18,420 lb
F_{34}	8,959 lb	8,980 lb
T_s	28,624 lb · in.	28,629 lb · in.

9.9 LINKAGE FORCE ANALYSIS BY THE METHOD OF VIRTUAL WORK

The methods of force analysis presented so far have been based on the principle of equilibrium of forces. Another method applicable for the analyses of linkages is that of *virtual work*, and it often results in much simpler solutions. This method is based on the principle that if a rigid body is in equilibrium under the action of external forces, the total work done by these forces is zero for a small displacement of the body.

To review the concept of work, consider Fig. 9.18, which shows a force **F** acting on a particle at point A. If the particle moves from point A to A' through a small distance δs, the work of force **F** during the displacement δs is

$$\delta U = F \, \delta s \cos \theta$$

As can be seen from the equation, the work done is the scaler product of the displacement and the component of the force in the direction of the displacement, and it can be positive, negative, or zero depending on angle θ. If θ is less than 90°, the force component and the displacement have the same sense and work is positive. If $\theta = 90°$, work is zero, and for θ greater than 90° but less than 270°, work is negative. If the preceding equation is compared with the vector equation

$$\mathbf{A} \cdot \mathbf{B} = AB \cos \theta$$

it can be seen that the equation for work can be written as the dot product of the force and displacement vectors as follows:

$$\delta U = \mathbf{F} \cdot \delta \mathbf{s} \tag{9.33}$$

The term *virtual work* is used for this method of analysis to indicate work that results from an infinitesimal displacement which is imaginary. Such a displacement is called a *virtual displacement* and labeled δs to distinguish it from an actual displacement ds. Although virtual displacements are imaginary, they must be consistent with the constraints of the mechansim under consideration. A virtual displacement may also be a measure of rotation and is labeled $\delta\theta$. The virtual work done by a torque T is therefore $\delta U = \mathbf{T} \cdot \delta\theta$.

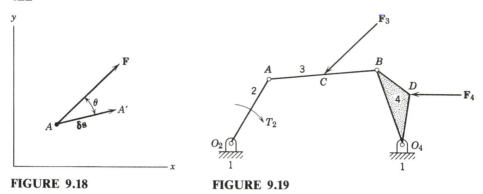

FIGURE 9.18　　　　　　　　　　　　　**FIGURE 9.19**

From the definition of virtual work, it follows that if a system that is in equilibrium under the action of external forces and torques is given a virtual displacement, the total virtual work must be zero. This concept may be expressed mathematically as follows:

$$\delta U = \Sigma\, \mathbf{F}_n \cdot \delta\mathbf{s}_n + \Sigma\, \mathbf{T}_n \cdot \delta\boldsymbol{\theta}_n = 0 \tag{9.34}$$

In the application of this equation, it must be remembered that the virtual displacements $\delta\mathbf{s}_n$ and $\delta\boldsymbol{\theta}_n$ must be consistent with the constraints of the mechanism. As an example of this, consider Fig. 9.19, in which a four-bar linkage is acted upon by forces $\mathbf{F}_3$ and $\mathbf{F}_4$ applied at points C and D, respectively, and it is required to determine torque T_2 necessary to maintain static equilibrium. If link 2 is given a virtual displacement $\delta\boldsymbol{\theta}_2$, the equations for $\delta\mathbf{s}_C$ and $\delta\mathbf{s}_D$ must be expressed as a function of $\delta\boldsymbol{\theta}_2$ in order to solve Eq. 9.34 for T_2.

The method of virtual work can also be applied to dynamic analyses if inertia forces and inertia torques are considered as applied forces and torques. Equation 9.34 can be modified for the dynamic case by dividing each term by dt. This is permissible because each virtual displacement takes place in the same interval of time. Making this change, one obtains

$$\Sigma\, \mathbf{F}_n \cdot \frac{\delta\mathbf{s}_n}{dt} + \Sigma\, \mathbf{T}_n \cdot \frac{\delta\boldsymbol{\theta}_n}{dt} = 0$$

and

$$\Sigma\, \mathbf{F}_n \cdot \mathbf{V}_n + \Sigma\, \mathbf{T}_n \cdot \boldsymbol{\omega}_n = 0 \tag{9.35}$$

Therefore, the virtual work of the external forces and torques is proportional to the velocity of the points of application of the forces on the links. Equation 9.35 can be expanded to give terms for applied forces and torques and for inertia forces and torques as follows:

$$\Sigma\, \mathbf{T}_n \cdot \boldsymbol{\omega}_n + \Sigma\, \mathbf{F}_n \cdot \mathbf{V}_n + \Sigma\, \mathbf{F}_{o_n} \cdot \mathbf{V}_{g_n} + \Sigma\, \mathbf{T}_{o_n} \cdot \boldsymbol{\omega}_n = 0 \tag{9.36}$$

where

$$\mathbf{F}_{o_n} \cdot \mathbf{V}_{g_n} = -\frac{W_n}{g} \mathbf{A}_{g_n} \cdot \mathbf{V}_{g_n} \qquad \text{(inertia force)}$$

$$\mathbf{T}_{o_n} \cdot \boldsymbol{\omega}_n = -I_n \boldsymbol{\alpha}_n \cdot \boldsymbol{\omega}_n \qquad \text{(inertia torque)}$$

After a velocity and acceleration analysis has been made, Eq. 9.36 can easily be solved for one unknown, which would usually be the torque required on the driving link to hold the mechanism in equilibrium.

In Eq. 9.36, only the virtual work done by the external forces and torques on a linkage appears. The internal forces between connecting links occur in pairs. These are equal in magnitude but opposite in sense so that their net work during any displacement is zero. For this reason, Eq. 9.36 cannot be used to evaluate bearing forces between connecting links.

Although analyses by the method of virtual work can be made graphically, an analytical solution is easier. In a graphical solution, if the forces and velocities are not both in the same direction, components of the forces must be used. In Eq. 9.36 with the terms expressed as dot products, the problem of direction is automatically taken care of.

An analytical solution of a force analysis by the method of virtual work is given in Example 9.4.

Example 9.4. Figure 9.20 shows the linkage of Fig. 9.10*a*. Use the method of virtual work to determine the torque T_2 necessary to hold the linkage in equilibrium. In the solution, all components have been taken relative to the *xy*-axes.

Solution. From the solution of Example 8.1,

$$\mathbf{V}_A = 5.2\mathbf{i} - 3.0\mathbf{j} \qquad \boldsymbol{\omega}_2 = -24\mathbf{k} \text{ rad/s}$$

$$\mathbf{V}_B = 5.2\mathbf{i} + 0.27\mathbf{j} \qquad \boldsymbol{\omega}_3 = 4.91\mathbf{k} \text{ rad/s}$$

$$\mathbf{A}_A = -72\mathbf{i} - 124.8\mathbf{j} \qquad \boldsymbol{\omega}_4 = 7.82\mathbf{k} \text{ rad/s}$$

$$\mathbf{A}_B = -88.1\mathbf{i} + 35.8\mathbf{j} \qquad \boldsymbol{\alpha}_3 = 241\mathbf{k} \text{ rad/s}^2$$

$$\boldsymbol{\alpha}_4 = -129\mathbf{k} \text{ rad/s}^2$$

By using the above values, the following velocities and accelerations of the centers of gravity of links 3 and 4 were determined. They are shown on the linkage in Fig. 9.20:

$$\mathbf{V}_{g_3} = 5.2\mathbf{i} - 1.37\mathbf{j} \qquad \mathbf{A}_{g_3} = -80.01\mathbf{i} - 44.53\mathbf{j}$$

$$\mathbf{V}_{g_4} = 3.27\mathbf{i} + 1.06\mathbf{j} \qquad \mathbf{A}_{g_4} = -62.17\mathbf{i} + 8.08\mathbf{j}$$

By writing Eq. 9.36 for the linkage of Fig. 9.20,

$$\mathbf{T}_2 \cdot \boldsymbol{\omega}_2 + \mathbf{F}_{o_3} \cdot \mathbf{V}_{g_3} + \mathbf{F}_{o_4} \cdot \mathbf{V}_{g_4} + \mathbf{T}_{o_3} \cdot \boldsymbol{\omega}_3 + \mathbf{T}_{o_4} \cdot \boldsymbol{\omega}_4 = 0$$

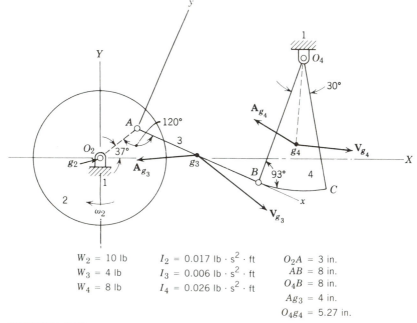

W_2 = 10 lb I_2 = 0.017 lb · s^2 · ft O_2A = 3 in.
W_3 = 4 lb I_3 = 0.006 lb · s^2 · ft AB = 8 in.
W_4 = 8 lb I_4 = 0.026 lb · s^2 · ft O_4B = 8 in.
                                                 Ag_3 = 4 in.
                                                 O_4g_4 = 5.27 in.

FIGURE 9.20

where

$$T_2 \cdot \omega_2 = (T_2\mathbf{k}) \cdot (-24\mathbf{k}) = -24T_2$$

$$F_{o_3} \cdot V_{g_3} = \left[\frac{-W_3}{g} A_{g_3} \right] \cdot V_{g_3}$$

$$= \frac{-4}{32.2} (-80.01\mathbf{i} - 44.53\mathbf{j}) \cdot (5.2\mathbf{i} - 1.37\mathbf{j})$$

$$= 44.13 \text{ ft} \cdot \text{lb/s}$$

$$F_{o_4} \cdot V_{g_4} = \left[\frac{-W_4}{g} A_{g_4} \right] \cdot V_{g_4}$$

$$= \frac{-8}{32.2} (-62.17\mathbf{i} + 8.08\mathbf{j}) \cdot (3.27\mathbf{i} + 1.06\mathbf{j})$$

$$= 48.32 \text{ ft} \cdot \text{lb/s}$$

$$T_{o_3} \cdot \omega_3 = (-I_3\alpha_3) \cdot \omega_3$$

$$= -0.006(241\mathbf{k}) \cdot (4.91\mathbf{k}) = -7.099 \text{ ft} \cdot \text{lb/s}$$

$$T_{o_4} \cdot \omega_4 = (-I_4\alpha_4) \cdot \omega_4$$

$$= -0.026(-129\mathbf{k}) \cdot (7.82\mathbf{k}) = 26.23 \text{ ft} \cdot \text{lb/s}$$

Therefore,

$$-24T_2 + 44.13 + 48.32 - 7.099 + 26.23 = 0$$

and

$$T_2 = 4.649 \text{ lb} \cdot \text{ft}$$

$$= 55.79 \text{ lb} \cdot \text{in.} \quad \text{(ccw)}$$

The value for T_2 agrees closely with the value found in Example 9.1.

9.10 LINKAGE FORCE ANALYSIS FROM DYNAMIC CHARACTERISTICS

In the preceding sections, linkages were considered which operated continuously with the driving link rotating at a known, generally uniform, angular velocity. From this, it was possible to make a velocity, acceleration, and inertia force analysis of the mechanism. Combining these forces with the static forces acting on the linkage, one was able to complete the analysis and determine the bearing loads.

Occasionally, the problem arises of having to determine the velocity and acceleration characteristics of a mechanism where the driving force is produced by a rapid release of energy as from a spring, solenoid, or air cylinder. A circuit breaker is such a mechanism, with its design depending upon the time required for a motion cycle and the applied force necessary to open the breaker.

In the solution of such problems, energy methods are generally preferred because of the ease with which they can be applied. Of the several methods that have been developed, the one by Quinn will be presented here. His method is based on the distribution of kinetic energy in a mechanism and is referred to as Quinn's *energy distribution theorem,* which states that "the percent of the total kinetic energy which the link of a mechanism contains will remain the same in any given position regardless of the speed." This theorem applies to mechanisms in which there is no change in the mass or moment of inertia of the links with speed, and in which a linear relationship exists between the velocities of the various links in a given position.

In applying this method, a convenient value is assumed for the velocity of the input link of a mechanism, and a velocity analysis is made including the determination of the velocities of the centers of gravity of the various links. From this, the kinetic energy of each link can be calculated, and the ratio of the kinetic energy of any link to the total for all the links is expressed as

$$\epsilon = \frac{KE}{\Sigma\, KE} \tag{9.37}$$

This ratio is known as an *energy contribution coefficient* and is a constant for any link in a particular phase regardless of its velocity. Values of ϵ can be calculated for various links over the range of several phases, and curves of ϵ versus crank angle or slider displacement plotted if desired.

In addition to the coefficient ϵ, which can easily be calculated for any link as shown above, it is necessary in making a dynamic analysis to know the variation of the external forces acting on the mechanism as well as the input velocity in some reference phase. The reference phase is usually the starting phase where the input velocity is zero. With this information, the actual velocity of the input link in a particular phase can be determined as follows:

1. Knowing the external forces in relation to the phase positions, calculate the work input to the mechanism between the starting phase and the phase under consideration.

2. Because a change in kinetic energy is equal to the work done, the actual change in kinetic energy of the mechanism from the starting phase to the phase in question is now known from item 1.

3. For the starting phase, the input velocity is zero so that the kinetic energy is also zero. Therefore, the kinetic energy of the mechanism for the phase in question is known from item 2.

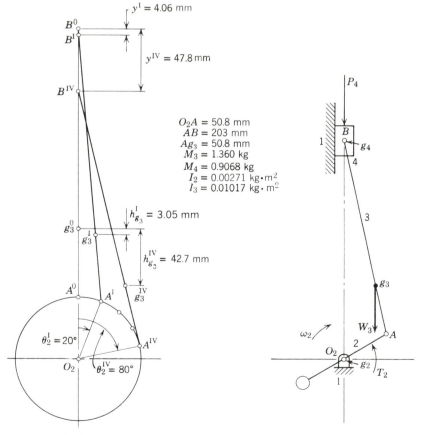

$O_2A = 50.8$ mm
$AB = 203$ mm
$Ag_3 = 50.8$ mm
$M_3 = 1.360$ kg
$M_4 = 0.9068$ kg
$I_2 = 0.00271$ kg·m^2
$I_3 = 0.01017$ kg·m^2

$y^I = 4.06$ mm

$y^{IV} = 47.8$ mm

$h_{g_3}^I = 3.05$ mm

$h_{g_3}^{IV} = 42.7$ mm

$\theta_2^I = 20°$

$\theta_2^{IV} = 80°$

FIGURE 9.21

4. After calculating the value of ϵ for the input link for the phase in question, determine the kinetic energy for the input link from item 3.

5. From item 4, calculate the velocity of the input link from the relation $KE = \frac{1}{2}I\omega^2$ for a rotating input link or from $KE = \frac{1}{2}MV^2$ if the input is a slider.

A sample problem using this method is given in Example 9.5.

From the above procedure, the velocity of the input link can be calculated for as many phases as necessary and a graph of input velocity versus displacement of the input link drawn. If many phases are to be analyzed, the calculations can easily be made and the graph plotted using a digital computer.

Example 9.5. For the slider-crank mechanism shown in Fig. 9.21, determine the angular velocity ω_2 of link 2 when the mechanism is in phase IV. The starting position is phase I where $\omega_2 = 0$. The torque T_2 on link 2 is constant at 10.2 N · m (ccw). The force **P** on piston 4 varies uniformly from 716 N in phase I to 160 N in phase IV.

Solution. The angular velocity ω_2' of link 2 is assumed to be 1 rad/s (cw), where the prime denotes an *assumed* value. From a velocity analysis of the linkage with $\omega_2' = 1$ rad/s, the velocities at phase IV are as follows:

$$V_A' = 0.0508 \text{ m/s} \qquad V_{g_3}' = 0.0508 \text{ m/s}$$

$$V_B' = 0.0521 \text{ m/s} \qquad \omega_3' = 0.049 \text{ rad/s} \qquad \text{(ccw)}$$

$$V_{BA}' = 0.00991 \text{ m/s}$$

By calculating the kinetic energy of the links based on the assumed value of $\omega_2' = 1$ rad/s,

$$KE_2' = \tfrac{1}{2}I_2(\omega_2')^2 = \tfrac{1}{2} \times 0.00271 \times 1$$

$$= 0.001355 \text{ N} \cdot \text{m}$$

$$KE_3' = \tfrac{1}{2}I_3(\omega_3')^2 + \tfrac{1}{2}M_3(V_{g_3}')^2$$

$$= \tfrac{1}{2} \times 0.01017 \times (0.049)^2 + \tfrac{1}{2} \times 1.360 \times (0.0508)^2$$

$$= 0.0000122 + 0.001755$$

$$= 0.001767 \text{ N} \cdot \text{m}$$

$$KE_4' = \tfrac{1}{2}M_4(V_B')^2 = \tfrac{1}{2} \times 0.9068 \times (0.0521)^2$$

$$= 0.001231 \text{ N} \cdot \text{m}$$

Therefore,

$$\Sigma \, KE' = 0.004353 \text{ N} \cdot \text{m}$$

From Eq. 9.37,

$$\epsilon_2 = \frac{KE_2'}{\Sigma \, KE'} = \frac{0.001355}{0.0043543} = 0.3112$$

The external work input to the mechanism by forces $\mathbf{P}_4$ and $\mathbf{W}_3$ and torque T_2 between phases I and IV is calculated as follows:

$$Wk_{P_4} = \tfrac{1}{2}(P_4^{IV} + P_4^{I})(y^{IV} - y^{I})$$

$$= \tfrac{1}{2}(160 + 716)\left(\frac{47.8 - 4.06}{1000}\right)$$

$$= 19.16 \text{ N} \cdot \text{m}$$

$$Wk_{W_3} = W_3(h_{g_3}^{IV} - h_{g_3}^{I})$$

$$= 1.360 \times 9.81\left(\frac{42.7 - 3.05}{1000}\right)$$

$$= 0.5290 \text{ N} \cdot \text{m}$$

$$Wk_{T_2} = -T_2(\theta_2^{IV} - \theta_2^{I})$$

$$= -10.2(80 - 20) \times \frac{\pi}{180}$$

$$= -10.68 \text{ N} \cdot \text{m}$$

Therefore,

$$\Sigma \, Wk = 9.009 \text{ N} \cdot \text{m}$$

From the fact that the change in kinetic energy must equal the work done,

$$\Sigma \, KE^{IV} - \Sigma \, KE^{I} = 9.009 \text{ N} \cdot \text{m}$$

but

$$\Sigma \, KE^{I} = 0$$

Therefore,

$$\Sigma \, KE^{IV} = 9.009 \text{ N} \cdot \text{m}$$

The actual value of ω_2 in phase IV is calculated from the relation

$$KE_2^{IV} = \tfrac{1}{2}I_2(\omega_2^{IV})^2$$

where

$$KE_2^{IV} = (\epsilon_2^{IV})(\Sigma \, KE^{IV})$$

$$= 0.3112 \times 9.009$$

$$= 2.804 \text{ N} \cdot \text{m}$$

Therefore,

$$\omega_2^{IV} = \left(\frac{2KE_2^{IV}}{I_2}\right)^{1/2} = \left(\frac{2 \times 2.804}{0.00271}\right)^{1/2}$$

$$= 45.49 \text{ rad/s}$$

9.11 LINKAGE FORCE ANALYSIS BY COMPLEX NUMBERS

Another analytical method of force analysis is to express vectors in complex form. This method is especially applicable when a complete cycle of a linkage is to be analyzed and computer facilities are available.

In Fig. 9.22a is shown a typical four-bar linkage in a given phase of the motion cycle. A shaft torque T_s acts on the drive link (link 2) at O_2. The accelerations A_g of the mass centers and the angular accelerations α of the moving links may be determined numerically by complex numbers as demonstrated in Chapter 8. The three inertia forces $\mathbf{F}_o$, which are related to the accelerations, represent the dynamic loading of the mechanism. The objective of the analysis is the determination of the bearing forces and the shaft torque which produce the dynamic loading.

Figure 9.22b shows the four-bar linkage with inertia force $\mathbf{F}_{o_3}$ as the only load vector acting so that the bearing forces and shaft torque to be determined are those related to $\mathbf{F}_{o_3}$ alone. Similar independent force analyses with $\mathbf{F}_{o_2}$ and $\mathbf{F}_{o_4}$ acting alone may be made, and eventually the resultant bearing forces and shaft torque may be obtained by superposition.

In the analysis with $\mathbf{F}_{o_3}$ acting alone, the free body to be considered first is that of link 3 shown in Fig. 9.22c. By assuming that the acceleration A_{g_3} (expressed as $A_{g_3}e^{i\beta_3}$) and the angular acceleration α_3 have been determined as in Chapter 8, the inertia force vector $\mathbf{F}_{o_3}$ may be determined from the following expression:

$$\mathbf{F}_{o_3} = (M_3 A_{g_3})e^{i(\beta_3 + \pi)} \qquad (9.38)$$

where $\beta_3 + \pi$ indicates that the sense of $\mathbf{F}_{o_3}$ is opposite to that of $\mathbf{A}_{g_3}$, which has the angular sense given by β_3. Because of the angular acceleration α_3, the line of action of $\mathbf{F}_{o_3}$ is offset $e_3 = I_3\alpha_3/F_{o_3}$ from the line of action of A_{g_3} as shown in Fig. 9.22b. For convenience in making calculations, the location of the line of action of $\mathbf{F}_{o_3}$ may be given by the distance l_3 shown in Fig. 9.22c:

$$l_3 = r_{g_3} + \frac{e_3}{\sin(\beta_3 - \theta_3)}$$

$$l_3 = r_{g_3} + \frac{I_3\alpha_3/F_{o_3}}{\sin(\beta_3 - \theta_3)} \qquad (9.39)$$

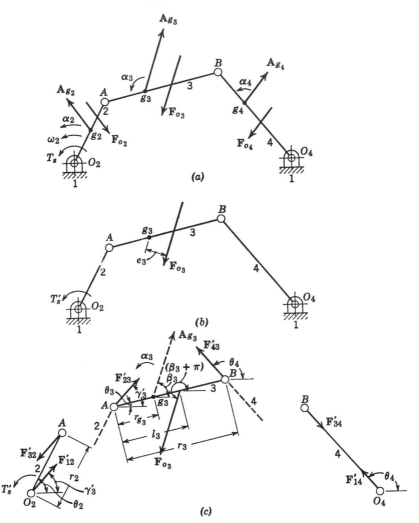

FIGURE 9.22

Figure 9.22c shows that three forces act on link 3, of which $\mathbf{F}_{o_3}$ is the known dynamic load and $\mathbf{F}'_{23}$ and $\mathbf{F}'_{43}$ are the unknown bearing forces to be determined. For static equilibrium of the forces, the following equations apply:

$$\mathbf{F}'_{23} + \mathbf{F}'_{43} + \mathbf{F}_{o_3} = 0$$

$$F'_{23}(e^{i\gamma'_3}) + F'_{43}(e^{i\theta_4}) + F_{o_3}(e^{i(\beta_3+\pi)}) = 0 \qquad \textbf{(9.40)}$$

Equating real and imaginary parts of the Eq. 9.40, we obtain

$$F'_{23}\cos\gamma'_3 + F'_{43}\cos\theta_4 + F_{o_3}\cos(\beta_3+\pi) = 0 \qquad \textbf{(9.41)}$$

$$F'_{23}\sin\gamma'_3 + F'_{43}\sin\theta_4 + F_{o_3}\sin(\beta_3+\pi) = 0 \qquad \textbf{(9.42)}$$

It may be seen that the number of unknowns to be determined is three, the magnitude of $\mathbf{F}'_{23}$ and its direction γ'_3, and the magnitude of $\mathbf{F}'_{43}$. The direction of $\mathbf{F}'_{43}$ is θ_4 and is known because link 4 is acted on by only two forces when considering $\mathbf{F}_{o_3}$ acting alone as shown in Fig. 9.22c. To determine the three unknowns, another equation is required in addition to Eqs. 9.41 and 9.42. The additional equation is one of equilibrium of moments, about either point A or B. By choosing point A, the sum of the moments about this point is required to be zero as follows:

$$F'_{43}r_3 \sin(\theta_4 - \theta_3) - F_{o_3}l_3 \sin(\beta_3 - \theta_3) = 0$$

$$F'_{43} = F_{o_3} \frac{l_3 \sin(\beta_3 - \theta_3)}{r_3 \sin(\theta_4 - \theta_3)} \tag{9.43}$$

On determining F'_{43} from Eq. 9.43, the real and imaginary components of F'_{23} may then be determined from Eqs. 9.41 and 9.42 as follows:

$$\mathcal{R} F'_{23} = F'_{23} \cos \gamma'_3 = -F'_{43} \cos \theta_4 - F_{o_3} \cos(\beta_3 + \pi) \tag{9.44}$$

$$\mathcal{I} F'_{23} = F'_{23} \sin \gamma'_3 = -F'_{43} \sin \theta_4 - F_{o_3} \sin(\beta_3 + \pi) \tag{9.45}$$

The symbols $\mathcal{R}$ and $\mathcal{I}$ indicate *real* and *imaginary* components of the vector $\mathbf{F}'_{23}$. The resultant of these components is the vector $\mathbf{F}'_{23}$, the magnitude of which may be determined as follows:

$$F'_{23} = \sqrt{(\mathcal{R}F'_{23})^2 + (\mathcal{I}F'_{23})^2} \tag{9.46}$$

The direction of $\mathbf{F}'_{23}$ is the angle γ'_3, which may be determined from the following expression:

$$\tan \gamma'_3 = \frac{\mathcal{I}F'_{23}}{\mathcal{R}F'_{23}} \tag{9.47}$$

Thus, the magnitudes and directions of the bearing forces at A and B may be calculated from the preceding equations. From the free body of link 4 shown in Fig. 9.22c, it is to be observed that the bearing force $\mathbf{F}'_{14}$ at O_4 is identical to the force $\mathbf{F}'_{43}$ because only two forces act on link 4. Similarly, since there are only two forces on link 2 as shown in the free body, the bearing force $\mathbf{F}'_{12}$ at O_2 is identical to $\mathbf{F}'_{23}$.

The final step of determining the shaft torque T'_s may be realized from the static equilibrium of couples acting on link 2:

$$\begin{aligned} T'_s &= -F'_{12}r_2 \sin(\theta_2 - \gamma'_3) \\ &= -F'_{23}r_2 \sin(\theta_2 - \gamma'_3) \end{aligned} \tag{9.48}$$

The preceding analysis has led to the determination of equations giving bearing forces and shaft torque due to the load $\mathbf{F}_{o_3}$. A similar analysis with only

$\mathbf{F}_{o_4}$ acting would yield another set of equations giving the influence of $\mathbf{F}_{o_4}$ on the bearing forces and shaft torque, and a third analysis would give the influence of $\mathbf{F}_{o_2}$. The resultant force at each of the bearings would be determined by superposition by summing the real and imaginary components calculated in the individual analyses. At bearing A, for example, the superposed resultant real and imaginary components of the bearing force are the sums $\Sigma \, \mathfrak{R} F_{23}$ and $\Sigma \, \mathfrak{I} F_{23}$, and

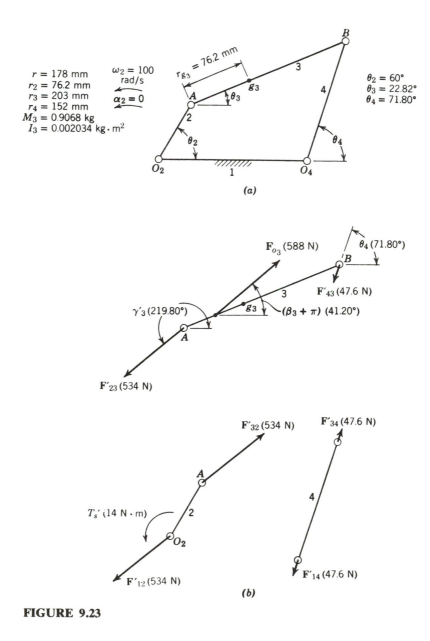

(a)

(b)

FIGURE 9.23

the resultant bearing force $\mathbf{F}_{23}$ may be determined from

$$F_{23} = \sqrt{(\Sigma \, \mathcal{R}F_{23})^2 + (\Sigma \, \mathcal{I}F_{23})^2} \qquad (9.49)$$

and the angle γ_3 of the resultant force from

$$\tan \gamma_3 = \frac{\Sigma \, \mathcal{I}F_{23}}{\Sigma \, \mathcal{R}F_{23}} \qquad (9.50)$$

Example 9.6. The drive link of the four-bar linkage shown in Fig. 9.23a rotates about O_2 at a constant angular velocity $\omega_2 = 100$ rad/s. Using the data given in Fig. 9.23a, determine the bearing forces and the shaft torque at O_2 due to the dynamic load $\mathbf{F}_{o_3}$ when the mechanism is in the phase where $\theta_2 = 60°$. The angular positions θ_3 and θ_4 of links 3 and 4 in this phase are the same as those determined in Chapter 2, Example 2.1, which refers to Fig. 2.4.

Solution. Preliminary to determining the inertia load $\mathbf{F}_{o_3}$, the following kinematic quantities are evaluated from Eqs. 8.84, 8.85, and 8.91:

$$\omega_3 = -10.16 \text{ rad/s} \qquad \text{(ccw)}$$

$$\omega_4 = 40.05 \text{ rad/s} \qquad \text{(ccw)}$$

$$\alpha_3 = 3361 \text{ rad/s}^2 \qquad \text{(ccw)}$$

The acceleration $\mathbf{A}_{g_3}$ of the mass center of link 3 may be determined as the sum of two acceleration vectors as follows:

$$\mathbf{A}_{g_3} = r_2(i\alpha_2 - \omega_2^2)e^{i\theta_2} + r_{g_3}(i\alpha_3 - \omega_3^2)e^{i\theta_3}$$

By noting that $\alpha_2 = 0$, this equation may be expanded as follows:

$$\mathbf{A}_{g_3} = (-r_2\omega_2^2 \cos \theta_2 - r_{g_3}\alpha_3 \sin \theta_3 - r_{g_3}\omega_3^2 \cos \theta_3)$$
$$+ i(-r_2\omega_2^2 \sin \theta_2 + r_{g_3}\alpha_3 \cos \theta_3 - r_{g_3}\omega_3^2 \sin \theta_3)$$

Substitution of given data yields the following numerical values:

$$\mathbf{A}_{g_3} = -487.6 - 426.9$$

The magnitude of $\mathbf{A}_{g_3}$ and its angular position β_3 are determined as follows:

$$|\mathbf{A}_{g_3}| = \sqrt{(-487.6)^2 + (-426.9)^2} = 648.1 \text{ m/s}^2$$

$$\tan \beta_3 = \frac{-426.9}{-487.6} = 0.8755 \qquad \text{(third quadrant)}$$

$$\beta_3 = 221.20°$$

The inertia force $\mathbf{F}_{o_3}$ may now be evaluated from Eq. 9.38:

$$\mathbf{F}_{o_3} = (M_3 A_{g_3}) e^{i(\beta_3 + \pi)}$$

$$= 0.9068(648.1) e^{i(221.20° + 180°)}$$

$$= 587.7 e^{i(41.20°)}$$

which indicates that the magnitude of the vector $\mathbf{F}_{o_3}$ is 587.7 N, and its angular position is 41.20°.

The location of the line of action of $\mathbf{F}_{o_3}$ is l_3 and may be determined from Eq. 9.39:

$$l_3 = r_{g_3} + \frac{I_3 \alpha_3}{F_{o_3} \sin (\beta_3 - \theta_3)}$$

$$= \frac{76.2}{1000} + \frac{(0.002034)(3361)}{(587.7)(\sin 198.38°)}$$

$$= 0.07620 - 0.03691$$

$$= 0.03929 \text{ m} = 39.3 \text{ mm}$$

It is to be noted that sin 198.38° is negative.

Equation 9.43 gives the magnitude of the bearing force $\mathbf{F}'_{43}$ at B. Substitution of numerical values in this equation yields the following:

$$F'_{43} = -47.6 \text{ N}$$

and substituting further in Eqs. 9.44 and 9.45 gives the numerical values of $\mathscr{R}F'_{23}$ and $\mathscr{I}F'_{23}$, from which the magnitude of the resultant $\mathbf{F}'_{23}$ is determined from Eq. 9.46 and its angle γ'_3 from Eq. 9.47:

$$F'_{23} = 534 \text{ N} \qquad \gamma'_3 = 219.80°$$

Finally, the shaft torque T'_s at O_2 is determined by substituting in Eq. 9.48:

$$T'_s = 14.0 \text{ N} \cdot \text{m}$$

In Fig. 9.23b are shown the forces on the several free bodies of the four-bar linkage, and these are shown to scale according to the preceding determined numerical values.

9.12   ENGINE FORCE ANALYSIS

In Fig. 9.24 is shown the slider-crank mechanism of a typical single-cylinder, four-stroke cycle internal-combustion engine. Shown also are the vectors which represent the principal loads on the mechanism: (a) the static gas load $\mathbf{P}$ on the piston and (b) the dynamic loads $\mathbf{F}_{o_4}$ and $\mathbf{F}_{o_3}$ acting on the piston and connecting rod, respectively. The inertia force $\mathbf{F}_{o_2}$ of the crank is zero since it is usual to counterweight the crankshaft such that the mass center is at the axis of rotation O_2. Thus, the crankshaft itself is nominally balanced so that $\mathbf{A}_{g_2}$ is zero. If the

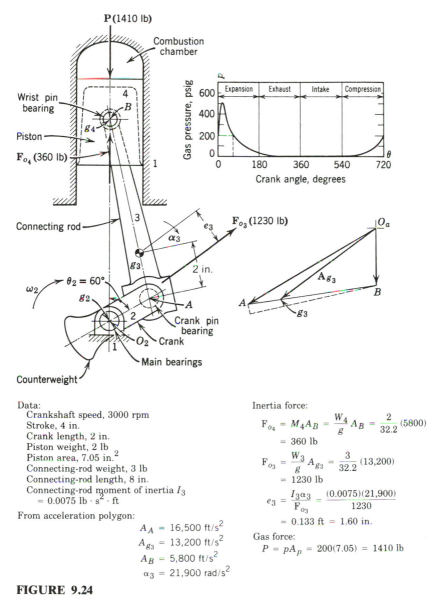

Data:
Crankshaft speed, 3000 rpm
Stroke, 4 in.
Crank length, 2 in.
Piston weight, 2 lb
Piston area, 7.05 in.^{2}
Connecting-rod weight, 3 lb
Connecting-rod length, 8 in.
Connecting-rod moment of inertia I_3
= 0.0075 lb $\cdot$ s^2 $\cdot$ ft

From acceleration polygon:
$$A_A = 16,500 \text{ ft/s}^2$$
$$A_{g3} = 13,200 \text{ ft/s}^2$$
$$A_B = 5,800 \text{ ft/s}^2$$
$$\alpha_3 = 21,900 \text{ rad/s}^2$$

Inertia force:
$$F_{o4} = M_4 A_B = \frac{W_4}{g} A_B = \frac{2}{32.2}(5800)$$
$$= 360 \text{ lb}$$
$$F_{o3} = \frac{W_3}{g} A_{g3} = \frac{3}{32.2}(13,200)$$
$$= 1230 \text{ lb}$$
$$e_3 = \frac{I_3 \alpha_3}{F_{o3}} = \frac{(0.0075)(21,900)}{1230}$$
$$= 0.133 \text{ ft} = 1.60 \text{ in.}$$

Gas force:
$$P = pA_p = 200(7.05) = 1410 \text{ lb}$$

FIGURE 9.24

analysis is made for constant rotative speed of the crank ($\alpha_2 = 0$), the inertia couple of the crankshaft is also zero. Gravity forces also act on the mechanism, but because the weights of the moving parts are small compared to the principal loads, these are usually neglected.

Also shown in Fig. 9.24 is a typical curve showing the variation of combustion chamber gas pressure in the four-stroke cycle corresponding with two revolutions of the crankshaft. Magnitudes of gas pressure are determined from a thermodynamic analysis or from experimental measurements of combustion chamber pressure. The gas force **P** on the piston is the product of gas pressure and piston head area.

Two force analyses of the engine mechanism will be made using (a) superposition solved graphically (Example 9.7) and (b) an analytical solution with unit vectors (Example 9.8). In the second case, it will not be necessary to use superposition because of certain simplifications that can be made and which are discussed in a later section.

Example 9.7. The crankshaft speed of the slider-crank engine shown in Fig. 9.24 is 3000 rpm. Using the data given, determine the loads on the mechanism when the crank is in the phase $\theta_2 = 60°$. From a force analysis of the mechanism, determine the forces transmitted through the wrist pin bearing, the crank pin bearing, and the main bearings. Determine also the crankshaft torque T_s.

Solution. As shown in Fig. 9.24, for the phase $\theta_2 = 60°$ the mechanism is in the expansion (power) stroke and the gas pressure is 200 psig. The corresponding gas load on the piston is $P = 1410$ lb. The inertia force $\mathbf{F}_{o_4}$ (360 lb) also acts on the piston, and its magnitude is determined from the product $M_4 A_B$. Inertia force $\mathbf{F}_{o_3}$ (1230 lb) of the connecting rod has the magnitude $M_3 A_{g_3}$, the direction of the acceleration $\mathbf{A}_{g_3}$ of the mass center of the rod, and a line of action offset e_3 because of the angular acceleration α_3.

In Fig. 9.25 is shown the force analysis of the mechanism in which $\mathbf{P}$, $\mathbf{F}_{o_4}$, and $\mathbf{F}_{o_3}$ are the known loads on the linkage. Superposition of forces is used to determine the unknown forces. In Fig. 9.25a forces in the linkage due to the loads $\mathbf{P}$ and $\mathbf{F}_{o_4}$ on the piston are determined, and in Fig. 9.25b the forces due to $\mathbf{F}_{o_3}$ on the connecting rod are determined. Finally, resultant forces are determined by superposition as shown in Fig. 9.25c. It is to be noted that friction forces are not included in the analysis; because of the pressure lubrication of the bearings and the cylinder wall, friction is assumed to be small and negligible.

By referring to Fig. 9.25a, $\mathbf{F}_4$ (1050 lb) is the resultant of the collinear forces $\mathbf{P}$ and $\mathbf{F}_{o_4}$. Beginning with the free body of the piston (link 4), three forces are shown concurrent at B. The direction of the connecting rod force $\mathbf{F}'_{34}$ on the piston is known since only two forces act on link 3. The direction of the cylinder wall force $\mathbf{F}'_{14}$ on the side of the piston is normal to the wall in the absence of friction, and the line of action of $\mathbf{F}'_{14}$ is through the point of concurrence at B. Beginning with the known force $\mathbf{F}_4$, the equilibrium force polygon shown is constructed to determine the magnitudes and senses of $\mathbf{F}'_{34}$ and $\mathbf{F}'_{14}$. The two collinear forces $\mathbf{F}'_{43}$ and $\mathbf{F}'_{23}$ on the free body of the connecting rod (link 3) are equal in magnitude to $\mathbf{F}'_{34}$ of the polygon. Also, the two parallel but noncollinear forces $\mathbf{F}'_{32}$ and $\mathbf{F}'_{12}$ on the free body of the crank (link 2) are equal in magnitude to $\mathbf{F}'_{34}$. Thus, all the unknown forces are determined from one force polygon. As shown on the free body of the crank, the shaft torque T'_s on the crankshaft at O_2 is the counterclockwise equilibrant of the couple formed by $\mathbf{F}'_{32}$ and $\mathbf{F}'_{12}$.

By referring to Fig. 9.25b, the known force on the linkage is $\mathbf{F}_{o_3}$ of the connecting rod. By isolating the connecting rod as a free body as shown, it may be seen that three concurrent forces act. The direction of the piston force $\mathbf{F}''_{43}$ on the rod is known since only two forces act on the piston, one of which must be normal to the piston side. The direction of the crank force $\mathbf{F}''_{23}$ on the rod is through the point k of concurrence determined from the intersection of the lines of action of $\mathbf{F}_{o_3}$ and $\mathbf{F}''_{43}$. Construction of the equilibrium force polygon shown determines the magnitudes and sense of $\mathbf{F}''_{43}$ and $\mathbf{F}''_{23}$. The two noncollinear

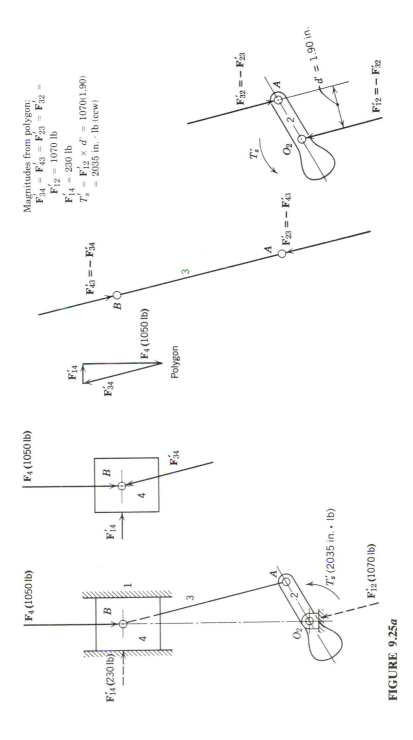

Magnitudes from polygon:

$\mathbf{F}'_{34} = \mathbf{F}'_{43} = \mathbf{F}'_{23} = \mathbf{F}'_{32} =$
$\quad \mathbf{F}'_{12} = 1070$ lb
$\quad \mathbf{F}'_{14} = 230$ lb
$\quad T'_s = \mathbf{F}'_{12} \times d' = 1070(1.90)$
$\quad\quad = 2035$ in. · lb (ccw)

$\mathbf{F}'_{32} = -\mathbf{F}'_{23}$

$\mathbf{F}'_{12} = -\mathbf{F}'_{32}$

$d' = 1.90$ in.

T'_s

O_2 2 A

$\mathbf{F}'_{43} = -\mathbf{F}'_{34}$

B 3 A

$\mathbf{F}'_{23} = -\mathbf{F}'_{43}$

$\mathbf{F}'_{14}$

$\mathbf{F}'_{34}$

$\mathbf{F}_4 \,(1050\,\text{lb})$

Polygon

$\mathbf{F}_4 \,(1050\,\text{lb})$

B 4 $\mathbf{F}'_{34}$

$\mathbf{F}'_{14}$

$\mathbf{F}_4 \,(1050\,\text{lb})$

B 1   3 A

4

$\mathbf{F}'_{14} \,(230\,\text{lb})$

O_2 2 $T'_s \,(2035\,\text{in. · lb})$

$\mathbf{F}'_{12} \,(1070\,\text{lb})$

FIGURE 9.25a

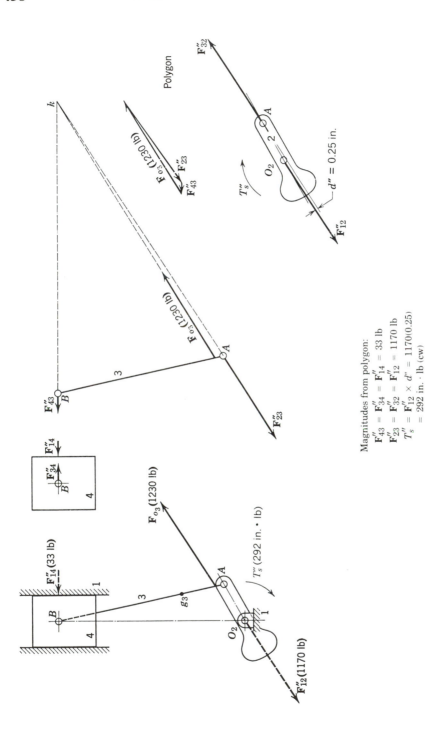

Magnitudes from polygon:
$$\mathbf{F}''_{43} = \mathbf{F}''_{34} = \mathbf{F}''_{14} = 33 \text{ lb}$$
$$\mathbf{F}''_{23} = \mathbf{F}''_{32} = \mathbf{F}''_{12} = 1170 \text{ lb}$$
$$T''_s = \mathbf{F}''_{12} \times d'' = 1170(0.25)$$
$$= 292 \text{ in.} \cdot \text{lb (cw)}$$

FIGURE 9.25b

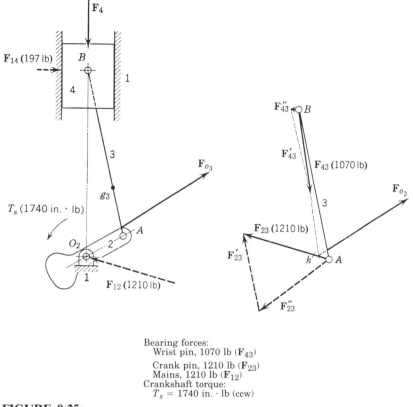

Bearing forces:
  Wrist pin, 1070 lb ($\mathbf{F}_{43}$)
  Crank pin, 1210 lb ($\mathbf{F}_{23}$)
  Mains, 1210 lb ($\mathbf{F}_{12}$)
Crankshaft torque:
  T_s = 1740 in. · lb (ccw)

FIGURE 9.25c

forces $\mathbf{F}''_{32}$ and $\mathbf{F}''_{12}$ on the free body of the crank are equal in magnitude to $\mathbf{F}''_{23}$. The shaft torque T''_s is the equilibrant of the couple of the forces on the crank. It is to be observed that the moment arm of the couple is small because the inertia force $\mathbf{F}_{o_3}$ very nearly acts through the crank pin center at A.

Although this analysis is primarily graphical, it is interesting to note that the polygon of Fig. 9.25a involving forces $\mathbf{F}_4$, $\mathbf{F}'_{14}$, and $\mathbf{F}'_{34}$ could have just as easily been treated analytically because the directions of the three forces can easily be written in unit vector form. The polygon of Fig. 9.25b with forces $\mathbf{F}_{o_3}$, $\mathbf{F}''_{23}$, and $\mathbf{F}''_{43}$ is more quickly solved graphically, however, because the directions of $\mathbf{F}_{o_3}$ and $\mathbf{F}''_{23}$ are not readily known in unit vector form without further calculations to determine angles.

The resultant forces obtained by superposition are shown in Fig. 9.25c. The free body of the connecting rod shows the resultant forces acting at the pin-connected ends of the rod. At the upper end of the rod is shown the resultant force $\mathbf{F}_{43}$ transmitted through the wrist pin bearing. $\mathbf{F}_{43}$ is the vector sum of $\mathbf{F}'_{43}$ and $\mathbf{F}''_{43}$. Similarly, the resultant force $\mathbf{F}_{23}$ transmitted through the crank pin bearing at A is the vector sum $\mathbf{F}'_{23}$ and $\mathbf{F}''_{23}$. It is to be observed that the connecting rod is in equilibrium under the action of these resultant forces and the inertia force $\mathbf{F}_{o_3}$ and that the three forces intersect at a common point k'. The resultant force through the main bearings is $\mathbf{F}_{12}$, which is identical to the force $\mathbf{F}_{23}$ through the crank-pin bearing. The crankshaft torque T_s is the algebraic sum of T'_s and T''_s.

9.13 DYNAMICALLY EQUIVALENT MASSES

Any rigid link in plane motion, having a mass M and a moment of inertia I, may be represented by an equivalent system of two point masses such that the inertia of the two masses is kinetically equivalent to the inertia of the link. In Fig. 9.26 is shown the inertia force $\mathbf{F}_o$ of a link, displaced a distance e from the mass center g by virtue of its angular acceleration α. Also shown are the two point masses M_P and M_Q, which are to be the equivalent of the link in order that the resultant of the inertia forces $\mathbf{F}_P = M_P\mathbf{A}_P$ and $\mathbf{F}_Q = M_Q\mathbf{A}_Q$ is equal to $\mathbf{F}_o = M\mathbf{A}_g$. Therefore,

$$\mathbf{F}_P + \mathbf{F}_Q = \mathbf{F}_o \qquad (9.51)$$

Although the proof is not undertaken here, it may be shown that to satisfy Eq. 9.51 the following three equivalents must be met:

1. *Equivalence of mass.* The sum of the point masses must be equal to the mass M of the link.

$$M_P + M_Q = M \qquad (9.52)$$

2. *Equivalence of mass center.* The mass center of the system of the two point masses must be at the mass center of the link. This requires that the point masses lie on a common link through g. This also requires that the sum of the moments of the point masses about g be zero.

$$M_P l_P - M_Q l_Q = 0 \qquad (9.53)$$

3. *Equivalence of moment of inertia.* The sum of the moments of inertia of

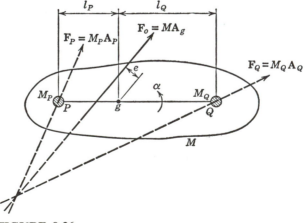

FIGURE 9.26

the point masses about g must be equal to the moment of inertia I of the link.

$$M_P l_P^2 + M_Q l_Q^2 = I \tag{9.54}$$

If a link is to be replaced by the equivalent system of two point masses, it is necessary to determine the four quantities of the system, the magnitudes of the two masses M_P and M_Q and the two distances l_P and l_Q. The last three equations given are not in the best form for determining these quantities. A more useful form of these equations may be derived as follows: By solving equations 9.52 and 9.53 simultaneously for M_P and M_Q, the following equations are obtained:

$$M_P = M \frac{l_Q}{l_P + l_Q} \tag{9.55}$$

$$M_Q = M \frac{l_P}{l_P + l_Q} \tag{9.56}$$

Then, substituting these equations in Eq. 9.54, the following is obtained:

$$M l_P l_Q = I$$

$$l_P l_Q = \frac{I}{M} \tag{9.57}$$

Since there are only three equations and four quantities are to be determined, it may be seen that one of the quantities must be arbitrarily chosen. Usually, one of the distances l_P or l_Q is so chosen, and the other is then calculated from Eq. 9.57. With the distances determined, the magnitudes of M_P and M_Q may then be calculated from Eqs. 9.55 and 9.56.

9.14 APPLICATION OF EQUIVALENT MASSES

Dynamically equivalent systems of two masses are most widely used in the analysis of automotive and aircraft piston engines, particularly with regard to connecting rods. Although applications of the method are made with approximations of small error, simplification in engine analysis is the primary advantage. Also, the method has influenced the design of counterweights on the crankshaft to reduce shaking of the engine.

Figure 9.27 shows a typical automotive connecting rod for which the center of gravity, weight, and moment of inertia are given. By arbitrarily locating one of the equivalent masses M_B at the wrist pin bearing B, one of the inertia forces is determined from the piston acceleration. The location of the second mass M_P is as shown and is calculated from Eq. 9.57 by solving for l_P. Because of the shape of the connecting rod, the center of gravity is near the crank pin center A

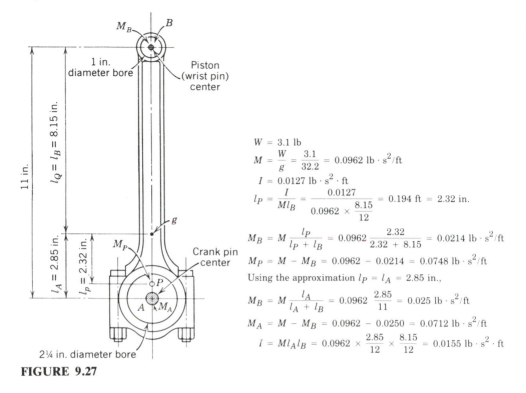

$$W = 3.1 \text{ lb}$$

$$M = \frac{W}{g} = \frac{3.1}{32.2} = 0.0962 \text{ lb} \cdot \text{s}^2/\text{ft}$$

$$I = 0.0127 \text{ lb} \cdot \text{s}^2 \cdot \text{ft}$$

$$l_P = \frac{I}{Ml_B} = \frac{0.0127}{0.0962 \times \frac{8.15}{12}} = 0.194 \text{ ft} = 2.32 \text{ in.}$$

$$M_B = M\frac{l_P}{l_P + l_B} = 0.0962 \frac{2.32}{2.32 + 8.15} = 0.0214 \text{ lb} \cdot \text{s}^2/\text{ft}$$

$$M_P = M - M_B = 0.0962 - 0.0214 = 0.0748 \text{ lb} \cdot \text{s}^2/\text{ft}$$

Using the approximation $l_P = l_A = 2.85$ in.,

$$M_B = M\frac{l_A}{l_A + l_B} = 0.0962 \frac{2.85}{11} = 0.025 \text{ lb} \cdot \text{s}^2/\text{ft}$$

$$M_A = M - M_B = 0.0962 - 0.0250 = 0.0712 \text{ lb} \cdot \text{s}^2/\text{ft}$$

$$I = Ml_Al_B = 0.0962 \times \frac{2.85}{12} \times \frac{8.15}{12} = 0.0155 \text{ lb} \cdot \text{s}^2 \cdot \text{ft}$$

FIGURE 9.27

as shown. Because of the nearness of P to A, the approximation may be made with little error that $l_P = l_A$. Thus, the second mass is at the crank pin center, and the inertia force may be determined from the acceleration of the crank pin, which is constant in all phases of the mechanism when operating at constant crankshaft speed. As shown in Fig. 9.27, the approximation results in a moment of inertia of the equivalent system slightly larger than the true moment of inertia of the link since Ml_Al_B is greater than Ml_Pl_B. The magnitudes of the masses M_A and M_B are calculated from Eqs. 9.55 and 9.56 using l_A and l_B.

Since M_A is a mass rotating about the crankshaft axis with constant centrifugal force (at constant crankshaft speed), counterweights attached to the crankshaft may be of such mass as to counterbalance M_A of the connecting rod as well as the mass of the crank.

9.15 ENGINE FORCE ANALYSIS USING POINT MASSES

In Fig. 9.28a is shown the internal-combustion engine mechanism with approximate kinetically equivalent point masses replacing the connecting rod. One of the point masses, M_{B_3}, is located at the wrist pin axis and the other, M_{A_3}, at the crank pin axis. Thus, the dynamic loading of the connecting rod is represented by the inertia force vectors $\mathbf{F}_{B_3}$ and $\mathbf{F}_{A_3}$, the magnitudes of which are $F_{B_3} = M_{B_3}A_B$ and $F_{A_3} = M_{A_3}A_A$. For all phases of the mechanism, the line of action of

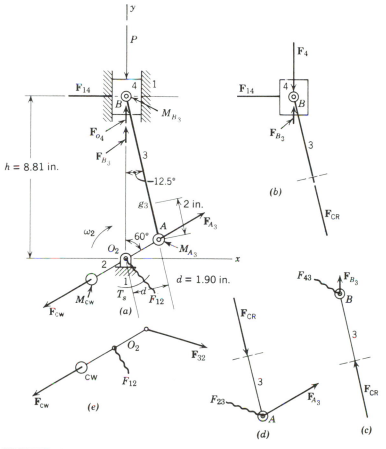

FIGURE 9.28

$\mathbf{F}_{B_3}$ is on the line of reciprocation of the wrist pin, and the force $\mathbf{F}_{A_3}$ is always radially outward on the crank line O_2A when the speed of the crank is uniform.

As shown in Fig. 9.28a, it is usual practice to add a mass M_{cw} to the counterweight of the crank so that an inertia force $\mathbf{F}_{CW}$ is induced to balance the inertia force $\mathbf{F}_{A_3}$ of the connecting-rod mass. By counterweighting in this manner, the masses rotating with the crank (crankshaft mass plus M_{A_3}) are balanced to put the mass center of the combination at O_2 so that no force from these masses acts on the main bearings.

It is to be observed that all the forces acting on the connecting rod, the inertia forces and the bearing forces, act at the ends of the rod at A and B. There are no transverse components of force between the ends of the rod to bend or shear the link, and therefore the member is in axial tension or compression. This is a result of assuming that M_{A_3} may be placed at the crank pin center A rather than at the correct point slightly removed from A. The fact that the connecting-rod force is axial in direction makes it possible to undertake the force analysis of the engine without superposition as illustrated in the following example.

Example 9.8. An analysis of the engine mechanism of Example 9.7 will now be made using approximate kinetically equivalent point masses for the connecting rod. In the solution, all components have been taken relative to the *xy*-axes.

Solution. From the data and solution of Example 9.7,

$$n = 3000 \text{ rpm}$$

$W_3 = 3$ lb	$A_A = 16{,}500$ ft/s^2
$l_3 = 8$ in.	$A_B = 5800$ ft/s^2
$l_A = 2$ in.	$F_{o_4} = 360$ lb
$l_B = 6$ in.	$P = 1410$ lb

With the data given above, the approximate kinetically equivalent masses of the connecting rod can be calculated as follows:

$$M_3 = \frac{W_3}{g} = \frac{3}{32.2} = 0.0933 \text{ lb} \cdot \text{s}^2/\text{ft}$$

$$M_{B_3} = M_3\left(\frac{l_A}{l_3}\right) = 0.0933(\tfrac{2}{8}) = 0.0233 \text{ lb} \cdot \text{s}^2/\text{ft}$$

$$M_{A_3} = M_3 - M_{B_3} = 0.0933 - 0.0233 = 0.0700 \text{ lb} \cdot \text{s}^2/\text{ft}$$

The inertia forces of the two ends of the connecting rod can also be calculated.

$$F_{B_3} = M_{B_3}A_B = 0.0233(5800) = 135 \text{ lb}$$

$$F_{A_3} = M_{A_3}A_A = 0.0700(16{,}500) = 1155 \text{ lb}$$

Figure 9.28*a* shows the forces acting on the mechanism. Of these forces, **P**, **F**$_{o_4}$, **F**$_{A_3}$, **F**$_{B_3}$, and **F**$_{cw}$ are known in magnitude, sense, and direction; force **F**$_{14}$ is known in direction only. Nothing is known about force **F**$_{12}$ except that it acts at point O_2. The equations for these forces, except **F**$_{12}$, can be written as follows:

$$\mathbf{F}_4 = \mathbf{P} + \mathbf{F}_{o_4} = -1410\mathbf{j} + 360\mathbf{j} = -1050\mathbf{j}$$

$$\mathbf{F}_{B_3} = 135\mathbf{j}$$

$$\mathbf{F}_{A_3} = 1155(\cos 30° \,\mathbf{i} + \sin 30° \,\mathbf{j})$$

$$= 1000.26\mathbf{i} + 577.50\mathbf{j}$$

$$\mathbf{F}_{cw} = -1000.26\mathbf{i} - 577.50\mathbf{j}$$

$$\mathbf{F}_{14} = F_{14}\mathbf{i}$$

In Fig. 9.28*b* is shown a free-body diagram of the piston and the upper part of the connecting rod. These members are acted upon by forces **F**$_4$, **F**$_{B_3}$, **F**$_{14}$, and **F**$_{CR}$. The force **F**$_{CR}$ is the force of the lower part of the connecting rod acting upon the upper part. Its direction is along the rod because the rod becomes a two-force member when equivalent masses are placed at points A and B. The equation for $\mathbf{F}_{CR}$ can be written as follows:

$$\mathbf{F}_{CR} = F_{CR}(-\sin 12.5°\, \mathbf{i} + \cos 12.5°\, \mathbf{j})$$

$$= -0.2164 F_{CR}\mathbf{i} + 0.9763 F_{CR}\mathbf{j}$$

Because the piston and the upper part of the rod are in equilibrium under the action of the four forces,

$$\mathbf{F}_4 + \mathbf{F}_{B_3} + \mathbf{F}_{14} + \mathbf{F}_{CR} = 0$$

Substituting the relations previously determined in the equation of equilibrium, one obtains

$$-1050\mathbf{j} + 135\mathbf{j} + F_{14}\mathbf{i} - 0.2164 F_{CR}\mathbf{i} + 0.9763 F_{CR}\mathbf{j} = 0$$

By summing **i** *components,*

$$F_{14}\mathbf{i} - 0.2164 F_{CR}\mathbf{i} = 0$$

By summing **j** *components,*

$$-1050\mathbf{j} + 135\mathbf{j} + 0.9763 F_{CR}\mathbf{j} = 0$$

$$F_{CR} = 937.2 \text{ lb}$$

Also,

$$F_{14}\mathbf{i} - 0.2164 \times 937.2\mathbf{i} = 0$$

$$F_{14} = 202.8 \text{ lb}$$

Therefore,

$$\mathbf{F}_{CR} = -0.2164 \times 937.2\mathbf{i} + 0.9763 \times 937.2\mathbf{j}$$

$$= -202.81\mathbf{i} + 915\mathbf{j}$$

$$\mathbf{F}_{14} = 202.8\mathbf{i}$$

Consider next Fig. 9.28c, which shows the upper part of the connecting rod acted upon by forces $\mathbf{F}_{B_3}$, $\mathbf{F}_{CR}$, and $\mathbf{F}_{43}$. The following equation of equilibrium can be written:

$$\mathbf{F}_{B_3} + \mathbf{F}_{CR} + \mathbf{F}_{43} = 0$$

where

$$\mathbf{F}_{43} = F_{43}(\lambda_x\mathbf{i} + \lambda_y\mathbf{j}) \qquad \text{(direction unknown; } \lambda \text{ is a unit vector in the direction of } \mathbf{F}_{43})$$

Substituting in the equation of equilibrium gives:

$$135\mathbf{j} - 202.81\mathbf{i} + 915\mathbf{j} + \lambda_x F_{43}\mathbf{i} + \lambda_y F_{43}\mathbf{j} = 0$$

By summing **i** *components,*

$$-202.81\mathbf{i} + \lambda_x F_{43}\mathbf{i} = 0$$

$$\lambda_x F_{43}\mathbf{i} = 202.81\mathbf{i}$$

By summing **j** *components,*

$$135\mathbf{j} + 915\mathbf{j} + \lambda_y F_{43}\mathbf{j} = 0$$

$$\lambda_y F_{43}\mathbf{j} = -1050\mathbf{j}$$

Therefore,

$$\mathbf{F}_{43} = 202.81\mathbf{i} - 1050\mathbf{j}$$

$$|\mathbf{F}_{43}| = 1069.4 \text{ lb}$$

Figure 9.28*d* shows a free-body diagram of the lower part of the connecting rod under action of forces $\mathbf{F}_{CR}$, $\mathbf{F}_{A_3}$, and $\mathbf{F}_{23}$. The following equation of equilibrium can be written:

$$\mathbf{F}_{CR} + \mathbf{F}_{A_3} + \mathbf{F}_{23} = 0$$

where

$$\mathbf{F}_{23} = F_{23}(\lambda_x\mathbf{i} + \lambda_y\mathbf{j}) \qquad \text{(direction unknown)}$$

and

$$\mathbf{F}_{CR} = 202.81\mathbf{i} - 915\mathbf{j} \qquad \text{(sense opposite to that of force } \mathbf{F}_{CR} \text{ acting on upper part of rod)}$$

Substituting in the equation of equilibrium gives:

$$202.81\mathbf{i} - 915\mathbf{j} + 1000.26\mathbf{i} + 577.50\mathbf{j} + \lambda_x F_{23}\mathbf{i} + \lambda_y F_{23}\mathbf{j} = 0$$

By summing **i** *components,*

$$202.81\mathbf{i} + 1000.26\mathbf{i} + \lambda_x F_{23}\mathbf{i} = 0$$

$$\lambda_x F_{23}\mathbf{i} = -1203.07\mathbf{i}$$

By summing **j** *components,*

$$-915\mathbf{j} + 577.50\mathbf{j} + \lambda_y F_{23}\mathbf{j} = 0$$

$$\lambda_y F_{23}\mathbf{j} = 337.5\mathbf{j}$$

Therefore,

$$\mathbf{F}_{23} = -1203.07\mathbf{i} + 337.5\mathbf{j}$$

$$|\mathbf{F}_{23}| = 1249.5 \text{ lb}$$

Figure 9.28*e* shows the crank and counterweight under action of forces $\mathbf{F}_{32}$, $\mathbf{F}_{cw}$, and $\mathbf{F}_{12}$. The equation of equilibrium is

$$\mathbf{F}_{32} + \mathbf{F}_{cw} + \mathbf{F}_{12} = 0$$

where

$$\mathbf{F}_{32} = -\mathbf{F}_{23} = 1203.07\mathbf{i} - 337.5\mathbf{j}$$

and

$$\mathbf{F}_{12} = F_{12}(\lambda_x\mathbf{i} + \lambda_y\mathbf{j}) \qquad \text{(direction unknown)}$$

Substituting in the above equation of equilibrium gives:

$$1203.07\mathbf{i} - 337.5\mathbf{j} - 1000.26\mathbf{i} - 577.50\mathbf{j} + \lambda_x F_{12}\mathbf{i} + \lambda_y F_{12}\mathbf{j} = 0$$

By summing $\mathbf{i}$ *components,*

$$1203.07\mathbf{i} - 1000.26\mathbf{i} + \lambda_x F_{12}\mathbf{i} = 0$$

$$\lambda_x F_{12}\mathbf{i} = -202.81\mathbf{i}$$

By summing $\mathbf{j}$ *components,*

$$-337.5\mathbf{j} - 577.50\mathbf{j} + \lambda_y F_{12}\mathbf{j} = 0$$

$$\lambda_y F_{12}\mathbf{j} = 915\mathbf{j}$$

Therefore,

$$\mathbf{F}_{12} = -202.81\mathbf{i} + 915\mathbf{j}$$

$$|\mathbf{F}_{12}| = 937.2 \text{ lb}$$

Comparing the equation for $\mathbf{F}_{12}$ with that for $\mathbf{F}_{CR}$ acting on the upper part of the rod, one can see that the equations are identical so that the two vectors are parallel and have the same sense and magnitude.

The shaft torque T_s necessary to hold link 2 in equilibrium can easily be calculated from the relation

$$T_s = F_{14}h = 202.8 \times 8.81 = 1786.8 \text{ lb} \cdot \text{in.} \qquad \text{(ccw)}$$

Torque T_s can also be obtained from the relation $F_{12}d$. The results of the analysis are as follows:

$$F_{14} = 203 \text{ lb}$$

$$F_{43} = 1069 \text{ lb} \qquad \text{(wrist pin force)}$$

$$F_{23} = 1250 \text{ lb} \qquad \text{(crank pin force)}$$

$$F_{12} = 937 \text{ lb} \qquad \text{(main-bearing force)}$$

$$T_s = 1787 \text{ lb} \cdot \text{in.} \qquad \text{(ccw)}$$

If these values are compared with those from Example 9.7 in which equivalent point masses were not used for the connecting rod, it will be seen that there is close agreement, with the exception of the force $\mathbf{F}_{12}$ at the main bearing. In this analysis, the magnitude and direction of $\mathbf{F}_{12}$ are quite different from those determined in Example 9.7. This difference, however, is not due to the use of equivalent masses but rather to the use of additional counterweight on the crank to balance inertia force $\mathbf{F}_{A_3}$; this partly unloads the main bearing.

It is interesting to note how easily the shaft torque T_s can be determined using the *method of virtual work*. This is illustrated in the following:

$$\mathbf{T}_2 \cdot \boldsymbol{\omega}_2 + \mathbf{F}_4 \cdot \mathbf{V}_B + \mathbf{F}_{B_3} \cdot \mathbf{V}_B + \mathbf{F}_{A_3} \cdot \mathbf{V}_A + \mathbf{F}_{\text{cw}} \cdot \mathbf{V}_{\text{cw}} = 0$$

$$\omega_2 = \frac{2\pi n}{60} = \frac{2\pi \times 3000}{60} = 314.16 \text{ rad/s} \quad (\text{cw})$$

$$|\mathbf{V}_B| = R\omega \left(\sin\theta + \frac{R}{2L}\sin 2\theta \right)$$

$$= \tfrac{2}{12} \times 314.16 \left(\sin 60° + \frac{2}{2 \times 8}\sin 120° \right)$$

$$= 51.01 \text{ ft/s}$$

Therefore,

$$\mathbf{V}_B = -51.01\mathbf{j}$$

$$\mathbf{F}_4 = \mathbf{P} + \mathbf{F}_{o_4} = -1050\mathbf{j}$$

$$\mathbf{F}_{B_3} = 135\mathbf{j}$$

$$\mathbf{T}_2 \cdot \boldsymbol{\omega}_2 = (T_2\mathbf{k}) \cdot (-314.16\mathbf{k}) = -314.16 T_2 \text{ ft} \cdot \text{lb/s}$$

$$\mathbf{F}_4 \cdot \mathbf{V}_B = (-1050\mathbf{j}) \cdot (-51.01\mathbf{j}) = 53{,}563.7 \text{ ft} \cdot \text{lb/s}$$

$$\mathbf{F}_{B_3} \cdot \mathbf{V}_B = (135\mathbf{j}) \cdot (-51.01\mathbf{j}) = -6886.8 \text{ ft} \cdot \text{lb/s}$$

$$\mathbf{F}_{A_3} \cdot \mathbf{V}_A = 0 \quad \text{(force and velocity vectors at right angles)}$$

$$\mathbf{F}_{\text{cw}} \cdot \mathbf{V}_{\text{cw}} = 0 \quad \text{(force and velocity vectors at right angles)}$$

Therefore,

$$-314.16 T_2 + 53{,}563.7 - 6886.8 = 0$$

and

$$T_2 = 148.58 \text{ lb} \cdot \text{ft}$$

$$= 1783 \text{ lb} \cdot \text{in.} \quad (\text{ccw})$$

This value agrees closely with the value found previously.

9.16 ENGINE BLOCK

In the foregoing discussion of the piston engine linkage, the frame of the engine, or the engine block, is considered the fixed member. However, in automotive installations, the engine block is supported on flexible mountings in order that a minimum of the unbalanced resultant force of the engine is transmitted to the engine supports. Figure 9.29 shows a free-body diagram of the engine block (link 1) fastened to the supporting link 0. The supporting link is shown rigid to illustrate the nature of the forces and moments which are transmitted to the supports. With flexible supports, the force system becomes one involving nonrigid members, and a vibration analysis is required.

The effects of the slider-crank mechanism of Fig. 9.28 is shown on the free-body diagram of the block in Fig. 9.29a, where F_{41} is the reactive force of the

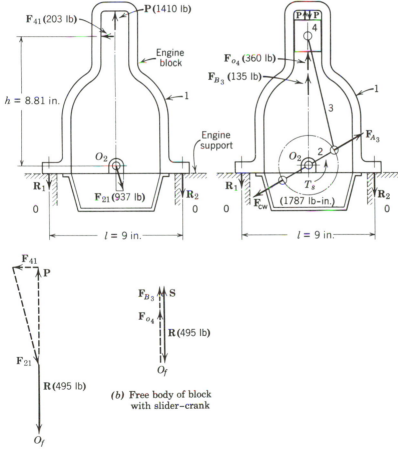

(a) Free body of engine block

(b) Free body of block with slider–crank

FIGURE 9.29

piston on the cylinder wall, and $\mathbf{F}_{21}$ is the force of the crankshaft on the main bearings fixed to the engine block. The force $\mathbf{P}$ by the gas pressure acts on the head of the block. These are known forces from the analysis made of the slider-crank mechanism. The reactive forces $\mathbf{R}_1$ and $\mathbf{R}_2$ of the engine supports are the unknowns to be determined.

The free-body diagram of Fig. 9.29b shows the forces acting on the combination of engine block with slider crank. Forces such as $\mathbf{F}_{41}$ and $\mathbf{F}_{14}$, as well as $\mathbf{F}_{21}$ and $\mathbf{F}_{12}$, are internal forces and are therefore not shown. The gas forces P are collinear, equal, and opposite and therefore do not affect $\mathbf{R}_1$ and $\mathbf{R}_2$. $\mathbf{F}_{cw}$ and $\mathbf{F}_{A_3}$ are also collinear, equal, and opposite. Those vectors shown on the free body that affect the forces $\mathbf{R}_1$ and $\mathbf{R}_2$ are the inertia forces $\mathbf{F}_{o_4}$ and $\mathbf{F}_{B_3}$ and the shaft torque T_s.

For static equilibrium of vertical forces, the resultant reactive force $\mathbf{R}$ (which is equal to $\mathbf{R}_1 + \mathbf{R}_2$) of the supports should be the same for both free bodies. The equilibrium force polygons for the two free bodies are shown in Fig. 9.29, and $\mathbf{R}$ is the same in both. From the polygon of Fig. 9.29b in particular, it may be seen that the resultant force $\mathbf{R}$ is equal in magnitude to the sum of the inertia forces $\mathbf{F}_{o_4}$ and $\mathbf{F}_{B_3}$ of the masses reciprocating with the wrist pin center. $\mathbf{R}$ is the equilibrant of the inertia forces. The opposite vector $\mathbf{S}$, which is the resultant of the inertia forces, is termed the *shaking force* since, if flexible supports are used, the engine block will be raised from its supports when the inertia forces are directed upward as shown and will be pressed against the supports when the inertia forces are directed downward in other phases of the cycle of the mechanism.

Thus, the reciprocating masses at the wrist pin center cause a vertical vibration or vertical shaking of the engine. It is to be observed that no resultant horizontal force acts on the engine, so there is no vibrational excitation in this direction. However, because of the shaft torque T_s, a couple acts on the engine, which, if mounted on flexible supports, excites an angular oscillation of the engine as the shaft torque changes in magnitude and sense during the engine cycle. Thus, T_s is a *shaking couple,* which is also transmitted to the engine supports and makes $\mathbf{R}_2$ greater than $\mathbf{R}_1$ as shown in Fig. 9.29. Calculations of the magnitudes of $\mathbf{R}_1$ and $\mathbf{R}_2$ are shown below Fig. 9.29 based on the free body of Fig. 9.29b.

9.17   ENGINE OUTPUT TORQUE

Of particular interest in engines is the shaft torque variation in the engine cycle corresponding to the 720° crank cycle. A plot of shaft torque versus crank angle θ shows a large variation in magnitude and sense of the torque, since by inspection of free bodies it may be seen that in some phases the torque is in the same sense as the crank motion and in other phases it is opposite to the crank motion. It would seem that the assumption of constant crank speed in the engine force analysis is invalid since a variation in torque would produce a variation in crank speed in the cycle. However, it is usual and necessary to fix a flywheel to the crankshaft. As shown in a following discussion, a flywheel of relatively small moment of inertia will reduce crank speed variations to negligibly small values (1 or 2% of crank speed). Because of its importance to flywheel design, an analytical method of evaluating output torque T as a function of crank angle θ is developed below. Output torque T and shaft torque T_s are the same in magnitude but of opposite sense. T_s is the torque on the mechanism as a free body and is the resisting torque of the load; T is the torque delivered to the flywheel and to the vehicle or load which the engine is driving. As shown in Example 9.8, the magnitude of the torque T_s may be calculated from either of two expressions: $T_s = F_{12}d$ or $T_s = F_{14}h$.

In Fig. 9.30 is shown the engine mechanism of Fig. 9.28a, in which the main bearing force $\mathbf{F}_{12}$ is known to be parallel to the connecting-rod axis because of the method of counterweighting the crankshaft. Considering the equilibrium of forces on the entire mechanism as a free body, it may be seen that the simple polygon of forces in Fig. 9.30 determines both $\mathbf{F}_{12}$ and $\mathbf{F}_{14}$ from the known collinear forces $\mathbf{P}$, $\mathbf{F}_{o_4}$, and $\mathbf{F}_{B_3}$ acting at the wrist pin center. The known forces may be combined and shown as the resultant force $\mathbf{F}_B$.

$$\mathbf{F}_B = \mathbf{P} + (\mathbf{F}_{o_4} + \mathbf{F}_{B_3}) \tag{9.58}$$

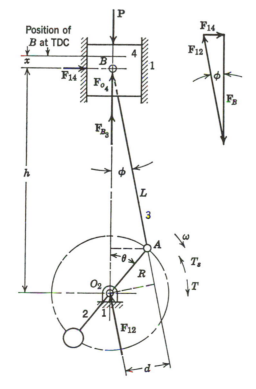

FIGURE 9.30

From the force polygon,

$$F_{12} = \frac{F_B}{\cos \phi} \tag{9.59}$$

$$F_{14} = F_B \tan \phi \tag{9.60}$$

Noting in Fig. 9.30 that $d = h \sin \phi$, we may evaluate the output torque T from either of the two expressions as follows:

$$T = F_{12}d = F_B \frac{h \sin \phi}{\cos \phi}$$

$$= F_B h \tan \phi \tag{9.61}$$

or

$$T = F_{14}h$$

$$= F_B h \tan \phi \tag{9.62}$$

Although the output torque is evaluated from different equations, Eqs. 9.61 and 9.62 show that they are identical. To determine torque as a function of crank angle θ from these equations, each of the right-hand factors (F_B, h, and ϕ) must be expressed as functions of θ. The first of these factors F_B depends on the several forces indicated in Eq. 9.58, each of which must also be expressed as functions of θ. An equation for the gas force P cannot be written directly as a function of θ since P is determined experimentally on an indicator diagram and is shown either versus θ as in Fig. 9.24 or versus piston position x in Fig. 9.30. Piston position is expressed as a function of θ in the following equations from Chapter 2:

$$x = R\left[1 - \cos \theta + \frac{L}{R}(1 - \cos \phi)\right]$$

$$\frac{x}{2R} = \frac{1}{2}\left[1 - \cos \theta + \frac{L}{R}(1 - \cos \phi)\right] \tag{9.63}$$

where R is the crank length and L is the connecting-rod length. The angle ϕ may be determined in terms of θ from the following relationship given by triangles in Fig. 9.30:

$$L \sin \phi = R \sin \theta$$

$$\phi = \sin^{-1}\left(\frac{R}{L} \sin \theta\right) \tag{9.64}$$

In Eq. 9.63, x is positive downward from the top dead center (T.D.C.) position of the piston and is expressed as a fraction of the stroke $2R$. The fraction gives the location on the abscissa of the indicator diagram at which the gas pressure is read for a given θ. Crank angle θ is positive clockwise in the sense of crank motion as shown in Fig. 9.30. For convenience, values of $x/2R$ for given angles θ have been calculated for a number of L/R ratios and are given in Table 9.2.

The sum of the inertia forces $\mathbf{F}_{O_4} + \mathbf{F}_{B_3}$ in Eq. 9.58 may be calculated from $(M_4 + M_{B_3})A_B$, in which A_B is the piston acceleration. As shown in Chapter 2, the piston acceleration is a function of θ in the following expression:

$$A_B = R\omega^2\left(\cos\theta + \frac{R}{L}\cos 2\theta\right)$$

$$\frac{A_B}{R\omega^2} = \left(\cos\theta + \frac{R}{L}\cos 2\theta\right) \qquad \textbf{(9.65)}$$

Values of $A_B/R\omega^2$ from Eq. 9.65 are tabulated in Table 9.2 for a number of L/R ratios as a function of θ. At the beginning of the stroke when θ is small, A_B is positive downward in the direction toward the crankshaft axis. However, since inertia force is opposite in sense to acceleration, $\mathbf{F}_{O_4}$ and $\mathbf{F}_{B_3}$ are shown opposite to $\mathbf{P}$ in Fig. 9.30.

Figure 9.31 shows the variation in one engine cycle of the resultant force or combined force $\mathbf{F}_B$. Also shown are the individual curves for gas force $\mathbf{P}$ and the inertia forces $\mathbf{F}_{O_4} + \mathbf{F}_{B_3}$. It may be seen from Table 9.2 that accelerations are positive near top dead center and are negative near the bottom of the stroke. Inertia forces are negative near the top of the stroke and positive near the bottom of the stroke as shown in Fig. 9.31.

The product $h \tan \phi$ in Eq. 9.62 may be expressed in terms of θ as follows. From triangles in Fig. 9.30,

$$h \tan \phi = (R \cos \theta + L \cos \phi)\tan \phi$$

$$= R \tan \phi\left(\cos\theta + \frac{L}{R}\cos\phi\right)$$

$$\frac{h}{R}\tan\phi = \tan\phi\left(\cos\theta + \frac{L}{R}\cos\phi\right) \qquad \textbf{(9.66)}$$

As shown in Table 9.2, values of $(h/R) \tan \phi$ are tabulated as a function of θ for various L/R ratios from Eqs. 9.66 and 9.64. It is to be observed that for downward strokes ($0° < \theta < 180°$), $(h/R) \tan \phi$ values are positive. Although they are not shown in the table, for upward strokes ($180° < \theta < 360°$), these values are negative because $\tan \phi$ is negative.

A typical curve of output torque of a single-cylinder engine for one complete cycle is shown in Fig. 9.32. Positive values of output torque are those which are

TABLE 9.2 Slider-Crank Functions for Output Torque, Piston Acceleration, and Piston Position

θ	$\frac{L}{R}=4.0$			$\frac{L}{R}=4.5$			$\frac{L}{R}=5.0$		
	$\frac{h}{R}\tan\phi$	$\frac{A_B{}^a}{R\omega^2}$	$\frac{x}{2r}$	$\frac{h}{R}\tan\phi$	$\frac{A_B}{R\omega^2}$	$\frac{x}{2R}$	$\frac{h}{R}\tan\phi$	$\frac{A_B}{R\omega^2}$	$\frac{x}{2R}$
0	0	1.2500	0	0	1.2222	0	0	1.2000	0
15	0.3214	1.1824	0.0212	0.3144	1.1583	0.0208	0.3088	1.1391	0.0204
30	0.6091	0.9910	0.0827	0.5968	0.9771	0.0809	0.5870	0.9660	0.0795
45	0.8341	0.7071	0.1779	0.8196	0.7071	0.1774	0.8081	0.7071	0.1716
60	0.9769	0.3750	0.2974	0.9642	0.3889	0.2921	0.9540	0.4000	0.2878
75	1.0303	0.0423	0.4298	1.0227	0.0666	0.4230	1.0169	0.0856	0.4177
90	1.0000	-0.2500	0.5635	1.0000	-0.2222	0.5563	1.0000	-0.2000	0.5505
105	0.9015	-0.4753	0.6886	0.9091	-0.4512	0.6819	0.9149	-0.4320	0.6765
120	0.7551	-0.6250	0.7974	0.7680	-0.6111	0.7921	0.7781	-0.6000	0.7878
135	0.5801	-0.7071	0.8851	0.5946	-0.7071	0.8815	0.6061	-0.7071	0.8787
150	0.3909	-0.7410	0.9487	0.4032	-0.7549	0.9469	0.4130	-0.7660	0.9455
165	0.1962	-0.7494	0.9872	0.2032	-0.7735	0.9867	0.2087	-0.7927	0.9863
180	0	-0.7500	1.0000	0	-0.7778	1.0000	0	-0.8000	1.0000

[a] When the value of $A_B/R\omega^2$ is negative, the sense of A_B is away from the crankshaft axis.

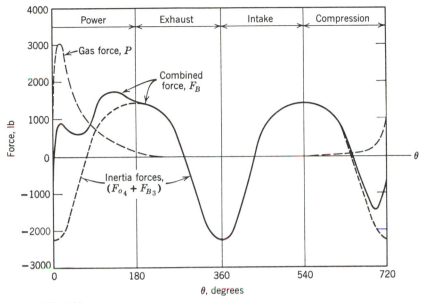

FIGURE 9.31

in the same sense as the crank motion. It may be seen that the torque changes sign where the torque is zero either because $(h/R) \tan \phi$ is zero or because the resultant force $\mathbf{F}_B$ is zero at the locations of θ shown in Fig. 9.31.

The dashed line of Fig. 9.32 represents the average torque T_{av} of one cylinder for one complete cycle. The work done on the mechanism by the gas force during the power stroke produces the average torque; without this work, the average

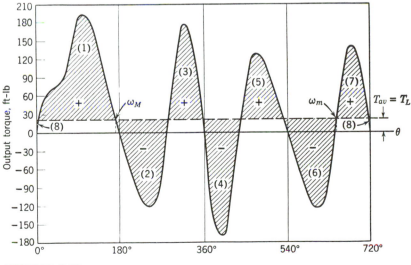

FIGURE 9.32

torque would be zero, and the changes of torque would be due to the inertia forces alone. Although the output torque in the crankshaft is greatly variable as shown in Fig. 9.32, with a flywheel attached to the shaft, the torque delivered after the flywheel is very nearly constant and equal to the average torque T_{av}. Under steady-state operation at a given crank speed, the torque T_{av} is equal to the resisting load torque T_L which the engine with flywheel is driving.

Horsepower output for one cylinder may be determined from the average torque output and the speed of the crankshaft:

$$\text{hp} = \frac{T_{av}\omega}{550} = \frac{T_{av}}{550}\frac{2\pi n}{60} = \frac{T_{av}n}{5250} \tag{9.67}$$

in which T_{av} is in foot-pounds and n is in revolutions per minute.[2] If friction is neglected in the torque analysis, the power given by Eq. 9.67 is very nearly equal to the indicated horsepower (ihp) as determined from the indicator diagram of gas pressure and stroke.

9.18 FLYWHEEL SIZE

As shown in Fig. 9.32, the output torque of the slider-crank mechanism is greater than the load torque for some portions of the engine cycle and is less in other parts of the cycle. Since the curve of Fig. 9.32 is a plot of torque versus θ, the shaded area represents work which either increases or decreases the kinetic energy of the system by causing an increase or decrease in crankshaft speed. The degree to which crank speed is increased or decreased depends on the inertia of the system since kinetic energy involves both mass, or moment of inertia, and speed. Control of the crank speed fluctuations is obtained primarily from a flywheel for which the moment of inertia may be calculated.

Figure 9.33 shows a single-cylinder engine with a flywheel. The free-body diagram of the flywheel shows the unbalance of torques acting on the flywheel to accelerate its angular motion. For output torque T of the slider crank greater than the load torque T_L, the equation of motion may be written:

$$T - T_L = I\alpha \tag{9.68}$$

in which I is the moment of inertia of the flywheel about the crank axis and α is in the sense of the resultant torque. Since $\alpha = (d\omega/dt)(d\theta/d\theta) = \omega(d\omega/d\theta)$, Eq.

---

[2]When working in SI units, the power is expressed in watts and is given by

$$\text{power} = T_{av}\omega \quad \text{(watts)}$$

where T_{av} is in newton · meters (N · m) and ω is in radians per second (rad/s).

9.68 may be rewritten:

$$T - T_L = I\omega \frac{d\omega}{d\theta}$$

$$(T - T_L)\, d\theta = I\omega\, d\omega$$

By integrating,

$$\int_{\theta\, at\, \omega_m}^{\theta\, at\, \omega_M} (T - T_L)\, d\theta = I \int_{\omega_m}^{\omega_M} \omega\, d\omega$$

$$= \tfrac{1}{2}I(\omega_M^2 - \omega_m^2) \tag{9.69}$$

In Eq. 9.69, the left-hand term is the work done on the flywheel and is represented by the shaded area of the torque diagrams of Figs. 9.32 and 9.33; the right-hand term is the corresponding change in kinetic energy of the flywheel due to its change in speed. Positive shaded areas of the torque diagram in Fig. 9.32 represent regions in the engine cycle where work is done to increase flywheel speed, and negative areas represent the work to decrease speed. Limits of θ on the integral of Eq. 9.69 are found by inspection so as to determine the greatest change in speed of the flywheel in the engine cycle where ω_M is the maximum and ω_m is the minimum angular velocity of the flywheel. The shaded loop in Fig. 9.32 having the greatest area would appear to represent the region of the greatest speed change. As shown, for a single-cylinder engine, the largest loop is in the power stroke, as would be expected because of the work done by the expanding

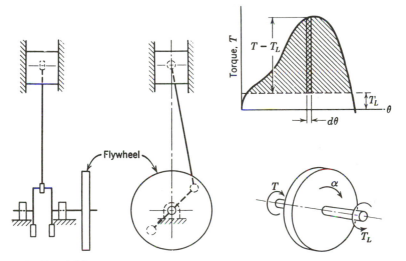

FIGURE 9.33

gas to speed up the engine. Thus, ω_M corresponds with θ at the end of the first loop. However, ω_m is not at the beginning of the first loop (1) but rather at the beginning of the seventh loop (7), since this loop is also positive and is nearly adjacent to the first loop except for the small negative loop (8) between the positive areas.

The locations of maximum and minimum crank speeds ω_M and ω_m on the torque diagram are not always easily determined by inspection. In such cases, a systematic arithmetic method may be used. For example, in Fig. 9.34, a torque diagram is shown with positive and negative areas above and below the average torque line. The relative magnitudes of the areas are given for the loops. If at the beginning of the first loop the speed is the datum value ω_0, then the speed at the end of the first loop is greater than ω_0 because of the positive area $A_1 = 7$. At the end of the second loop, which is negative, the speed is lower than at the end of the first loop but greater than ω_0 because the algebraic sum of the first two areas is positive: $A_1 + A_2 = 7 - 4 = 3$. At the end of each loop in Fig. 9.34 is shown the sum of the areas from the beginning of the first loop $(A_1 + A_2 + \cdots + A_n)$. The sum of the areas of all of the loops must equal zero since the average torque line is established at such a position that the sum of positive areas above the average torque line equals the sum of the negative areas below the average torque line. The maximum value of the sums gives the location of ω_M, which is the maximum speed greater than ω_0. As shown in Fig. 9.34, the location of ω_M is at the end of the first loop where the greatest sum is $+7$. Similarly, the minimum value of the sums $(-2$ at the end of the fourth loop) gives the location of ω_m, which is the speed most below ω_0. The algebraic sum of the areas between the locations of ω_m and ω_M represents the work done by torque to change the flywheel kinetic energy from a minimum to a maximum value.

The integral expression of Eq. 9.69 may be represented by the area A:

$$A = \int_{\theta \text{ at } \omega_m}^{\theta \text{ at } \omega_M} (T - T_L)\, d\theta \qquad (9.70)$$

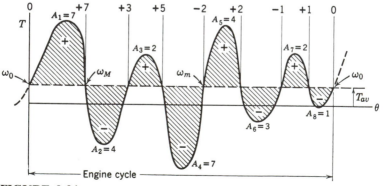

FIGURE 9.34

which is the algebraic summation of areas of loops in the cycle giving the greatest change in flywheel speed. The net positive area in square inches may be obtained by use of a planimeter and converted to foot-pounds of work A with the proper use of scales for the diagram. Substitution in Eq. 9.69 gives

$$A = \tfrac{1}{2}I(\omega_M^2 - \omega_m^2) \tag{9.71}$$

In multicylinder engines, the firing order and crank arrangement are such that the pulses of torque from the power strokes of the individual cylinders are uniformly distributed throughout the engine cycle of 720°. In a six-cylinder engine, for example, the cranks are spaced at 120° of crank angle (720/6) so that a power stroke begins every 120° of crankshaft rotation. The resultant torque curve, obtained by the superposition of the torque curves of the individual cylinders in proper phase relationship, is shown in Fig. 9.35. As shown, the loops of the torque curve are uniform in the sense that the loops are of the same form every 120°. The dashed line of average torque is located to make each positive loop equal in area to a negative loop. Therefore, in Eq. 9.70, A is determined from the area of any individual loop. The minimum speed ω_m is at the end of every negative loop, and ω_M is at the end of every positive loop.

To determine the required moment of inertia I of the flywheel, a coefficient of fluctuation K is assigned so that the fluctuation, or difference of maximum and minimum speeds, is a small fraction of the average design speed ω_{av} of the engine.

$$K = \frac{\omega_M - \omega_m}{\omega_{av}} \tag{9.72}$$

The average speed ω_{av} is

$$\omega_{av} = \frac{\omega_M + \omega_m}{2} \tag{9.73}$$

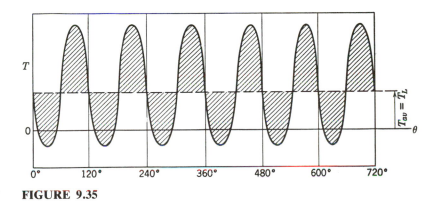

FIGURE 9.35

Equation 9.71 may be rewritten

$$A = \tfrac{1}{2}I(\omega_M + \omega_m)(\omega_M - \omega_m)$$

$$= \tfrac{1}{2}I(2\omega_{av})(K\omega_{av})$$

$$A = IK\omega_{av}^2$$

$$= IK \frac{4\pi^2 n^2}{(60)^2}$$

Solving for I gives

$$I = 91 \frac{A}{Kn^2} \tag{9.74}$$

Equation 9.74 gives the required moment of inertia of the flywheel for the average speed n in revolutions per minute at which the torque analysis is made; the units of I are lb · s² · ft, and those of A are lb · ft.

When using SI units, Eq. 9.74 becomes

$$I = \frac{A}{k\omega_{av}^2}$$

where the units of ω are rad/s, the units of I are kg · m², and the units of A are N · m.

Example 9.9. Determine the required moment of inertia of a flywheel for a single-cylinder engine for which Fig. 9.32 is the torque diagram at a speed of 3300 rpm. The maximum allowable fluctuation in speed in the engine cycle is 40 rpm.

Solution. From Fig. 9.32,

Area of first loop	= +0.762 in.²
Area of seventh loop	= +0.265
	1.027
Negative area of eighth loop	= −0.007
	1.020 in.²

In Fig. 9.32, the torque scale is 120 ft · lb/in. and the angular scale is $(\tfrac{8}{9})\pi$ rad/in.; therefore, each square inch of the torque diagram represents 335 ft · lb of work.

$$A = 1.020(335) = 342 \text{ ft} \cdot \text{lb}$$

$$K = \frac{40}{3300} = 0.01212$$

$$I = 91 \frac{A}{Kn^2} = \frac{(91)(342)}{0.01212 \times (3300)^2}$$

$$= 0.236 \text{ lb} \cdot \text{s}^2 \cdot \text{ft}$$

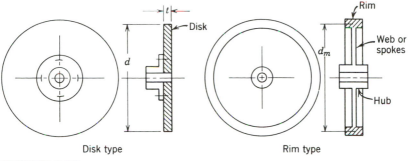

Disk type   Rim type

FIGURE 9.36

Flywheels are usually disks in automotive engines and rim type in steam engines or punch presses (Fig. 9.36). In automotive installations, a thin disk of large diameter gives the lowest weight for the moment of inertia required. However, a compromise as to diameter must be made because of the stresses in the material caused by centrifugal force and because of space and road clearance requirements. Large flywheels of the rim type in steam engines and punch presses are limited in diameter primarily by allowable stresses due to centrifugal force.

For disk-type flywheels, the moment of inertia and mass are related as follows:

$$I = \frac{Mr^2}{2} = \frac{Md^2}{8} = \frac{Wd^2}{8g} \tag{9.75}$$

in which r and d are radius and diameter, respectively, of the disk.

Example 9.10. Determine the mass M and thickness t of a steel disk-type flywheel 0.3048 m in diameter to give $I = 0.320$ kg · m. The density of steel is $\rho = 7.80 \times 10^3$ kg/m^3.

Solution

$$I = \frac{Md^2}{8}$$

$$M = \frac{8I}{d^2} = \frac{8 \times 0.320}{0.3048^2} = 27.6 \text{ kg}$$

$$M = (\text{vol. of disk})\rho = \left(\frac{\pi d^2 t}{4}\right)\rho$$

$$t = \frac{4M}{\pi d^2 \rho}$$

$$= \frac{4 \times 27.6}{\pi (0.3048)^2 \times (7.80 \times 10^3)}$$

$$= 0.0485 \text{ m} = 48.5 \text{ mm}$$

It may be seen that a small speed fluctuation may be obtained with a reasonable weight of flywheel. However, as Eq. 9.74 indicates, a larger flywheel is required at low speeds, although A may be much less at low speeds.

For rim-type flywheels, $I = Mk^2$, in which k is the radius of gyration. It is sufficiently accurate to assume that the mean radius r_m of the rim is equal to k.

$$I = Mr_m^2 = \frac{W}{4g} d_m^2 \qquad (9.76)$$

The solution of Eq. 9.76 for W gives only the weight of the rim. The weights of the hub, web, or spokes also contribute a small amount to the moment of inertia of the flywheel, with the result that the speed fluctuation is somewhat less than the assigned value.

Components of an engine installation other than the flywheel may contribute flywheel effect. The crankshaft and the equivalent mass of the connecting rod at the crank pin act as a flywheel. In automotive installations during road operation with clutch engaged, the rotating parts of the driving system as well as the car itself serve to reduce engine speed fluctuation to the degree that almost no flywheel is required. However, the idling condition at low speed in automotive engines is the prime condition for automotive flywheel design. Flywheel effect is important also in maintaining motor spin during starting. Aircraft piston engines are normally without a flywheel because of the large flywheel effect of propellers and propeller reduction gears. In the design of all reciprocating machinery such as diesel–electric systems, compressors, steam winches operated by donkey engines, quick-return mechanisms, motorcycles, outboard motors, and punch presses, flywheel effect is required, but the degree to which moment of inertia must be added in a flywheel depends on the requirements of the installation.

9.19 FORCES ON GEAR TEETH

For gears in mesh, the line of transmission of force is along the line of action, which is always normal to the contacting tooth surfaces as the teeth traverse the arc of action. As shown in Fig. 9.37, the line of action of the tooth force $\mathbf{F}$ is at the pressure angle ϕ to the tangent of the pitch circles. The tooth of the driver shown in Fig. 9.37 is in contact with a tooth of the driven gear at the pitch point. In this position, the teeth are in the state of pure rolling, and no friction due to relative sliding exists. At other positions in the arc of action, relative sliding exists and the resultant force on the gear tooth is inclined to the line of action by the angle of friction. In a force analysis of mechanisms with gears, the friction angle may be neglected with little error in the determination of the magnitude of tooth force.

If two sets of gear teeth are in contact, the transmitted force is divided between the two sets of teeth. The free-body diagram of the driving gear, for example, would show two tooth forces, both of which act along the line of action. The resultant of the two forces, equal to the transmitted force, also acts along the line of action. The proportion of transmitted force carried by each tooth depends on the accuracy with which the gear teeth are in mesh, which in turn depends on the accuracy of manufacture of the tooth forms. Since one tooth is likely to carry more force than the other, it is usual to assume that one tooth carries the full transmitted force.

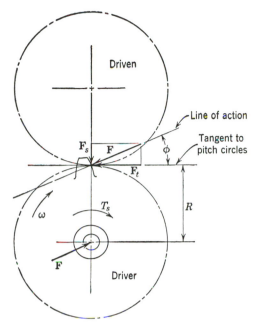

FIGURE 9.37

As shown in Fig. 9.37, the tooth force **F** is represented by the components **F**$_t$ and **F**$_s$, of which **F**$_t$ is called the tangential force and **F**$_s$ the separating force; $F_t = F \cos \phi$ and $F_s = F \sin \phi$. In many problems, the shaft torque T_s applied to the driving gear is known, and F_t may be determined from the equilibrium of moments about the shaft axis:

$$T_s = F_t R \tag{9.77}$$

where R is the pitch radius of the gear. The transmitted force **F** may be determined from

$$F = \frac{F_t}{\cos \phi} = \frac{T_s}{R \cos \phi} \tag{9.78}$$

Equation 9.78 shows that, for a given torque applied to the gear, the tooth force **F** increases with pressure angle. The separating force **F**$_s$ also increases with pressure angle. It may be seen that **F**$_t$ acts to shear and bend the tooth and that **F**$_s$ acts to compress the tooth. The transmitted force **F** causes high local stresses in the material in the vicinity of contact on the tooth face.

Example 9.11. It is desired to determine the tooth forces acting on the several gears of the planetary gear train shown in Fig. 9.38a. One hundred horsepower is transmitted by the gear train at constant speed. The sun gear (link 4) rotating clockwise at $n_4 = 2000$ rpm is the input side of the train, and the carrier (link 2) rotating clockwise at $n_2 = 667$ rpm is the output side. The shaft torque T_{s_4} acting on the sun gear is the driving torque,

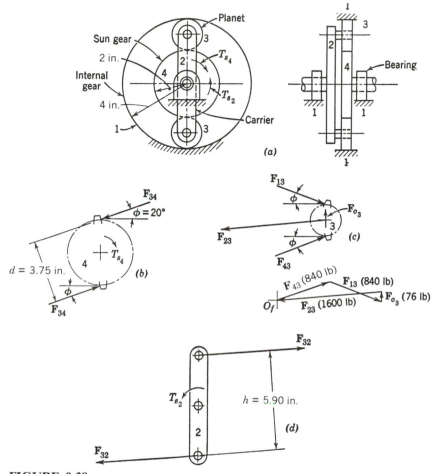

FIGURE 9.38

and T_{s_2} acting on the carrier is the resisting shaft torque of the load. Spur gears with involute shaped teeth at 20° pressure angle are used in the gear train.

Solution. Carefully drawn free-body diagrams of the individual links as shown in Fig. 9.38 aid in the determination of the forces acting on each link for static equilibrium. Inertia forces are zero for the sun gear and carrier as well as for the internal gear since the accelerations of the mass centers of these elements are zero; the inertia torques are also zero since the train operates at constant angular velocity and zero angular acceleration. Insofar as the planets are concerned, centrifugal inertia forces act because of the centripetal acceleration of the centers of the planets. By assuming a planet weight of 2 lb, the centrifugal force acting is

$$F_{o_3} = M_3 A_{g_3} = \left(\frac{2}{32.2}\right)\left(\frac{3}{12}\right)\left(\frac{2\pi 667}{60}\right)^2$$

$$= 76 \text{ lb}$$

The free-body diagram of the sun gear in Fig. 9.38*b* shows the driving torque T_{s_4} and two tooth forces $\mathbf{F}_{34}$, which are shown along the line of action for 20° pressure angle. Since the power transmitted and the speed of the gear are known, T_{s_4} may be calculated as follows:

$$\text{hp} = \frac{T_{s_4}\omega_4}{550}$$

$$T_{s_4} = \frac{550 \text{ hp}}{\omega_4} = \frac{550 \text{ hp}}{2\pi n_4/60}$$

$$= \frac{(550)(100)}{2\pi 2000/60}$$

$$= 262 \text{ ft} \cdot \text{lb} \qquad (\text{cw})$$

Since two couples in equilibrium act on link 4,

$$F_{34}d = T_{s_4}$$

$$F_{34} = \frac{T_{s_4}}{d} = \frac{262}{3.75/12} = 840 \text{ lb}$$

The free-body diagram of the planet in Fig. 9.38*c* shows that four forces act on the link, of which the forces $\mathbf{F}_{43}$ and $\mathbf{F}_{o_3}$ are known. The direction and sense of the tooth force $\mathbf{F}_{13}$ may be ascertained by considering the moment equilibrium of forces acting about the planet center. If $\mathbf{F}_{13}$ also acts at the pressure angle of 20°, then $\mathbf{F}_{13}$ and $\mathbf{F}_{43}$ are equal in magnitude to satisfy equilibrium of moments about the planet center. The force $\mathbf{F}_{23}$ of the carrier acting on the planet is the remaining unknown, which may be determined from the force polygon in Fig. 9.38*c* or analytically as follows:

The equation of equilibrium can be written from the free-body diagram of the planet as

$$\mathbf{F}_{13} + \mathbf{F}_{o_3} + \mathbf{F}_{43} + \mathbf{F}_{23} = 0$$

where

$$\mathbf{F}_{13} = F_{13}(\cos 20° \, \mathbf{i} - \sin 20° \, \mathbf{j})$$

$$= 0.9397F_{13}\mathbf{i} - 0.3420F_{13}\mathbf{j}$$

$$\mathbf{F}_{o_3} = 76.0\mathbf{j}$$

$$\mathbf{F}_{43} = F_{43}(\cos 20° \, \mathbf{i} + \sin 20° \, \mathbf{j})$$

$$= 840 \times 0.9397\mathbf{i} + 840 \times 0.3420\mathbf{j}$$

$$= 789.35\mathbf{i} + 287.28\mathbf{j}$$

But,

$$|\mathbf{F}_{13}| = |\mathbf{F}_{43}|$$

Therefore,

$$\mathbf{F}_{13} = 789.35\mathbf{i} - 287.28\mathbf{j}$$

Also,

$$\mathbf{F}_{23} = F_{23}(\lambda_x\mathbf{i} + \lambda_y\mathbf{j}) \qquad \text{(direction unknown)}$$

Substituting in the equation of equilibrium gives:

$$789.35\mathbf{i} - 287.28\mathbf{j} + 76.0\mathbf{j} + 789.35\mathbf{i} + 287.28\mathbf{j} + \lambda_xF_{23}\mathbf{i} + \lambda_yF_{23}\mathbf{j} = 0$$

By summing **i** *components,*

$$789.35\mathbf{i} + 789.35\mathbf{i} + \lambda_xF_{23}\mathbf{i} = 0$$

$$\lambda_xF_{23}\mathbf{i} = -1578.70\mathbf{i}$$

By summing **j** *components,*

$$76.0\mathbf{j} + \lambda_yF_{23}\mathbf{j} = 0$$

$$\lambda_yF_{23}\mathbf{j} = -76.0\mathbf{j}$$

Therefore,

$$\mathbf{F}_{23} = -1578.70\mathbf{i} - 76.0\mathbf{j}$$

$$|\mathbf{F}_{23}| = 1580.53 \text{ lb}$$

From the free-body diagram of the carrier in Fig. 9.38*d*, the shaft torque T_{s_2} may be determined from the equilibrium of moments about the carrier axis:

$$T_{s_2} = F_{32}h$$

$$= 1600 \times \frac{5.9}{12}$$

$$= 787 \text{ ft} \cdot \text{lb} \qquad \text{(ccw)}$$

T_{s_2} may also be determined from the transmitted horsepower and the speed of the carrier:

$$\text{hp} = \frac{T_{s_2}\omega_2}{550}$$

$$T_{s_2} = \frac{550 \text{ hp}}{\omega_2} = \frac{550 \text{ hp}}{2\pi n_2/60}$$

$$T_{s_2} = \frac{(550)(100)}{2\pi 667/60}$$

$$= 787 \text{ ft} \cdot \text{lb} \qquad \text{(ccw)}$$

The foregoing solution indicates that the tooth force to be expected is 840 lb. However, a subtlety exists in interpreting the effect of the inertia force $\mathbf{F}_{o_3}$ on forces. In the solution above, it was assumed that the carrier constrained the planet center to remain at the meshing pressure angle of 20° at the pitch point by providing the reaction to the planet inertia force $\mathbf{F}_{o_3}$. However, if a large clearance exists for the pin connecting the planet to the carrier, then the carrier cannot provide the reaction to $\mathbf{F}_{o_3}$, in which case the planet will displace slightly toward the internal gear in such a way that it meshes with the internal gear at a pressure angle slightly less than 20° and with the sun gear at a pressure angle slightly more than 20°. The result is that $\mathbf{F}_{43}$ will be somewhat greater than 840 lb and $\mathbf{F}_{13}$ somewhat less.

An interesting extension of this example is to consider the effect of replacing the spur gears in the planetary drive of Fig. 9.38 with helical gears while keeping the center distances, gear ratios, and power transmitted the same. If the helical gears have a normal pressure angle ϕ_n of 20° and a helix angle ψ of 30°, the following relations can be developed for planet 3 assuming that the normal force in the plane of rotation $\mathbf{F}_{13}$ remains at 840 lb. By referring to Fig. 9.39,

$$\tan \phi = \frac{\tan \phi_n}{\cos \psi} = \frac{\tan 20°}{\cos 30°}$$

$$\phi = 22.8°$$

$$\mathbf{F}_{13} = F_{13}(\cos \phi \, \mathbf{i} - \sin \phi \, \mathbf{j})$$

$$= 840(\cos 22.8° \, \mathbf{i} - \sin 22.8° \, \mathbf{j})$$

$$= 774.39\mathbf{i} - 325.45\mathbf{j}$$

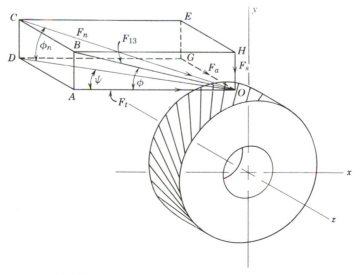

FIGURE 9.39

Therefore,

$$\mathbf{F}_t = 774.39\mathbf{i}$$

$$\mathbf{F}_s = -325.45\mathbf{j}$$

and

$$|\mathbf{F}_a| = F_t \tan \psi = 774.39 \tan 30°$$

$$= 447.09 \text{ lb}$$

$$\mathbf{F}_a = 447.09\mathbf{k}$$

Therefore,

$$\mathbf{F}_n = 774.39\mathbf{i} - 325.45\mathbf{j} + 447.09\mathbf{k}$$

$$|\mathbf{F}_n| = 951.57 \text{ lb}$$

It is left to the reader to determine how this change will affect the bearing force $\mathbf{F}_{23}$.

9.20 CAM FORCES

At high cam speeds, the force transmitted at the contact of cam and follower is high and may cause serious wear of the contacting surfaces. Figure 9.40 shows a disk cam with radial roller follower. Two phases of the cam are shown as it rotates counterclockwise at a uniform speed n_c of 8550 rpm. In Fig. 9.40a, the cam is in such a phase that the acceleration $\mathbf{A}_f$ of the follower is away from the cam. In this phase, the inertia force $\mathbf{F}_f$ of the follower is such that even without the force $\mathbf{S}$ of the compressed spring, the follower is held in contact with the cam. However, in Fig. 9.40b, the phase of the cam is such that a high downward acceleration $\mathbf{A}_f$ of the follower is present. In this instance, the follower inertia force $\mathbf{F}_f$ is great enough to cause the follower to leave the cam unless a force $\mathbf{S}$ equal to $\mathbf{F}_f$ is applied by the spring. By assuming that the weight of the follower including roller, stem, and spring is 1 lb,

$$F_f = M_f A_f = \left(\frac{1}{32.2}\right)(101,000) = 3140 \text{ lb}$$

and the required spring force is $S = 3140$ lb. Since $\mathbf{S}$ and $\mathbf{F}_f$ are equal forces but opposite in sense, the force $\mathbf{F}$ acting at the contacting surfaces of cam and roller is zero. Since the spring is compressed $\delta = \frac{1}{2}$ in. corresponding to the lift of the cam, the required spring constant k is

$$k = \frac{S}{\delta} = \frac{3140}{1/2} = 6280 \text{ lb/in.}$$

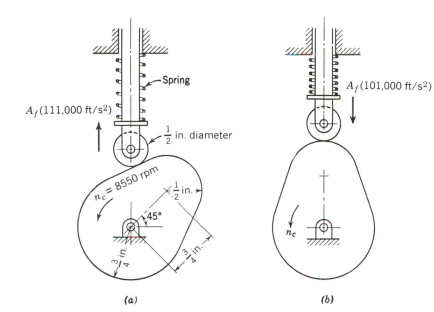

(a)

(b)

(c)

FIGURE 9.40

By returning to the phase of the cam shown in Fig. 9.40a, it may be observed that the force **F** at the contacting surfaces will be high because it represents a reaction to both the inertia force $\mathbf{F}_f$ and the spring force **S** as shown in the free-body diagram of the follower in Fig. 9.40c. **F** is normal to the surfaces of contact, and the angle which **F** makes with the direction of motion of the follower is the pressure angle ϕ. For static equilibrium, the summation of forces in the direction of follower displacement is zero:

$$F \cos \phi - F_f - S = 0$$

$$F = \frac{F_f + S}{\cos \phi} \tag{9.79}$$

Inertia force $\mathbf{F}_f$ is

$$F_f = M_f A_f = \left(\frac{1}{32.2}\right)(111,000) = 3450 \text{ lb}$$

The spring is compressed $\frac{1}{8}$ in.

$$S = k\delta = 6280 \times \tfrac{1}{8} = 785 \text{ lb}$$

The pressure angle $\phi = 25°$ ($\cos 25° = 0.907$).
  From Eq. 9.79,

$$F = \frac{3450 + 785}{0.907} = 4670 \text{ lb}$$

A surface force $F = 4670$ lb and spring constant $k = 6280$ lb/in. are high for a cam of the size shown. A stress analysis would show that the speed of the cam should be limited to a smaller value. Valve cams in automotive installations are a constant challenge to the mechanical designer because of the demand for increased engine speeds.

In Fig. 9.40c is also shown the free-body diagram of the cam. The shaft torque T_s may be determined from the couple formed by **F** and the shaft reaction equal and opposite to **F**. Since the center of gravity g_c of the cam is not at the axis of rotation, a centrifugal inertia force for the cam should be shown. However, since the cam rotates at constant speed, the cam inertia force does not influence the calculation of the shaft torque T_s but does enter into the calculation of the resultant shaft reaction.

9.21 GYROSCOPIC FORCES

In vehicles having engines with rotating parts of high moment of inertia, gyroscopic forces are in action when the vehicle is changing direction of motion. In automotive vehicles undergoing roadway turns at high velocity, gyroscopic forces act on such spinning parts as crankshaft, flywheel, clutch, transmission gears, propeller shaft, and wheels. Engine parts as well as the propeller and the gear reduction system of an airplane are under the action of gryscopic effects in turns and pullouts. Locomotives and ships are similarly affected.

Figure 9.41 shows a rigid body spinning at a constant angular velocity ω about a spin axis through the mass center. The angular momentum **H** of the spinning body is represented by a vector whose magnitude is $I\omega$, in which I is the moment of inertia of the body about the spin axis and axis through the mass center. Although the angular momentum of the body is in a plane parallel to the planes of motion of the individual particles of the body, it is represented by a vector normal to the plane of motion as shown. The sense of the vector is determined by the right-hand screw rule in which the arrowhead of the vector is in the sense of the advance of a right-hand screw turned in the sense of the angular velocity ω of the body. The length of the vector represents the magnitude of the angular momentum.

From the study of mechanics, it is known that the rate of change of angular momentum with respect to time is proportional to an applied torque T as determined from the following equation of motion:

$$T = I\alpha = I\frac{d\omega}{dt} = \frac{d}{dt}(I\omega)$$

Also,

$$H = I\omega$$

Therefore,

$$T = \frac{dH}{dt} \qquad \textbf{(9.80)}$$

In the case shown in Fig. 9.41, a torque applied in the plane of motion of the spinning body in the sense of ω increases the angular momentum at a given rate which may be shown as an increase in the length of the vector.

In the foregoing discussion, the spin axis was considered fixed. If the spin axis is made to change angular position as in a vehicle traversing a plane curved path as shown in Fig. 9.42a, gyroscopic action results. For constant ω, the mag-

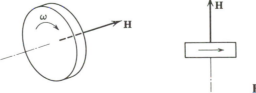

FIGURE 9.41

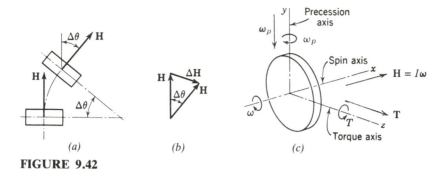

(a)   (b)   (c)

FIGURE 9.42

nitude of the angular momentum remains constant for an angular displacement $\Delta\theta$ of the spin axis as shown by the vectors. However, a change in angular momentum exists because of the change in direction of the momentum as shown by the polygon of free vectors in Fig. 9.42b. For a small value of $\Delta\theta$, the magnitude of the change in angular momentum $\Delta\mathbf{H}$ is

$$\Delta H = (I\omega)\,\Delta\theta$$

The rate of change of angular momentum with respect to time is

$$\frac{dH}{dt} = \lim_{\Delta t \to 0} \frac{\Delta H}{\Delta t} = \lim_{\Delta t \to 0} (I\omega)\frac{\Delta\theta}{\Delta t} = I\omega\frac{d\theta}{dt}$$

Therefore,

$$\frac{dH}{dt} = I\omega\omega_p$$

and

$$T = I\omega\omega_p \tag{9.81}$$

where $\omega_p = d\theta/dt$ is the angular velocity of precession of the spin axis, or the rate at which the spin axis displaces angularly.

The magnitude of the torque $\mathbf{T}$ which is associated with the precession of the spin axis can easily be determined from Eq. 9.81, and it is now necessary to determine its direction. Referring again to Fig. 9.42b, one can see that as Δt approaches zero, the vector $\Delta\mathbf{H}$ becomes normal to the vector $\mathbf{H}$, which has the same direction as the spin axis. Therefore, the torque vector $\mathbf{T}$ will also be normal to the spin axis in the plane of $\mathbf{H}$ and $\Delta\mathbf{H}$. Figure 9.42c shows the x-axis as the spin axis and the y-axis as the precession axis. The z-axis becomes the torque axis because the direction of torque **T** is normal to the spin axis and lies in the *xz*-plane. From the orientation of the spin, precession, and torque axes, it can be seen that Eq. 9.81 can be written in the following vector form:

$$\mathbf{T} = \boldsymbol{\omega}_p \times I\boldsymbol{\omega} \qquad (9.82)$$

The applied torque **T** in Eq. 9.82 is a couple referred to as the *gyroscopic couple*. Because this couple has the same direction as $\Delta\mathbf{H}$, the couple lies in the *xy*-plane and represents a torque applied to the body about the *z*- or torque axis. Thus, it may be seen that to cause precession of a spinning body, a torque must be applied to the body in a plane normal to the plane in which the spin axis is precessing.

The flywheel of an automotive engine is an example of a spinning body that is subject to the gyroscopic couple in roadway turns at high vehicular speed. As shown in Fig. 9.43, the flywheel of a single-cylinder engine is fixed to the crankshaft, which in turn is supported by the main bearings. The crankshaft and the equivalent mass of the connecting rod are also spinning masses which may be considered as part of the flywheel. Forces $\mathbf{F}_{12}$, which represent the gyroscopic couple, are applied to the crankshaft by the bearings. These forces are superposed on the forces produced by the operation of the slider-crank mechanism. Other bearing forces which are induced by the turning vehicular motion are those resulting from centrifugal force while the vehicle is in the curved path.

Example 9.12. For the single-cylinder engine of Fig. 9.43, determine the bearing forces $\mathbf{F}_{12}$ caused by the gyroscopic action of the flywheel of Example 9.9 as the engine vehicle traverses a 1000-ft-radius curve at 60 mph (88 ft/s) in a turn to the right. The engine speed is 3300 rpm and is turning clockwise viewed from the front of the engine.

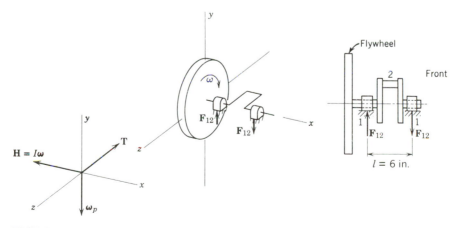

FIGURE 9.43

Solution

$$I \text{ of flywheel} = 0.236 \text{ lb} \cdot \text{s}^2 \cdot \text{ft}$$

$$\omega = \frac{2\pi n}{60} = \frac{2\pi(3300)}{60} = 346 \text{ rad/s}$$

$$\omega_p = \frac{V}{R} = \frac{88}{1000} = 0.088 \text{ rad/s}$$

$$\mathbf{T} = \boldsymbol{\omega}_p \times I\boldsymbol{\omega}$$

where

$$\boldsymbol{\omega}_p = -0.088\mathbf{j}$$

$$I\boldsymbol{\omega} = (0.236)(-346\mathbf{i}) = -81.66\mathbf{i}$$

Therefore,

$$\mathbf{T} = (-0.088\mathbf{j}) \times (-81.66\mathbf{i})$$
$$= -7.18\mathbf{k} \text{ ft} \cdot \text{lb} \qquad (\text{cw})$$

and

$$F_{12} = \frac{T}{l} = \frac{7.18}{6/12} = 14.4 \text{ lb}$$

With the direction of torque **T** clockwise, the sense of force $\mathbf{F}_{12}$ applied to the crankshaft by the bearing will be up at the rear bearing and down at the front bearing as shown in Fig. 9.43.

As may be seen from Example 9.12, the gyroscopic forces on the bearings are small compared to those due to slider-crank action. These forces are greater in sharper roadway turns. Gyroscopic forces on bearings supporting clutch, transmission gears, and drive shaft are also small because of the low moment of inertia of the parts. However, the moment of inertia of the front wheel may be great enough to apply appreciable gyroscopic forces to the ball joints and the steering mechanism. The gyroscopic forces of the spinning parts of the engine that are transmitted to the car body have the effect of raising the front end of the car on its suspension as it traverses a curve to the right.

The gyroscopic forces of heavy flywheels of shipboard engines may be of large magnitude. Metal airplane propellers of large diameter cause high gyroscopic couples in some maneuvers as shown in the following example.

Example 9.13. Determine the gyroscopic couple of a 3.05-m-diameter solid aluminum alloy four-bladed propeller in which each blade has a mass of 18.1 kg. The test maneuver of the airplane is a power-on flat spin in which the propeller speed is 1500 rpm and the rotation of the flat spin is 1 rad/s. The radius of gyration k of the propeller with respect to the propeller axis is approximately one-half of the propeller radius.

Solution

$$k = r_m = \tfrac{1}{2}(3.05) = 1.525 \text{ m}$$

$$I = Mk^2 = 18.1 \times 1.525^2$$

$$= 42.1 \text{ kg} \cdot \text{m}^2$$

$$\omega = \frac{2\pi n}{60} = \frac{2\pi(1500)}{60} = 157 \text{ rad/s}$$

$$\omega_p = 1 \text{ rad/s}$$

$$T = I\omega\omega_p = 42.1 \times 157 \times 1$$

$$= 6610 \text{ N} \cdot \text{m}$$

The effect of the couple is to impose a large load on bearings supporting the propeller shaft as well as to impose large bending moments on the individual blades near the propeller hub. The gyroscopic effect is great enough to affect the maneuver by raising or lowering the nose of the airplane depending on the senses of the propeller spin and the precession.

It is a characteristic of the gyroscope that a gyroscopic couple must be applied to cause precession. In many instrument applications such as the gyrocompass and artificial horizon used in aircraft, precession is undesirable and great care is taken to reduce the gyroscopic couple to a minimum as the vehicle in which it is mounted undergoes turns that would cause precession. The resistance of a gyroscope to precession becomes greater as $I\omega$ increases; a high moment of inertia and high spin velocity give it the characteristic of "rigidity" against precessing in space. Rigidity is the desired characteristic in the gyrocompass, which provides a fixed datum required for navigational purposes. Although the gyroscope is mounted in low-friction bearings in such a manner that the vehicle's turning transmits a minimum of gyroscopic couple, some torque is nevertheless applied by friction and the gyroscope must periodically be reset to the desired datum position. The rigidity characteristic in gyroscopes is also utilized in control equipment. In naval gun directors, the gyroscope provides a datum during pitching and rolling of the ship, and an electrical signal may be transmitted to machinery which holds gun positions relative to the gyroscope rather than relative to the ship. Gyroscopes in automatic pilots control the flight position of aircraft by transmitting signals to the control surfaces as wind currents and other disturbances cause the aircraft to yaw, pitch, and roll.

9.22 MOMENT-OF-INERTIA DETERMINATION

In the foregoing discussions of force analyses, the moments of inertia I of the individual links were known or assumed. The designer or analyst of a machine is often confronted with the need for determining moment of inertia. Formulas are available in handbooks and textbooks on mechanics for the determination of the moment of inertia of bodies having simple geometrical forms such as cylinders, disks, and bars and tubes of round and rectangular cross section. Many machine elements such as gears, pulleys, flywheels, gyroscopes, rotors, and shafts are simple enough in form that determination of moments of inertia by formula is quite accurate. Although calculations of I for links of more complex forms such as connecting rods, crankshafts, planet carriers, and odd-shaped cams may be made by considering the complicated forms as composites of simpler forms, the determinations are less accurate. If parts are available, moments of inertia may be determined experimentally in most cases. One of the most useful experimental methods is to mount the part in such a way that it may oscillate as a pendulum and to observe the period of oscillation, which is a function of the moment of inertia of the pendulum.

Figure 9.44 shows a pendulum suspended from the knife edge at O so that O is the axis of rotation about which the pendulum oscillates from θ_1 to $-\theta_1$. The mass center g is at a distance l_o from O. Two forces act on the pendulum: the force of gravity W and the supporting force of the knife edge. The following equation of motion is written using the moment center O:

$$\Sigma \, T_O = I_O \alpha$$

$$-Wl_O \sin \theta = I_O \alpha = I_O \frac{d^2\theta}{dt^2} \qquad \textbf{(9.83)}$$

where I_O is the moment of inertia of the pendulum about the axis through O. The moment T_O depends upon the position θ of the pendulum from the vertical. Since α is in the same sense as increasing values of θ, the minus sign of Eq. 9.83 indicates that T_O is in the opposite sense to α. For small oscillations of the pendulum, $\theta = \sin \theta$ may be assumed with little error. Thus,

$$\frac{d^2\theta}{dt^2} = -\frac{Wl_O}{I_O} \theta \qquad \textbf{(9.84)}$$

Equation 9.84 is a differential equation, which on double integration yields

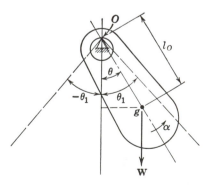

FIGURE 9.44

an equation relating time t and θ:

$$t = \sqrt{I_O/Wl_O}\ \cos^{-1}\left(\frac{\theta}{\theta_1}\right) \tag{9.85}$$

The two constants of integration are evaluated for the conditions $\omega = d\theta/dt = 0$ at $t = 0$, and $\theta = \theta_1$ at $t = 0$. Since the time of the oscillation is measured from the starting position $\theta = \theta_1$, the time to reach the vertical position is determined by substituting $\theta = 0$:

$$t = \frac{\pi}{2}\sqrt{I_O/Wl_O} \tag{9.86}$$

The period τ of the pendulum, or the time for one complete oscillation, is four times the time given by Eq. 9.86:

$$\tau = 2\pi\sqrt{I_O/Wl_O}$$

$$I_O = \frac{\tau^2}{4\pi^2}Wl_O \tag{9.87}$$

where I_O is the moment of inertia about the axis through O. Usually, the moment of inertia I about the axis through the mass center is wanted and may be determined from the parallel axis theorem:

$$I_O = I + Ml_O^2$$

$$I = I_O - Ml_O^2$$

$$= \left(\frac{\tau^2}{4\pi^2}Wl_O\right) - \left(\frac{W}{g}l_O^2\right)$$

$$I = Wl_O\left(\frac{\tau^2}{4\pi^2} - \frac{l_O}{g}\right) \tag{9.88}$$

Thus, I may be determined from Eq. 9.88 by experimentally noting the time for a large number of oscillations of a part suspended as a pendulum. A connecting rod, for example, may be suspended on a knife edge from either the wrist pin bore or the crank pin bore. The quantity in parentheses in Eq. 9.88 approaches zero as l_O becomes large because the two terms are nearly equal in magnitude. Under these conditions, the accuracy of determining I depends on the accuracy of measuring both l_O and τ. Accuracy is greatly increased by making l_O a small measurable value other than zero. Thus, accuracy is better for the case in which the connecting rod is suspended from the end closest to the center of gravity. It should be mentioned, however, that it is often difficult to obtain an

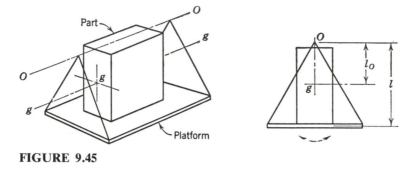

FIGURE 9.45

accurate time of oscillation if the point about which the body is swung is too close to the center of gravity.

The moment of inertia of a part may also be determined experimentally by mounting the part on a pendulum made of a lightweight platform suspended by chords as shown in Fig. 9.45. To determine the moment of inertia of the part about the centroidal axis g–g, the part is oriented such that g–g is directly below and parallel to the suspension axis O–O. The period for small oscillations is observed by counting oscillations for a time of several minutes. The following equation, which determines I of the part, is a modification of Eq. 9.88, in which the second term accounts for the effect of the platform; τ_b represents the period of the platform without the part:

$$I = W l_O \left(\frac{\tau^2}{4\pi^2} - \frac{l_O}{g} \right) + \frac{W_b l}{4\pi^2} (\tau^2 - \tau_b^2) \qquad (9.89)$$

where
$\quad W$ = weight of part
$\quad W_b$ = weight of platform
$\quad l_O$ = distance from O to center of gravity of part
$\quad l$ = distance from O to center of gravity of platform
$\quad \tau$ = period of platform with part
$\quad \tau_b$ = period of platform alone

To allow for accuracy in determining I when using Eq. 9.89, the length of the suspension should be such that l_O is as small as possible but accurately measurable.

A third method for determining I is to orient the part on an equilateral triangular (or round) platform suspended as shown in Fig. 9.46 and to observe the period of the apparatus as a torsional pendulum oscillating about axis g–g. To determine the moment of inertia of the part about the centroidal axis g–g, the part is oriented such that g–g is parallel to the three vertical suspension chords and is equidistant r from the three chords.

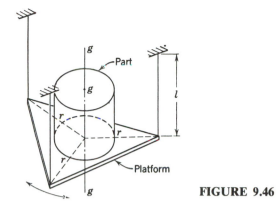

FIGURE 9.46

The moment of inertia I of the part is determined from the following equation in which the second term accounts for the effect of the platform:

$$I = \frac{Wr^2\tau^2}{4\pi^2 l} + \frac{W_b r^2}{4\pi^2 l}(\tau^2 - \tau_b^2) \qquad (9.90)$$

Problems

9.1. The rotor of a jet compressor has blades 100 mm long mounted on a 900-mm-diameter hub. Assuming the configuration of Fig. 9.1 and that $b = 70$ mm, $t = 6$ mm, and $w = 0.0272 \times 10^6$ N/m³, determine the allowable rotative speed of the rotor for an allowable maximum blade stress of 690×10^5 N/m². The blades are aluminum.

9.2. The blades of one of the stages of a jet engine compressor are 4 in. long and are mounted on a rotor hub 36 in. in diameter. Assuming that the blades have the configuration of Fig. 9.1, that $b = 3$ in., and $t = \frac{1}{4}$ in. and that the rotor speed is 8000 rpm, determine the force exerted on the hub by the centrifugal force of an individual blade and the corresponding stress at the base of the blade for (a) a steel blade ($w = 0.285$ lb/in.³) and for (b) an aluminum alloy blade ($w = 0.10$ lb/in.³).

9.3. Consider a blade of the type shown in Fig. 9.1 in which the cross-sectional area of the base is A_b and that of the tip is A_t. Assuming a uniform taper of cross-sectional area with radius, derive an expression for the stress s_b due to centrifugal force at the base of the blade in terms of the taper ratio $k = A_t/A_b$.

9.4. Assuming that a fan is simulated by the configuration of Fig. 9.2, determine the maximum blade stress (N/m²) due to centrifugal force for a 260-mm-diameter fan. The fan has eight blades 1.07 mm thick mounted on a hub 60 mm in diameter rotating at 3600 rpm ($w = 0.0769 \times 10^4$ N/m³).

9.5. Assume that a 15-ft-diameter solid aluminum alloy propeller rotating at 1200 rpm has the uniform blade shape shown in Fig. 9.3. The blade length is 5 ft, $b = 8$ in., and $t = 1$ in., and the blade angle is 20°. Determine the tensile force on the shank by the centrifugal force on the blade and the corresponding twisting moment on the shank ($w = 0.10$ lb/in.³).

9.6. In Fig. 9.47, link 2 rotates about a fixed axis at O_2. For the data given, determine the inertia force vector $\mathbf{F}_o$ and show it in its proper position on a scale drawing of link 2. Show also the resultant force vector $\mathbf{R}$ representing the forces which produce the angular motions shown.

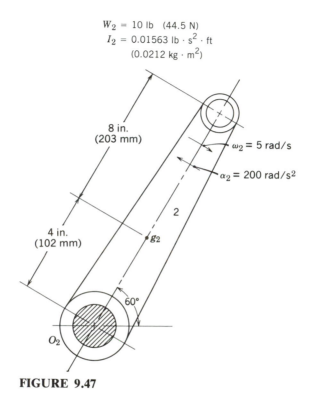

$$W_2 = 10 \text{ lb} \quad (44.5 \text{ N})$$
$$I_2 = 0.01563 \text{ lb} \cdot \text{s}^2 \cdot \text{ft}$$
$$(0.0212 \text{ kg} \cdot \text{m}^2)$$

8 in.
(203 mm)

$\omega_2 = 5 \text{ rad/s}$

$\alpha_2 = 200 \text{ rad/s}^2$

2

4 in.
(102 mm)

g_2

60°

O_2

FIGURE 9.47

9.7. For the mechanism shown in Fig. 9.48, determine the magnitudes, directions, senses, and locations of the inertia forces acting on links 2, 3, and 4. Show the results on a scale drawing of the mechanism. Draw the given acceleration polygon to scale for use in determining unknown accelerations.

9.8. In the four-bar mechanism of Fig. 9.49, the center of gravity of link 3 is coincident with the centroid of the rectangle shown. From the given information, determine the inertia force of link 3, and show it as a vector in its correct relationship to the mechanism in the phase shown.

9.9. The link in Fig. 9.50, rotating about the fixed center O_2, is in motion in such a way that the center of gravity is accelerating in the direction shown and its normal component $A_{g_2}^n = 2000 \text{ ft/s}^2$. Using vector polygons, determine the force $\mathbf{F}_A$ and the reactive force at O_2 that are producing the motion of the link.

9.10. For the mechanism of Fig. 9.51, determine the forces $\mathbf{F}_{14}$ and $\mathbf{F}_{12}$ due to the action of the inertia force $\mathbf{F}_{o_4}$. Also determine the shaft torque T_s applied to link 2 at O_2. Draw the mechanism to scale and show the answers in their proper locations.

9.11. Given the mechanism shown in Fig. 9.52 and its acceleration polygon, calculate $\mathbf{F}_{o_3}$, and show it on the configuration diagram in its correct location.

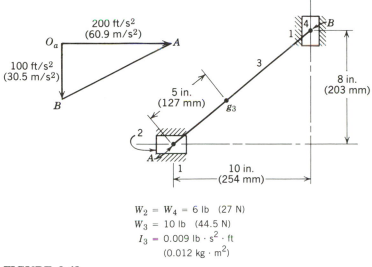

$$W_2 = W_4 = 6 \text{ lb} \quad (27 \text{ N})$$
$$W_3 = 10 \text{ lb} \quad (44.5 \text{ N})$$
$$I_3 = 0.009 \text{ lb} \cdot s^2 \cdot \text{ft}$$
$$(0.012 \text{ kg} \cdot m^2)$$

FIGURE 9.48

9.12. For the linkage shown in Fig. 9.53 with its acceleration polygon, calculate $\mathbf{F}_{o_3}$ and show it on the configuration diagram in its correct location.

9.13. In the crank shaper mechanism of Fig. 9.54, the tool holder 6 is acted upon by a static force of 100 lb as the tool cuts the work. Using force polygons, determine the forces acting on the bearings at A, B, C, O_2, and O_4 due to the tool force. Also determine the shaft torque T_s applied to link 2 at O_2. Draw the free-body diagram of each link (except link 1), and show the forces acting to scale. Also show T_s on link 2.

9.14. A 335-N cutting force acts on the Whitworth mechanism as shown in Fig. 9.55. From a static force analysis of the mechanism, determine the forces acting on the bearings due to $\mathbf{F}_6$ and the shaft torque T_s applied to link 2 at O_2.

9.15. For the mechanism shown in Fig. 9.56, determine the forces acting on the bearings due to $\mathbf{F}_6$ and the torque T_s on the shaft at O_2.

$O_2A = 3$ in. (76.2 mm)	$W_3 = 16.1$ lb (71.6 N)	
$AB = 12$ in. (305 mm)	$I_3 = 0.50$ lb $\cdot$ s^2 $\cdot$ ft	
$O_4B = 7$ in. (178 mm)	(0.678 kg $\cdot$ m^2)	
$AD = 3$ in. (76.2 mm)		

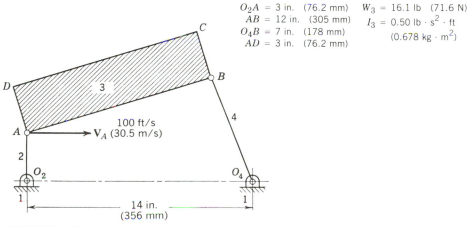

FIGURE 9.49

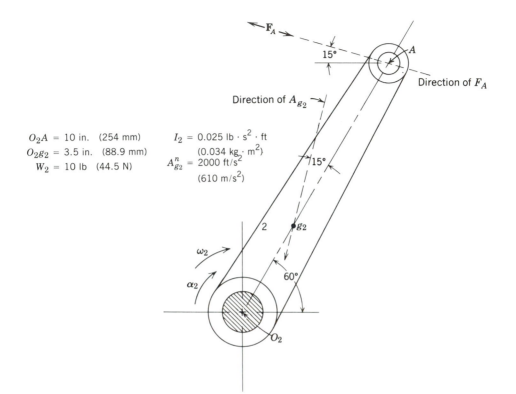

$O_2A = 10$ in. (254 mm) $I_2 = 0.025$ lb · s^2 · ft
$O_2g_2 = 3.5$ in. (88.9 mm)   (0.034 kg · m^2)
$W_2 = 10$ lb (44.5 N) $A_{g_2}^n = 2000$ ft/s^2
                                (610 m/s^2)

FIGURE 9.50

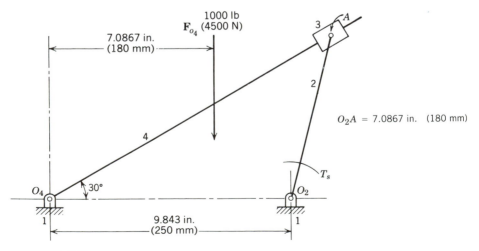

1000 lb
$\mathbf{F}_{o_4}$ (4500 N)

7.0867 in.
(180 mm)

$O_2A = 7.0867$ in. (180 mm)

T_s

30°

9.843 in.
(250 mm)

FIGURE 9.51

482

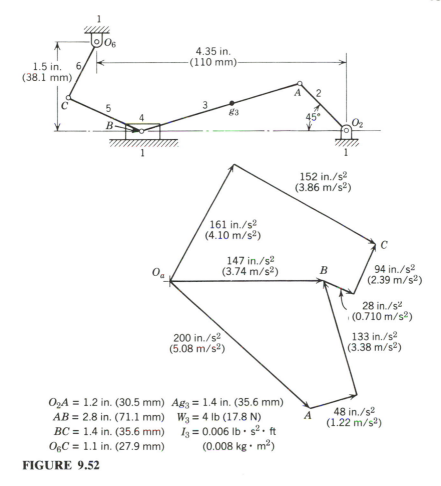

O₂A = 1.2 in. (30.5 mm)   Ag₃ = 1.4 in. (35.6 mm)
AB = 2.8 in. (71.1 mm)   W₃ = 4 lb (17.8 N)
BC = 1.4 in. (35.6 mm)   I₃ = 0.006 lb · s² · ft
O₆C = 1.1 in. (27.9 mm)   (0.008 kg · m²)

FIGURE 9.52

9.16. Refer to Fig. 9.57. Using superposition, construct force polygons and determine the forces on the bearings at A, B, C, and O_2 to maintain static equilibrium. Determine also the shaft torque T_s at O_2 of the driving link.

9.17. The flyball governor of Fig. 9.58 rotates about the Y–Y-axis at a constant angular velocity. The spring exerts a force of 100 lb to balance the inertial force on the balls. Determine the rotation speed (rpm) of the governor in the position shown. Each ball weighs 3.22 lb.

9.18. The tensioning mechanism of Fig. 9.59 is shown in both the open and closed positions. **P** is the force applied to the handle, and **Q** represents the tension in the cable. By eye, sketch the force polygons for both the open and the closed positions, and show that the ratio of Q/P becomes infinite when the points A, B, and C are on a straight line.

9.19. Determine force **Q** which must be applied to link 6 in Fig. 9.60 to maintain static equilibrium of the system under the action of force **P**.

9.20. For the mechanism of Fig. 9.61, determine the force **Q** necessary to maintain static equilibrium of the system under the action of the force $P = 1000$ lb.

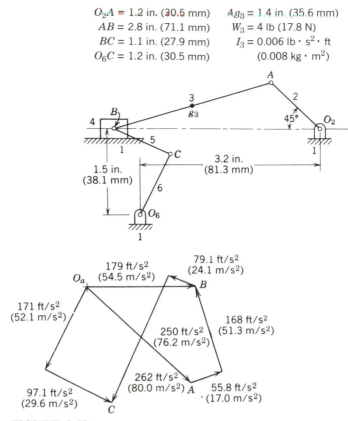

$O_2A = 1.2$ in. (30.5 mm) $Ag_3 = 1.4$ in. (35.6 mm)
$AB = 2.8$ in. (71.1 mm) $W_3 = 4$ lb (17.8 N)
$BC = 1.1$ in. (27.9 mm) $I_3 = 0.006$ lb $\cdot$ s$^2 \cdot$ ft
$O_6C = 1.2$ in. (30.5 mm) (0.008 kg $\cdot$ m^2)

3.2 in.
(81.3 mm)

1.5 in.
(38.1 mm)

79.1 ft/s^2
(24.1 m/s^2)

179 ft/s^2
(54.5 m/s^2)

171 ft/s^2
(52.1 m/s^2)

168 ft/s^2
(51.3 m/s^2)

250 ft/s^2
(76.2 m/s^2)

262 ft/s^2
(80.0 m/s^2) A

55.8 ft/s^2
(17.0 m/s^2)

97.1 ft/s^2
(29.6 m/s^2)

FIGURE 9.53

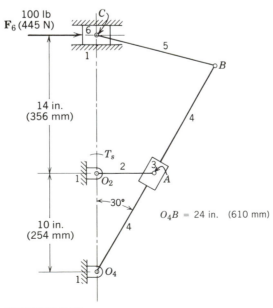

100 lb
F_6 (445 N)

14 in.
(356 mm)

T_s

$O_4B = 24$ in.   (610 mm)

10 in.
(254 mm)

30°

FIGURE 9.54

484

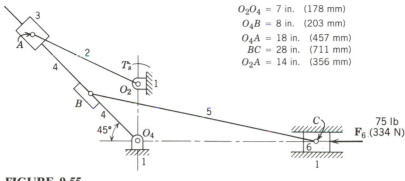

$$O_2O_4 = 7 \text{ in. } (178 \text{ mm})$$
$$O_4B = 8 \text{ in. } (203 \text{ mm})$$
$$O_4A = 18 \text{ in. } (457 \text{ mm})$$
$$BC = 28 \text{ in. } (711 \text{ mm})$$
$$O_2A = 14 \text{ in. } (356 \text{ mm})$$

FIGURE 9.55

9.21. Refer to Fig. 9.62. Given the resistance **P** = 5338 N, determine the force which must be applied at M in the direction shown to maintain static equilibrium.

9.22. The four-bar mechanism of Fig. 9.63 is driven at O_2 at a constant angular velocity of 500 rad/s. From the data given, make a complete dynamic analysis including a kinematic analysis, inertia force determinations, and a force analysis.

9.23. From the slider-crank data given in Fig. 9.64, make a complete dynamic analysis including a kinematic analysis, inertia force determinations, and a force analysis.

9.24. The Scotch yoke mechanism is often utilized in actuating small vibration tables as shown in Fig. 9.65. The motion is simple harmonic. Determine the maximum force on the bearing when the crank of length $e = \frac{1}{8}$ in. rotates at 6000 rpm giving a vibration frequency of 6000 cycles/min. Include the inertia effects of all parts inducing force on the bearing.

9.25. From the data given for the donkey engine of Fig. 9.66, make a force analysis and determine the forces on the bearings at O_2, A, and B.

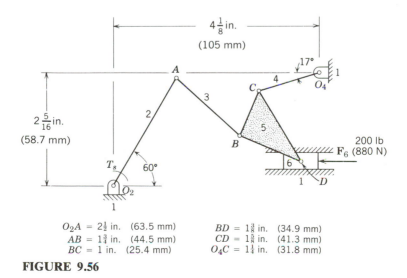

$$O_2A = 2\frac{1}{2} \text{ in. } (63.5 \text{ mm})$$
$$AB = 1\frac{3}{4} \text{ in. } (44.5 \text{ mm})$$
$$BC = 1 \text{ in. } (25.4 \text{ mm})$$

$$BD = 1\frac{3}{8} \text{ in. } (34.9 \text{ mm})$$
$$CD = 1\frac{5}{8} \text{ in. } (41.3 \text{ mm})$$
$$O_4C = 1\frac{1}{4} \text{ in. } (31.8 \text{ mm})$$

FIGURE 9.56

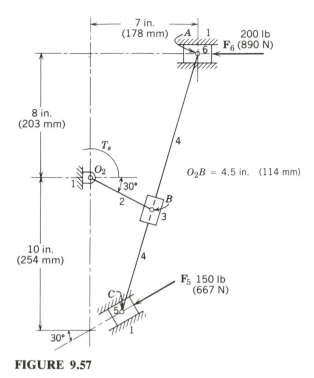

FIGURE 9.57

9.26. In Fig. 9.67a, the uniform steel bar is in motion and the acceleration of points A and B are as shown. Determine the following, giving magnitude, direction, and sense: (a) transverse and longitudinal distribution of acceleration; (b) transverse and longitudinal distribution of inertia force (for steel $w = 0.0769 \times 10^4$ N/m³); (c) transverse and longitudinal resultant inertia forces; show line of action; (d) resultant inertia force; show line of action. Determine the same quantities for the nonuniform steel bar of Fig. 9.67b.

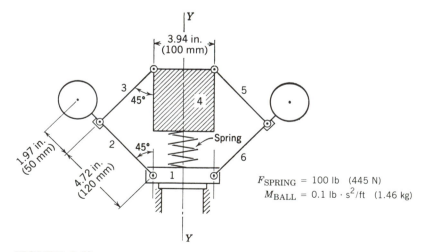

$F_{\text{SPRING}} = 100$ lb   (445 N)

$M_{\text{BALL}} = 0.1$ lb · s²/ft   (1.46 kg)

FIGURE 9.58

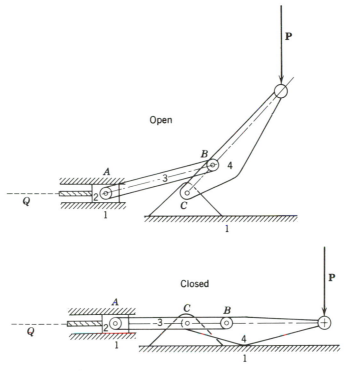

FIGURE 9.59

9.27. Determine the shaft torque T_s that must be applied to link 2 on Fig. 9.63 to hold the linkage in equilibrium. Use the method of virtual work.

9.28. Using the method of virtual work, calculate the crankshaft torque T_s necessary to hold the linkage in equilibrium for the slider-crank mechanism of Fig. 9.68. The gas force may be assumed zero.

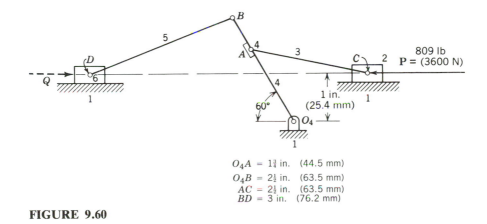

$$O_4A = 1\tfrac{3}{4} \text{ in.} \quad (44.5 \text{ mm})$$
$$O_4B = 2\tfrac{1}{2} \text{ in.} \quad (63.5 \text{ mm})$$
$$AC = 2\tfrac{1}{2} \text{ in.} \quad (63.5 \text{ mm})$$
$$BD = 3 \text{ in.} \quad (76.2 \text{ mm})$$

FIGURE 9.60

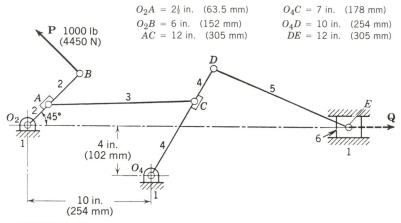

FIGURE 9.61

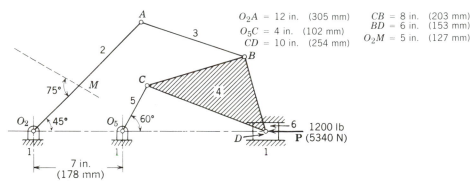

FIGURE 9.62

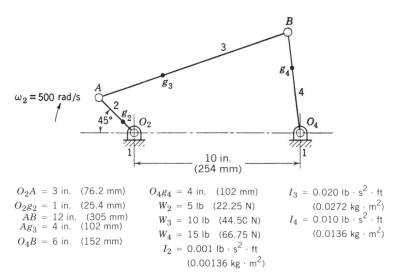

FIGURE 9.63

488

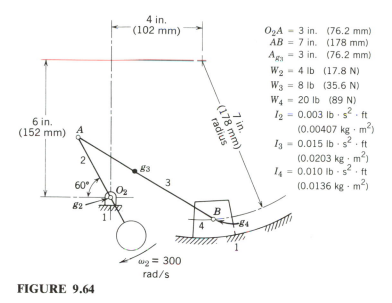

FIGURE 9.64

9.29. For the mechanism of Fig. 9.64, determine the crankshaft torque T_s necessary to hold the linkage in equilibrium. Use the method of virtual work.

9.30. Calculate the shaft torque T_s which must be applied to link 2 of Fig. 9.66 to hold the linkage in equilibrium. Use the method of virtual work.

9.31. Use the method of virtual work to determine the crankshaft torque T_s necessary to hold the linkage in equilibrium for the two-cylinder engine shown in Fig. 9.69.

9.32. For the slider-crank mechanism shown in Fig. 9.21, a spring provides the driving force on piston 4 with a spring rate of 100 lb/in. of deflection. Assume that the spring is compressed 1.75 in. from its free length when piston 4 is at its starting position at phase I and that ω_2 is zero at this position. Determine ω_2 at position III if T_2 is constant at 50 lb · in. (ccw) and $\theta_2^{III} = 60°$.

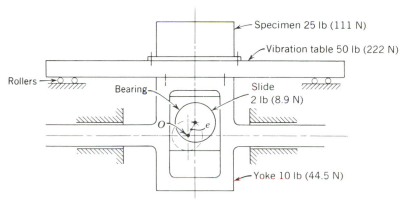

FIGURE 9.65

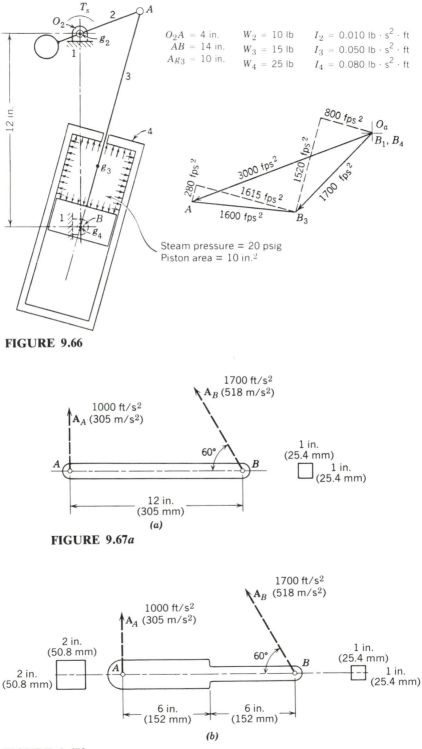

O_2A = 4 in. W_2 = 10 lb I_2 = 0.010 lb · s² · ft
AB = 14 in. W_3 = 15 lb I_3 = 0.050 lb · s² · ft
Ag_3 = 10 in. W_4 = 25 lb I_4 = 0.080 lb · s² · ft

Steam pressure = 20 psig
Piston area = 10 in.²

FIGURE 9.66

1700 ft/s²
$\mathbf{A}_B$ (518 m/s²)

1000 ft/s²
$\mathbf{A}_A$ (305 m/s²)

60°

A   B

1 in.
(25.4 mm)

1 in.
(25.4 mm)

12 in.
(305 mm)

(a)

FIGURE 9.67a

1700 ft/s²
$\mathbf{A}_B$ (518 m/s²)

1000 ft/s²
$\mathbf{A}_A$ (305 m/s²)

2 in.
(50.8 mm)

2 in.
(50.8 mm)

60°

A   B

1 in.
(25.4 mm)

1 in.
(25.4 mm)

6 in.
(152 mm)

6 in.
(152 mm)

(b)

FIGURE 9.67b

490

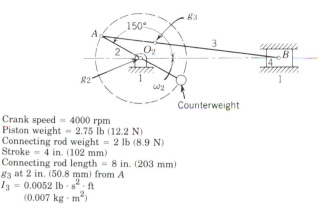

Crank speed = 4000 rpm
Piston weight = 2.75 lb (12.2 N)
Connecting rod weight = 2 lb (8.9 N)
Stroke = 4 in. (102 mm)
Connecting rod length = 8 in. (203 mm)
g_3 at 2 in. (50.8 mm) from A
$I_3 = 0.0052$ lb · s^2 · ft
    (0.007 kg · m^2)

FIGURE 9.68

9.33. For the link shown in Fig. 9.70 M_Q, l_Q, and l_P are known. (*a*) Determine an equation for M_P in terms of the above values. (*b*) Calculate the moment of inertia of the link if $l_Q = 76$ mm, $l_P = 51$ mm, and $M_Q = 0.073$ kg.

9.34. For the link shown in Fig. 9.71, the two point masses at A and B are intended to be kinetically equivalent. Determine whether they are kinetically equivalent.

9.35. A link is shown in Fig. 9.72 that has its mass divided between points A and B in the manner shown. Determine whether the two masses are kinetically equivalent.

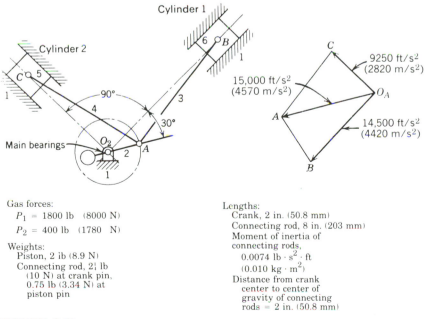

Gas forces:
    P_1 = 1800 lb (8000 N)
    P_2 = 400 lb (1780 N)

Weights:
    Piston, 2 lb (8.9 N)
    Connecting rod, 2¼ lb
        (10 N) at crank pin,
        0.75 lb (3.34 N) at
        piston pin

Lengths:
    Crank, 2 in. (50.8 mm)
    Connecting rod, 8 in. (203 mm)
    Moment of inertia of
    connecting rods,
        0.0074 lb · s^2 · ft
        (0.010 kg · m^2)
    Distance from crank
    center to center of
    gravity of connecting
    rods = 2 in. (50.8 mm)

FIGURE 9.69

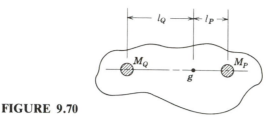

FIGURE 9.70

9.36. For the link shown in Fig. 9.73, determine the location of W_2 and the magnitudes of W_1 and W_2 so that they are kinetically equivalent. Determine also the magnitudes, directions, and senses of the inertia forces of the point masses and show them on a drawing of the link to scale.

9.37. The single-cylinder, four-stroke engine of Fig. 9.68 is shown in the intake phase in which the gas force may be assumed zero. From the data given, determine the following: (a) the velocity and acceleration polygons; (b) the true kinetically equivalent masses of the connecting rod locating one of them at point B; (c) the approximate kinetically equivalent masses locating one at B and the other at A; (d) the inertia forces $\mathbf{F}_{o_4}$, $\mathbf{F}_{B_3}$, $\mathbf{F}_{A_3}$, and show them on the diagram to scale using the masses of section (c); (e) using the free-body diagram of the complete mechanism (excluding link 1), determine $\mathbf{F}_{14}$ and $\mathbf{F}_{12}$ due to the inertia forces. Determine the crankshaft torque T_s. Assume that the counterweight balances the crank and the equivalent masses at A.

9.38. For the two-cylinder, 90° V engine shown in Fig. 9.69, determine the resultant force on the main bearings due to the gas forces and inertia forces. The center of gravity of the crankshaft is at O_2. However, there is no counterweight to counterbalance the connecting

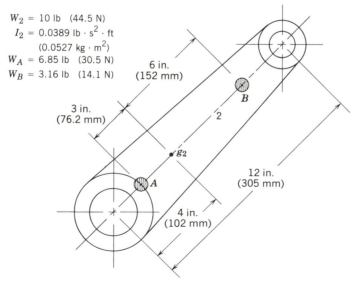

FIGURE 9.71

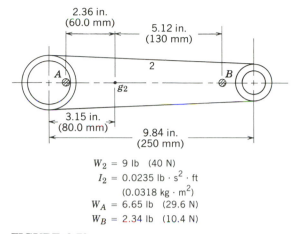

$W_2 = 9$ lb (40 N)
$I_2 = 0.0235$ lb · s^2 · ft
(0.0318 kg · m^2)
$W_A = 6.65$ lb (29.6 N)
$W_B = 2.34$ lb (10.4 N)

FIGURE 9.72

rod equivalent masses at the crank pin. Determine the crankshaft torque T_s. Show your answers on the layout of the mechanism.

9.39. In Fig. 9.74, two free-body diagrams are shown of a single-cylinder engine in which the rotating masses are counterbalanced. Figure 9.74a is a free-body diagram of the moving parts of the slider crank, and Fig. 9.74b is a free-body diagram of the engine block together with the slider crank. Show vectors of the forces and torques acting on the free bodies when the engine operates at constant crank speed and is under gas pressure during the power stroke. Explain each vector.

9.40. Sketch a free-body diagram of the engine block of the two-cylinder 90° V engine of Fig. 9.69 and show vectors of the forces acting. Explain each vector.

9.41. Using Table 9.2, determine the instantaneous output torque of the single-cylinder engine shown in Fig. 9.68. Zero gas pressure may be assumed since the phase shown is in the intake stroke. Inertia forces act.

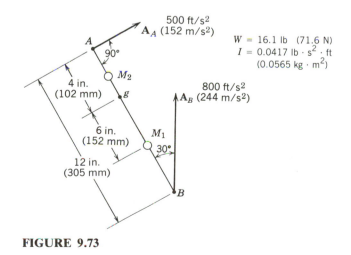

FIGURE 9.73

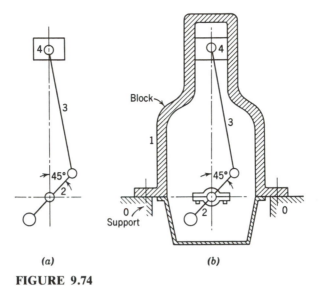

(a) (b)

FIGURE 9.74

9.42. By the analytical method and Table 9.2, determine the instantaneous output torque of the two-cylinder 90° V engine of Fig. 9.69 for a crank speed of 4000 rpm. The acceleration polygon of Fig. 9.69 does not apply.

9.43. In Fig. 9.75 is shown a two-cylinder four-cycle engine with cranks 90° apart. Using Table 9.2, determine and plot the output torque for one cylinder for each 15° crank angle

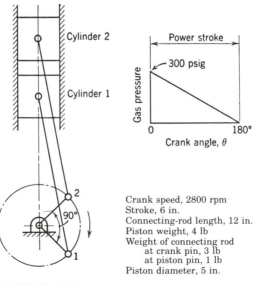

Crank speed, 2800 rpm
Stroke, 6 in.
Connecting-rod length, 12 in.
Piston weight, 4 lb
Weight of connecting rod
    at crank pin, 3 lb
    at piston pin, 1 lb
Piston diameter, 5 in.

FIGURE 9.75

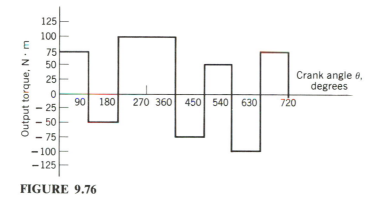

FIGURE 9.76

of the 720° cycle. Plot the output torque of the engine versus crank angle by superposing the torque curves of the two cylinders with a phase angle of 90°. Assume that the gas pressure during the power stroke varies as shown in Fig. 9.75 and that the gas pressure is zero for the other strokes. Assume also that all masses rotating with the crank are counterbalanced.

9.44. The torque output diagram of a single-cylinder engine is shown in Fig. 9.76. Determine: (*a*) the average output torque and the kilowatt output for the engine which operates at 3500 rpm; (*b*) the locations of the crank angles at which the crankshaft speed is a maximum and a minimum during the engine cycle; and (*c*) calculate the work done to change the speed from the minimum to the maximum.

9.45. Assume that the torque output diagram of Fig. 9.77 is for the first cylinder of a two-cylinder in-line engine with cranks at 180°. On this diagram, superimpose the same diagram for the second cylinder. Determine the locations of maximum and minimum crankshaft speeds in terms of the crank angle of the first cylinder.

9.46. If each square inch of a torque diagram represents 375 ft · lb of work, the area between the points of ω_M and ω_m is 1.20 in.², the engine speed is 3500 rpm, and the maximum allowable fluctuation of speed in the engine cycle is 35 rpm, determine the

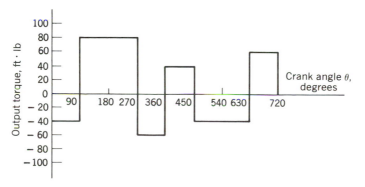

FIGURE 9.77

following: (*a*) the moment of inertia of a steel disk-type flywheel; (*b*) the weight and thickness of the flywheel if the diameter is 15 in. ($w = 490$ lb/ft³).

9.47. A single-cylinder engine has a 230-mm mean diameter rim-type flywheel weighing 200 N. The engine operates at 3000 rpm and has an allowable fluctuation of speed in the engine cycle of 30 rpm. Determine the total work output in N · m.

9.48. The torque output diagram of Fig. 9.76 is for a single-cylinder engine at 3000 rpm. Determine the weight of a steel disk-type flywheel required to limit the crank speed to 10 rpm above and 10 rpm below the average speed of 3000 rpm. The outside diameter of the flywheel is 250 mm. Determine also the weight of the rim of a rim-type steel flywheel of 250 mm mean diameter for the same allowable fluctuation in speed.

9.49. In the punch press shown in Fig. 9.78*a*, a slider-crank mechanism with flywheel is used to punch holes in steel plates. A hole is punched for each revolution of the flywheel, which operates at an average speed of 300 rpm. The load *P* on the punch during punching is the force necessary to shear the plate, and the variation of punching force with shearing deformation of the plate is shown in Fig. 9.78*b*. For the maximum size of plate and hole to be punched, it is estimated that 3200 ft · lb of work is required to punch the hole and that the punching is accomplished in one-sixth of a revolution of the flywheel. Figure 9.78*c* shows the torque diagram for one cycle of punching. Determine the following: (*a*) the average crank torque for one cycle; (*b*) the horsepower required of the motor; (*c*) the required moment of inertia of the flywheel for a minimum speed of 280 rpm just after punching.

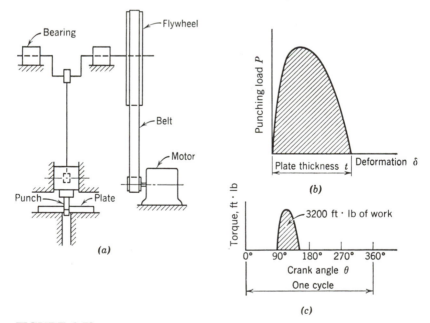

FIGURE 9.78

9.50. Gears are normally used to drive crank shaper machinery as shown in Fig. 9.79. Gears 2 and 3 are in mesh on standard pitch circles and have 20° stub teeth. Determine the tooth force on gears 2 and 3 and the shaft torque at O_2 to maintain static equilibrium of the mechanism acted upon by the known cutting tool force of 500 lb.

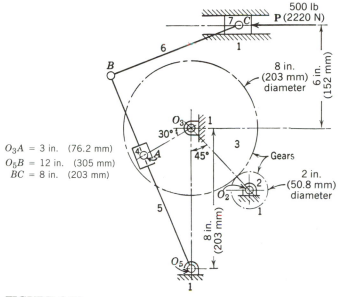

FIGURE 9.79

9.51. For the mechanism of Fig. 9.80, determine the force at the pitch point of the meshed gears due to the inertia force of link 4. The pressure angle of the gear teeth is 20°. The mechanism is driven by a shaft at O_2.

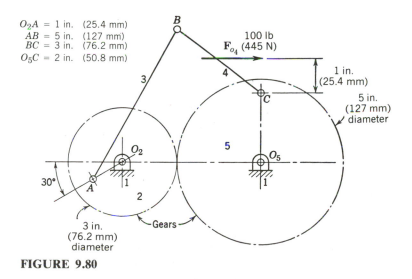

FIGURE 9.80

9.52. The gear and rack of Fig. 9.81 are in static equilibrium under the static force **P**. Note that since there is no bearing support of the gear at O, the force **P** causes a tooth of the gear to bear against two teeth on the rack. The resultant of the two forces on the

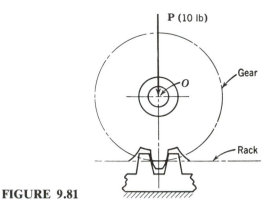

FIGURE 9.81

gear tooth must be equal, opposite, and collinear with **P** for static equilibrium of the gear. If the gears are of $14\frac{1}{2}°$ tooth form, determine the force of contact on one of the rack teeth.

9.53. Make a force analysis of the mechanism of Fig. 9.82, which is in static equilibrium under the action of the force **P**. The shaft at O_2 is capable of producing a resisting torque. Is it necessary to know the pressure angle of the gear teeth to determine T_s? Why? Sketch an enlarged view of the contact of the teeth, and show force vectors.

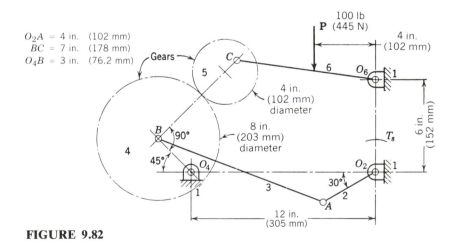

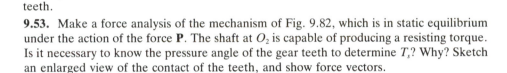

$O_2A = 4$ in. (102 mm)
$BC = 7$ in. (178 mm)
$O_4B = 3$ in. (76.2 mm)

FIGURE 9.82

9.54. A disk cam rotating at 200 rpm lifts a radial roller follower with simple harmonic motion through a maximum displacement of 2.5 in. while the cam rotates through 90°. There is then a dwell of the follower for 180° of cam rotation, followed by a return of the follower in simple harmonic motion in the remaining 90° of cam rotation. For a follower weight of 8 lb, determine the inertia force of the follower for each 15° of cam rotation and plot the results. Determine and plot the resultant of the weight force and inertia force. Determine the required spring constant of a spring which will maintain contact of the follower with the cam throughout the cam rotation.

9.55. A spring-loaded radial roller follower of a disk cam is lifted through a distance of 50 mm with simple harmonic motion. The weight of the follower including the weight of the spring is 45 N. Because of its very low spring constant, the spring is assumed to apply a constant force of 25 N. Determine the maximum speed of the cam so that the follower does not lose contact with the cam.

9.56. In a Geneva mechanism, the pin of the driving wheel exerts a contacting force against the slot of the driven wheel. Determine this force for the following data: driver speed, 400 rpm (constant); center distance of the wheels, 4 in.; pin location, on 3-in. radius; number of slots, 4; phase, pin radius at 30° from line of wheel centers; weight of driven wheel, 1 lb; radius of gyration of driven wheel, 1 in. Determine the pin force due to inertia of the driven wheel, and determine the torque on the driving shaft.

9.57. The gyroscope as shown in Fig. 9.83 is often used to demonstrate gyroscopic precession due to the action of the gravitational force. Determine the angular velocity of precession of a gyroscope of 16.1 lb weight spinning at 6000 rpm. The radius r is 4 in. and a is 17.9 in.

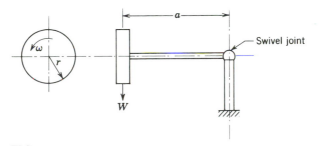

FIGURE 9.83

9.58. The rotor of a jet airplane engine is supported by two bearings as shown in Fig. 9.84. The rotor assembly including compressor, turbine, and shaft is 6672 N in weight and has a radius of gyration of 229 mm. Determine the maximum bearing force as the airplane undergoes a pullout on a 1830-m-radius curve at a constant airplane speed of 966 km/h and an engine rotor speed of 10,000 rpm. Include the effect of centrifugal force due to the pullout as well as the gyroscopic effect.

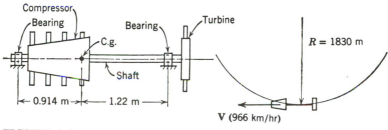

FIGURE 9.84

9.59. In the bevel gear planetary gear train of Fig. 9.85, the carrier (link 4) rotates at 1200 rpm. Determine the force on the bearings of planet (link 3) produced by gyroscopic action; I about the spin axis of the planet is 0.060 lb · s² · in.

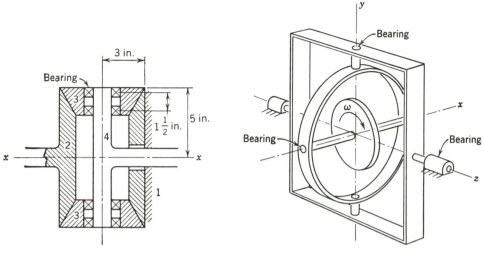

FIGURE 9.85   **FIGURE 9.86**

9.60. In Fig. 9.86 is shown the gimbal mounting of a gyroscope used in instrument applications to maintain a fixed axis in space. Low-friction bearings are used to minimize the precession of the gyro. Through which bearings must friction torque be applied to cause precession of the x-axis in the xz-plane? If the gyro spins at 10,000 rpm and I of the gyro is 0.001 lb · s² · ft, how much friction torque must be applied continuously to cause a precession at 1°/h?

9.61. To determine the moment of inertia I about an axis through the mass center of the connecting rod of Fig. 9.87a, the rod is suspended as a pendulum and the period of small

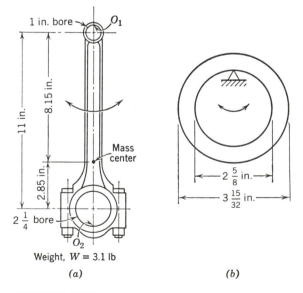

FIGURE 9.87

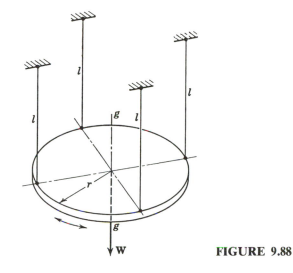

FIGURE 9.88

oscillations is observed. (*a*) When suspended on a knife edge at O_1, 59 oscillations are counted in 60 s. Determine I. Calculate the percent difference in I if 58 oscillations are counted in 60 s. (*b*) When suspended at O_2, 66 oscillations are counted in 60 s. Determine I, and calculate the percent difference in I if only 65 oscillations are counted in 60 s. Which suspension gives the greater accuracy in determining I? Why?

9.62. In an experiment, the ring of Fig. 9.87*b* makes 107 oscillations in 1 min when supported as shown. The ring weighs 1.203 lb. Determine the moment of inertia I of the ring about an axis through its center of gravity (*a*) theoretically and (*b*) from the experimental data. Compute the percent error in I based on the theoretical value.

9.63. In Fig. 9.88 is shown a solid, thin disk suspended as a torsional pendulum by four weightless strings. Derive an expression for the determination of I of the disk about the g–g-axis in terms of the disk weight W, l, r, and the torsional period τ of small oscillations.

9.64. The slider crank of Fig. 9.89 is operated at a uniform crank speed of 200 rad/s.

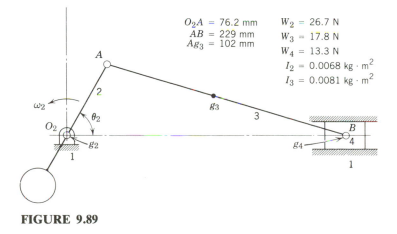

$O_2A = 76.2$ mm $W_2 = 26.7$ N
$AB = 229$ mm $W_3 = 17.8$ N
$Ag_3 = 102$ mm $W_4 = 13.3$ N
$I_2 = 0.0068$ kg $\cdot$ m^2
$I_3 = 0.0081$ kg $\cdot$ m^2

FIGURE 9.89

The lengths, the centers of gravity, the weights, and the moments of inertia I of the links are given. Using the analytical method of complex numbers, determine the numerical values of the following for the phase $\theta_2 = 60°$: (a) magnitude and angle of the inertia forces $\mathbf{F}_{o_3}$ and $\mathbf{F}_{o_4}$; (b) magnitude and angle of the bearing force $\mathbf{F}'_{34}$ due to inertia force $\mathbf{F}_{o_4}$; (c) magnitude and angle of the bearing force $\mathbf{F}''_{34}$ due to inertia force $\mathbf{F}_{o_3}$; (d) magnitude and angle of the resultant force $\mathbf{F}_{34}$. For kinematic equations, refer to section 8.26.

Chapter Ten

Balance of Machinery

10.1 INTRODUCTION

As discussed in Chapter 9, the inertia forces of the slider-crank mechanism in an engine cause shaking of the engine block. Shaking forces in machines due to inertia forces may be minimized by balancing inertia forces in opposition to each other in such a way that little or no force is transmitted to the machine supports.

In Fig. 10.1, for example, the rotating mass M without counterbalance induces a shaking force equal to the inertia force **F** that is transmitted to the bearings and the supports. Because of the rotation, the shaking force has the characteristics of forced vibration at a circular frequency ω. The degree to which

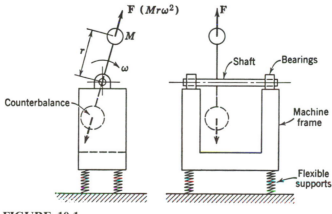

FIGURE 10.1

the forced vibration is undesirable depends upon the frequency of the forced vibration and the natural frequency of the flexible members through which force is transmitted such as the shaft or the supports. If conditions are near resonance, amplitudes of vibration may become large enough to cause discomfort as in an automobile or may cause failure of the shaft, the bearings, or the supports. As shown in Fig. 10.1, the shaking force may be minimized by counterbalancing in such a manner that the resultant of the inertia forces of the mass M and the counterbalance is zero.

In the subsequent discussions, methods are shown for determining the requirements of balance in (a) systems of masses rotating about a common axis and (b) systems of reciprocating masses. In piston engines, both systems are present, the crankshaft being a system of rotating masses and the pistons being one of reciprocation. In addition, a method is shown for balancing a four-bar linkage.

10.2   BALANCE OF ROTORS

Figure 10.2 shows a rigid rotor consisting of a system of three masses rotating in a common *transverse* plane about the axis O–O. A fourth mass is to be added to the system so that the sum of the inertia forces (shaking force) is zero and

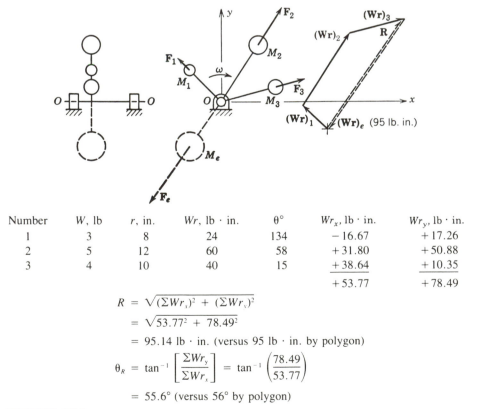

Number	W, lb	r, in.	Wr, lb · in.	$\theta°$	Wr_x, lb · in.	Wr_y, lb · in.
1	3	8	24	134	− 16.67	+ 17.26
2	5	12	60	58	+ 31.80	+ 50.88
3	4	10	40	15	+ 38.64	+ 10.35
					+ 53.77	+ 78.49

$$R = \sqrt{(\Sigma Wr_x)^2 + (\Sigma Wr_y)^2}$$

$$= \sqrt{53.77^2 + 78.49^2}$$

$$= 95.14 \text{ lb · in. (versus 95 lb · in. by polygon)}$$

$$\theta_R = \tan^{-1}\left[\frac{\Sigma Wr_y}{\Sigma Wr_x}\right] = \tan^{-1}\left(\frac{78.49}{53.77}\right)$$

$$= 55.6° \text{ (versus 56° by polygon)}$$

FIGURE 10.2

balance is achieved. For constant ω, the inertia force for any given mass M is $F = Mr\omega^2$ with direction and sense radially outward. For balance, the vector sum, taken either graphically or analytically, of the inertia forces of the system is zero:

$$\Sigma \, \mathbf{F} = \Sigma \, (\mathbf{M}r\omega^2) = \Sigma \left(\frac{\mathbf{W}}{g} \, \mathbf{r}\omega^2 \right) = \frac{\omega^2}{g} \, \Sigma \, (\mathbf{Wr}) = 0$$

$$\Sigma \, (\mathbf{Wr}) = 0 \qquad\qquad\qquad (10.1)$$

Since for all masses, ω^2/g is constant, balance is achieved if Eq. 10.1 is satisfied; $\mathbf{Wr}$ for each mass is a vector in the same direction and sense as the inertia force. In Fig. 10.2, values of Wr for the three known masses are tabulated, and the Wr value of the fourth mass is to be determined to satisfy Eq. 10.1 for balance. As shown in the vector polygon, the resultant $\mathbf{R}$ represents the unbalance of the three masses. The unbalance can also be determined analytically by summing x- and y-components around the vector loop. Both methods give $(Wr)_e =$ 95 lb · in. In Fig. 10.2, the balancing weight W_e is a 10-lb weight at $r_e = 9.5$ in., although any arbitrary value of W_e or r_e may be selected. Without the balancing mass, the resultant force in the rotating system is $R\omega^2/g$, which causes a bending of the shaft and exerts forces on the bearings supporting the shaft; in Fig. 10.2, the left bearing would carry a greater part of the unbalanced load. With the balancing mass added, the shaft bending and bearing loads are reduced to a minimum. Any number of masses rotating in a common radial plane may be balanced with a single mass.

For the case in which the rotating masses of a rigid rotor lie in a common *axial* plane as in Fig. 10.3, the inertia forces are parallel vectors. Balance of inertia forces is achieved in this case as in the previous case by satisfying Eq. 10.1. However, balance of the *moments* of inertia forces is also required. In the system of Fig. 10.2, moment equilibrium is inherent since the inertia force vectors are concurrent. In Fig. 10.3, however, the inertia forces are not concurrent when viewed in the axial plane. Thus, for balance of moments, the moments of the inertia forces about an arbitrarily chosen axis normal to the axial plane must be zero:

$$\Sigma \, (Fa) = \Sigma \left(\frac{W}{g} \, r\omega^2 a \right) = \frac{\omega^2}{g} \, \Sigma \, (Wra) = 0$$

$$\Sigma \, (Wra) = 0 \qquad\qquad\qquad (10.2)$$

in which a is the moment arm of any given inertia force.

The magnitude of the resultant force $\mathbf{R}$ of the three unbalanced masses in Fig. 10.3 is the algebraic sum as well as the vector sum of $\mathbf{Wr}$ terms of the three masses since the inertia force vectors are parallel. As shown in the table of Fig. 10.3, upward Wr values are taken as positive. The line of action of $\mathbf{R}$ is determined using the principle of moments in which moments are taken about the moment center O. The distance a_R from the moment center O locates the line of action

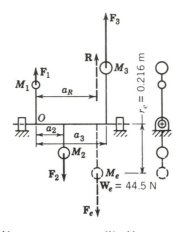

Number	W, N	r, m	Wr, N · m	a, m	Wra, N · m²
1	22.3	0.178	+ 3.9694	0	0
2	44.5	0.127	− 5.6515	0.127	− 0.7177
3	44.5	0.254	+ 11.303	0.305	+ 3.4474
			+ 9.6209		+ 2.7297

$$R = \Sigma(Wr) = 9.6209 \text{ N} \cdot \text{m}$$
$$W_c r_c = -R = -9.6209 \text{ N} \cdot \text{m}$$
$$a_R = \frac{\Sigma(Wra)}{\Sigma(Wr)} = \frac{2.7297}{9.6209} = 0.2837 \text{ m} = 284 \text{ mm}$$

FIGURE 10.3

of **R**. As shown in the table of Fig. 10.3, counterclockwise values of Wra are positive. To satisfy Eqs. 10.1 and 10.2 for balance, the equilibrant $(\mathbf{Wr})_e$ is equal, opposite, and collinear with **R**. In Fig. 10.3, a 44.5-N weight at $r_e = 0.216$ m is shown as the balancing weight.

In some instances, as shown in Fig. 10.4, the resultant of the system of masses to be balanced is a couple. The resultant force **R** for the two equal masses in Fig. 10.4 is zero. However, because the inertia forces of the two masses are not collinear, an unbalanced couple exists. To meet the requirements of moment balance, *two* additional masses are needed to provide a balancing couple.

In the foregoing cases, balancing requirements are met by determining the minimum number of additional masses to achieve balance. Often, more than the minimum number is used. For example, in Fig. 10.3, the single counterbalancing mass is added to reduce shaking forces to zero and to remove load from the bearings supporting the shaft. However, the shaft is under the action of bending, which in some cases may be severe. Balance may also be achieved by providing a counterbalance opposite each mass, a total of three counterbalances, with the advantage that shaft bending is reduced to near zero. As shown in Fig. 10.5a, crankshafts are frequently balanced by counterbalancing each crank separately to reduce shaft bending. Greater total weight is a disadvantage in utilizing large numbers of counterweights. As shown for the crankshaft in Fig. 10.5b, the sym-

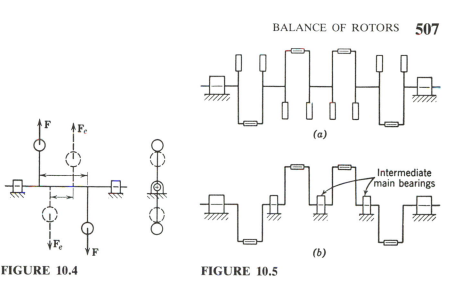

FIGURE 10.4   FIGURE 10.5

metrical distribution of cranks provides balance without the addition of counterweights; but to reduce shaft bending, intermediate main bearings are added.

The most general case of distribution of rotating masses on a rigid rotor is
that in which the masses lie in various transverse and axial planes as in Fig. 10.6.
As in the foregoing cases, Eq. 10.1 must be satisfied for balance of inertia forces.
As shown in Fig. 10.6b, the resultant **R** of the three unbalanced masses of Fig.
10.6a is obtained from a vector polygon. Although it would appear that a single
balancing mass would satisfy Eq. 10.1, a consideration of moment balance shows
that a minimum of two balancing masses is required.

In Fig. 10.6a, the transverse plane A–A is arbitrarily chosen about which
moments of the inertia forces are evaluated. It may be seen that the moments
of the various individual forces are in different axial planes. For moment balance,
the *vector* sum of the moments of the forces must be zero:

$$\Sigma \, (\mathbf{Wra}) = 0 \qquad\qquad (10.3)$$

Equation 10.3 is similar to Eq. 10.2, except that a vector sum is indicated rather
than an algebraic sum. Since, in the general case, the resultant unbalanced moment is in a different axial plane from the resultant **R** of unbalanced forces, a
single balancing mass does not satisfy both Eqs. 10.1 and 10.3.

In Fig. 10.6c is shown the vector polygon of moments taken about the
transverse plane A–A. Plane B–B is chosen as a transverse plane in which a
balancing mass M_b is to be placed to achieve balance of moments. Magnitudes
of the moment vectors are tabulated as shown. Although moment vectors are
usually represented in direction and sense according to the right-hand screw rule,
in Fig. 10.6c they are shown in the same direction and sense as the inertia forces.
In Fig. 10.6c, the known moment vectors $(\mathbf{Wra})_2$ and $(\mathbf{Wra})_3$ are laid off first,
and the closing side $(\mathbf{Wra})_b$ determines the required moment vector for balance.
The direction of $(\mathbf{Wra})_b$ shows the axial plane in which M_b is to be placed. As

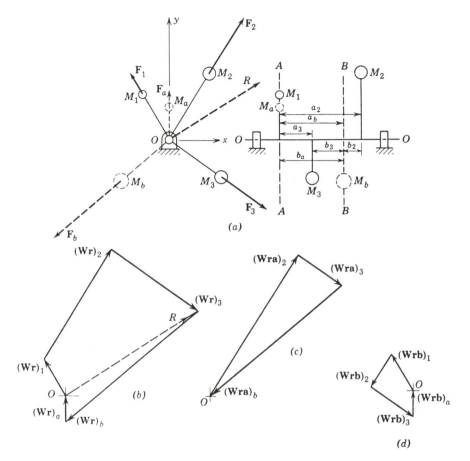

Number	W, lb	r, in.	Wr, lb · in.	a, in.	Wra, lb · in.^{2}	b, in.	Wrb, lb. in.^{2}
1	5	8	40	0	0	10	400
2	10	12	120	13	1560	−3	−360
3	10	10	100	5	500	5	500
a	5*	5	25*	0	0	10	250*
b	16.3*	10	163*	10	1630*	0	0

$$(Wr)_b = \frac{(Wra)_b}{a_b} = \frac{1630}{10} = 163 \quad \text{lb · in.}$$

$$(Wr)_a = \frac{(Wrb)_a}{b_a} = \frac{250}{10} = 25 \quad \text{lb · in.}$$

*From polygons and calculations.

FIGURE 10.6

shown, the magnitude of the force vector $(\mathbf{Wr})_b$ is calculated from $(Wra)_b/a_b$ and laid off on the force polygon of Fig. 10.6b. For balance of forces, a second mass M_a is required to close the force polygon as indicated by $(\mathbf{Wr})_a$. $(\mathbf{Wr})_a$ and $(\mathbf{Wr})_b$ form the equilibrant of **R**. By placing M_a in plane A–A such that it has zero moment about plane A–A, the moment vector polygon (Fig. 10.6c) for balance is unchanged. Thus, both Eqs. 10.1 and 10.3 are satisfied. As indicated in the table in Fig. 10.6, a 5-lb weight at $r_a = 5$ in. in plane A–A and a 16.3-lb weight at $r_b = 10$ in. in plane B–B are used for balance. Figure 10.6a shows the axial planes of the balancing masses as determined from the directions of $(\mathbf{Wr})_a$ and $(\mathbf{Wr})_b$ in Fig. 10.6b.

In Fig. 10.6d is shown the moment vector polygon in which moments are taken about plane B–B to determine the moment vector $(\mathbf{Wrb})_a$ due to M_a in plane A–A. The vector $(\mathbf{Wr})_a$ obtained from this polygon is the same as in the previous solution. As shown in the table and in Fig. 10.6d, the sense of $(\mathbf{Wrb})_2$ is negative since M_2 is on the opposite side of plane B–B from M_1 and M_3.

The problem of Fig. 10.6 may be solved using any two of the three vector polygons shown. Also, for the general case of Fig. 10.6, any number of masses may be balanced by a minimum of two masses placed in any two arbitrarily selected transverse planes such as A–A and B–B.

The balance of the shaft and weights shown in Fig. 10.6 which was solved graphically was also solved analytically. The angles are $\theta_1 = 120°$, $\theta_2 = 59°$, and $\theta_3 = 324°$. By considering Fig. 10.6b the following equation was written:

$$(\mathbf{Wr})_1 + (\mathbf{Wr})_2 + (\mathbf{Wr})_3 = \mathbf{R}$$

This equation was solved to give $R = 145.8$ lb $\cdot$ in. and $\theta_R = 32.7°$. From Fig. 10.6c,

$$(\mathbf{Wra})_2 + (\mathbf{Wra})_3 + (\mathbf{Wra})_b = 0$$

This was solved to give $(Wra)_b = 1596$ lb $\cdot$ in.^{2}, compared to 1630 lb $\cdot$ in.^{2} found graphically from the polygon. Using this value with $a_b = 10$ in. and $r_b = 10$ in. gives $(Wr)_b = 159.6$ lb $\cdot$ in. and $W_b = 15.96$ lb, compared to 16.3 lb found graphically.

By returning to Fig. 10.6b,

$$(\mathbf{Wr})_a + \mathbf{R} + (\mathbf{Wr})_b = 0$$

The direction of $(\mathbf{Wr})_a$ is along the y-axis (i.e., 90°) and the y-components of **R** and $(\mathbf{Wr})_b$ are known. The equation can therefore be solved to give $(Wr)_a = 25.54$ lb $\cdot$ in. and $W_a = 5.11$ lb. The graphical solution gives 25 lb $\cdot$ in. and 5 lb, respectively, for these values.

As shown in Fig. 10.7, the cranks of six- and eight-cylinder in-line engines are arranged so that crankshaft balance is obtained by symmetry, although the individual crank masses (including equivalent connecting-rod masses) are in different axial planes.

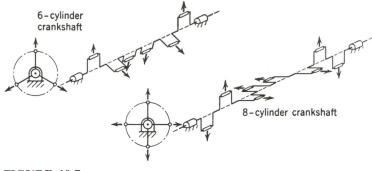

6-cylinder crankshaft

8-cylinder crankshaft

FIGURE 10.7

10.3 DYNAMIC AND STATIC BALANCE

The requirements for balance of rigid rotors as illustrated in section 10.2 are those for *dynamic* balance or balance due to the action of inertia forces. *Static* balance is a balance of forces due to the action of gravity. Figure 10.8 shows a rigid rotor with the shaft laid on horizontal parallel ways. Under the action of gravity, the rotor will not roll if it is in static balance regardless of the angular position of the rotor. The requirement for static balance is that the center of gravity of the system of masses be at the axis $O–O$ of rotation. For the center of gravity to be at $O–O$ of Fig. 10.8, the moments of the masses about the x-axis and the y-axis, respectively, must be zero.

$$\Sigma \ (Wr \sin \theta) = 0 \qquad\qquad (10.4)$$

$$\Sigma \ (Wr \cos \theta) = 0 \qquad\qquad (10.5)$$

By referring to Fig. 10.2, it may be seen that Eqs. 10.4 and 10.5 for static equilibrium also apply for dynamic balance of inertia forces. In the vector polygon of Fig. 10.2, the vertical components of forces are represented in Eq. 10.4 and the horizontal components in Eq. 10.5. Thus, if the conditions for dynamic balance are met, the conditions for static balance are also met. This is true also for

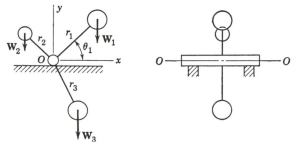

FIGURE 10.8

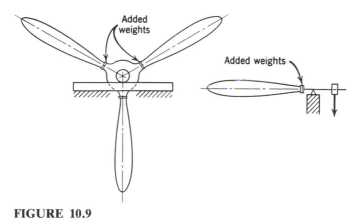

FIGURE 10.9

the rotors of Figs. 10.3, 10.4, and 10.6. However, it is not true that if a rotor is statically balanced, it is also dynamically balanced. The rotor of Fig. 10.4, for example, is statically balanced without the balancing masses, but it is not dynamically balanced because of the unbalance of moments in the axial plane. Thus, static balance fails to indicate moment balance required for the dynamic case. A static balance is a reliable test of dynamic balance only in the case of Fig. 10.2, where all masses lie in a common transverse plane and a dynamic unbalance of moment is unlikely.

The use of horizontal parallel ways as in Fig. 10.8 is a simple method for "shop" balancing or "production" balancing of rotors having masses in a common radial plane. As shown in Fig. 10.9, airplane propellers are tested for dynamic balance in this manner from a test for static balance. A high degree of balance is achieved by adding washers to the propeller hub as shown. Also, as shown, washers are added to the blade shank in balancing the individual propeller blades against a standard balancing moment. Balancing can also be accomplished by removing metal by drilling at 180° opposite to where material is to be added. This should be avoided in some applications such as propeller balancing because of the stress concentration caused by holes. Rotors having the general shape of a thin disk such as gears, pulleys, wheels, cams, fans, flywheels, and impellers are often balanced statically.

10.4 BALANCING MACHINES

Although the dynamic balance of a rotor is properly met in the design of the rotor, some unbalance, however small, results from the manufacture of the rotor. Carefully machined parts are likely to be in better balance than cast parts. In many instances, it is more economical to allow an unbalance caused by manufacture and to balance the part by adding or removing material as indicated by a balancing machine. Commercial balancing machines are available which enable balancing of parts at mass production rates of manufacture.

The degree to which a rotor is to be dynamically balanced depends upon the speed at which the rotor is to operate. A small unbalance of mass may be tolerable at low speed since the inertia force representing unbalance may be small, but because the unbalanced force increases as the square of the speed, the unbalance transmitted to the bearings may be large at high speed. The rotor of a jet engine operating at greater than 10,000 rpm, for example, must be balanced to a high degree. For such rotors, the individual compressor and turbine blades are balanced in pairs at opposite locations on the rotor in such a way that, if one blade is damaged, the pair is replaced to restore balance.

The principle on which dynamic balancing machines are based is shown in Fig. 10.10. The rotor to be dynamically balanced is supported on flexible springs and rotated at the speed the rotor is to be used. As shown in Fig. 10.10, the springs permit only a lateral oscillation of the rotor under the action of the unbalanced force **F**. If there is also an unbalanced moment in the rotor, the amplitudes of oscillation of the two springs will be different and, in some cases, opposite in sign. The amplitudes of oscillation of each spring are measured with a highly sensitive electronic pickup which is calibrated to show the amount of unbalance. The machine also indicates the angular position of the unbalance on the rotor by imparting a signal at the instant the rotating force vector is horizontal and the amplitude is maximum. After the amount and angular position of the unbalance is read, the rotor is removed from the machine and material is added by soldering masses to it or is removed by drilling holes. As shown in Fig. 10.10, material is usually added or removed at two specific locations where it is not injurious to the rotor surface. In electric motor armatures, for example, it is not always possible to add or remove material in the region of the electrical windings. Long rotors such as armatures, crankshafts, and jet engine rotors are balanced in machines of this kind.

It is often necessary to balance a rotor that is too large to be handled in a balancing machine or to balance a rotor assembled in its own bearings. Also, when a unit is being rebuilt, it may be impractical to transport the rotor back to

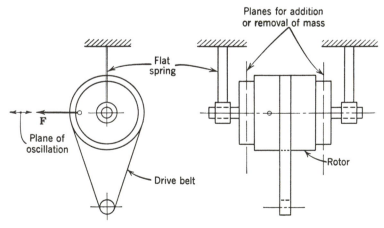

FIGURE 10.10

the shop for rebalancing. In such cases *field balancing* must be used. The initial development of this method of balancing was carried out by Thearle.[1] More recent descriptions of this method are given by Hirschhorn[2] and by Shigley and Uicker.[3]

10.5 BALANCE OF RECIPROCATING MASSES

As shown in Fig. 9.29, the shaking of a piston engine is due primarily to the inertia forces of the reciprocating masses located at the wrist pin. The masses rotating with the crankshaft are normally balanced, and they do not transmit a shaking force to the engine block. As shown in the slider-crank free body of Fig. 10.11, the effect of the inertia force **F** of the reciprocating masses is the transmission of force to the engine block at the cylinder wall and at the main bearings. The vertical component of the main bearing force $\mathbf{F}_{12}^y$ and the inertia force **F** are equal, opposite, and collinear. The horizontal component of the bearing force $\mathbf{F}_{12}^x$ and the cylinder wall force $\mathbf{F}_{14}$ are equal and opposite and form a couple $\mathbf{F}_{14}h$ since they are not collinear. The effect of the reciprocating masses on the engine block, as shown in the free body of the block is a shaking force $\mathbf{S} = \mathbf{F}$ and a shaking couple $\mathbf{F}_{41}h$. Since both the shaking force and the shaking couple change in magnitude and sense during the engine cycle, forced vibrations are imposed

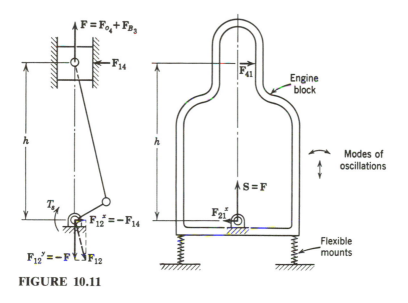

FIGURE 10.11

[1] F. I. Thearle, "Dynamic Balancing in the Field," *Trans. ASME, Journal of Engineering for Industry*, 1934, p. 745.

[2] J. Hirschhorn, *Dynamics of Machinery*, Barnes and Noble, New York, 1968, p. 348.

[3] J. J. Shigley and J. J. Uicker, *Theory of Machines and Mechanisms*, McGraw-Hill, New York, 1980, p. 497.

on the engine block. If the engine block is mounted on flexible supports, the mode of oscillation of the block imposed by **S** is an up-and-down mode; the shaking couple produces a rotational oscillation or lateral rocking.

In Chapter 9, it was shown that the gas forces do not contribute to the vertical shaking but that they do produce a shaking couple as do the inertia forces of the reciprocating masses. Since output torque and shaking couple are equal (except in sense), the output torque diagrams of Chapter 9 show the variations of the shaking couple in the engine cycle due both to gas force and to inertia forces of the reciprocating masses.

In the following discussion, it is shown that the resultant shaking force on an engine block may be reduced to zero in some instances by combining several slider cranks to form a multicylinder engine in which the individual shaking forces balance one another. The resultant shaking couple of a multicylinder engine, however, is not reduced to zero as shown by the torque diagram of the six-cylinder engine in Fig. 9.35. However, by the proper design of the flexible sup-

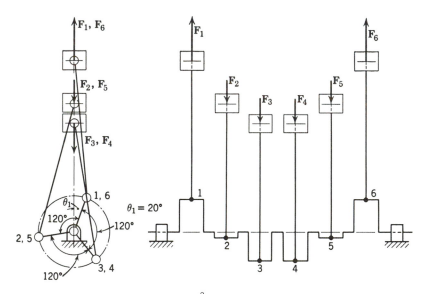

$$M = 0.0777 \text{ lb} \cdot \text{sec}^2/\text{ft} \qquad n = 3000 \text{ rpm}$$
$$R = 2 \text{ in.} \qquad R/L = \tfrac{1}{4}$$

Number	θ, °	$\cos \theta$	$\cos 2\theta$	$R/L(\cos 2\theta)$	$\cos \theta + R/L$ $(\cos 2\theta)$	F, lb
1	20	+0.904	+0.766	+0.191	+1.131	+1450
2	260	−0.174	−0.940	−0.235	−0.409	−525
3	140	−0.766	+0.174	+0.044	−0.722	−925
4	140	−0.766	+0.174	+0.044	−0.722	−925
5	260	−0.174	−0.940	−0.235	−0.409	−525
6	20	+0.940	+0.766	+0.191	+1.131	+1450
		$\Sigma = 0$	$\Sigma = 0$	$\Sigma = 0$	$\Sigma = 0$	$\Sigma = 0$

FIGURE 10.12

ports connecting the engine block to the supporting frame, the oscillations due to the shaking couple may be isolated from the frame for certain shaking couple frequencies.

Figure 10.12 shows a typical arrangement of cranks in a six-cylinder in-line engine. In this engine, the cranks are fixed at 120° to each other as shown and all slider-crank parts are the same as to size, shape, and weight. As shown in the table of Fig. 10.12, the inertia force **F** of the individual reciprocating masses are calculated from the following equation:

$$F = MA_B = MR\omega^2\left(\cos\theta + \frac{R}{L}\cos 2\theta\right)$$

in which M is the combined masses M_4 and M_{B_3} for a single cylinder, R is the crank length, L is the connecting-rod length, ω is the crank angular velocity, and θ is the crank angle from T.D.C. (top dead center).

As the table in Fig. 10.12 shows, the arrangement of cranks for a six-cylinder engine is such that the resultant of the six inertia forces is zero for the position of the crankshaft given by $\theta_1 = 20°$. It may be shown that the resultant is zero for all positions of the crankshaft. Thus, no shaking force is transmitted to the main bearings supporting the crankshaft or to the engine block. The six-cylinder engine is well known for its inherent balance of reciprocating masses. The single-, two-, three-, four-, and five-cylinder engines of the in-line type are not inherently balanced against shaking by reciprocating masses as is the six-cylinder engine.

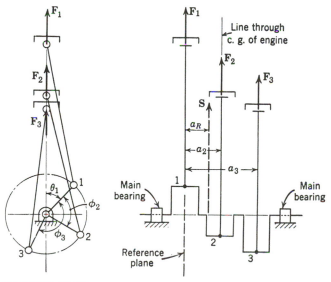

FIGURE 10.13

10.6 ANALYTICAL DETERMINATION OF UNBALANCE

Analytical methods are available for the determination of the unbalance or the shaking force of a multicylinder engine. The method leads to simple algebraic expressions which give the magnitude and sense of unbalance as a function of crank position θ_1. Crank position of a multicylinder engine in the engine cycle is given by the crank angle θ_1 of the first cylinder as shown in Fig. 10.13. In automotive engines, the first cylinder is at the front end, and θ_1 is measured clockwise in the direction of rotation when viewed from the front end.

The following analytical derivation applies only to in-line types of engines in which the cylinders are in line on the same side of the crankshaft. The reciprocating mass M and the R/L ratio is the same for each cylinder.

As shown in Fig. 10.13, θ_1 of the first crank locates the clockwise position of the crankshaft in the engine cycle, ϕ_2 and ϕ_3 are the fixed angles of cranks 2 and 3, respectively, measured clockwise from crank 1. Although three cylinders are shown in Fig. 10.13, any number of cylinders may be considered. The inertia force F of any given cylinder at θ is

$$F = MR\omega^2\left(\cos\theta + \frac{R}{L}\cos 2\theta\right)$$

$$= MR\omega^2\cos\theta + M\frac{R^2}{L}\omega^2\cos 2\theta \qquad (10.6)$$

The two right-hand terms in Eq. 10.6 are the first two terms of a series, the remaining terms of which are usually considered negligible. The first term (first harmonic) is known as the primary force F_p and the second term (second harmonic) as the secondary force F_s. Thus,

$$F = F_p + F_s \qquad (10.7)$$

where $F_p = MR\omega^2\cos\theta$ and $F_s = M(R^2/L)\omega^2\cos 2\theta$.

The summation of the inertia forces of a multicylinder engine is the resultant force or shaking force S, which represents the unbalance:

$$S = \Sigma F = \Sigma F_p + \Sigma F_s \qquad (10.8)$$

In some engines, the primary forces may be balanced although the secondary forces may not. The reverse may also be the case. The equation of unbalance of primary forces is developed as follows, in which $\theta = \theta_1 + \phi$.

$$\Sigma F_p = \Sigma MR\omega^2\cos\theta$$

$$= MR\omega^2\Sigma\cos\theta$$

$$= MR\omega^2\Sigma\cos(\theta_1 + \phi)$$

$$= MR\omega^2\Sigma[(\cos\theta_1)(\cos\phi) - (\sin\theta_1)(\sin\phi)]$$

Since $\cos \theta_1$ and $\sin \theta_1$ are constant for all terms of the summation,

$$\Sigma F_p = MR\omega^2[(\cos \theta_1) \Sigma (\cos \phi) - (\sin \theta_1) \Sigma (\sin \phi)] \qquad (10.9)$$

The equation of unbalance of secondary forces is similar in form.

$$\Sigma F_s = M \frac{R^2}{L} \omega^2[(\cos 2\theta_1) \Sigma (\cos 2\phi) - (\sin 2\theta_1) \Sigma (\sin 2\phi)] \qquad (10.10)$$

It may be seen from Eqs. 10.9 and 10.10 that, for any given arrangement of cranks in a multicylinder engine, the angles ϕ are known such that $\Sigma \cos \phi$, $\Sigma \sin \phi$, $\Sigma \cos 2\phi$, and $\Sigma \sin 2\phi$ may be evaluated, and the equations of unbalance become functions only of θ_1. It may also be seen that for balance, or zero shaking force, the following summations must all be zero:

$$\Sigma \cos \phi = 0$$

$$\Sigma \sin \phi = 0$$

$$\Sigma \cos 2\phi = 0$$

$$\Sigma \sin 2\phi = 0$$

Another mode of shaking must be considered for multicylinder engines. By viewing the engine of Fig. 10.13 from the side, it may be seen that the line of action of the resultant shaking force in the axial plane may not lie on a line of symmetry between the main bearings. Moreover, the line of action of the resultant **S** may be shifting axially in the axial plane as a function of θ_1. In this event, the engine oscillates in an end-over-end rotational mode. The line of action of **S** may be determined from the principle of moments in terms of a primary moment C_p and a secondary moment C_s, in which moments are taken with respect to a reference plane at the first cylinder. In Fig. 10.13, a is the distance from the reference plane to the line of action of the inertia force of any given cylinder.

$$C_p = \Sigma F_p a = MR\omega^2 \Sigma (a \cos \theta)$$

$$= MR\omega^2[(\cos \theta_1) \Sigma (a \cos \phi) - (\sin \theta_1) \Sigma (a \sin \phi)] \qquad (10.11)$$

and

$$C_s = \frac{MR^2\omega^2}{L} [(\cos 2\theta_1) \Sigma (a \cos 2\phi) - (\sin 2\theta_1) \Sigma (a \sin 2\phi)] \qquad (10.12)$$

$$C = C_p + C_s \qquad (10.13)$$

The distance a_R of the line of action of the shaking force **S** may be determined from the resultant moment C about the reference plane as follows:

$$a_R = \frac{C}{S} \tag{10.14}$$

In certain cases, the shaking force S is zero, indicating a balance of inertia forces, but the resultant moment C is not. In this case, the resultant is a couple C in the axial plane, which produces an end-over-end axial shaking couple. In some cases, a_R is not a function of θ_1 but is constant. In this case, if constant a_R places the line of action of S other than through the center of gravity of the engine, an end-over-end shaking couple exists.

Example 10.1. Determine the unbalance S of the reciprocating masses of the conventional four-cylinder engine shown in Fig. 10.14, in which the cranks are at 180°. Determine also the unbalance of the axial shaking couple.

Solution. The fixed angles ϕ are shown in Fig. 10.14. It should be noted that, although ϕ_1 and ϕ_4 are zero, their cosine functions are unity and must be taken into account in the equations that determine unbalance. The following summations are made to determine the constants which appear in the equations of unbalance:

$$\Sigma \cos \phi = 1 - 1 - 1 + 1 = 0$$

$$\Sigma \sin \phi = 0 + 0 + 0 + 0 = 0$$

$$\Sigma \cos 2\phi = 1 + 1 + 1 + 1 = 4$$

$$\Sigma \sin 2\phi = 0 + 0 + 0 + 0 = 0$$

By referring to Eqs. 10.9 and 10.10, it may be seen that the primary forces are balanced and that the secondary forces are not:

$$\Sigma F_p = MR\omega^2[(\cos \theta_1)(0) - (\sin \theta_1)(0)] = 0$$

$$\Sigma F_s = \frac{MR^2\omega^2}{L} [(\cos 2\theta_1)(4) - (\sin 2\theta_1)(0)]$$

$$\Sigma F_s = 4\frac{MR^2\omega^2}{L} \cos 2\theta_1$$

$$S = \Sigma F_p + \Sigma F_s$$

$$S = 4\frac{MR^2\omega^2}{L} \cos 2\theta_1 = \frac{MR^2}{L}(2\omega)^2 \cos 2\theta_1 \tag{10.15}$$

Equation 10.15, which gives the shaking force of the conventional four-cylinder engine as a function of θ_1, is shown plotted in Fig. 10.15. It may be seen that the shaking-force curve is a simple harmonic curve whose circular frequency 2ω is twice the speed of the crankshaft.

The following summations give the constants which apply in the moment equations

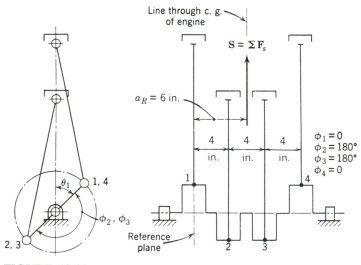

FIGURE 10.14

10.11 and 10.12:

$$\Sigma\ (a\ \cos\ \phi) = 0(1) + 4(-1) + 8(-1) + 12(1) = 0$$
$$\Sigma\ (a\ \sin\ \phi) = 0(0) + 4(0) + 8(0) + 12(0) = 0$$
$$\Sigma\ (a\ \cos\ 2\phi) = 0(1) + 4(1) + 8(1) + 12(1) = 24$$
$$\Sigma\ (a\ \sin\ 2\phi) = 0(0) + 4(0) + 8(0) + 12(0) = 0$$

By referring to Eqs. 10.11 and 10.12, it may be seen that a secondary moment C_s exists about the reference plane and that the primary moments are zero:

$$C = C_p + C_s = 24\ \frac{MR^2\omega^2}{L}\ \cos\ 2\theta_1$$

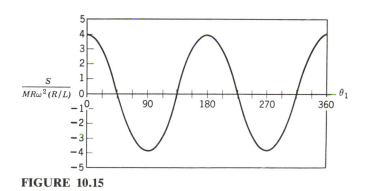

FIGURE 10.15

The line of action of the shaking force **S** is determined as follows:

$$a_R = \frac{C}{S} = \frac{24(MR^2\omega^2/L)\cos 2\theta_1}{4(MR^2\omega^2/L)\cos 2\theta_1}$$

$$a_R = 6 \text{ in.}$$

The line of action of the shaking force is constant since a_R is not a function of θ_1. Also, by assuming that the line of action of **S** is through the center of gravity of the engine at $a_R = 6$ in., no axial shaking couple exists.

As shown, the only unbalance is a shaking force due to the secondary forces which tend to cause an up-and-down vibration of the engine. A common device used to balance the secondary forces of a four-cylinder engine is the Lanchester balancer shown in Fig. 10.16. This type of balancer is currently being used by one automobile manufacturer in its large four-cylinder engines and is known as *silent shaft*. The balancer consists of two meshed gears with eccentric masses as shown. The pitch point of the meshing gears is directly under the centerline of the engine such that the resultant inertia force of the rotating masses counterbalances the shaking force **S**. A crossed helical gear on the crankshaft drives the balancing gears at twice the crankshaft speed in order that the balancing forces be of the same circular frequency as the unbalanced secondary forces. In cases where unbalance of primary forces only exists, the balancing gears rotate at crankshaft speed.

A five-cylinder engine as shown in Fig. 10.17 has been developed for diesel engines. While shaking forces are balanced, the primary and secondary shaking couples are not balanced as will be shown in the following example.

Example 10.2. Analyze the balance of the five-cylinder engine shown in Fig. 10.17 relative to shaking forces and shaking couples.

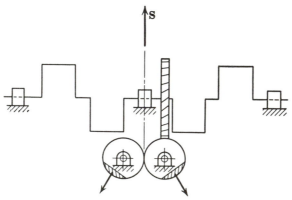

FIGURE 10.16

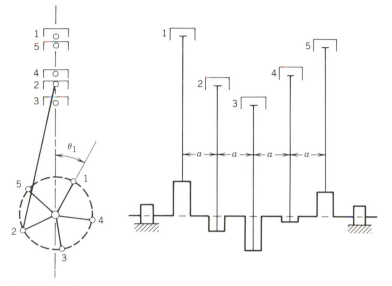

FIGURE 10.17

Solution. From Fig. 10.17 the fixed angles ϕ are

$\phi_1 = 0°$

$\phi_2 = 216°$

$\phi_3 = 144°$

$\phi_4 = 72°$

$\phi_5 = 288°$

Shaking forces:

From Eqs. 10.9 and 10.10,

$$\Sigma F_p = MR\omega^2[(\cos \theta_1) \Sigma (\cos \phi) - (\sin \theta_1) \Sigma (\sin \phi)]$$

$$\Sigma F_s = \frac{MR^2\omega^2}{L} [(\cos 2\theta_1) \Sigma (\cos 2\phi) - (\sin 2\theta_1) \Sigma (\sin 2\phi)]$$

$$\Sigma \cos \phi = \cos 0° + \cos 216° + \cos 144° + \cos 72° + \cos 288°$$

$$= 1 - 0.8090 - 0.8090 + 0.3090 + 0.3090$$

$$= 0$$

$$\Sigma \sin \phi = \sin 0° + \sin 216° + \sin 144° + \sin 72° + \sin 288°$$

$$= 0 - 0.5878 + 0.5878 + 0.9511 - 0.9511$$

$$= 0$$

$$\Sigma \cos 2\phi = \cos 0° + \cos 432° + \cos 288° + \cos 144° + \cos 576°$$

$$= 1 + 0.3090 + 0.3090 - 0.8090 - 0.8090$$

$$= 0$$

$$\Sigma \sin 2\phi = \sin 0° + \sin 432° + \sin 288° + \sin 144° + \sin 576°$$

$$= 0 + 0.9511 - 0.9511 + 0.5878 - 0.5878$$

$$= 0$$

Therefore, as can be seen above, the primary and secondary forces are balanced.

Shaking couples:

From Eqs. 10.11 and 10.12,

$$C_p = MR\omega^2[(\cos \theta_1) \Sigma (a \cos \phi) - (\sin \theta_1) \Sigma (a \sin \phi)]$$

$$C_s = \frac{MR^2\omega^2}{L} [(\cos 2\theta_1) \Sigma (a \cos 2\phi) - (\sin 2\theta_1) \Sigma (a \sin 2\phi)]$$

$$\Sigma (a \cos \phi) = 0(1) + a(-0.8090) + 2a(-0.8090) + 3a(0.3090)$$
$$+ 4a(0.3090)$$
$$= -0.264a$$

$$\Sigma (a \sin \phi) = 0(1) + a(-0.5878) + 2a(+0.5878) + 3a(0.9511)$$
$$+ 4a(-0.9511)$$
$$= -0.3633a$$

$$\Sigma (a \cos 2\phi) = 0(1) + a(0.3090) + 2a(0.3090) + 3a(-0.8090)$$
$$+ 4a(-0.8090)$$
$$= -4.7360a$$

$$\Sigma (a \sin 2\phi) = 0(0) + a(0.9511) + 2a(-0.9511) + 3a(0.5878)$$
$$+ 4a(-0.5878)$$
$$= -1.5389a$$

Therefore,

$$C_p = MR\omega^2[(\cos \theta_1)(-0.264a) - (\sin \theta_1)(-0.3633a)]$$

$$= MR\omega^2a[-0.264 \cos \theta_1 + 0.3633 \sin \theta_1]$$

and

$$C_s = \frac{MR^2\omega^2}{L} [(\cos 2\theta_1)(-4.7360a) - (\sin 2\theta_1)(-1.5389a)]$$

$$= \frac{MR^2\omega^2a}{L} [-4.7360 \cos 2\theta_1 + 1.5389 \sin 2\theta_1]$$

Example 10.3. Show that the conventional six-cylinder engine of Fig. 10.12 is in balance according to Eqs. 10.9, 10.10, 10.11, and 10.12. Assume that cylinders are b in. apart.

Solution. To show that the six-cylinder engine is in balance, it is necessary to show that the following summations are zero. From Fig. 10.12, the fixed angles ϕ are

$$\phi_1 = \phi_6 = 0°$$

$$\phi_2 = \phi_5 = 240°$$

$$\phi_3 = \phi_4 = 120°$$

$$\Sigma \cos \phi = 1 - \tfrac{1}{2} - \tfrac{1}{2} - \tfrac{1}{2} - \tfrac{1}{2} + 1 = 0$$

$$\Sigma \sin \phi = 0 - \frac{\sqrt{3}}{2} + \frac{\sqrt{3}}{2} + \frac{\sqrt{3}}{2} - \frac{\sqrt{3}}{2} + 0 = 0$$

$$\Sigma \cos 2\phi = 1 - \tfrac{1}{2} - \tfrac{1}{2} - \tfrac{1}{2} - \tfrac{1}{2} + 1 = 0$$

$$\Sigma \sin 2\phi = 0 + \frac{\sqrt{3}}{2} - \frac{\sqrt{3}}{2} - \frac{\sqrt{3}}{2} + \frac{\sqrt{3}}{2} + 0 = 0$$

$$\Sigma (a \cos \phi) = 0(1) + b(-\tfrac{1}{2}) + 2b(-\tfrac{1}{2}) + 3b(-\tfrac{1}{2}) + 4b(-\tfrac{1}{2})$$
$$+ 5b(1) = 0$$

$$\Sigma (a \sin \phi) = 0(0) + b\left(-\frac{\sqrt{3}}{2}\right) + 2b\left(\frac{\sqrt{3}}{2}\right) + 3b\left(\frac{\sqrt{3}}{2}\right)$$
$$+ 4b\left(-\frac{\sqrt{3}}{2}\right) + 5b(0) = 0$$

$$\Sigma (a \cos 2\phi) = 0(1) + b(-\tfrac{1}{2}) + 2b(-\tfrac{1}{2}) + 3b(-\tfrac{1}{2}) + 4b(-\tfrac{1}{2}) + 5b(1) = 0$$

$$\Sigma (a \sin 2\phi) = 0(0) + b\left(\frac{\sqrt{3}}{2}\right) + 2b\left(-\frac{\sqrt{3}}{2}\right) + 3b\left(-\frac{\sqrt{3}}{2}\right)$$
$$+ 4b\left(\frac{\sqrt{3}}{2}\right) + 5b(0) = 0$$

Substitution of the preceding summations in Eqs. 10.9, 10.10, 10.11, and 10.12 shows that there is no resultant shaking force and no resultant axial moment; this signifies a balance of inertia forces of the six reciprocating masses.

The usual straight-eight engine consists of a combination of two four-cylinder engines at 90° crank angle as shown in Fig. 10.18. One of the four-cylinder engines is split with two cylinders at the front end and two at the rear, and the second four-cylinder engine is in the center. As shown in Example 10.1, the four-cylinder engine is unbalanced as to secondary forces. For the split four-cylinder engine, the unbalance is

$$S_1 = 4 \frac{MR^2\omega^2}{L} \cos 2\theta_1$$

The shaking force of the middle set of four cylinders in terms of θ_3 of the first cylinder in the middle set is

$$S_2 = 4\,\frac{MR^2\omega^2}{L}\cos 2\theta_3$$

However, since $\theta_3 = \theta_1 + 270°$,

$$S_2 = 4\,\frac{MR^2\omega^2}{L}\cos 2(\theta_1 + 270°)$$

$$= -4\,\frac{MR^2\omega^2}{L}\cos 2\theta_1$$

Since $S_1 = -S_2$, the resultant shaking force is zero. Also, since the lines of action of $\mathbf{S}_1$ and $\mathbf{S}_2$ are coincident at the center of the engine, there is no axial shaking couple. Thus, like the six-cylinder engine, the straight-eight of Fig. 10.18 is a balanced engine.

Small engines with one, two, or three cylinders are used in a multitude of applications such as outboard motors, mowers, and garden machinery. Air compressors and compressors for spraying are reciprocating machines of one, two, and three cylinders. The balance of reciprocating masses in these machines is poor, as the equations of unbalance would show. The three-cylinder machine with cranks at 120° is in balance as to shaking force, but an axial shaking couple exists. Because comfort is unimportant in small-engine installations, except possibly with outboard motors, unbalance is tolerable. If the expense is warranted, small engines may be mounted on springs or rubber mounts to isolate the machine vibrations from the frame on which the engine is supported.

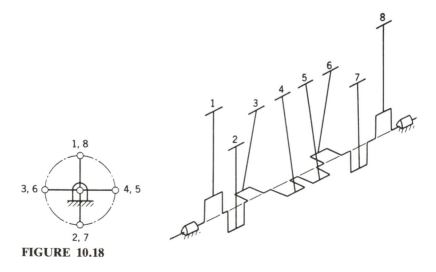

FIGURE 10.18

10.7 FIRING ORDER

In multicylinder engines, the crank arrangements are such that there is a smooth distribution of torque in the engine cycle as well as a balance of inertia forces of the reciprocating masses. For example, in the four-cylinder engine, a power stroke begins every 180° of crank angle in the following order of cylinder numbers: 1–3–4–2. In the six-cylinder engine, a power stroke begins every 120° of crank angle with a 1–5–3–6–2–4 firing order. The eight-cylinder engine fires every 90° of crank angle. In this discussion of firing order, only the four-stroke cycle engine is considered, where one power stroke per cylinder occurs for every 720° of crank rotation. Four events take place in the 720° cycle, which are intake, compression, power or expansion, and exhaust.

10.8 V ENGINES

As shown in Fig. 10.19, the V engine consists of two in-line engines in which the crankshaft is common to both engines. The axial planes in which the two sets of pistons reciprocate intersect at the crankshaft axis and form a V of angle β. In automotive installations, V-6 and V-8 engines are common in which β is either 60° or 90°. Although V-12 engines are no longer used in conventional cars, they are used in a few sport cars. Small engines and compressors are often V-2 or V-4.

In Fig. 10.19 is shown a common arrangement of cylinders used in the V-8 engine in which the cranks are at 90° and β is 90°. The engine consists of two four-cylinder in-line engines or two "banks" of four cylinders each. As shown, the connecting rods of each pair of cylinders, one from each bank, are side by side on a common crank, or "throw." It may be seen that the side-by-side arrangement introduces a small axial couple. In some instances, this couple is minimized by reversing the side-by-side position of some pairs of cylinders from those of other pairs. In other instances, the connecting rods are offset, with the cylinders in the same transverse plane but with the crank pin ends of the connecting rods set side by side. In the following analysis of balance of the V-8 engine, the effect of the side-by-side arrangement on balance is neglected.

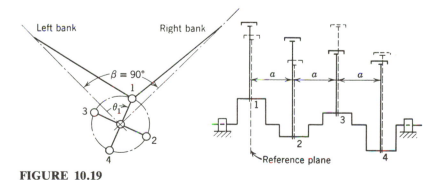

FIGURE 10.19

Since Eqs. 10.9, 10.10, 10.11, and 10.12 apply to in-line engines only, each bank of the V-8 engine may be analyzed separately for balance or unbalance. Any unbalanced force or couple of a given bank is in the axial plane in which the cylinders reciprocate. The resultant unbalance of the complete engine is determined from the vector sum of the unbalance of the two banks.

The following summations apply to either bank since the fixed angles ϕ are the same for both banks:

$$\phi_1 = 0 \qquad \phi_2 = 90° \qquad \phi_3 = 270° \qquad \phi_4 = 180°$$

$$\Sigma \cos \phi = 0 \qquad\qquad \Sigma \sin \phi = 0$$

$$\Sigma \cos 2\phi = 0 \qquad\qquad \Sigma \sin 2\phi = 0$$

$$\Sigma (a \cos \phi) = -3a \qquad\qquad \Sigma (a \sin \phi) = -a$$

$$\Sigma (a \cos 2\phi) = 0 \qquad\qquad \Sigma (a \sin 2\phi) = 0$$

As may be seen from the summations, there is no resultant shaking force **S** for either bank since the primary and secondary forces are balanced. However, as shown by the summations, a resultant moment due to primary forces exists; since **S** is zero, the resultant moment is manifest as an axial shaking couple. For the left bank, the axial shaking couple C_L may be evaluated from Eq. 10.11 in terms of θ_1 measured from T.D.C. of the left bank.

$$C_L = MR\omega^2[-3a \cos \theta_1 + a \sin \theta_1] \qquad\qquad \textbf{(10.16)}$$

For the right bank, the crank angle of the first cylinder is $-(\beta - \theta_1) = \theta_1 - \beta = \theta_1 - 90°$. The axial shaking couple C_R for the right bank is

$$C_R = MR\omega^2[-3a \cos (\theta_1 - 90) + a \sin (\theta_1 - 90)]$$

$$= MR\omega^2[-3a \sin \theta_1 - a \cos \theta_1] \qquad\qquad \textbf{(10.17)}$$

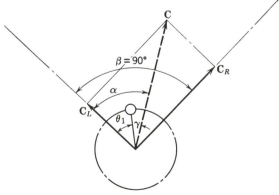

FIGURE 10.20

Since the couples $\mathbf{C}_L$ and $\mathbf{C}_R$ are in axial planes at 90° as shown in Fig. 10.20, the magnitude of the resultant couple C is

$$C = \sqrt{C_L^2 + C_R^2} \tag{10.18}$$

Substitution of the values of C_L and C_R from Eqs. 10.16 and 10.17 gives

$$C = \sqrt{10}\, MR\omega^2 a \tag{10.19}$$

It may be seen that the resultant unbalanced couple of the V-8 engine is independent of θ_1 and is therefore constant in magnitude for all angular positions of the crankshaft. The axial plane in which the resultant couple $\mathbf{C}$ lies is given by the angle α measured clockwise from the plane of the left bank as shown in Fig. 10.20.

$$\tan \alpha = \frac{C_R}{C_L} \tag{10.20}$$

It may be seen that α is a function of θ_1 and that the vector $\mathbf{C}$ rotates with the engine crankshaft. The angle that $\mathbf{C}$ makes with the first crank is γ, which may be determined as follows, since $\alpha = \theta_1 + \gamma$:

$$\tan (\theta_1 + \gamma) = \frac{C_R}{C_L}$$

$$\frac{\tan \theta_1 + \tan \gamma}{1 - \tan \theta_1 \tan \gamma} = \frac{C_R}{C_L}$$

$$\tan \gamma = \frac{C_R - C_L \tan \theta_1}{C_L + C_R \tan \theta_1} \tag{10.21}$$

Substitution of the values of C_L and C_R from Eqs. 10.16 and 10.17 gives

$$\tan \gamma = \frac{-1}{-3}$$

$$\gamma = 198.43° \qquad \text{(third quadrant)} \tag{10.22}$$

As the crankshaft turns in the clockwise sense, the resultant axial shaking couple $\mathbf{C}$ acts in an axial plane which leads the axial plane of the first crank by a constant angle $\gamma = 198.43°$, or leads the fourth crank by 18.43°. Figure 10.21 shows the resultant unbalanced axial couple $\mathbf{C}$ in its correct position with respect to cranks 1 and 4. As shown, the engine is completely balanced by the introduction of a balancing couple $\mathbf{C}_e$, the equilibrant of $\mathbf{C}$, in the form of two counterweights such that

$$C_e = F_e l = -\sqrt{10}\, MR\omega^2 a$$

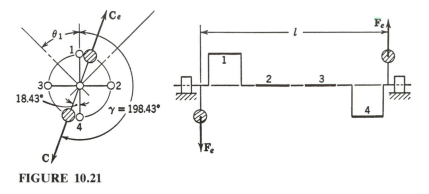

FIGURE 10.21

Figure 10.22 shows the arrangement of a V-6 engine with the angle β between banks of cylinders of 90°. V-6 engines with β of 60° are also produced.

An analysis of the balance of the engine shown in Fig. 10.22 follows with the development of the equations for the primary and secondary shaking forces and shaking couples.

RIGHT BANK	LEFT BANK
Cylinders 1, 3, 5	Cylinders 2, 4, 6
(No. 1 reference cylinder)	(No. 2 reference cylinder)
$\phi_1 = 0°$	$\phi_2 = 0°$
$\phi_3 = 240°$	$\phi_4 = 240°$
$\phi_5 = 120°$	$\phi_6 = 120°$

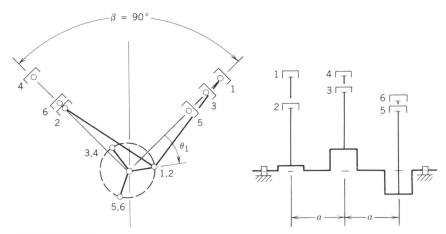

FIGURE 10.22

Shaking Forces

From Eqs. 10.9 and 10.10,

$$\Sigma F_p = MR\omega^2[(\cos\theta_1)\,\Sigma\,(\cos\phi) - (\sin\theta_1)\,\Sigma\,(\sin\phi)]$$

$$\Sigma F_s = \frac{MR^2\omega^2}{L}[(\cos 2\theta_1)\,\Sigma\,(\cos 2\phi) - (\sin 2\theta_1)\,\Sigma\,(\sin 2\phi)]$$

Right Bank

$$\Sigma\cos\phi = \cos 0° + \cos 240° + \cos 120°$$
$$= 1 - 0.5 - 0.5$$
$$= 0$$
$$\Sigma\sin\phi = \sin 0° + \sin 240° + \sin 120°$$
$$= 0 - 0.866 + 0.866$$
$$= 0$$
$$\Sigma\cos 2\phi = \cos 0° + \cos 480° + \cos 240°$$
$$= 1 - 0.5 - 0.5$$
$$= 0$$
$$\Sigma\sin 2\phi = \sin 0° + \sin 480° + \sin 240°$$
$$= 0 + 0.866 - 0.866$$
$$= 0$$

Left Bank

$$\Sigma\cos\phi = 0$$
$$\Sigma\sin\phi = 0$$
$$\Sigma\cos 2\phi = 0$$
$$\Sigma\sin 2\phi = 0$$

Therefore, the primary and secondary shaking forces are balanced for both banks.

Shaking Couples

From Eqs. 10.11 and 10.12,

$$C_p = MR\omega^2[(\cos\theta_1)\,\Sigma\,(a\cos\phi) - (\sin\theta_1)\,\Sigma\,(a\sin\phi)]$$

$$C_s = \frac{MR^2\omega^2}{L}[(\cos 2\theta_1)\,\Sigma\,(a\cos 2\phi) - (\sin 2\theta_1)\,\Sigma\,(a\sin 2\phi)]$$

Right Bank (θ_1 reference angle):

$$\Sigma \ (a \cos \phi) = 0(1) - a(0.5) - 2a(0.5) \qquad = -1.5a$$
$$\Sigma \ (a \sin \phi) = 0(0) - a(0.866) + 2a(0.866) = +0.866a$$
$$\Sigma \ (a \cos 2\phi) = 0(1) - a(0.5) - 2a(0.5) \qquad = -1.5a$$
$$\Sigma \ (a \sin 2\phi) = 0(0) + a(0.866) - 2a(0.5) \qquad = -0.866a$$

Left Bank (ϕ_2 reference angle):

$$\Sigma \ (a \cos \phi) = -1.5a$$
$$\Sigma \ (a \sin \phi) = +0.866a$$
$$\Sigma \ (a \cos 2\phi) = -1.5a$$
$$\Sigma \ (a \sin 2\phi) = -0.866a$$

Right Bank

Therefore,

$$C_p = MR\omega^2[(\cos \theta_1)(-1.5a) - (\sin \theta_1)(0.866a)]$$
$$= MR\omega^2 a[(-1.5 \cos \theta_1) - (0.866 \sin \theta_1)]$$

and

$$C_s = \frac{MR^2\omega^2}{L}[(\cos 2\theta_1)(-1.5a) - (\sin 2\theta_1)(-0.866a)]$$
$$= \frac{MR^2\omega^2 a}{L}[(-1.5 \cos 2\theta_1) + (0.866 \sin 2\theta_1)]$$

Left Bank (where $\theta_2 = \theta_1 + 90°$)

$$C_p = MR\omega^2 a[-1.5 \cos \theta_2 - 0.866 \sin \theta_2]$$
$$= MR\omega^2 a[-1.5 \cos (\theta_1 + 90°) - 0.866 \sin (\theta_1 + 90°)]$$
$$= MR\omega^2 a[-1.5 (-\sin \theta_1) - 0.866 (\cos \theta_1)]$$

Therefore,

$$C_p = MR\omega^2 a[1.5 \sin \theta_1 - 0.866 \cos \theta_1]$$
$$C_s = \frac{MR^2\omega^2 a}{L}[-1.5 \cos 2\theta_2 + 0.866 \sin 2\theta_2]$$

$$C_s = \frac{MR^2\omega^2 a}{L}[-1.5\cos(2\theta_1 + 180°) + 0.866\sin(2\theta_1 + 180°)]$$

$$= \frac{MR^2\omega^2 a}{L}[-1.5(-\cos 2\theta_1) + 0.866(-\sin 2\theta_1)]$$

Therefore,

$$C_s = \frac{MR^2\omega^2 a}{L}[1.5\cos 2\theta_1 - 0.866\sin 2\theta_1]$$

As can be seen from the analysis above, the primary and secondary shaking forces are balanced for each bank. However, the primary and secondary couples are badly out of balance for both banks, and there is no way of easily correcting this as was done for the V-8 engine.

In addition to the question of the balance of the inertia forces and couples of the V-6 engine, it is interesting to consider the development of that engine and the problems encountered with the output torque variation due to crank configuration.

A V-6 engine was developed in 1962 by Buick with a three-throw crankshaft spaced 120° apart as shown in Fig. 10.22. This design with a firing order of 1–6–5–4–3–2 gives uneven intervals of crankshaft rotation of 150°–90°–150°–90°–150°–90° between cylinder firings which results in a high fluctuating output torque. This engine was discontinued in 1967.

As a result of the fuel crisis in 1973–1974, the Buick V-6 engine was reintroduced in 1975 as a means of improving fuel economy. The smoothness of the engine was improved by *splitting* each crank pin by an included angle of 30°. The throw was advanced 15° for the cylinders of one bank and retarded 15° for the cylinders of the other bank. This produced equal intervals of 120° timing between cylinders.[4] The result was a smoother-running engine than was possible with the earlier crankshaft design.

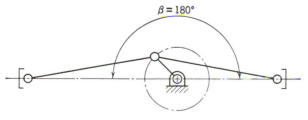

$\beta = 180°$

FIGURE 10.23

[4]D. M. Manner and R. A. Miller, "Buick's New Even Firing 90° V-6 Engine," SAE Paper 770821, Detroit, MI, September 1977.

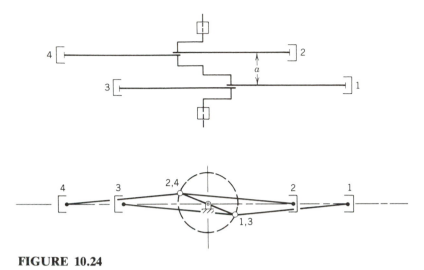

FIGURE 10.24

10.9 OPPOSED ENGINES

As shown in Fig. 10.23, the opposed engine consists of two banks of cylinders, or two in-line engines on opposite sides of the crankshaft in a common horizontal plane. The opposed engine is a special case of the V engine in which $\beta = 180°$ and the determination of balance or unbalance may be made as in V engines. The resultant shaking force **S** and the resultant unbalanced axial moment C lie

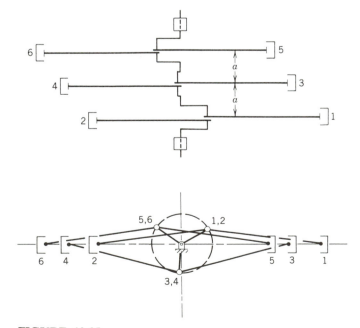

FIGURE 10.25

in the horizontal plane. Opposed engines of four and six cylinders are commonly used in automotive and light-aircraft applications. An eight-cylinder opposed engine is also available for aircraft use. Figures 10.24 and 10.25 show the cylinder arrangement for the four-cylinder and six-cylinder opposed engines.

10.10 BALANCE OF FOUR-BAR LINKAGES

A simple method for force balancing four-bar linkages can be developed from the theory of dynamically equivalent masses given in Chapter 9. In general, the shaking moment will not be balanced by this method.

Consider the four-bar linkage O_2ABO_4 shown in Fig. 10.26. Links 2 and 4 move in pure rotation and may be balanced by adding appropriate counterweights. The only remaining unbalance is due to coupler link mass M_3. It will be assumed that the mass center of coupler link 3 lies along the link centerline $\overline{AB}$. This assumption can always be satisfied by adding mass to the opposite side of the centerline to bring the mass center onto the centerline. The coupler link mass M_3 is now divided into point masses M_A and M_B located at pivots A and B. Recall from Chapter 9 that, for dynamically equivalent systems, these masses must satisfy three requirements:

1. Equivalent total mass

$$M_A + M_B = M_3$$

2. Equivalent mass center

$$M_A l_A - M_B l_B = O$$

3. Equivalent moment of inertia

$$M_A l_A^2 + M_B l_B^2 = I$$

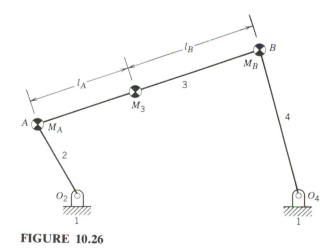

FIGURE 10.26

Having selected the locations for both M_A and M_B, it is clear that these three equations cannot be simultaneously satisfied. By satisfying only the first two of these equations, a link is obtained which is an inertia force equivalent to the original link, although it is not an inertia torque equivalent. The linkage may thus be balanced for shaking forces, but not for shaking moment. Solving the first two equations above simultaneously for M_A and M_B gives

$$M_A = \frac{l_B}{l_A + l_B} M_3 \qquad (10.23)$$

$$M_B = \frac{l_A}{l_A + l_B} M_3 \qquad (10.24)$$

Balancing of the shaking forces is completed by adding counterweights to balance M_A on link 2 and M_B on link 4. The net acceleration of the combined mass center of the three moving links has been reduced to zero. The mechanism is now perfectly balanced with respect to the shaking forces in all positions of the linkage. Note that, since mass has been added to the system, the driving torque and the internal joint forces must be recalculated.

It should be mentioned that Eqs. 10.23 and 10.24 are similar to those used to divide the mass of the engine connecting rod shown in Fig. 9.27. This gave an approximate kinetically equivalent system with part of the connecting rod mass concentrated at the wrist pin center and part at the crank pin center.

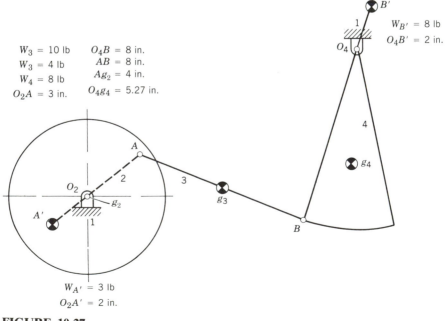

$W_3 = 10$ lb $\quad O_4B = 8$ in.
$W_3 = 4$ lb $\quad\ AB = 8$ in.
$W_4 = 8$ lb $\quad Ag_2 = 4$ in.
$O_2A = 3$ in. $\quad O_4g_4 = 5.27$ in.

$W_{B'} = 8$ lb
$O_4B' = 2$ in.

$W_{A'} = 3$ lb
$O_2A' = 2$ in.

FIGURE 10.27

Example 10.4. It is required to balance the shaking forces in the mechanism of Example 9.1 which is shown again here as Fig. 10.27. Determine the locations and amounts of the weights to be added.

Solution. The input link O_2A is balanced as given. The output link O_4B is balanced by adding weight to link 4 opposite g_4. One possibility is to add 8 lb at 5.27 in. The weight of link 3, W_3, is separated into W_A at point A and W_B at point B, as follows:

$$W_A = \frac{l_B}{l_A + l_B} W_3 = \frac{4 \text{ in.}}{4 \text{ in.} + 4 \text{ in.}} 4 \text{ lb}$$

$$= 2 \text{ lb}$$

$$W_B = \frac{l_A}{l_A + l_B} W_3 = \frac{4 \text{ in.}}{4 \text{ in.} + 4 \text{ in.}} 4 \text{ lb}$$

$$= 2 \text{ lb}$$

These weights are next counterbalanced by adding weights on the opposite sides of pivots O_2 and O_4. The locations for these counterwieghts have been chosen to be A' and B' both located 2 in. from their respective ground pivots O_2 and O_4. The counterweights to be added, W_{ACW} and W_{BCW}, are determined as follows:

$$(\overline{O_2A})W_A = (\overline{O_2A'})W_{ACW}$$

$$W_{ACW} = \frac{\overline{O_2A}}{\overline{O_2A'}} W_A$$

$$= \frac{3 \text{ in.}}{2 \text{ in.}} 2 \text{ lb}$$

$$= 3 \text{ lb}$$

$$(\overline{O_4B})W_B = (\overline{O_4B'})W_{BCW}$$

$$W_{BCW} = \frac{\overline{O_4B}}{\overline{O_4B'}} W_B$$

$$= \frac{8 \text{ in.}}{2 \text{ in.}} 2 \text{ lb}$$

$$= 8 \text{ lb}$$

The position of these counterweights are shown on Fig. 10.27.

Problems

10.1. The rigid rotor of Fig. 10.28 shown with three masses is to be balanced by the addition of a fourth mass. Determine the required weight and angular position of the balancing mass, which is to be located at $r_4 = 10$ in. Show answers on a scale drawing of the rotor.

10.2. For the rigid rotor of Fig. 10.29, determine the bearing reactions at A and B for a rotor speed of 2000 rpm.

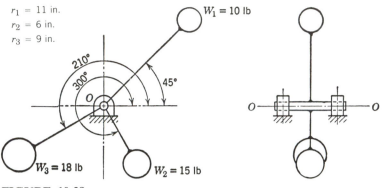

$r_1 = 11$ in.
$r_2 = 6$ in.
$r_3 = 9$ in.

$W_1 = 10$ lb

$210°$
$300°$
$45°$

$W_3 = 18$ lb

$W_2 = 15$ lb

FIGURE 10.28

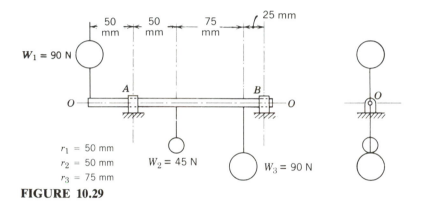

50 mm
50 mm
75 mm
25 mm

$W_1 = 90$ N

A
B
O — O

$r_1 = 50$ mm
$r_2 = 50$ mm
$r_3 = 75$ mm

$W_2 = 45$ N

$W_3 = 90$ N

FIGURE 10.29

10.3. Determine the bearing reactions of the rigid rotor of Fig. 10.30 for a rotor speed of 1200 rpm. Determine the mass or masses which should be added to the rotor at a radius of 2 in. in order that the bearing reactions be due only to the weight of the rotor. Show results using one mass and two masses.

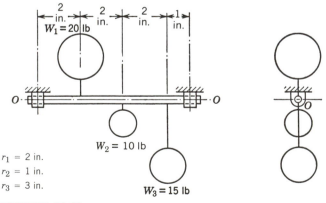

2 in.
2 in.
2 in.
1 in.

$W_1 = 20$ lb

O — O

$W_2 = 10$ lb

$r_1 = 2$ in.
$r_2 = 1$ in.
$r_3 = 3$ in.

$W_3 = 15$ lb

FIGURE 10.30

10.4. For the rigid rotor of Fig. 10.31 shown with two masses, determine the weights W_A and W_B in planes A–A and B–B, respectively, which put the rotor in dynamic balance for a rotor speed of 500 rpm. Determine also the angular positions of the balancing weights.

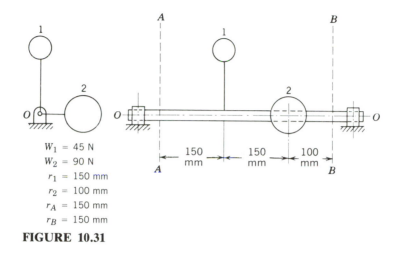

W_1 = 45 N
W_2 = 90 N
r_1 = 150 mm
r_2 = 100 mm
r_A = 150 mm
r_B = 150 mm

FIGURE 10.31

10.5. Weights W_1 and W_2 of the rotor in Fig. 10.32 rotate in the transverse planes shown. Determine the weights W_3 and W_4 in planes 3 and 4, respectively, which give dynamic rotating balance. Show the correct angular positions of W_3 and W_4.

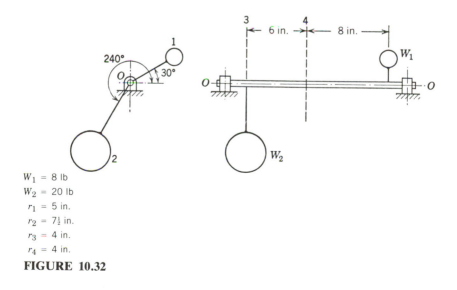

W_1 = 8 lb
W_2 = 20 lb
r_1 = 5 in.
r_2 = $7\frac{1}{2}$ in.
r_3 = 4 in.
r_4 = 4 in.

FIGURE 10.32

10.6. The crankshaft of Fig. 10.33 has four equal cranks at 90° and spaced 100 mm apart. Each crank is equivalent to 18 N at a radial distance of 50 mm. Calculate the bearing reactions due to the inertia forces if the shaft is run at 3000 rpm. Balance this system with

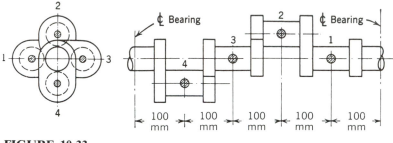

FIGURE 10.33

two weights W_A and W_B in the planes of W_1 and W_3, respectively, and at a radial distance of 50 mm. Determine W_A and W_B, and show their positions.

10.7. The shaking force produced by a given unbalance in a rotor increases with the rotative speed of the rotor. An unbalance of 1 oz at an eccentricity of 1 in. ($Wr = 1$ oz · in.) may be small at a low speed and large at a high speed. Calculate the inertia force of 1 oz at 1 in. for speeds in 1000-rpm increments to 10,000 rpm, and plot a curve of inertia force versus speed.

10.8. A jet engine rotor weighs 6700 N. Determine the amount that the center of gravity of the rotor mass may be eccentric from the axis of rotation to produce an inertia force equal to the weight of the rotor at speeds of 1000, 5000, and 10,000 rpm.

10.9. The degree of unbalance permitted in rotors is often specified by limiting the centripetal acceleration of the center of gravity of the rotor to $g/4$. Determine the eccentricity which produces this amount of acceleration at 5000 rpm, and give the permissible amount of unbalance in N · m (Wr) for a 4500 N rotor.

10.10. The rotor with the steel gears shown in Fig. 10.34 was dynamically balanced in a balancing machine by the addition of the clay masses shown on the periphery of the gears. However, the balancing is to be achieved by drilling holes in the webs of the gears at the diameters shown. Determine the size and location of the holes for dynamic balance.

10.11. In the Scotch yoke mechanism shown in Fig. 10.35, the yoke (link 4) is in simple harmonic motion when the crank of length R rotates at constant angular velocity ω_2. Write a mathematical expression for shaking force due to the reciprocating mass M of the yoke.

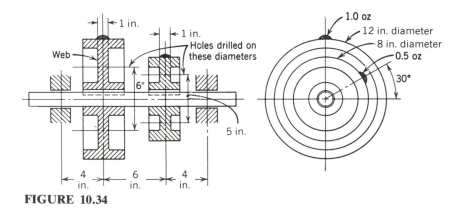

FIGURE 10.34

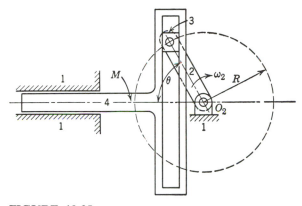

FIGURE 10.35

10.12. For the Scotch yoke mechanism of Fig. 10.35, sketch the supporting frame (link 1) of the mechanism as a free body and show vectors of forces and couples imposed on the frame by the moving parts of the mechanism in approximately the phase shown. Link 2 is driven at constant angular velocity by torque applied at O_2. Designate the shaking force and the shaking couple.

10.13. In Fig. 10.36, the weight of the reciprocating weight is 28.6 N at D and 14.3 N at C. Determine the resultant unbalanced force due to the reciprocating masses for the phase shown; $R\omega^2 = 305$ m/s².

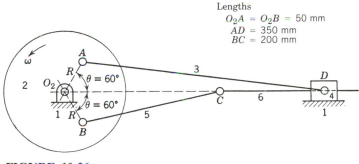

Lengths
$O_2A = O_2B = 50$ mm
$AD = 350$ mm
$BC = 200$ mm

FIGURE 10.36

10.14. In the three-cylinder radial engine shown in Fig. 10.37, all of the connecting rods are attached to a common crank. The reciprocating mass of each cylinder is M_r, and the equivalent mass of each connecting rod at the crank pin is M_A. M_r and M_A are equal. The mass center of the crank is at O, but there is no counterweight to balance the masses M_A at the crank pin. Calculate inertia forces, and, using force polygons, determine the resultant shaking force **S** on the engine for the phase shown when the crank speed is such that $M_rR\omega^2 = 1000$ lb; $R/L = \frac{1}{4}$. Show **S** as a vector on the drawing of the mechanism.

10.15. Using the data of Problem 10.14 and Eq. 9.62, calculate the shaking couple produced by the reciprocating weights of the three-cylinder engine of Fig. 10.37 when $\theta_1 = 30°$ and $R = 3$ in.

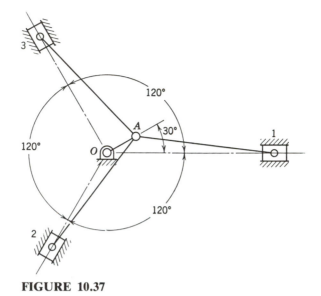

FIGURE 10.37

10.16. Draw a sketch of the engine block of the three-cylinder radial engine of Fig. 10.37, and show vectors of forces imposed on the block by the reciprocating inertia forces of the slider-crank mechanisms.

10.17. Figure 10.38 shows the four-cylinder mechanism of Fig. 10.14 in the engine block and shows the shaking force **S** at the centerline of the four cylinders. M_e is the mass of the complete engine including the block, and the center of gravity is located c distance from the line of action of **S**. The engine is supported by motor mounts having spring constants k. Due to the reciprocating shaking force, the engine vibrates. For the displacements x and ϕ shown, write the equations of motion $\Sigma\, F_x = M_e A = M_e\,(d^2x/dt^2)$ and $\Sigma\, T = I\alpha = I\,(d^2\phi/dt^2)$.

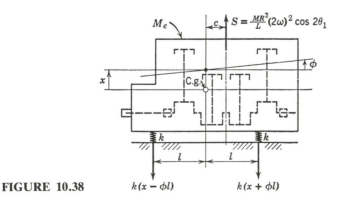

FIGURE 10.38

10.18. By the analytical method, determine the equations of unbalance of reciprocating masses for the two-cylinder engine of Fig. 10.39 in which the cranks are at 90°. Determine

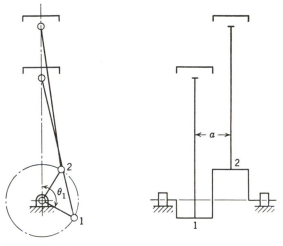

FIGURE 10.39

equations for S and a_R in terms of θ_1. Determine S and a_R for $\theta_1 = 30°$, $MR\omega^2 = 8900$ N, $R/L = \frac{1}{4}$, and $a = 100$ mm.

10.19. Work Problem 10.18 for a two-cylinder engine with cranks at 180° instead of 90°.

10.20. For the three-cylinder engine of Fig. 10.40 with cranks as shown, determine the equation of the unbalanced shaking force S of reciprocating masses in terms of θ_1. Determine also the equation for the distance a_R of the line of action of S from the plane of cylinder 1. Plot curves of S versus θ_1 and a_R versus θ_1 for a complete engine cycle using the following data: Reciprocating weight, 3.22 lb; crank speed, 3000 rpm; stroke, 4 in.; $R/L = \frac{1}{4}$; distance between cylinders, 4 in.

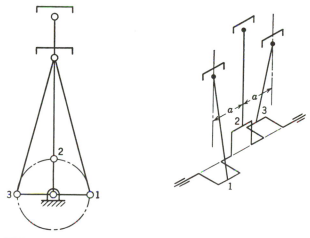

FIGURE 10.40

10.21. The equations of unbalance of reciprocating masses for the conventional four-cylinder engine are developed in Example 10.1. The straight eight-cylinder engine in Fig. 10.41 consists of two four-cylinder engines in tandem with their crank throw planes at 90°. Determine the magnitude and direction of the resultant shaking force or couple for the phase shown ($\theta_1 = 0$).

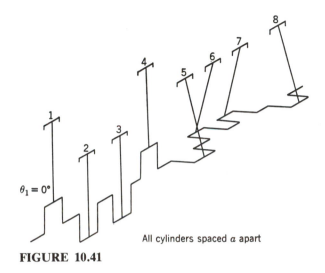

All cylinders spaced a apart

FIGURE 10.41

10.22. For the two-cylinder 90° V engine shown in Fig. 10.42, derive the following equations of unbalance as a function of θ_1: resultant primary force $\mathbf{F}_p$, resultant secondary force $\mathbf{F}_s$, resultant shaking force $\mathbf{S}$, direction of shaking force, and the distance a_R from the plane of the first cylinder to the line of action of $\mathbf{S}$. Using the equations, determine $\mathbf{S}$ and a_R for $\theta_1 = 60°$. $MR\omega^2 = 1$, $R/L = \frac{1}{4}$, $a = 100$ mm.

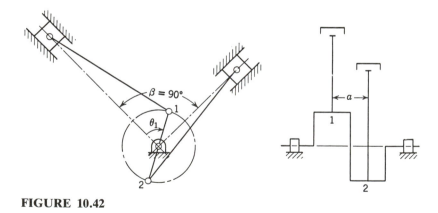

FIGURE 10.42

10.23. For the four-cylinder opposed engine of Fig. 10.43, derive the following equations (in terms of θ_1) of unbalance due to reciprocating masses: primary force $\mathbf{F}_p$, secondary force $\mathbf{F}_s$, shaking force $\mathbf{S}$, and the distance a_R from the plane of cylinder 1 to the line of action of $\mathbf{S}$. Evaluate $\mathbf{S}$ and a_R for $\theta_1 = 90°$, assuming $MR\omega^2$ and the distance between sets of cylinders are unity; $R/L = \frac{1}{4}$. For what angle or angles θ_1, if any, will the resultant primary force be zero?

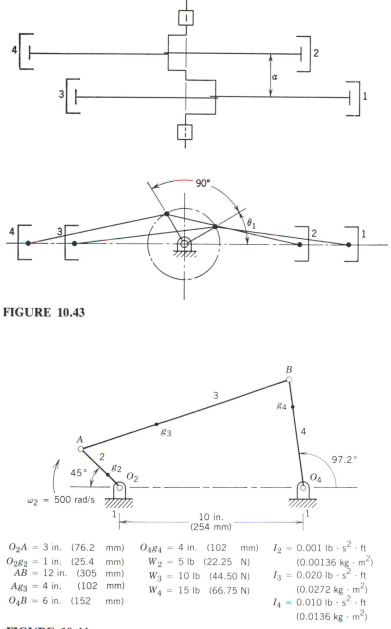

FIGURE 10.43

FIGURE 10.44

O_2A = 3 in.	(76.2 mm)	O_4g_4 = 4 in.	(102 mm)	I_2 = 0.001 lb · s² · ft	
O_2g_2 = 1 in.	(25.4 mm)	W_2 = 5 lb	(22.25 N)	(0.00136 kg · m²)	
AB = 12 in.	(305 mm)	W_3 = 10 lb	(44.50 N)	I_3 = 0.020 lb · s² · ft	
Ag_3 = 4 in.	(102 mm)	W_4 = 15 lb	(66.75 N)	(0.0272 kg · m²)	
O_4B = 6 in.	(152 mm)			I_4 = 0.010 lb · s² · ft	
				(0.0136 kg · m²)	

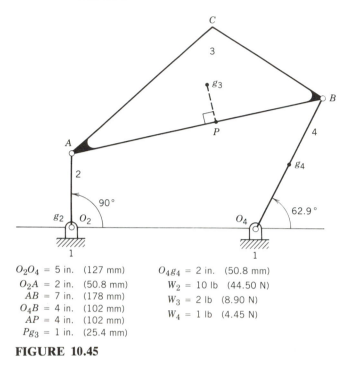

O_2O_4 = 5 in. (127 mm) O_4g_4 = 2 in. (50.8 mm)
O_2A = 2 in. (50.8 mm) W_2 = 10 lb (44.50 N)
AB = 7 in. (178 mm) W_3 = 2 lb (8.90 N)
O_4B = 4 in. (102 mm) W_4 = 1 lb (4.45 N)
AP = 4 in. (102 mm)
Pg_3 = 1 in. (25.4 mm)

FIGURE 10.45

10.24. For the six-cylinder opposed engine of Fig. 10.25, derive the equations for the shaking forces and the shaking couples in terms of M, R, ω, a, and L as a function of θ_1.

10.25. For the V-6 engine of Fig. 10.22, determine the effect on the inertia forces and inertia couples of "splitting" the crank pins.

10.26. Determine the locations and amounts of the weights to be added to balance the shaking forces in the four-bar linkage of Fig. 10.44.

10.27. Determine the locations and amounts of the weights to be added to balance the shaking forces in the four-bar linkage of Fig. 10.45.

Chapter Eleven

Introduction
to Synthesis

In the study of mechanisms so far, the proportions of a linkage have been given and the problem has been to analyze the motion produced by the linkage. It is quite a different matter, however, to start with a required motion and to try to proportion a mechanism to give this motion. This procedure is known as the *synthesis of mechanisms*. As has been mentioned earlier, designing a cam from the required displacement diagram is the only problem in synthesis that can be solved every time. In the application of synthesis to the design of a mechanism, the problem divides itself into three parts: (*a*) the type of mechanism to be used, (*b*) the number of links and connections needed to produce the required motion, and (*c*) the proportions or lengths of the links necessary. These divisions are often respectively referred to as *type*, *number*, and *dimensional* synthesis.

The designer generally relies on intuition and experience as a guide to type and number synthesis. Very little supporting theory is available in these areas. For this reason, the designer should be familiar with the capabilities and typical applications of a variety of mechanisms, including gears, belts and pulleys, chain drives, cams, and linkages.

In contrast to type and number synthesis, a great deal of theoretical background exists for dimensional synthesis of mechanisms. This chapter is primarily devoted to introducing some of this theory, particularly as it applies to the dimensional synthesis of linkages.

In the application of synthesis, one factor that must be continually kept in mind is that of the accuracy required of the mechanism. Sometimes, it is possible to design a linkage that will theoretically generate a given motion. Often, however, the designer must be satisfied with an approximation to the given motion. The difference between the motion that is desired and the motion that is actually produced is known as *structural error*. In addition, there are errors due to manufacture. The error resulting from tolerances in the lengths of the links and bearing clearances is referred to as *mechanical error*. Methods of calculating mechanical error are given by Hartenberg and Denavit[1] and by Garrett and Hall.[2]

In the early development of synthesis, graphical methods played a predominant role. This may have stemmed from the fact that some of the early methods were undoubtedly trial-and-error methods, which later developed into more rational procedures. With the continued development of synthesis, a number of analytical methods have been introduced. In this chapter, a variety of graphical and analytical methods are presented to illustrate the principles involved, the difficulties encountered, and the application of the methods.

11.1 CLASSIFICATION OF KINEMATIC SYNTHESIS PROBLEMS

Experience gained over a number of years has shown that problems in kinematic synthesis can generally be placed in one of three categories, namely, function generation, path generation, and body guidance.

Function generation most often involves coordinating the angular orientations of two links within a mechanism. A disk cam with an oscillating follower is one mechanism commonly used for function generation. The angular orientation of the follower is specified as a function of the rotation angle of the cam. The synthesis problem, as discussed in Chapter 3, is to find the shape of the cam surface given the follower displacements.

Another mechanism commonly used for function generation is the four-bar linkage, as shown in Fig. 11.1. Here, the synthesis problem is to find the dimensions of the linkage required to produce a specified functional relationship between the input angle θ and the output angle ψ. Function generation problems may involve translational as well as rotational inputs and outputs. For example, slider-crank linkages and cam and reciprocating follower mechanisms are used for linear-to-angular or angular-to-linear function generation. An example of a

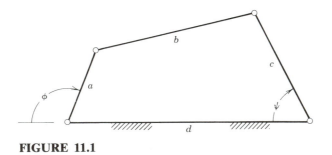

FIGURE 11.1

[1]R. S. Hartenberg and J. Denavit, *Kinematic Synthesis of Linkages*, McGraw-Hill, New York, 1964.

[2]R. E. Garrett and A. S. Hall, "Effect of Tolerance and Clearance in Linkage Design," *Trans. ASME, Journal of Engineering for Industry*, **91**(1), February 1969.

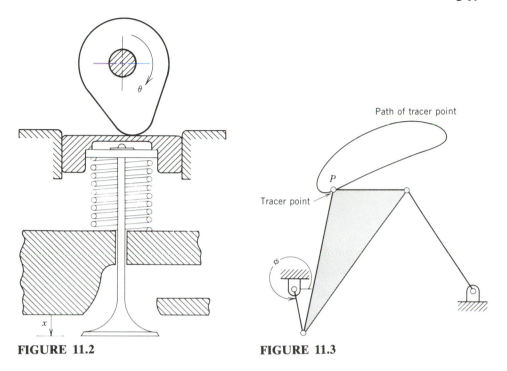

FIGURE 11.2   **FIGURE 11.3**

mechanism used for angular-to-linear function generation is the valve train in an internal-combustion engine. Figure 11.2 shows a valve being directly actuated by an overhead cam. The linear motion of the valve x must be an accurately defined function of the cam rotation angle θ.

In *path generation*, a mechanism is required to guide a point (called the tracer point) along a specified path. The path of one such tracer point is shown in Fig. 11.3. An example of a typical application of path generation is found in the knitting motion of an industrial loom. Figure 11.4 shows a latch needle hook about to pick up a strand of woven fabric and pull it into place. A plan view of the required teardrop-shaped path and the mechanism subsequently designed to produce this path is shown in Fig. 11.5.

Quite often in path generation, the motion of the point along its path must be coordinated with the motion of the input link. In other words, at specified values of the input angle ϕ, the tracer point in Fig. 11.3 is required to be at specified locations along its path. This type of problem is called path generation with prescribed input timing.

In *body guidance*, both the position of a point within a moving body and the angular orientation of the body are specified. Cam and follower mechanisms, simple gears, belts and pulleys, and similar devices are not capable of general body guidance, since points on the links of these mechanisms move either on a circular arc or along a straight line. For this same reason, the links connected to the ground of a four-bar linkage (links a and c in Fig. 11.6) cannot be used for

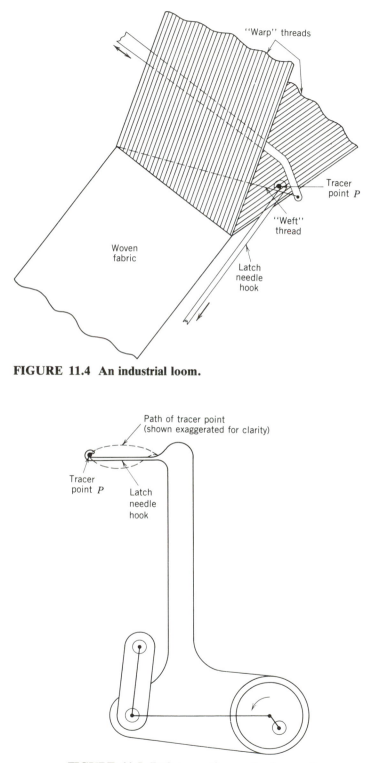

FIGURE 11.4 An industrial loom.

FIGURE 11.5 Path generating mechanism used in the loom of Fig. 11.4.

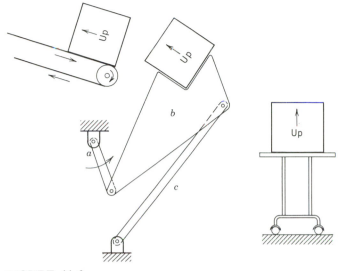

FIGURE 11.6

body guidance. The coupler link (link *b* in Fig. 11.6), however, does move with general rigid-body motion. Hence, the four-bar linkage is the simplest device capable of general body guidance. An example of a typical body guidance problem is shown in Fig. 11.6. Here, a carton is being automatically loaded from a conveyor onto a cart. During motion, the carton is securely held to the coupler link so that both carton and coupler link undergo the same rotations and translations.

As previously mentioned, a large majority of kinematic synthesis problems can be classified as either function generation, body guidance, or path generation. However, the reader should not be deceived into thinking that all problems fall naturally into one of these categories. An example of a problem that cannot easily be placed in one of the standard categories has been presented by Chuang and Waldron.[3]

The reader should also be aware that it is sometimes necessary to specify higher-order motion properties in mechanism synthesis. For example, the designer may wish to synthesize a function-generating four-bar linkage in which the angular position, velocity, and acceleration of the output link are specified in terms of the position, velocity, and acceleration of the input link.

11.2 SPACING OF ACCURACY POINTS FOR FUNCTION GENERATION

In designing a mechanism to generate a particular function, it is usually impossible to accurately produce the function at more than a few points. These points are known as *accuracy points*, or *precision points*, and must be so located as to

---

[3]J. C. Chuang and K. J. Waldron, "Synthesis with Mixed Motion and Path Generation Position Specifications," *Trans. ASME, Journal of Mechanisms, Transmission and Automation in Design*, **105**(4), December 1983.

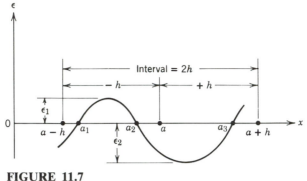

FIGURE 11.7

minimize the error generated between these points. As previously mentioned, the error produced is a structural error which may be expressed as follows:

$$\epsilon = f(x) - g(x)$$

where

$f(x)$ = desired function

$g(x)$ = function actually produced

In Fig. 11.7 is shown a plot of the variation in structural error as a function is generated over an interval $2h$ with the center of the interval at $x = a$. The error is zero at points a_1, a_2, and a_3, which are the accuracy points mentioned above. From this figure, it can be seen that the maximum error ϵ_1 produced by the mechanism in going from point a_1 to point a_2 is considerably smaller than the maximum error ϵ_2 produced in going from a_2 to a_3. Total structural error will be approximately minimized when these two errors are made equal. By using a theory developed by Chebyshev,[4] it is possible to locate the points a_1, a_2, and a_3

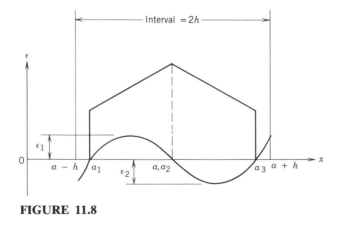

FIGURE 11.8

[4]R. S. Hartenberg and J. Denavit, *Kinematic Synthesis of Linkages*, McGraw-Hill, New York, 1964.

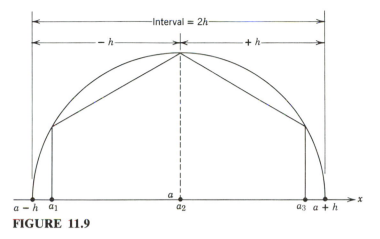

FIGURE 11.9

of Fig. 11.7 in such a manner that ϵ_1 is approximately equal to ϵ_2. Figure 11.8 shows this arrangement, and Fig. 11.9 illustrates the method of locating the three accuracy points with Chebyshev spacing. A semicircle is drawn on the x-axis with a radius h and center at point a. Half of a regular polygon is then inscribed in the semicircle so that two of its sides are perpendicular to the x-axis. Lines drawn perpendicularly to the x-axis from the vertices of the half-polygon determine the accuracy points a_1, a_2, and a_3. Figure 11.10 shows the construction for four accuracy points. It can be seen that for three accuracy points, the polygon is a hexagon and for four accuracy points, it is an octagon. In other words, the number of sides of the polygon is twice the number of accuracy points desired.

In general, the Chebyshev points may be calculated from the following equation:

$$a_j = a - h \cos \left[\frac{\pi(j - \frac{1}{2})}{n} \right] \qquad j = 1, 2, \ldots, n \qquad \textbf{(11.1)}$$

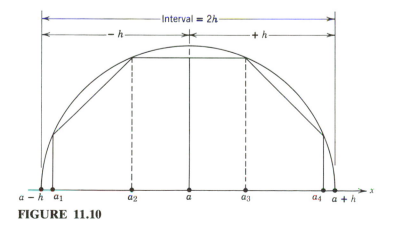

FIGURE 11.10

where

   n = number of accuracy points to be determined

   a_j = Chebyshev points

   a = center point of the interval

   h = half the width of the interval

11.3 ANALYTICAL DESIGN OF A FOUR-BAR LINKAGE AS A FUNCTION GENERATOR

It is often necessary to design a linkage to generate a given function, for example, $y = \log x$. Figure 11.11 shows a four-bar linkage arranged to generate the function $y = f(x)$ over a limited range. As link OA moves between the limits ϕ_1 and ϕ_n with the input x, link BC gives the value of $y = f(x)$ between the limits ψ_1 and ψ_n. It can be seen that in the linkage, there are three independent side ratios that define the proportions of the linkage. Also to be considered is the range (and scale factors) of ϕ and ψ and the initial angles ϕ_1 and ψ_1. In all, there are seven variables that must be considered in designing the linkage to generate $y = f(x)$. The magnitude of the task of synthesizing this function is immediately apparent.

A method has been developed by Freudenstein[5] by which a four-bar linkage can be designed to generate a function which is accurate at a finite number of points called *precision points*. The function is generated in an approximate sense between these points. In other words, the ideal function and the function actually generated will agree only at the precision points. Between these points, the actual function will differ from the ideal by an amount depending upon the distance between the points and upon the nature of the ideal function. Referring again to Fig. 11.11, the function would therefore only be exact at ψ_1 and ψ_n and at a specific number of points in between.

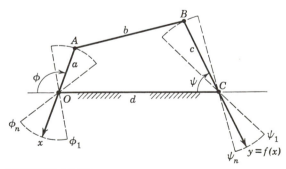

FIGURE 11.11

[5]F. Freudenstein, "Approximate Synthesis of Four-Bar Linkages," *Trans. ASME, Journal of Engineering for Industry,* **77**(6), 1955, p. 853.

In developing Freudenstein's method, the first step is to determine the relation between ϕ and ψ using the minimum number of side ratios. This relation can be derived considering Fig. 11.12, where a line parallel to link OA has been drawn from point B and a line parallel to link AB has been drawn from point O to give the parallelogram $OABD$. The links form a closed loop, and the sum of the x-components of lengths a, b, c must equal length d. In equation form,

$$a \cos(\pi - \phi) + b \cos \alpha + c \cos \psi = d \qquad \text{(11.2)}$$

By applying the law of cosines to triangle DOC,

$$e^2 = b^2 + d^2 - 2bd \cos \alpha \qquad \text{(11.3)}$$

Also from triangle DBC,

$$e^2 = a^2 + c^2 - 2ac \cos(\phi - \psi) \qquad \text{(11.4)}$$

Solving Eqs. 11.3 and 11.4 for $b \cos \alpha$ gives

$$b \cos \alpha = \frac{b^2 + d^2 - a^2 - c^2 + 2ac \cos(\phi - \psi)}{2d} \qquad \text{(11.5)}$$

By substituting Eq. 11.5 into Eq. 11.2 and letting $\cos(\pi - \phi) = -\cos \phi$,

$$a^2 - b^2 + c^2 + d^2 + 2ad \cos \phi - 2cd \cos \psi = 2ac \cos(\phi - \psi) \qquad \text{(11.6)}$$

By dividing by $2ac$,

$$\frac{a^2 - b^2 + c^2 + d^2}{2ac} + \frac{d}{c} \cos \phi - \frac{d}{a} \cos \psi = \cos(\phi - \psi) \qquad \text{(11.7)}$$

By letting

$$R_1 = \frac{d}{c}$$

$$R_2 = \frac{d}{a} \qquad \text{(11.8)}$$

$$R_3 = \frac{a^2 - b^2 + c^2 + d^2}{2ac}$$

Eq. 11.7 becomes

$$R_1 \cos \phi - R_2 \cos \psi + R_3 = \cos(\phi - \psi) \qquad \text{(11.9)}$$

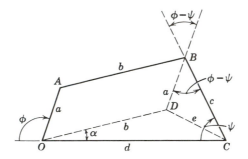

FIGURE 11.12

where R_1, R_2, and R_3 are three independent side ratios. Equation 11.9 gives the simplest relation possible between ϕ and ψ.

By use of Eq. 11.9, the method will now be extended to cover the design of a linkage to generate a function which is exact at three points. For greater accuracy, four- and five-point approximations have been developed. However, these systems are much more complicated and will not be included here.

The pairs of angles (ϕ, ψ) that correspond to the precision points are substituted into Eq. 11.9, which gives three simultaneous equations. The side ratios can then be determined from the solution of these equations. If the linkage is to pass through (ϕ_1, ψ_1), (ϕ_2, ψ_2), and (ϕ_3, ψ_3), then

$$R_1 \cos \phi_1 - R_2 \cos \psi_1 + R_3 = \cos(\phi_1 - \psi_1)$$

$$R_1 \cos \phi_2 - R_2 \cos \psi_2 + R_3 = \cos(\phi_2 - \psi_2) \qquad \textbf{(11.10)}$$

$$R_1 \cos \phi_3 - R_2 \cos \psi_3 + R_3 = \cos(\phi_3 - \psi_3)$$

In solving the simultaneous equations 11.10, let

$$\cos \phi_1 - \cos \phi_2 = w_1$$

$$\cos \phi_1 - \cos \phi_3 = w_2$$

$$\cos \psi_1 - \cos \psi_2 = w_3$$

$$\cos \psi_1 - \cos \psi_3 = w_4$$

$$\cos(\phi_1 - \psi_1) - \cos(\phi_2 - \psi_2) = w_5$$

$$\cos(\phi_1 - \psi_1) - \cos(\phi_3 - \psi_3) = w_6$$

Then,

$$R_1 = \frac{w_3 w_6 - w_4 w_5}{w_2 w_3 - w_1 w_4}$$

$$R_2 = \frac{w_1 w_6 - w_2 w_5}{w_2 w_3 - w_1 w_4} \qquad \textbf{(11.11)}$$

$$R_3 = \cos(\phi_i - \psi_i) + R_2 \cos \psi_i - R_1 \cos \phi_i \qquad \text{where } i = 1, 2, \text{ or } 3$$

From these side ratios, the lengths of the links can be determined from Eqs. 11.8. In determining the lengths of links a and c, a negative sign must be interpreted in a vector sense when drawing the linkage.

Example 11.1. Let it be required to proportion a four-bar linkage to generate $y = x^{1.5}$, where x varies between 1.0 and 4.0. Use Chebyshev spacing, and let $\phi_s = 30°$, $\Delta\phi = 90°$, $\psi_s = 90°$, and $\Delta\psi = 90°$. Assume $d = 1.000$ in.

$x_s = 1.0 \qquad y_s = 1.0$

$x_f = 4.0 \qquad y_f = 8.0$

The accuracy points are determined from the Chebyshev spacing, as shown in Fig. 11.13, and are calculated as follows:

$x_1 = 2.5 - 1.5\cos 30° = 1.201 \qquad y_1 = 1.317$

$x_2 = 2.50 \qquad\qquad\qquad\qquad y_2 = 3.96$

$x_3 = 2.5 + 1.5\cos 30° = 3.799 \qquad y_3 = 7.40$

$\phi_1 = \phi_s + \dfrac{x_1 - x_s}{x_f - x_s}\Delta\phi = 30 + \dfrac{1.201 - 1.0}{4.0 - 1.0} \times 90 = 36.03°$

$\phi_2 = \phi_1 + \dfrac{x_2 - x_1}{x_f - x_s}\Delta\phi = 36.03 + \dfrac{2.50 - 1.20}{3} \times 90 = 75.03°$

$\phi_3 = \phi_1 + \dfrac{x_3 - x_1}{x_f - x_s}\Delta\phi = 36.03 + \dfrac{3.799 - 1.20}{3} \times 90 = 114.0°$

$\psi_1 = \psi_s + \dfrac{y_1 - y_s}{y_f - y_s}\Delta\psi = 90 + \dfrac{1.317 - 1.0}{8.0 - 1.0} \times 90 = 94.08°$

$\psi_2 = \psi_1 + \dfrac{y_2 - y_1}{y_f - y_s}\Delta\psi = 94.08 + \dfrac{3.96 - 1.32}{7} \times 90 = 128.02°$

$\psi_3 = \psi_1 + \dfrac{y_3 - y_1}{y_f - y_s}\Delta\psi = 94.08 + \dfrac{7.40 - 1.32}{7} \times 90 = 172.25°$

$w_1 = \cos\phi_1 - \cos\phi_2 = 0.8087 - 0.2583 = 0.5504$

$w_2 = \cos\phi_1 - \cos\phi_3 = 0.8087 + 0.4067 = 1.2154$

$w_3 = \cos\psi_1 - \cos\psi_2 = -0.0713 + 0.6159 = 0.5446$

$w_4 = \cos\psi_1 - \cos\psi_3 = -0.0713 + 0.9909 = 0.9196$

$w_5 = \cos(\phi_1 - \psi_1) - \cos(\phi_2 - \psi_2) = 0.5292 - 0.6019 = -0.0727$

$w_6 = \cos(\phi_1 - \psi_1) - \cos(\phi_3 - \psi_3) = 0.5292 - 0.5262 = 0.003$

$R_1 = \dfrac{w_3 w_6 - w_4 w_5}{w_2 w_3 - w_1 w_4} = \dfrac{(0.545)(0.003) - (0.920)(-0.073)}{(1.215)(0.545) - (0.550)(0.920)}$

$R_1 = 0.440$

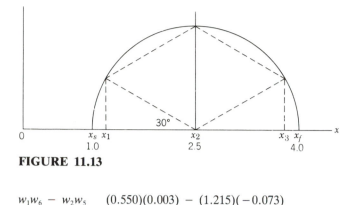

FIGURE 11.13

$$R_2 = \frac{w_1 w_6 - w_2 w_5}{w_2 w_3 - w_1 w_4} = \frac{(0.550)(0.003) - (1.215)(-0.073)}{(1.215)(0.545) - (0.550)(0.920)}$$

$$R_2 = 0.578$$

$$R_3 = \cos(\phi_1 - \psi_1) + R_2 \cos \psi_1 - R_1 \cos \phi_1$$

$$= 0.5292 + (0.578)(-0.0713) - (0.440)(0.8087)$$

$$= 0.132$$

From Eqs. 11.8 with $d = 1.000$ in.,

$$a = \frac{d}{R_2} = \frac{1.000}{0.578} = 1.730 \text{ in.}$$

$$c = \frac{d}{R_1} = \frac{1.000}{0.440} = 2.273 \text{ in.}$$

$$b = (a^2 + c^2 + d^2 - 2acR_3)^{1/2}$$

$$= [1.730^2 + 2.273^2 + 1.00^2 - 2(1.730)(2.273)(0.132)]^{1/2}$$

$$= 2.850 \text{ in.}$$

A sketch of the linkage $OABC$ is shown in Fig. 11.14.

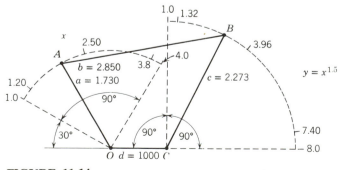

FIGURE 11.14

11.4 CURVE MATCHING APPLIED TO THE DESIGN OF A FOUR-BAR LINKAGE AS A FUNCTION GENERATOR

Another method of synthesis using displacement equations has been developed based on the work of Raven.[6] Consider the four-bar linkage as shown in Fig. 11.15, and let it be required to have θ_4 vary as a function of θ_2. A vector equation in terms of complex numbers can be written for the linkage as follows:

$$\mathbf{r}_B = \mathbf{r}_2 + \mathbf{r}_3 = \mathbf{r}_1 + \mathbf{r}_4$$

$$= r_2 e^{i\theta_2} + r_3 e^{i\theta_3} = r_1 + r_4 e^{i\theta_4} \qquad (11.12)$$

The lengths of the links can be made nondimensional by letting

$$R_2 = \frac{r_2}{r_1} \qquad R_3 = \frac{r_3}{r_1} \qquad R_4 = \frac{r_4}{r_1}$$

Equation 11.12 may therefore be written

$$R_2 e^{i\theta_2} + R_3 e^{i\theta_3} = 1 + R_4 e^{i\theta_4} \qquad (11.13)$$

By writing Eq. 11.13 in terms of real and imaginary parts,

$$R_2(\cos \theta_2 + i \sin \theta_2) + R_3(\cos \theta_3 + i \sin \theta_3) = 1 + R_4(\cos \theta_4 + i \sin \theta_4)$$

By separating the real and imaginary parts and solving for $R_3 \cos \theta_3$ and $R_3 \sin \theta_3$,

$$R_3 \cos \theta_3 = 1 + R_4 \cos \theta_4 - R_2 \cos \theta_2 \qquad \text{(real)} \qquad (11.14)$$

$$R_3 \sin \theta_3 = R_4 \sin \theta_4 - R_2 \sin \theta_2 \qquad \text{(imaginary)}$$

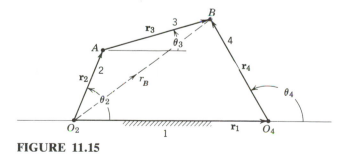

FIGURE 11.15

[6]F. H. Raven, "Position, Velocity, and Acceleration Analysis and Kinematic Synthesis of Plane and Space Mechanisms by a Generalized Procedure Called the Method of Independent Position Equations," L.C. Card No. 58–58, University Microfilms, Ann Arbor, MI, 1958.

The unknown angle θ_3 can be eliminated from Eqs. 11.14 by squaring the real and imaginary parts and adding the two parts:

$$1 + R_2^2 - R_3^2 + R_4^2 = 2R_2 \cos \theta_2 - 2R_4 \cos \theta_4 + 2R_2R_4 \cos(\theta_4 - \theta_2) \quad \textbf{(11.15)}$$

By expanding the term $\cos(\theta_4 - \theta_2)$ and rearranging, Eq. 11.15 can be written as

$$1 + R_2^2 - R_3^2 + R_4^2 = 2R_2 \cos \theta_2 + 2R_4(R_2 \cos \theta_2 - 1) \cos \theta_4$$
$$+ 2R_2R_4 \sin \theta_2 \sin \theta_4 \quad \textbf{(11.16)}$$

By solving Eq. 11.16 for θ_4,

$$\sin(\theta_4 + \beta) = \frac{1 + R_2(R_2 - 2 \cos \theta_2) - R_3^2 + R_4^2}{2R_4 \sqrt{1 + R_2(R_2 - 2 \cos \theta_2)}}$$

where

$$\beta = \tan^{-1} \frac{R_2 \cos \theta_2 - 1}{R_2 \sin \theta_2} \quad \textbf{(11.17)}$$

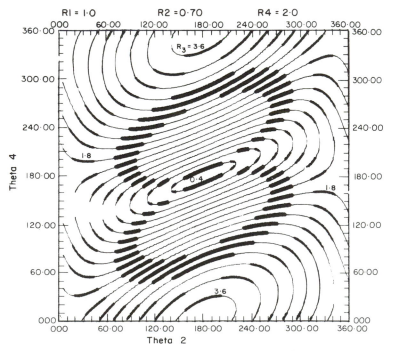

FIGURE 11.16 Reprinted with permission from R. S. Brown and H. H. Mabie, "Application of Curve Matching to Designing Four-Bar Mechanisms," *Journal of Mechanisms,* **5, 1970, p. 566.**

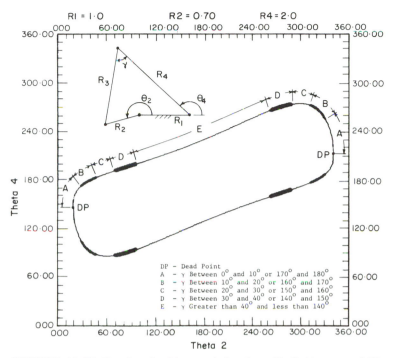

FIGURE 11.17 Reprinted with permission from R. S. Brown and H. H. Mabie, "Application of Curve Matching to Designing Four-Bar Mechanisms," *Journal of Mechanisms,* **5, 1970, p. 567.**

From the complexity of Eq. 11.17, it is obvious that some means other than direct substitution must be employed to proportion the linkage to generate θ_4 as a desired function of θ_2. A method[7] that has been successful is to plot a series of curves of constant R_3 for θ_4 versus θ_2 with R_2 and R_4 given values. Such curves are known as *displacement curves*. To select a linkage to generate a given function, the desired relation of θ_4 versus θ_2 is first plotted on transparent paper, and this curve is then superimposed on the displacement curves. The displacement curve which best fits the desired curve gives the approximate proportions of the linkage. Figure 11.16 shows an example of displacement curves plotted by computer with $R_1 = 1.0$, $R_2 = 0.7$, and $R_4 = 2.0$.

The variation in the width of the lines in Fig. 11.16 indicates values of transmission angles according to the legend given in Fig. 11.17 where only one displacement curve ($R_3 = 1.6$) is shown from Fig. 11.16.

To have a workable system, it is of course necessary to have plots of displacement curves for many combinations of R_2, R_3, and R_4. This system of synthesis is known as *curve matching*, and examples of this method are given by Brown and Mabie.

[7]R. S. Brown and H. H. Mabie, "Application of Curve Matching to Designing Four-Bar Mechanisms," *Journal of Mechanisms,* **5**, pp. 563–575, 1970.

11.5 GRAPHICAL DESIGN OF A FOUR-BAR LINKAGE AS A FUNCTION GENERATOR

There are many graphical methods of synthesis that have been developed. One method is presented here and others are given in an excellent work by Professor A. S. Hall of Purdue University.

The method to be discussed is one by which the proportions of a four-bar linkage can be found to give a required input-to-output motion at three positions. Figure 11.18 shows the layout where link 2 of known length passes through positions A_1, A_2, and A_3 and drives link 4 (or a pointer attached to it) through the angular positions B_1, B_2, and B_3. The distance O_2O_4 is also known, and it is required to find the lengths of links 3 and 4.

The easiest way to handle the problem is to invert the mechanism so that link 4 is fixed instead of link 1. As the mechanism passes through its cycle, it is evident that point O_2 will trace a circle about point O_4 and that point A will trace a circle about point B. Locating the center of the latter circle determines the position of point B and therefore the lengths of links 3 and 4.

Figure 11.19 shows the graphical construction for determining point B. Link 4 is considered fixed, and link 1 rotates counterclockwise about point O_4 through angles α' and β' which are equal but opposite in direction to α and β. Point O_2 moves through two positions O_2' and O_2'' while point A moves to A_2' and A_3' (the rotated positions of A_2 and A_3). Point A_2' is the intersection of the arc of radius O_2A swung about point O_2' and the arc of radius O_4A_2 swung about O_4. Point A_3' can similarly be determined using the arc of radius O_2A about point O_2'' and the arc of radius O_4A_3 about O_4. With points A_1, A_2', and A_3' available, the perpendicular bisectors of A_1A_2' and $A_2'A_3'$ can be drawn. Their intersection gives point B.

It should be mentioned that although a geometrical solution is possible, there is no way of telling before a layout is made whether the solution will give a practical mechanism. It must be examined for dead points, reversals, and mechanical advantage. If the solution is impractical, the length or position of link 2 or the length of link 1 must be changed and another trial made.

This method can also be applied to a three-dimensional linkage.

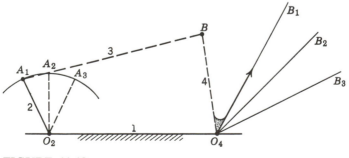

FIGURE 11.18

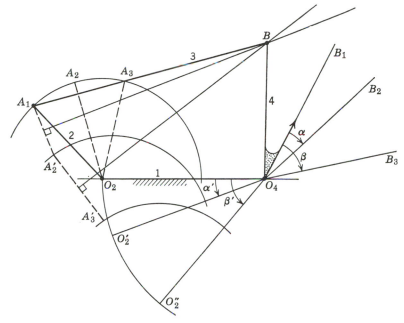

FIGURE 11.19

11.6 GRAPHICAL DESIGN OF A FOUR-BAR LINKAGE FOR BODY GUIDANCE

It has already been mentioned that linkages can be synthesized to generate only a small number of theoretically exact positions; these are the so-called precision positions. In general, the more precision positions a designer specifies, the more difficult the solution becomes. A four-bar linkage can be synthesized to satisfy a theoretical maximum of five precision positions of body guidance, but this number is seldom attainable in practice. Four-precision-position synthesis procedures are widely used in computer-aided mechanism design software,[8,9,10,11] but these methods are often impractical for hand calculation or graphical layout.

Three-precision-position synthesis procedures, on the other hand, are read-

---

[8]A. J. Rubel and R. E. Kaufman, "KINSYN III: A New Human-Engineered System for Interactive Computer-Aided Design of Planar Linkages," *Trans. ASME, Journal of Engineering for Industry*, **99**(2), May 1977.

[9]A. G. Erdman and J. E. Gustafson, "LINCAGES: Linkage Interactive Computer Analysis and Graphically Enhanced Synthesis Package," ASME Design Engineering Technical Conference, Paper No. 77-DET-5, September 1977.

[10]O. Sivertsen and A. Myklebust, "MECSYN: An Interactive Computer Graphics System for Mechanism Synthesis by Algebraic Means," ASME Design Engineering Technical Conference, Paper No. 80-DET-68, September, 1980.

[11]J. C. Chuang, R. T. Strong, and K. J. Waldron, "Implementation of Solution Rectification Techniques in an Interactive Linkage Synthesis Program," *Trans. ASME, Journal of Mechanical Design*, July 1981.

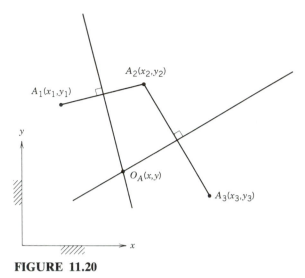

FIGURE 11.20

ily handled both graphically and analytically. They are sufficient to solve a wide class of industrial problems, and they give the designer a strong insight into the synthesis process. As with function generation, the three discrete precision positions may sometimes serve as an approximation to a continuous sequence of positions.

Before discussing the graphical three-position synthesis technique, it is nec-

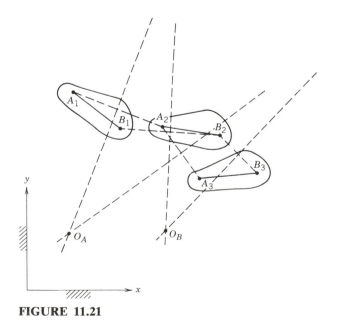

FIGURE 11.21

essary for the reader to recall how to graphically find the center of a circle defined by three points, such as points A_1, A_2, and A_3 in Fig. 11.20. The following procedure is suggested:

1. Draw any two of the three line segments $\overline{A_1A_2}$, $\overline{A_2A_3}$, or $\overline{A_1A_3}$. (Fig. 11.20 shows the procedure using segments $\overline{A_1A_2}$ and $\overline{A_2A_3}$.)
2. Find the perpendicular bisector of each line segment.
3. The intersection of the perpendicular bisectors of the two segments locates the center of the circle labeled O_A in the figure.

In retrospect, there is actually no need to draw in the line segments in step 1. These simply serve as a visual aid in finding the perpendicular bisectors. Once the reader is comfortable with the procedure, this step may be omitted.

Returning to the synthesis problem, consider the three positions of a rigid planar body containing the points A and B, as shown in Fig. 11.21. The three positions of point A are labeled A_1, A_2, and A_3, and these define a circle centered at the point labeled O_A. It is evident that a rigid link pinned to the body at point A and pinned to ground at point O_A will guide point A through its three positions. Similarly, the three positions of point B labeled B_1, B_2, and B_3 define a circle centered at O_B. A rigid link pinned to the body at point B and pinned to ground at point O_B will guide point B through its three positions. This construction has formed the four-bar linkage O_A–A–B–O_B which guides the body through the three specified positions. Figure 11.22 shows the linkage in these three positions.

It is important to realize that any point in the body may be selected as a moving pivot location. Generally speaking, only two guiding links can be used in any one mechanism. The addition of a third guiding link results in a structure that can be assembled at the precision positions but cannot move between them.

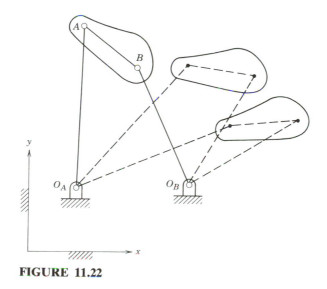

FIGURE 11.22

11.7 ANALYTICAL DESIGN OF A FOUR-BAR LINKAGE FOR BODY GUIDANCE

The graphical synthesis procedure described in the previous section is reasonably accurate in most cases and provides valuable insight into the synthesis process. At times, however, the accuracy of graphical methods is insufficient. This often occurs when it is necessary to graphically find the intersection of two lines that are nearly parallel. In addition, graphical methods may become tedious when a large number of trials are needed to find an acceptable solution mechanism. For these reasons, it is desirable to have an analytical solution that can be programmed on a computer or hand calculator. Although several solution methods are possible, the approach taken here will be to analytically find a point equidistant from the three locations of the moving pivot point.

Consider again the three locations of the point A labeled A_1, A_2, and A_3 in Fig. 11.20. The distance between point A_1 at x_1, y_1 and point O_A at x, y is given by

$$|\overline{A_1O_A}| = [(x_1 - x)^2 + (y_1 - y)^2]^{1/2} \tag{11.18}$$

Likewise, the distances $|\overline{A_2O_A}|$ and $|\overline{A_3O_A}|$ are given by

$$|\overline{A_2O_A}| = [(x_2 - x)^2 + (y_2 - y)^2]^{1/2} \tag{11.19}$$

$$|\overline{A_3O_A}| = [(x_3 - x)^2 + (y_3 - y)^2]^{1/2} \tag{11.20}$$

For O_A to be the center of the circle passing through points A_1, A_2, and A_3, these distances must be equal. This requirement is satisfied by making $|\overline{A_1O_A}| = |\overline{A_2O_A}|$ and $|\overline{A_2O_A}| = |\overline{A_3O_A}|$, or, upon expansion,

$$[(x_1 - x)^2 + (y_1 - y)^2]^{1/2} = [(x_2 - x)^2 + (y_2 - y)^2]^{1/2} \tag{11.21}$$

$$[(x_2 - x)^2 - (y_2 - y)^2]^{1/2} = [(x_3 - x)^2 + (y_3 - y)^2]^{1/2} \tag{11.22}$$

Squaring both sides of these two equations and simplifying gives the following two equations in the unknowns x and y:

$$2x(x_2 - x_1) + x_1^2 - x_2^2 + 2y(y_2 - y_1) + y_1^2 - y_2^2 = 0 \tag{11.23}$$

$$2x(x_3 - x_2) + x_2^2 - x_3^2 + 2y(y_3 - y_2) + y_2^2 - y_3^2 = 0 \tag{11.24}$$

These two equations can be solved simultaneously to find the coordinates x, y of the point O_A representing the intersections of the two midnormal lines. This intersection then becomes the location of the fixed pivot guiding point A through its three positions. The length of the guiding link can be determined from any of the three equations 11.18, 11.19, or 11.20. Other moving pivot points such as point B may be used in a similar fashion to find additional fixed pivot points and constraining links. The graphical and analytical body guidance synthesis procedures will now be illustrated by way of a numerical example.

Example 11.2. In designing pressure-sealing or thermal-sealing doors, it is sometimes necessary to reduce the clearance surrounding the door to less than that which could be obtained using a conventional hinge. One possible solution is to design a four-bar linkage that guides the door in and out with little rotation until it clears the surrounding structure, after which it swings fully open to one side. Figure 11.23 shows three positions of such a door undergoing this type of motion. Use both graphical and analytical techniques to find a four-bar linkage with moving pivots at points A and B that guides the body through these three positions.

Solution. The graphical solution is shown in Fig. 11.24. The fixed pivot locations are measured to be $O_A(-0.22, 1.5)$, $O_B(0.29, 1.5)$.

The coordinates of the points needed in the analytical solution are given as follows:

$A_1(2, 1)$ $\quad$ $B_1(3, 1)$

$A_2(2, 2)$ $\quad$ $B_2(3, 2)$

$A_3(1.5, 3)$ $\quad$ $B_3(1.5, 4)$

Substituting the coordinates of points A_1, A_2, and A_3 in synthesis equations 11.23 and 11.24 gives

$$2y(2 - 1) + (1)^2 - (2)^2 + 2x(2 - 2) + (2)^2 - (2)^2 = 0$$

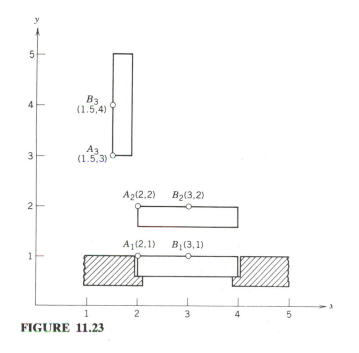

FIGURE 11.23

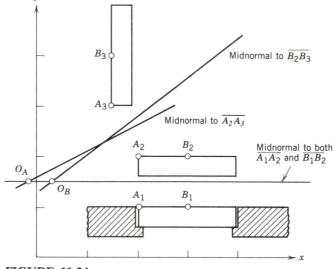

FIGURE 11.24

and

$$2y(3 - 2) + (2)^2 - (3)^2 + 2x(1.5 - 2) + (2)^2 - (1.5)^2 = 0$$

or, upon simplification,

$$y = 1.5$$

and

$$y = 0.5x + 1.625$$

Simultaneous solution of these equations gives $x = -0.25$ and $y = 1.5$, which are the coordinates of the fixed pivot point O_A. This process is repeated using the points B_1, B_2, and B_3 to find the point B_O. Substituting the coordinates of the points B_1, B_2, and B_3 in synthesis equations 11.23 and 11.24 gives

$$2y(2 - 1) + (1)^2 - (2)^2 + 2x(3 - 3) + (3)^2 - (3)^2 = 0$$

and

$$2y(4 - 2) + (2)^2 - (4)^2 + 2x(1.5 - 3) + (3)^2 - (1.5)^2 = 0$$

Solving these two equations gives $x = 0.25$ and $y = 1.5$ as the coordinates of the point O_B. The graphical and analytical results are seen to agree.

The resulting mechanism is shown attached to the door in the initial position in Fig. 11.25. The positions of the mechanism links in the second and third position are shown as dashed lines representing the centerlines of the links. Although the mechanism can be assembled in each of the three positions, it is not able to move between positions 1 and

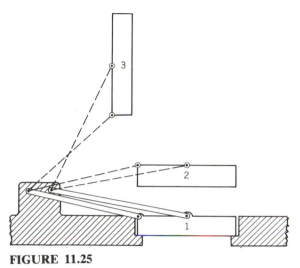

FIGURE 11.25

2. For this reason, it is not a workable solution. The problem is that positions 1 and 3 lie on one branch of the mechanism, and position 2 lies on the other branch of the mechanism. This problem, known as *branch defect*, is discussed in greater detail in section 11.10. To obtain a workable solution, one must try selecting different moving pivot locations or try altering the required positions of the door.

11.8 ANALYTICAL SYNTHESIS USING COMPLEX NUMBERS

The algebraic method of linkage synthesis described in the previous section is easily developed because it follows directly from the graphical procedure. Unfortunately, however, it is difficult to generalize this approach to allow for the specification of other free-choice parameters. Also, it assumes that the location of points in the moving body are known in the fixed frame of reference when in fact their location must usually be determined beforehand in a separate calculation. The complex-number approach overcomes both of these disadvantages.

Figure 11.26 shows a body undergoing general planar motion. Attached to and moving with the body is the moving coordinate system labeled o, u, v. The fixed, or reference, coordinate system is labeled O, x, y. The position and orientation of the moving coordinate system are known with respect to the fixed coordinate system, that is, $re^{i\theta}$ and α are known. Now consider point A attached to the moving body. In the previous section, it was assumed that the location of this point was known in the fixed O, x, y-coordinate system. In practice, however, the location of this point is usually known in the moving o, u, v-coordinate system. In other words, $pe^{i\beta}$ is known, and the problem is to find $qe^{i\gamma}$ (or equivalently x_A, y_A). In complex polar form, the unknown vector is given by

$$qe^{i\gamma} = re^{i\theta} + pe^{i(\alpha+\beta)}$$
$$= re^{i\theta} + pe^{i\alpha}e^{i\beta} \qquad\qquad \textbf{(11.25)}$$

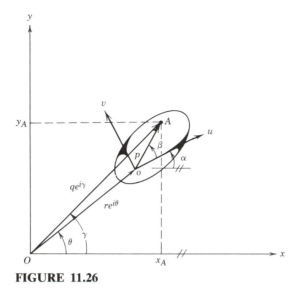

FIGURE 11.26

By separating this into real and imaginary parts, the coordinates of point A in the fixed reference system may be expressed as

$$x_A = r \cos \theta + p \cos(\alpha + \beta)$$
$$= r \cos \theta + p(\cos \alpha \cos \beta - \sin \alpha \sin \beta) \qquad \textbf{(11.26)}$$
$$y_A = r \sin \theta + p \sin(\alpha + \beta)$$
$$= r \sin \theta + p(\cos \alpha \sin \beta - \sin \alpha \cos \beta)$$

Note in Eq. 11.25 that multiplying the vector $pe^{i\alpha}$ by $e^{i\beta}$ has the effect of rotating it by an amount β in a right-hand sense (ccw) about the z-axis. The ease of rotating vectors in the plane is one of the principal advantages of the complex-number approach.

The synthesis methods presented in the two previous sections were based on what is often called the dyadic approach, where each constraining link of a mechanism is determined separately. The term *dyad* refers to a two-link chain composed of the guided body and the constraining link. The following complex-number synthesis formulation is also based on this method.

Consider the moving body shown in the 1st and jth positions in Fig. 11.27. Attached to the body is the moving o, u, v-coordinate system whose position $re^{i\theta}$ and orientation α have been specified in the two positions 1 and j with respect to the fixed O, x, y-coordinate system. Point A is the location of the moving revolute joint; its position in the o, u, v-coordinate system is given by $pe^{i\gamma}$, which is a constant. Point O_A is the location of the fixed revolute joint. The constraining link (the link from O_A to A) is defined by a constant-length vector $se^{i\beta}$ pointing from O_A to A. Note that the angular displacement β_j is shown negative because

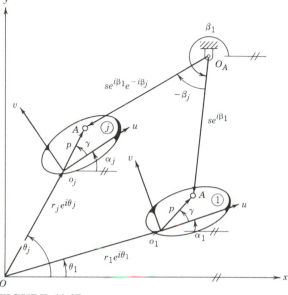

FIGURE 11.27

this is a clockwise rotation. The synthesis equation is obtained by summing the vectors forming a closed loop as follows:

$$r_1 e^{i\theta_1} + p e^{i\gamma} e^{i\alpha_1} - s e^{i\beta_1} + s e^{i\beta_1} e^{-i\beta_j} - p e^{i\gamma} e^{i\alpha_j} - r_j e^{i\theta_j} = 0 \qquad (11.27)$$

Grouping terms and rearranging this result leads to

$$p e^{i\gamma}(e^{i\alpha_1} - e^{i\alpha_j}) + s e^{i\beta_1}(e^{-i\beta_j} - 1) = r_j e^{i\theta_j} - r_1 e^{i\theta_1} \qquad (11.28)$$

In the body guidance problem, r_1, r_j, α_1, and α_j will be specified. The remaining variables, p, γ, s, β_1, and β_j, will be unknown. If three rigid-body positions have been specified, Eq. 11.28 will be written twice, namely, for $j = 2$ and $j = 3$, as follows:

$$p e^{i\gamma}(e^{i\alpha_1} - e^{i\alpha_2}) + s e^{i\beta_1}(e^{-i\beta_2} - 1) = r_j e^{i\theta_2} - r_1 e^{i\theta_1} \qquad (11.29)$$

$$p e^{i\gamma}(e^{i\alpha_1} - e^{i\alpha_3}) + s e^{i\beta_1}(e^{-i\beta_3} - 1) = r_j e^{i\theta_3} - r_1 e^{i\theta_1} \qquad (11.30)$$

This pair of vector equations is equivalent to four scalar equations containing six unknowns (p, γ, s, β_1, β_2, and β_3). It should therefore be possible to select any two unknowns and solve for the remaining four. Although this was not stated explicitly, in the graphical and analytical methods presented in the previous two sections, these free choices were always taken to be p and γ. The complex-number approach allows any two of the six unknowns to be selected as free-choice parameters. It is possible, for example, to select the angular displacements β_2 and

β_3 and thus to coordinate the input rotations with the motion of the body. Another useful combination of free-choice parameters is s and β_1. This combination completely determines the vector $se^{i\beta_1}$ and therefore determines the location of the fixed revolute joint.

Solution of the above equations (Eqs. 11.29 and 11.30) can be carried out analytically or numerically. Numerical techniques can easily be extended to four- or five-position synthesis, whereas the corresponding analytical solutions become quite involved.[12]

11.9 DESIGN OF A FOUR-BAR LINKAGE AS A PATH GENERATOR USING COGNATES

Three-position synthesis of a path-generating four-bar linkage is an easy matter when coordination of the motion along the path with the rotation of the input link is not required. This becomes a less restrictive case of the body guidance problem discussed in the previous sections. The positions of a point within the body are specified, but the orientations of the body are not specified. The designer can therefore arbitrarily select body orientations at the three prescribed positions and proceed as if synthesizing a body guidance mechanism. Different body orientations can be used to produce different solution mechanisms.

The problem of three-precision-position path generation is somewhat more difficult when the position of the tracer point must be coordinated with the rotations of the input link. One method for solving this problem involves a remarkable and useful concept known as the Roberts–Chebyshev cognate linkages. These cognate linkages are different in appearance, but they have closely related geometrical properties. Perhaps the best known of these properties is given in the following statement of the Roberts–Chebyshev[13] theorem: *Three different planar four-bar linkages will generate identical coupler point* (tracer point) *curves.* Thus, for any four-bar linkage, there are two related cognate linkages that will trace the same path. These cognates can be constructed as shown in Fig. 11.28. The original linkage O_A–A–B–O_B is shown in solid lines. The coupler link, containing tracer point P, is a triangle defined by the angles α, β, and γ. The two cognate linkages O_A–A_1–C_1–O_C and O_B–B_2–C_2–O_C are shown in dashed lines. Tracer point P is common to all three linkages, and the three coupler links form similar triangles.

The cognate linkages may be constructed using the following procedure:

1. From the original linkage, complete the parallelograms $O_A A P A_1$ and $O_B B P B_2$.
2. Construct the coupler triangles of the cognate linkages noting carefully the positions of the angles. Angle γ will always be at points C_1 and C_2. The cognate with the fixed pivot O_A will have the angle β at point P and, similarly, the cognate with fixed pivot O_B will have angle α at point P.

---

[12]G. N. Sandor and A. G. Erdman, *Advanced Mechanism Design: Analysis and Synthesis*, Prentice-Hall, Englewood Cliffs, NJ, 1984.
[13]R. S. Hartenberg and J. Denavit, *Kinematic Synthesis of Linkages*, McGraw-Hill, New York, 1964.

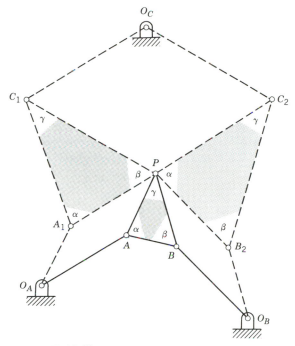

FIGURE 11.28

3. Complete the parallelogram $PC_1O_CC_2$, locating the third fixed pivot point O_C.
4. As a check on this construction, pivot point O_C can be located in another way. Points O_AO_B and O_C should form a triangle similar to traingle ABP, with angle α at O_A, angle β at O_B, and angle γ at O_C.

It is to be noted that, while the path of point P is identical in all three of these mechanisms, the rotations of the coupler links will be different.

Now, a key observation is made. Note from the parallelograms that the cognate links O_AA_1 and O_BB_2 undergo the same rotations as the coupler link in the original mechanism. Links O_AA_1 and O_BB_2 are, however, grounded links which could serve as input links for the first and second cognates, respectively. This suggests the following procedure for path generation synthesis with prescribed timing:

1. Transform the path generation problem into a body guidance problem by assigning the required input link rotations to the moving body.
2. Solve this new body guidance problem either graphically, using the method of section 11.6, or analytically, using the methods of section 11.7 or section 11.8.
3. Construct the two cognate linkages. Both of these satisfy the original path

generation with prescribed input-timing problem. The coordinated input link will be $O_A A_1$ for the first cognate and $O_B B_2$ for the second cognate.

It should be pointed out that cognate linkages are also useful in solving path generation problems when input link timing is not required. Suppose, for example, a mechanism has been synthesized which produces the desired path but other problems, such as poor transmission angle or unacceptable dynamic characteristics, make the solution unacceptable. The cognate mechanism may produce an acceptable solution, since the cognate linkages will trace the same path but will have different kinematic and dynamic characteristics.

11.10 PRACTICAL CONSIDERATIONS IN MECHANISM SYNTHESIS (*MECHANISM DEFECTS*)

The synthesis methods discussed in this chapter will always result in mechanisms that can reach the specified precision positions. Knowing this, many designers have proceeded to build prototype mechanisms only to find that the mechanism they synthesized is unable to satisfy the kinematic design requirements. This happens because several important factors have not been considered in the synthesis process. Specifically, three types of problems, or "defects," occur which can render a mechanism kinematically unsuitable for the design task. These are known as branch defect, order defect, and Grashof defect. Each of these will be discussed below in some detail.

Branch defect is perhaps the most perplexing problem to those who are unaware of it. Upon building a prototype of the synthesized mechanism, the

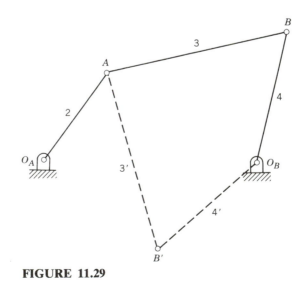

FIGURE 11.29

designer may find it satisfies only a portion of the precision positions. To understand this phenomenon, consider the four-bar linkage $O_A-A-B-O_B$ shown in solid lines in Fig. 11.29. It is evident that, without moving the input link $O_A A$, it is possible to assemble the mechanism in another configuration, namely, $O_A-A-B'-O_B$. These two distinct configurations are called the branches of the mechanism. Once the mechanism is assembled in one branch, it cannot move into the other branch, except by physically taking it apart and reassembling it in this other branch. Unfortunately, the synthesis techniques cannot differentiate between the two branches. For this reason, the mechanism must be checked after synthesis to see if all the precision positions lie in one branch. If not, the mechanism suffers from branch defect and is unsuitable.

Consider, for example, the three body positions shown in Fig. 11.30. Selecting the moving pivots at A and B and using the previously described synthesis techniques gives the ground pivots O_A and O_B. The extreme positions of link $O_B B$ in each branch are shown in Fig. 11.30. These are easily determined by graphically finding the mechanism positions where links $O_A A$ and AB are collinear. From this, it can be seen that positions 1 and 2 lie in one branch and position 3 lies in the other branch. The mechanism therefore cannot move through all three positions in a continuous motion cycle.

A simple analytical test also exists for determining whether or not a mechanism suffers from branch defect. Let μ be the angle from link $O_B B$ to link AB measured clockwise about point B. This angle is shown in Fig. 11.30 with subscripts 1, 2, and 3 to indicate the position of the mechanism being considered. The mechanism will be free from branch defect if, in all precision positions, either

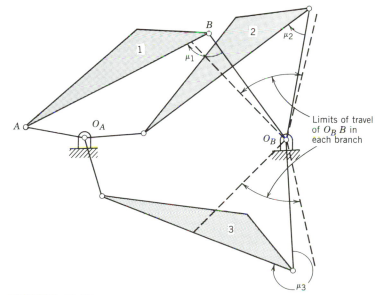

FIGURE 11.30

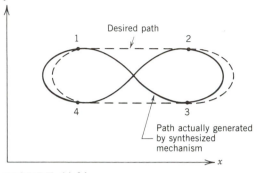

FIGURE 11.31

$0 < \mu < \pi$ or, in all positions, $\pi < \mu < 2\pi$. In the present example, μ_1 and μ_2 are between zero and π, but μ_3 is between π and 2π. Position 3 is thus seen to be in a different branch than positions 1 and 2.

The second type of defect to be considered is known as order defect. This type of defect only occurs in path generation and body guidance and only occurs when more than three precision positions have been specified. Consider, for example, the four positions of a point along its path, as shown in Fig. 11.31. The designer would like these points to be traversed in the order 1–2–3–4, as would be the case for a mechanism generating the path shown in dashed lines. Unfortunately, the synthesized mechanism may instead generate the figure-8 path shown as a solid line in Fig. 11.31. In this case, the positions cannot be generated in the order 1–2–3–4 (try moving along the path to see this), and the mechanism is said to suffer from order defect. The reader may verify that, regardless of the path, three positions can always be generated in the order 1–2–3. To accomplish this, however, it may be necessary to reverse the direction the path is being traversed or, in other words, to reverse the direction of input link rotation.

The last type of defect to be considered, and perhaps the most easily understood, is the so-called Grashof defect. Recall from Chapter 2, section 2.2, that Grashof's law predicts the relative rotatability of links within a four-bar linkage. Quite often, it is necessary to drive a linkage from a continuously rotating input source, such as an electric motor. In such a case, the input link of the mechanism would be required to rotate fully or, in the notation of section 2.2, a type 1 Grashof linkage would be required. If the synthesized mechanism is not of the correct Grashof type, the mechanism is said to suffer from Grashof defect. A more rigorous discussion of branch, order, and Grashof defects is given by Waldron and Stevensen.[14]

The three types of defects discussed in this section result from purely ki-

---

[14]K. J. Waldron and E. N. Stevensen, Jr., "Elimination of Branch, Grashof, and Order Defects in Path-Angle Generation and Function Generation Synthesis," *Trans. ASME, Journal of Mechanical Design*, **101**(3), July 1979.

nematic considerations; they depend only on the geometry of the mechanism. A number of additional problems may also render a mechanism unsuitable for the design task. Among these are excessive forces, stresses or deflections, imbalance, and vibration. These and other important topics are treated in Chapters 8, 9, and 10.

Problems

11.1. Using Freudenstein's method, determine the proportions of a four-bar linkage to generate $y = \tan x$ when x varies between $0°$ and $45°$. Use Chebyshev spacing. Let $\phi_s = 45°$, $\Delta\phi = 90°$, $\psi_s = 90°$, and $\Delta\psi = 90°$. Make a sketch of the linkage letting the ground link d be 1.00 in.

11.2. Using Freudenstein's method, determine the proportions of a four-bar linkage to generate $y = \log_{10} x$ when x varies between 1 and 10. Use Chebyshev spacing. Let $\phi_s = 45°$, $\Delta\phi = 60°$, $\psi_s = 135°$, and $\Delta\psi = 90°$. Make a sketch of the linkage letting the ground link d be 50 mm and check for dead points.

11.3. Using the methods of complex variables, derive Eq. 11.6 of Freudenstein's method.

11.4. The crank-shaper mechanism shown in Fig. 11.32 can be used as a function generator to give θ_4 as a function of θ_2. Using complex variables, prove that the relation between θ_4 and θ_2 is given by $\cos\theta_4 + R_2 \sin(\theta_2 - \theta_4) = 0$, where $R_2 = r_2/O_2O_4$.

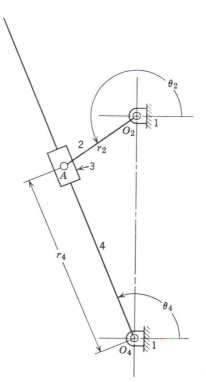

FIGURE 11.32

11.5. Using the relation given in Problem 11.4 for the crank-shaper mechanism of Fig. 11.32, plot θ_4 versus θ_2 for constant values of R_2 of $\frac{1}{2}$, 1, and 2. Let θ_2 and θ_4 both vary from $-90°$ to $270°$.

11.6. In a four-bar linkage, the length of link 2 is 38.0 mm and it is to rotate clockwise from its initial position (position 1) of 30° above the horizontal to 60° (position 2) and to 90° (position 3). As link 2 rotates from position 1 to position 2, link 4 rotates 13°. As link 2 goes from position 2 to position 3, link 4 rotates 20°. If the length of link 1 (O_2O_4) is 51.0 mm, determine graphically the lengths of links 3 and 4. Check the operation of the linkage by drawing it in position 2 and 3.

11.7. In a four-bar linkage, the length of link 2 is $1\frac{1}{2}$ in., and it is to rotate clockwise from its initial position (position 1) of 60° above the horizontal to 90° (position 2) and to 120° (position 3). As link 2 rotates from position 1 to position 2, link 4 rotates 10°. As link 2 goes from position 2 to position 3, link 4 rotates 15°. If the length of link 1 (O_2O_4) is 2 in., determine graphically the length of links 3 and 4. Check the operation of the linkage by drawing it in positions 2 and 3.

11.8. The maximum load the tower crane of Fig. 11.33 can lift without being toppled over is proportional to the moment arm the load acts through. This is given by $l \cos \theta$. Since l is a constant value, the crane operator only requires a readout of $\cos \theta$ on the instrument panel to make decisions about lifting a known load. The present crane design has a flexible cable transmitting the rotation angle θ to a gauge in the cab. This gauge, shown in Fig. 11.34, is marked off to read $\cos \theta$ from 0 to 1 in 0.1 increments. Unfortunately, this nonlinear scale is sometimes hard to read. Synthesize a function-generating four-bar linkage that will give an approximately linear output scale for $\cos \theta$ with θ as the input. Use a range of 90° for the input and 120° for the output.

11.9. In practice, the motion of the latch needle hook described in the knitting machine example of section 11.1 must also be coordinated with the motion of the input link. This is because the motion of the hook must be in correct phase with other machine motions

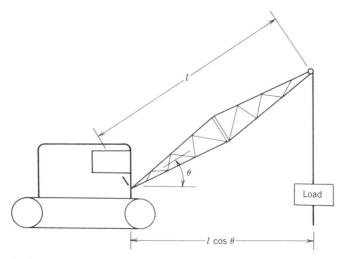

FIGURE 11.33

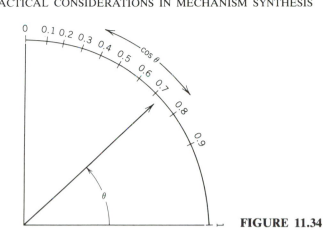

FIGURE 11.34

which are driven from the same shaft. The desired precision path points and the input link rotations are listed below, and are also shown in Fig. 11.35.

POINT	INPUT ANGLE
(5, 5)	0°
(8, 6)	60°
(12, 5)	120°

Use the method of cognates to synthesize a path-generating linkage satisfying these requirements.

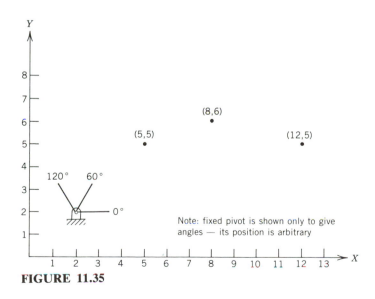

FIGURE 11.35

11.10.[15] A weather radar antenna has been designed to fit within the wing cavity of a single-engine aircraft. The antenna must scan from side to side, with the radar waves passing through a radar-transparent material that also serves as the leading edge of the wing. Unfortunately, the metal wing ribs required for structural support severely limit the antenna's field of view, as shown by the scan angle in Fig. 11.36. Presently, the maximum

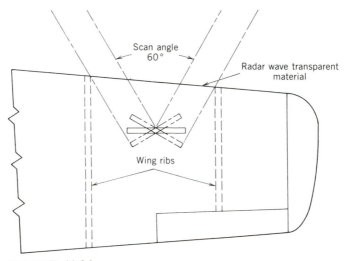

FIGURE 11.36

scan is approximately 60°. A new design has been proposed that would improve the field of view by translating the antenna to the right as it scans to the left, and vice versa. Synthesize a mechanism to guide the antenna through the three positions shown in Fig. 11.37 using (a) graphical methods; (b) analytical methods. Using the complex number notation of section 11.8 and of Fig. 11.27, the position of the antenna is given by

$$r_1 e^{i\theta_1} = 5e^{i(180°)}, \qquad r_2 = 0, \qquad r_3 e^{i\theta_3} = 5e^{i(0°)}$$

$$\alpha_1 = -45°, \qquad \alpha_2 = 0°, \qquad \text{and } \alpha_3 = 45°$$

The wing ribs are located at $x = 8$ and $x = -8$, the leading edge of the wing is at $y = 6$, and the trailing edge of the wing is at $y = -14$. Use the following moving pivot points, expressed in the uv-coordinate system:

Point A: $\qquad pe^{i\gamma} = 3.06\ e^{i(122°)}$

Point B: $\qquad pe^{i\gamma} = 2.05\ e^{i(117°)}$

---

[15]This problem was adapted from the paper by A. Myklebust, C. F. Reinholtz, W. H. Frances, and M. J. Keil, "Design of a Radar Guidance Mechanism Using MECSYN-ANIMEC," ASME paper No. 84-DET-139, 1984.

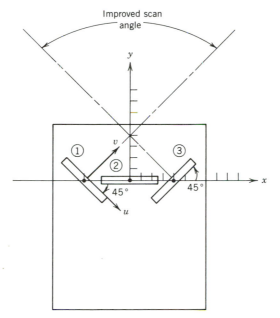

FIGURE 11.37

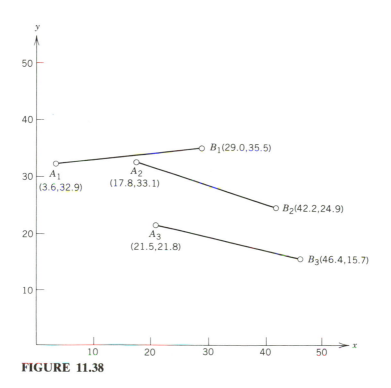

FIGURE 11.38

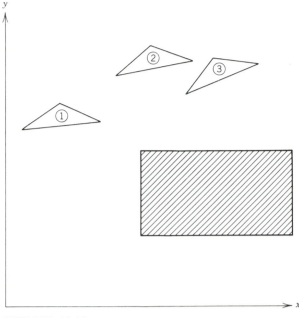

FIGURE 11.39

11.11. Synthesize a four-bar linkage to move the rod AB in Fig. 11.38 through the positions 1, 2, and 3. Use endpoints A and B as moving pivot points.

11.12. Synthesize a four-bar linkage to approximately generate the path $y = x^{1/2}$, where x varies from 0 to 1. The precision positions should be at $x = 0.07$, $x = 0.50$, and $x = 0.93$.

11.13. Synthesize a four-bar linkage to generate the same precision points specified in

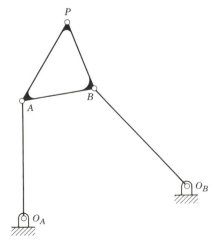

FIGURE 11.40

problem 11.12 while also having equal 15° input link rotations between positions 1 and 2 and between positions 2 and 3.

11.14. Synthesize a four-bar linkage to generate the three rigid-body positions shown in Fig. 11.39. As an additional requirement, try to find a solution linkage whose fixed pivots are within the shaded region.

11.15. Find the two cognate linkages to the linkage shown in Fig. 11.40.

11.16. Show that the linkage synthesized in Example 11.2 (page 565) is subject to branch defect.

Chapter Twelve

Spatial Mechanisms and Robotics

12.1 INTRODUCTION

Until recently, there has been little interest in the design and analysis of spatial mechanisms and robotic manipulators. Operations too difficult to automate with planar mechanisms were relegated to humans, often without regard to the boring or dangerous nature of the work. With the extraordinary advances in computers and electronics, many of the obstacles that once hindered the design and use of spatial devices have been removed. There is a growing awareness that more complex manufacturing operations can and must be automated if increased productivity and improved product quality are to be realized. Machine designers are recognizing that spatial mechanisms and robots are no longer novel devices of strictly theoretical interest; they are now viable alternatives which must be evaluated in the design process.

The material in this chapter is an introduction to the subjects of spatial mechanisms and robotics. The intent is to describe some of the devices that are available and the tasks these devices are able to perform, and to provide the basic tools needed for design and analysis. With this knowledge, the designer should be able to decide if a spatial mechanism or a robotic manipulator is appropriate for the particular task at hand. It should be cautioned, however, that the need to use spatial devices will be the exception rather than the rule. As with any other component of a machine, the use of these devices must be justified in terms of cost and performance.

As discussed in Chapter 2, the links of a planar mechanism are constrained to move in a single plane or in parallel planes. As a result, their motion can always be displayed graphically in the plane of the paper. Spatial mechanisms move in three-dimensional space. Consequently, their motion cannot be fully displayed by a single-view drawing. Although it is possible to graphically design and analyze some spatial mechanisms using a set of projected views, this process is tedious and the results are often inaccurate. For this reason, most of the material in this chapter is based on analytical vector and matrix formulations rather than graphical layouts.

12.2 MOBILITY

In developing Grubler's equation of mobility in Chapter 2, each link of a mechanism was assumed to be constrained to planar motion and thus to have three degrees of freedom (two translations and one rotation). In space, each link will have six degrees of freedom (three translations and three rotations). Therefore, connecting two spatial links with a joint having one degree of freedom, such as a revolute joint, has the effect of removing five degrees of freedom. Similarly, connecting two links with a two-degree-of-freedom joint has the effect of removing four degrees of freedom, and so forth. One link of a spatial mechanism will have all six degrees of freedom removed because it is fixed to ground. The total mobility of a system of n interconnected spatial links is therefore given by the following equation, often known as the Kutzbach equation:

$$M = 6(n - 1) - 5f_1 - 4f_2 - 3f_3 - 2f_4 - f_5 \qquad (12.1)$$

where

M = mobility, or number of degrees of freedom

n = total number of links, including the ground

f_1 = number of one-degree-of-freedom joints

f_2 = number of two-degree-of-freedom joints

f_3 = number of three-degree-of-freedom joints

f_4 = number of four-degree-of-freedom joints

f_5 = number of five-degree-of-freedom joints

In planar mechanisms, only four types of joints, or pairs, are commonly used: (1) the revolute joint, (2) the prismatic joint, (3) the rolling contact joint, and (4) the cam, or gear, joint. In the case of revolute, prismatic, and rolling contact joints, each has one degree of freedom, while the cam, or gear, joint has two degrees of freedom. Many other joint types are possible in spatial mechanisms. The most common of these include the screw, or helical, joint (one degree of freedom), the cylindric joint (two degrees of freedom), the spheric, or ball-and-socket, joint (three degrees of freedom), and the spatial cam joint (five degrees of freedom). These joints and the relative motion they permit are illustrated in Fig. 12.1.

The Kutzbach equation should be thought of as an aid in predicting the mobility of a device rather than a rule defining the mobility. Many useful mechanisms are known for which this equation will not work. As an example, the Kutzbach equation will be applied to the planar four-bar linkage shown in Fig.

Joint type	Physical form	Schematic representation		Degrees of freedom
		Plane	Space	
Revolute (R)	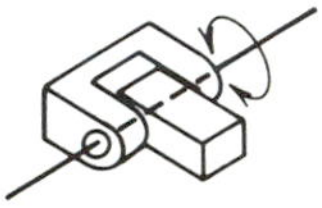			1
Cylindric (C)	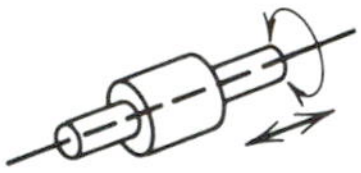			2
Prismatic (P)	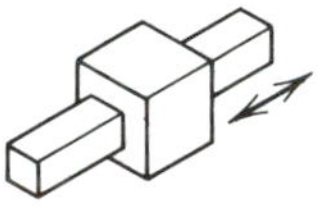			1
Spheric (S)				3
Screw, or helical (H)	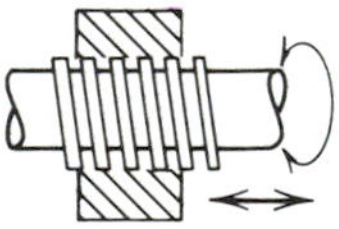			1
Spatial cam	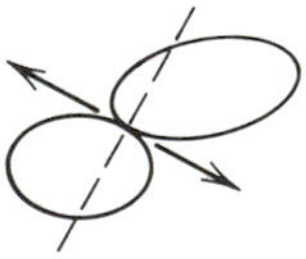			5 (sliding along the common tangent plane and rotation about the contact point)

FIGURE 12.1

12.2*a* as if it were a spatial device. There are four links and four revolute joints, and so the mobility is predicted to be

$$M = 6(4 - 1) - 5(4) = -2$$

This result is obviously incorrect. The problem is that the revolute joints all have the same orientation. As a result, they are not all able to remove five degrees of freedom. To see this, apply the Kutzbach equation to the device shown in Fig.

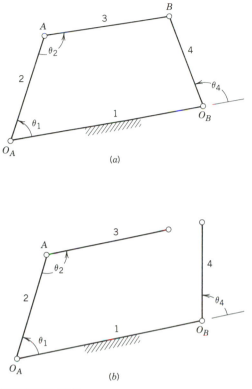

FIGURE 12.2

12.2*b*, where the revolute joint at point *B* has been eliminated from the four-bar linkage:

$$M = 6(4 - 1) - (5)3 = 3$$

This result is correct. The three degrees of freedom could, for example, be the three angles θ_1, θ_2, and θ_3. Now, observe that links 2 and 3 are constrained to move only in the plane of the paper. Connecting them with a planar revolute joint removes only two additional degrees of freedom (two relative translations in the plane). The joint at *B* also restricts the three motions out of the plane, but this is a redundant constraint and should not be considered in calculating the total mobility of the device.

The preceding example gives insight into one situation where the Kutzbach equation may fail to correctly predict mobility, namely, when two or more joint axes within a mechanism are parallel. Other special cases are known to occur when joint axes intersect or are perpendicular. Unfortunately, there is no set of rules which can be used to predict all the special situations that arise. By definition, a spherical mechanism results when all the joint axes intersect at a point. Hooke's coupling (the universal joint) is one example of a spherical mechanism.

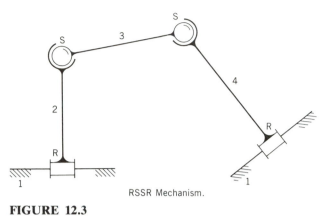

RSSR Mechanism.

FIGURE 12.3

All links in a spherical mechanism are constrained to move on the surface of a sphere (or on concentric spheres). Since this is a two-dimensional surface, Grubler's equation may be applied directly to spherical mechanisms.

Another device that apparently violates the Kutzbach equation is the revolute–spheric–spheric–revolute (RSSR) mechanism shown in Fig. 12.3. Application of Eq. 12.1 gives

$$M = 6(4 - 1) - 5(2) - 3(2) = 2$$

This result is correct but misleading. One degree of freedom is the idle rotation of the coupler SS link about its own axis. Specifying the position of one RS link will determine the position of the other RS link. In other words, this device has one degree of freedom as a function generator. The RSSR mechanism is perhaps the most commonly used spatial mechanism.

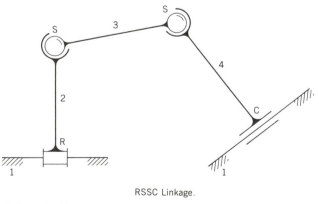

RSSC Linkage.

FIGURE 12.4

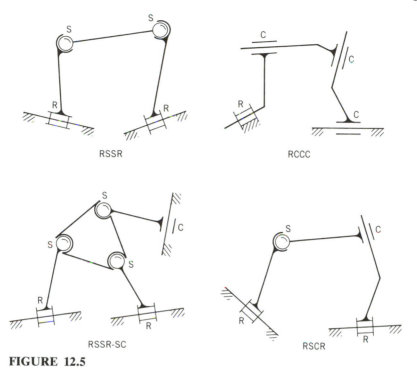

FIGURE 12.5

Example 12.1. Determine the mobility of the RSSC linkage shown in Fig. 12.4. Applying the Kutzbach equation (Eq. 12.1) gives

$$M = 6(4 - 1) - 5(1) - 4(1) - 3(2) = 3$$

Although the SS link again contains an idle degree of freedom, two degrees of freedom remain between the input and output links. In its general form, this will not be a useful mechanism.

Figure 12.5 shows several of the better-known and more commonly studied spatial linkages.

12.3 DESCRIBING SPATIAL MOTIONS

One of the most difficult problems encountered in extending planar kinematics to three dimensions is describing the angular displacements of rigid bodies. Rigid-body displacements, both planar and spatial, can always be expressed as the sum of two basic components: the angular displacement of the body plus the linear displacement of a reference point within the body. Describing the linear displacement of a point within a body is an easy matter. In the plane, it will be expressed as a two-component vector; and in space, it will be expressed as a three-component vector. Planar angular displacements of a body are also relatively easy to accomplish using matrix or complex-number operators. Spatial angular displacements, however, present a much greater challenge. In the discussion that follows, the concept of a planar rotation matrix operator is first developed. This concept is then used to develop a general spatial rotation matrix operator.

Consider, for example, the two positions of the planar body shown in Fig. 12.6. The vectors $\mathbf{p}_1$ and $\mathbf{v}_1$ locating the body in the first position are given. Also given are the linear displacement $\mathbf{d}_{12}$ of point p and the rotation α of the body about the z-axis. The problem is to find the vector $\mathbf{q}_2$ locating the displaced position of point q. The following vector relationships are easily deduced:

$$\mathbf{q}_1 = \mathbf{p}_1 + \mathbf{v}_1 \tag{12.2}$$

$$\mathbf{q}_2 = \mathbf{p}_2 + \mathbf{v}_2 \tag{12.3}$$

$$\mathbf{q}_2 = \mathbf{p}_1 + \mathbf{d}_{12} + \mathbf{v}_2 \tag{12.4}$$

Unfortunately, the vector $\mathbf{v}_2$ is not known directly. It can, however, be expressed in terms of $\mathbf{v}_1$ and α by using complex numbers as follows:

$$\mathbf{v}_2 = \mathbf{v}_1 e^{i\alpha} \tag{12.5}$$

where the operator $e^{i\alpha}$ rotates $\mathbf{v}_1$ by an amount α in a right-hand (ccw) sense. Expanding this result into real and imaginary parts by using the identity $e^{i\alpha} = \cos \alpha + i \sin \alpha$ gives

$$\mathbf{v}_2 = (v_{1x} + i\, v_{1y})(\cos \alpha + i \sin \alpha)$$

$$= (v_{1x} \cos \alpha - v_{1y} \sin \alpha) + i(v_{1x} \sin \alpha + v_{1y} \cos \alpha) \tag{12.6}$$

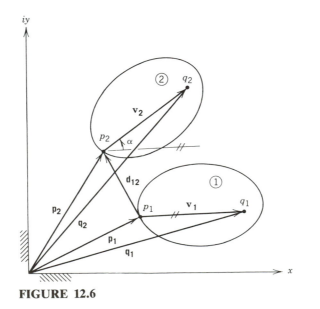

FIGURE 12.6

This same result can be expressed in matrix form as follows:

$$
\begin{bmatrix} v_{2x} \\ v_{2y} \end{bmatrix} = \begin{bmatrix} \cos \alpha & -\sin \alpha \\ \sin \alpha & \cos \alpha \end{bmatrix} \begin{bmatrix} v_{1x} \\ v_{1y} \end{bmatrix}
\tag{12.7}
$$

or, in shorthand form,

$$
\mathbf{v}_2 = [R]\mathbf{v}_1
$$

where R is the *plane rotation matrix* transforming the vector from the first orientation to the second. This same transformation (rotation about the z-axis by an amount α) can also be expressed in three-dimensional form as follows:

$$
\begin{bmatrix} v_{2x} \\ v_{2y} \\ v_{2z} \end{bmatrix} = \begin{bmatrix} \cos \alpha & -\sin \alpha & 0 \\ \sin \alpha & \cos \alpha & 0 \\ 0 & 0 & 1 \end{bmatrix} \begin{bmatrix} v_{1x} \\ v_{1y} \\ v_{1z} \end{bmatrix}
\tag{12.8}
$$

or, in shorthand form,

$$
\mathbf{v}_2 = [R_{\alpha,z}]\mathbf{v}_1
$$

Equation 12.8 actually forms one component of a three-dimensional rigid-body rotation. The other two components are rotations about the y-axis and about the x-axis. Rotating the vector $\mathbf{v}_1$ about the y-axis by an amount β to a new position $\mathbf{v}_2'$ gives

$$
\begin{bmatrix} v_{2x}' \\ v_{2y}' \\ v_{2z}' \end{bmatrix} = \begin{bmatrix} \cos \beta & 0 & \sin \beta \\ 0 & 1 & 0 \\ -\sin \beta & 0 & \cos \beta \end{bmatrix} \begin{bmatrix} v_{1x} \\ v_{1y} \\ v_{1z} \end{bmatrix}
\tag{12.9}
$$

or

$$
\mathbf{v}_2' = [R_{\beta,y}]\mathbf{v}_1
$$

Rotating the vector $\mathbf{v}_1$ about the x-axis by an amount γ to a new position $\mathbf{v}_2''$ gives

$$
\begin{bmatrix} v_{2x}'' \\ v_{2y}'' \\ v_{2z}'' \end{bmatrix} = \begin{bmatrix} 1 & 0 & 0 \\ 0 & \cos \gamma & -\sin \gamma \\ 0 & \sin \gamma & \cos \gamma \end{bmatrix} \begin{bmatrix} v_{1x} \\ v_{1y} \\ v_{1z} \end{bmatrix}
\tag{12.10}
$$

or

$$
\mathbf{v}_2'' = [R_{\gamma,x}]\mathbf{v}_1
$$

All spatial rotations may be defined in terms of the three basic planar transformations of Eqs. 12.8, 12.9, and 12.10. However, great care must be taken in defining the order in which these transformations are to occur, since rotations of a body in space are not commutative. To show this, consider a rectangular body initially lying in the yz-plane, as depicted in Figs. 12.7a and 12.7b. Figure 12.7a shows a sequence of 90° (ccw) rotations in the order α, β, γ (i.e., 90° rotation about the z-axis followed by 90° rotation about the y-axis followed by

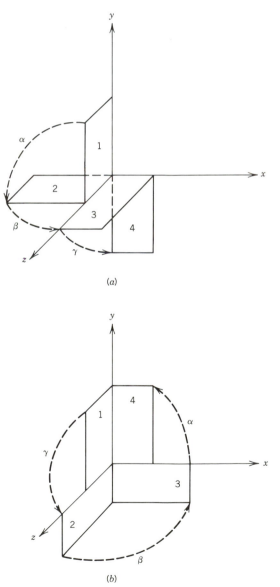

(a)

(b)

FIGURE 12.7

90° rotation about the x-axis). Figure 12.7b also shows a sequence of 90° (ccw) rotations, but this time in the order γ, β, α (i.e., 90° rotation about the x-axis followed by 90° rotation about the y-axis followed by 90° rotation about the z-axis). Clearly, these two sets of rotations are not equivalent. Spatial rotations can be defined using any order of the three basic rotations. Once a particular order is selected, however, it must be adhered to. In this text, rotations will be taken in the following order: (1) α about the z-axis, (2) β about the y-axis, and (3) γ about the x-axis. A general spatial rotation of the vector $\mathbf{v}_1$ to some new position $\mathbf{v}_2$ can now be expressed in terms of the three basic rotation matrices as follows:

$$\mathbf{v}_2 = [R_{\gamma,x}][R_{\beta,y}][R_{\alpha,z}]\mathbf{v}_1 \qquad (12.11)$$

Note the order in which the rotations must be performed. Vector $\mathbf{v}_1$ is first rotated by an amount α about the z-axis followed by a rotation β about the y-axis followed by a rotation γ about the x-axis. Combining the three basic rotation matrices into a single spatial rotation matrix by performing the successive matrix multiplications leads to

$$\mathbf{v}_2 = \begin{bmatrix} C\alpha C\beta & -S\alpha C\beta & S\beta \\ S\alpha C\gamma + C\alpha S\beta S\gamma & C\alpha C\gamma + S\alpha S\beta S\gamma & -C\beta S\gamma \\ S\alpha S\gamma - C\alpha S\beta C\gamma & C\alpha S\gamma + S\alpha S\beta C\gamma & C\beta C\gamma \end{bmatrix} \mathbf{v}_1 \qquad (12.12)$$

where S and C represent sine and cosine, respectively. This result may also be written in condensed form as

$$\mathbf{v}_2 = [R_{\gamma,\beta,\alpha}]\mathbf{v}_1 \qquad (12.13)$$

In the preceding equations (Eqs. 12.12 and 12.13), the three scalar parameters α, β, and γ completely determine the angular displacement from position 1 to position 2. Nevertheless, it is usually more convenient to work with the 3×3 nine-component rotation matrix when describing spatial angular displacements. Although the rotation matrix describing a given angular displacement is unique, several other choices are possible for the three independent scalar parameters used to determine the rotation matrix. The preceding discussion was based on a sequence of rotations about a right-hand set of Cartesian axes. A more useful but less obvious method is to define a single rotation about an axis in space.

It often happens that one of the links of a mechanism or manipulator rotates about a known axis which is not parallel to any of the Cartesian coordinate axes. It would be advantageous to be able to describe link rotations directly in terms of an axis direction and the angle of rotation about this axis. It can be shown that a body undergoing a finite angular displacement has within it a line which remains stationary during the rotation. In other words, given the finite angular displacement of a body, it will always be possible to find an axis which, at least momentarily, can be considered a fixed axis of rotation. The orientation of the rotation axis will be expressed by the unit vector **u** having components u_x, u_y, and u_z. By definition, the magnitude of a unit vector must equal 1:

$$u_x^2 + u_y^2 + u_z^2 = 1$$

This shows that only two scalar components of the vector **u** are independent. The rotation θ of the body about this axis is the third scalar quantity defining the angular displacement. To express the total angular displacement in terms of the basic rotation matrices, it is necessary to align one of the Cartesian coordinate axes along the vector **u**. A method for accomplishing this is shown in Fig. 12.8. Begin by rotating the vector **u** by an amount $-\beta$ about the y-axis. This is followed by a rotation γ about the x-axis. The z-axis will then be aligned with the rotated vector **u**. Now, perform the desired rotation θ about the z-axis. Following this, the vector **u** is returned to its original position by rotating it $-\gamma$ about the x-axis and β about the y-axis. This series of transformations can be written in matrix form as follows:

$$\mathbf{v}_2 = [R_{\beta,y}][R_{-\gamma,x}][R_{\theta,z}][R_{\gamma,x}][R_{-\beta,y}]\mathbf{v}_1 \qquad \textbf{(12.14)}$$

Carrying out the matrix multiplication and making the following substitutions

$$\sin \gamma = u_y$$

$$\cos \gamma \sin \beta = u_x$$

$$\cos \gamma \cos \beta = u_z$$

leads to

$$\mathbf{v}_2 = \begin{bmatrix} u_x^2 V\theta + C\theta & u_x u_y V\theta - u_z S\theta & u_x u_z V\theta + u_y S\theta \\ u_x u_y V\theta + u_z S\theta & u_y^2 V\theta + C\theta & u_y u_z V\theta - u_x S\theta \\ u_x u_z V\theta - u_y S\theta & u_y u_z V\theta + u_x S\theta & u_z^2 V\theta + C\theta \end{bmatrix} \mathbf{v}_1 \qquad \textbf{(12.15)}$$

where

$$V\theta = 1 - \cos \theta$$

$$C\theta = \cos \theta$$

$$S\theta = \sin \theta$$

This may be written in the condensed form

$$\mathbf{v}_2 = [R_{\theta,\mathbf{u}}]\mathbf{v}_1$$

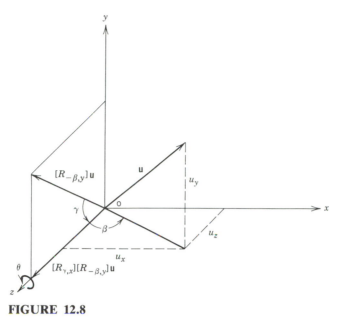

FIGURE 12.8

The rotation matrix $[R_{\theta,\mathbf{u}}]$ is called the *axis rotation matrix*. It is used extensively in the analysis and synthesis of spatial mechanisms. It should be pointed out that several other possibilities exist for selecting the three scalar quantities describing spatial angular displacements. Nevertheless, it must be emphasized that the rotation matrix describing a given angular displacement is unique, even though a variety of methods exist for selecting the three independent angular-motion parameters.

Example 12.2. Figure 12.9 shows a three-link spatial chain with two revolute joints. Link 1 is the ground, and links 2 and 3 are moving links. The following vectors are defined in the initial position:

$\mathbf{u}_1$ = unit vector along the revolute joint axis at A

$\quad = 0\mathbf{i} + 0\mathbf{j} + 1\mathbf{k}$

$\mathbf{u}_2$ = unit vector along the revolute joint axis at B

$\quad = 0\mathbf{i} - 1\mathbf{j} + 0\mathbf{k}$

$\mathbf{v}_1$ = vector along link 2 from A to B

$\quad = 10\mathbf{i} + 0\mathbf{j} + 0\mathbf{k}$

$\mathbf{v}_2$ = vector along link 3 from B to C

$\quad = 0\mathbf{i} + 0\mathbf{j} - 10\mathbf{k}$

$\mathbf{q}$ = vector from the origin to point C

$\quad = \mathbf{v}_1 + \mathbf{v}_2 = 10\mathbf{i} + 0\mathbf{j} - 10\mathbf{k}$

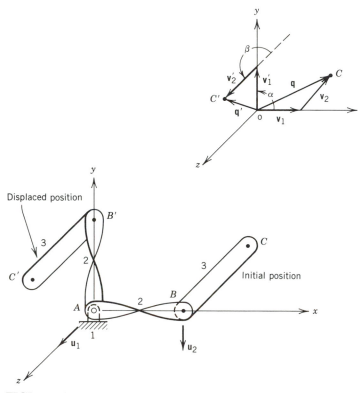

FIGURE 12.9

For a rotation $\alpha = 90°$ about $\mathbf{u}_1$ and $\beta = 180°$ about $\mathbf{u}_2$, find the displaced position of point C.

Solution. By denoting the displaced positions of the above vectors with a prime (i.e., $\mathbf{v}_1'$, $\mathbf{v}_2'$, $\mathbf{q}'$), the new position of point C is given by

$$\mathbf{q}' = [R_{\alpha,\mathbf{u}_1}](\mathbf{v}_1 + \mathbf{v}_2')$$

where

$$\mathbf{v}_2' = [R_{\beta,\mathbf{u}_2}]\mathbf{v}_2$$

Substituting $\theta = \beta = 180°$ and $u_x = 0$, $u_y = -1$, $u_z = 0$ into Eq. 12.15 gives

$$[R_{\beta,\mathbf{u}_2}] = \begin{bmatrix} -1 & 0 & 0 \\ 0 & 1 & 0 \\ 0 & 0 & -1 \end{bmatrix}$$

and

$$\mathbf{v}_2' = \begin{bmatrix} -1 & 0 & 0 \\ 0 & 1 & 0 \\ 0 & 0 & -1 \end{bmatrix} \begin{bmatrix} 0 \\ 0 \\ -10 \end{bmatrix} = \begin{bmatrix} 0 \\ 0 \\ 10 \end{bmatrix}$$

Substituting $\theta = \alpha = 90°$ and $u_x = 0$, $u_y = 0$, $u_z = 1$ into Eq. 12.15 gives

$$[R_{\alpha, \mathbf{u}_1}] = \begin{bmatrix} 0 & -1 & 0 \\ 1 & 0 & 0 \\ 0 & 0 & 1 \end{bmatrix}$$

and

$$\mathbf{q}' = \begin{bmatrix} 0 & -1 & 0 \\ 1 & 0 & 0 \\ 0 & 0 & 1 \end{bmatrix} \begin{bmatrix} 10 + 0 \\ 0 + 0 \\ 0 + 10 \end{bmatrix} = \begin{bmatrix} 0 \\ 10 \\ 10 \end{bmatrix}$$

or, written in unit vector form,

$$\mathbf{q}' = 0\mathbf{i} + 10\mathbf{j} + 10\mathbf{k}$$

Note that, in this example, both rotations occurred about Cartesian coordinate axes, so either of the two types of rotation matrices defined above could have been used.

12.4 KINEMATIC ANALYSIS OF SPATIAL MECHANISMS

The wide variety of spatial mechanisms makes it difficult to develop a unified method of analysis that applies to all cases. Most spatial mechanisms currently believed to be of practical importance can be directly analyzed by using vector loop closure equations or constraint equations in conjunction with the axis rotation matrix.

As an example, consider the RSSR spatial linkage shown in Fig. 12.10. The mechanism is described in its initial position by link vectors $\mathbf{r}_1$, $\mathbf{r}_2$, $\mathbf{r}_3$, and $\mathbf{r}_4$ and by the joint axes unit vectors $\mathbf{u}_2$ and $\mathbf{u}_4$. Additionally, unit vector $\mathbf{s}_3$ is directed along link 3, so that

$$\mathbf{r}_3 = r_3 \mathbf{s}_3 \tag{12.16}$$

where r_3 is the magnitude of vector $\mathbf{r}_3$. The loop closure equation in the initial position is

$$\mathbf{r}_1 + \mathbf{r}_4 - \mathbf{r}_2 - \mathbf{r}_3 = 0 \tag{12.17}$$

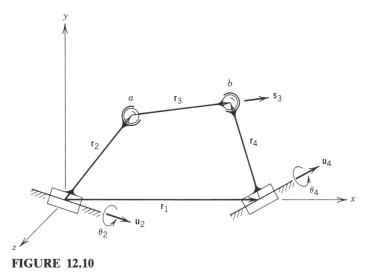

FIGURE 12.10

After a specified rotation θ_2 of the input link (link 2) about $\mathbf{u}_2$, the link vectors are displaced to some new positions $\mathbf{r}'_2$, $\mathbf{r}'_3$, and $\mathbf{r}'_4$. The loop closure equation for the displaced position is

$$\mathbf{r}_1 + \mathbf{r}'_4 - \mathbf{r}'_2 - \mathbf{r}'_3 = 0 \tag{12.18}$$

Note that link 1 is fixed to ground and does not displace. Link vectors $\mathbf{r}'_2$ and $\mathbf{r}'_4$ can now be expressed in terms of the known starting vectors and the rotation angles θ_2 and θ_4 by using the axis rotation matrix:

$$\mathbf{r}'_2 = [R_{\theta_2, \mathbf{u}_2}]\mathbf{r}_2 \tag{12.19}$$

$$\mathbf{r}'_4 = [R_{\theta_4, \mathbf{u}_4}]\mathbf{r}_4 \tag{12.20}$$

If the initial position of the mechanism and the input link rotation θ_2 are given, the vector $\mathbf{r}'_2$ can be calculated from Eq. 12.19. The position of the output link vector $\mathbf{r}'_4$ is only a function of the rotation angle θ_4. The displaced coupler link vector $\mathbf{r}'_3$ can be expressed in terms of the known vector magnitude r_3 and the unknown orientation of the displaced unit vector $\mathbf{s}'_3$ as follows:

$$\mathbf{r}'_3 = r_3\mathbf{s}'_3 \tag{12.21}$$

Substituting the results of Eqs. 12.19, 12.20, and 12.21 into Eq. 12.18 gives

$$\mathbf{r}_1 + [R_{\theta_4, \mathbf{u}_4}]\mathbf{r}_4 - [R_{\theta_2, \mathbf{u}_2}]\mathbf{r}_2 - r_3\mathbf{s}'_3 = 0 \tag{12.22}$$

This is a single vector equation (or, equivalently, three scalar equations) in the unknowns θ_4 (a scalar quantity) and $\mathbf{s}'_3$ (a unit vector). The unit vector $\mathbf{s}'_3$ may be expressed in terms of its components s'_{3x}, s'_{3y}, and s'_{3z}. It is most convenient to treat these as three independent unknowns in the analysis and then to include the following unit magnitude equation:

$$(s'_{3x})^2 + (s'_{3y})^2 + (s'_{3z})^2 = 1 \qquad \text{(12.23)}$$

Equations 12.22 and 12.23 form a set of four scalar equations in the four unknowns θ_4, s'_{3x}, s'_{3y}, and s'_{3z}. This set of equations may be easily solved using numerical techniques such as the Newton–Raphson root-finding method discussed in Chapter 2. Since these equations are somewhat complicated, it is probably best to obtain numerical approximations to the partial derivatives needed in the Newton–Raphson scheme rather than calculating them analytically.

It is also possible to develop a closed-form solution to the above equations. To do this, it is first necessary to express the axis rotation matrix $[R_{\theta,u}]$ as an explicit function of $\cos \theta$ and $\sin \theta$ as follows:

$$[R_{\theta,u}] = -[P_u][P_u] \cos \theta + [P_u] \sin \theta + [Q_u] \qquad \text{(12.24)}$$

where

$$[P_u] = \begin{bmatrix} 0 & -u_z & u_y \\ u_z & 0 & -u_x \\ -u_y & u_x & 0 \end{bmatrix}$$

and

$$[Q_u] = \begin{bmatrix} u_x^2 & u_x u_y & u_x u_z \\ u_x u_y & u_y^2 & u_y u_z \\ u_x u_z & u_y u_z & u_z^2 \end{bmatrix}$$

Using this form of the rotation matrix in place of $[R_{\theta_4,u_4}]$ in Eq. 12.22 results in a set of four nonlinear equations. These can be solved by substituting $\sin \theta_4 = (1 - \cos^2 \theta_4)^{1/2}$ or by using the tangent-half-angle substitution described in Appendix 1.

12.5 KINEMATIC SYNTHESIS OF SPATIAL MECHANISMS

Although there are many more possible spatial mechanisms than planar mechanisms, the basic tasks of kinematic synthesis (i.e., function generation, path generation, and body guidance) remain the same. In addition, the dyad-based approach to synthesis described for planar mechanisms in Chapter 11 can be extended to include most spatial mechanisms of practical importance. In this section, body guidance synthesis equations will be developed for the SS and RS spatial dyads. As with planar mechanisms, the function generation problem can be solved as an inversion of the body guidance problem. The path generation problem can be solved as a body guidance problem by arbitrarily assuming body orientations associated with the tracer point.

Dyadic synthesis equations for planar linkages were derived in section 11.7 by recognizing that the distance between the fixed and moving revolute joints must be constant. Synthesis equations for the SS spatial dyad are derived in exactly the same way, that is, by expressing the requirement that the SS link be of constant length. The SS dyad is shown in its 1st and jth position in Fig. 12.11. The constant-length equation is most easily written in the following vector dot product form:

$$(\mathbf{a}_1 - \mathbf{a}_0) \cdot (\mathbf{a}_1 - \mathbf{a}_0) = (\mathbf{a}_j - \mathbf{a}_0) \cdot (\mathbf{a}_j - \mathbf{a}_0) \qquad j = 2, 3, \ldots, n \qquad \textbf{(12.25)}$$

The positions $\mathbf{o}_j$ and orientations $[R_j]$, $j = 1, 2, \ldots, n$, of the body are given quantities, and the synthesis problem is to find the vectors $\mathbf{a}_0$ and $\mathbf{a}_1$ defining the dyad in its starting position. The moving spheric joint undergoes the same rotations and translations as the moving body. Its position is therefore given by

$$\mathbf{a}_j = \mathbf{o}_j + [R_j](\mathbf{o}_1 - \mathbf{a}_1) \qquad \textbf{(12.26)}$$

Substituting this expression for $\mathbf{a}_j$ into Eq. 12.25 results in a set of $n - 1$ equations

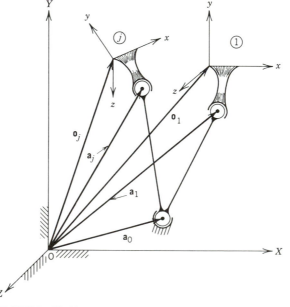

FIGURE 12.11

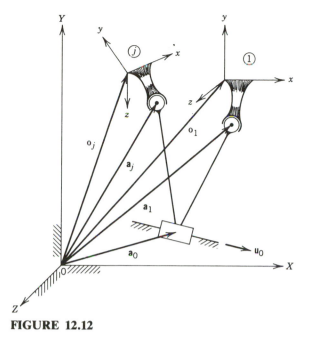

FIGURE 12.12

in the six unknown components of $\mathbf{a}_0$ and $\mathbf{a}_1$. Therefore, a maximum of seven positions can be satisfied using the SS dyad.

The RS dyad shown in Fig. 12.12 must also satisfy the constant link length condition of Eq. 12.25. In addition, it must satisfy the requirement that the spheric joint lie in a plane perpendicular to the axis of the revolute joint in all positions of the mechanism. This requirement can be expressed by the following equation:

$$\mathbf{u}_0 \cdot (\mathbf{a}_j - \mathbf{a}_0) = 0 \qquad j = 1, 2, \ldots, n \qquad \textbf{(12.27)}$$

The expression for $\mathbf{a}_j$ from Eq. 12.26 is substituted into Eqs. 12.25 and 12.27. This leads to a set of $2n - 1$ design equations in the eight unknown components of $\mathbf{a}_0$, $\mathbf{a}_1$, and $\mathbf{u}_0$ ($\mathbf{u}_0$ is a unit vector containing only two independent scalar unknowns). The maximum number of positions that can be satisfied using the RS dyad is four. This will result in a system of seven equations in eight unknowns, with one unknown to be selected arbitrarily.

All other spatial dyads may be synthesized by mathematically expressing the physical constraints imposed by the links. For a complete description of these other dyads, the reader is referred to the text by Suh and Radcliffe.[1]

[1]C. H. Suh and C. W. Radcliffe, *Kinematics and Mechanisms Design*, Wiley, New York, 1978.

12.6 INTRODUCTION TO ROBOTIC MANIPULATORS

The word *robot* is taken from the Czechoslovakian word meaning serf, or worker. In English, the word has come to mean a machine which can be programmed to perform a variety of tasks. Some robots are able to make decisions during operation; these are commonly called intelligent robots. A clear distinction should be made between devices such as cams and linkages, which are designed to perform a single, repetitive task, and robots, which can be programmed to perform many different tasks. Because of this basic difference, operations that use robots are sometimes referred to as *flexible* automation, and operations that use devices such as cams and linkages are referred to as *fixed* automation. A *manipulator* acts as an arm and often, although not always, resembles a human arm.

Industrial robotic manipulators are often classified by the number and types of joints they contain and by the total resulting number of degrees of freedom they possess. Most industrial manipulators contain only revolute (turning) and prismatic (sliding) joints.

Perhaps the simplest robot geometry is found in the Cartesian or *xyz*-manipulator, such as the one shown in Fig. 12.13. Here, the first two joints are prismatic joints which locate the hand in the *xy*-reference frame. The third joint is also a prismatic joint which moves the hand in the *z*-direction (i.e., normal to the *xy*-plane). The fourth joint is a revolute joint whose axis is parallel to the *z*-axis. This robot is quite useful for flat-surface assembly operations, such as placing chips on a circuit board.

Another common manipulator geometry is based on cylindrical coordinates. The variables of this coordinate system are *h* (height), θ (rotation), and *r* (reach). A typical industrial robot using this geometry is shown in Fig. 12.14. The fourth

FIGURE 12.13 (Courtesy of Seiko Instruments, Inc.)

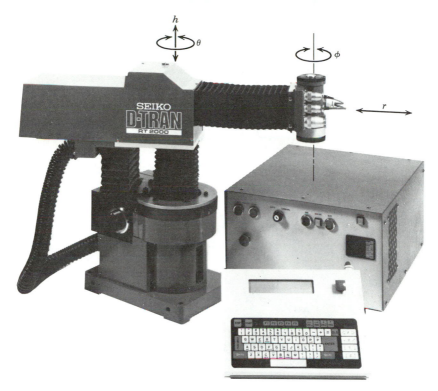

FIGURE 12.14 (Courtesy of Seiko Instruments, Inc.)

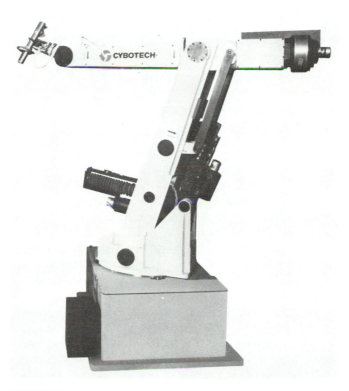

FIGURE 12.15 (Courtesy of Cybotech Corporation.)

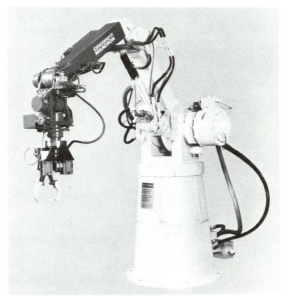

FIGURE 12.16 (Courtesy of Cincinnati Milacron.)

joint of this manipulator is once again a revolute joint which allows rotation about a vertical (z) axis.

Other industrial robotic manipulators are shown in Fig. 12.15 and Fig. 12.16. The geometries of these devices are more complex, and as a result, they are able to perform tasks requiring more general types of motion.

12.7 KINEMATICS OF ROBOTIC MANIPULATORS

While the broad field of robotics draws upon many disciplines, perhaps none is of more fundamental importance than kinematics. In the design of robotic manipulators, it is up to the kinematician to determine the number and types of joints and the link dimensions required to produce a given motion. The engineer involved in the selection and implementation of robotic manipulators must have a clear understanding of the motions a given manipulator is able to produce. Some robotic manipulators are designed to perform only simple planar tasks, while others may perform complex spatial tasks. The first key to understanding the motion a given manipulator can produce is an appreciation of the concept of mobility.

Consider a single link rotating about a fixed pivot, as shown in Fig. 12.17. Rigidly attached to the end of this link is a so-called end effector, or hand, which may be a tool or a gripping device. The location of the end effector is given by the coordinates x_P, y_P of its center point P. Specifying the angle θ_1 completely determines the location of every point in this link, including point P. This simple device has one degree of freedom and may be considered a robotic manipulator

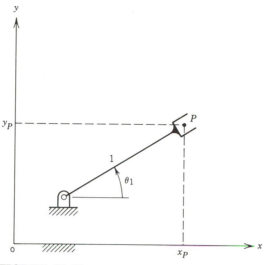

FIGURE 12.17

with mobility equal to 1. Obviously, the types of tasks this "robot" can perform are quite limited. In the most general planar manipulator, the operator should be able to freely specify both the position x_P, y_P and the orientation θ_1 of the end effector. In the single-link planar manipulator of Fig. 12.17, only one of these parameters may be selected independently.

Now, consider the two-link planar manipulator of Fig. 12.18. This device has two independent input parameters, θ_1 and θ_2, and therefore has two degrees

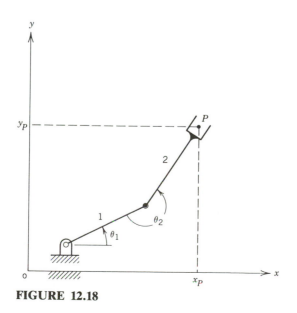

FIGURE 12.18

of freedom. In this case, the operator has independent control over two of the three end effector parameters x_P, y_P, θ_2.

Clearly, the three-link planar manipulator of Fig. 12.19 is the simplest device capable of producing general planar motion. By proper selection of the three parameters θ_1, θ_2, and θ_3, the end effector can theoretically be made to assume any planar position and orientation. There are, however, several practical kinematic considerations which greatly complicate this issue.

The actual working area of a planar manipulator is limited by the length of its links and the range of motion of its joints. For example, in the manipulator shown in Fig. 12.19, points cannot be reached which are a greater distance than the sum of the link lengths away from the fixed pivot. It must also be recognized that the relative joint angles θ_1, θ_2, and θ_3 are typically controlled by electric, hydraulic, or pneumatic actuators acting between successive links. Often, these devices cannot produce a full 360° of rotation. This further limits the actual work area of the manipulator. This suggests that in some instances more than three joints (i.e., more than three degrees of freedom) may be desirable in a planar manipulator. These additional degrees of freedom are sometimes referred to as the *dexterity* of the manipulator. Dexterity may also allow the manipulator to maneuver around obstacles within the work area.

The preceding discussion was directed toward planar manipulators containing only revolute joints. Many of these same concepts also apply to spatial manipulators and to manipulators containing other joint types. To move with general spatial motion, a manipulator must possess a minimum of six degrees of freedom. Possible robot configurations can be determined using the Kutzbach mobility equation (Eq. 12.1) with $M = 6$:

$$M = 6 = 6(n - 1) - 5f_1 - 4f_2 - 3f_3 - 2f_4 - f_5 \tag{12.28}$$

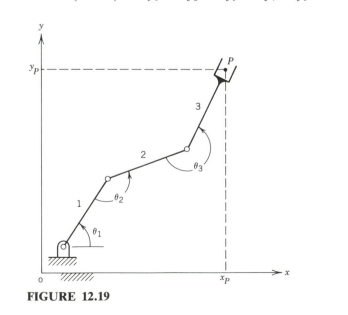

FIGURE 12.19

In most cases, only single-degree-of-freedom (f_1) joints are used in robots. It is possible to use joints with a greater number of degrees of freedom, but these are difficult to actuate. Thus, by considering only one-degree-of-freedom joints, the mobility equation becomes

$$M = 6 = 6(n - 1) - 5f_1 \tag{12.29}$$

or

$$2 = n - \left(\frac{5}{6}\right)f_1$$

Several interesting solutions to Eq. 12.29 exist. The first of these is the case where $n = 2$ (two links) and $f_1 = 0$ (zero joints). One link will be the ground, and the other link will float freely without attachment to ground. This may at first seem absurd; but, in fact, a spacecraft or helicopter is exactly this type of a robot. It is not possible to construct a robot with $M = 6$ and $n = 3, 4, 5$, or 6 because the resulting number of joints will not be an integer. The simplest robot having $M = 6$ and all links physically connected will contain seven links (one fixed and six moving) and six one-degree-of-freedom joints.

It is not possible to use only prismatic joints in constructing a general planar or spatial robot. In the planar robot shown in Fig. 12.20, note that at least one revolute joint is necessary to provide the rotational degree of freedom. A minimum of three revolute joints are necessary in the general spatial robot. Also

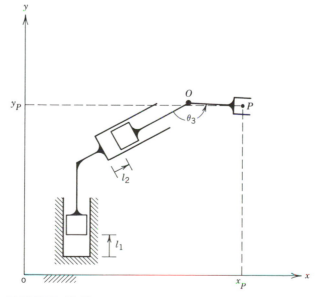

FIGURE 12.20

note in Fig. 12.20 that if the axis of the two sliding joints were made parallel, the manipulator would possess only two degrees of freedom. This can be seen by observing that in such a case point O would move along a straight line rather than in two-dimensional space. In the manipulator of Fig. 12.20, the axes of the two prismatic joints can only become parallel if they are so assembled. Figure 12.21, however, shows a manipulator that operates with three degrees of freedom except when $\theta_2 = 0°$ or $180°$. At these angles, one degree of freedom is lost due to the instantaneous geometry, and the manipulator is said to be in a *singular position*. It would be impossible to independently specify both the angular velocity of the end effector and the y-direction velocity of point P. This problem is easy to avoid when working with simple planar manipulators. Manipulators capable of general spatial motion, however, must possess a minimum of six degrees of freedom, and the control of instantaneous losses of mobility becomes much more complex.

To control the motion of a robotic manipulator, the designer must be able to determine the position, velocity, and acceleration of the end effector given the position, velocity, and acceleration of each joint actuator. This is sometimes referred to as the *forward* kinematics problem. Many industrial robotic manipulators are configured in a single, open-loop chain. In this case, the position of the end effector is found by adding the link vectors from ground to the end effector. For example, the location of point P in the planar manipulator of Fig. 12.22 is given by

$$\mathbf{S} = l_1 e^{i\theta_1} + l_2 e^{i\psi_2} + l_3 e^{i\psi_3} \tag{12.30}$$

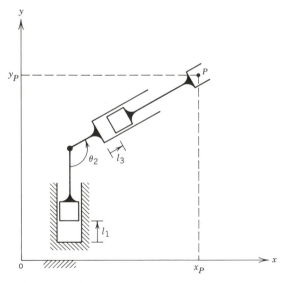

FIGURE 12.21

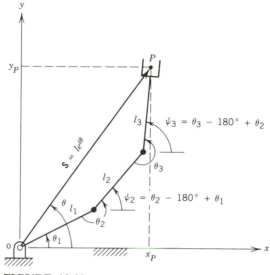

FIGURE 12.22

where

$$\psi_2 = \theta_2 - (\pi - \theta_1)$$

$$\psi_3 = \theta_3 - (\pi - \psi_2)$$

and where l_1, l_2, and l_3 are the link lengths. The orientation of the end effector is simply ψ_3, which is the angle that link 3 forms with the horizontal. The velocity and acceleration equations are found by differentiating Eq. 12.30 with respect to time, as follows:

$$\mathbf{V} = \frac{d\mathbf{S}}{dt} = i(l_1\dot{\theta}_1 e^{i\theta_1} + l_2\dot{\psi}_2 e^{i\psi_2} + l_3\dot{\psi}_3 e^{i\psi_3})$$

$$\mathbf{A} = l_1(i\ddot{\theta}_1 - \dot{\theta}_1^2)e^{i\theta_1} + l_2(i\ddot{\psi}_2 - \dot{\psi}_2^2)e^{i\psi_2} + l_3(i\ddot{\psi}_3 + \dot{\psi}_3^2)e^{i\psi_3} \qquad \textbf{(12.31)}$$

Obviously, when sliding joints are present, the variable link lengths will also become functions of time. The forward kinematics problem for spatial manipulators is also accomplished by serial addition of the vectors from ground out to the end effector. In fact, this procedure has already been demonstrated in Example 12.2 using a three-link spatial chain.

In the forward kinematics problem just discussed, the joint variables are known, and the motion of the end effector is to be determined. A second and much more difficult problem is to find the values of the joint variables and their derivatives given the required motion (position, velocity, acceleration) of the end effector. This is often called the *backward*, or *inverse*, kinematics problem. Consider, for example, the planar three-link manipulator of Fig. 12.22. The position and orientation of the end effector are given by $le^{i\theta}$ and ψ_3, respectively. With these values specified, the inverse kinematics problem is to find the unknown joint variables θ_1 and ψ_2 which will produce this position. This is not a trivial problem, even for the simple case presented here. In fact, careful study shows this to be exactly the same problem as the kinematic analysis of a planar four-bar linkage. Since the position of point P is known, for the purpose of analysis this may be considered to be a second ground pivot. Specifying the angle ψ_3 is tantamount to specifying the input angle of the four-bar linkage. The solution procedure for the four-bar linkage is presented using the law of cosines in Chapter 2, section 2.1. An alternate solution based on complex number methods is presented in Appendix 1.

Problems

12.1. Calculate the mobility of the devices shown in Fig. 12.5 using the Kutzbach mobility equation.

12.2. Calculate the mobility of the device shown in Fig. 12.23. What would be the mobility if link 4 were fixed to ground?

12.3. Calculate the mobility of the device shown in Fig. 12.24.

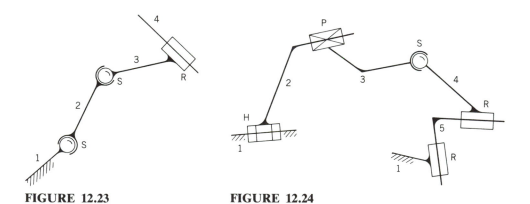

FIGURE 12.23   **FIGURE 12.24**

12.4. Calculate the mobility of the device shown in Fig. 12.25 where the subdevice ABCD is a planar four-bar linkage.

12.5. In practice, it is impossible to ensure that all the revolute joint axes in a planar four-bar linkage are perfectly parallel. As a result, "planar" four bars are actually spatial devices, and the Kutzbach mobility equation will predict them to be structures. Is the mobility equation incorrect in this case, or are there other factors that must be considered when dealing with "planar" devices?

12.6. Explain what will happen to the RCCC mechanism shown in Fig. 12.5 if all the joint axes become parallel.

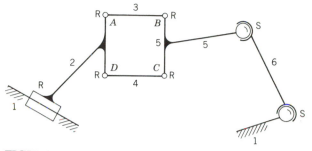

FIGURE 12.25

12.7. A very useful property of rotation matrices (both planar and spatial) is that they are *orthogonal*, meaning the matrix inverse is equal to the matrix transpose. For the planar rotation matrix defined in Eq. 12.7, show that $[R]^{-1} = [R]^T$.

12.8. Figure 12.26 shows the initial position (solid lines) and the final position (dashed lines) of a three-link spatial chain. The following data are given for the initial position:

$$
\begin{array}{cccc}
 & \mathbf{i} & \mathbf{j} & \mathbf{k} \\
\mathbf{u}_1 = & 1 & 0 & 0 \\
\mathbf{u}_2 = & 0 & 0 & 1 \\
\mathbf{p} = & 0 & 5 & 5 \\
\mathbf{q} = & 0 & 7 & 5
\end{array}
$$

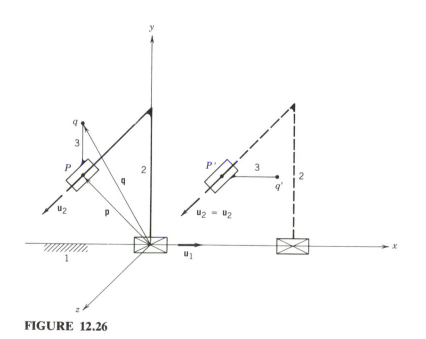

FIGURE 12.26

The prismatic joint undergoes a displacement d = 5 units, and the revolute joint undergoes a displacement α = $-90°$. Using the relationships

$$\mathbf{p'} = \mathbf{p} + (d)\mathbf{u}_1$$

$$\mathbf{q'} = [R_{\alpha,\mathbf{u}_2}](\mathbf{q} - \mathbf{p}) + \mathbf{p'}$$

find $\mathbf{q'}$, the displaced position of point q.

12.9. For the three-link chain analyzed in Example 12.2, show that the links return to their original position when α = 360° and β = 360°.

12.10. For the four-link spatial chain shown in Fig. 12.27, find the displaced position of point q given the joint displacements α = 90°, β = 45°, and γ = 180°. The following vector quantities are given in the initial position:

	i	j	k
$\mathbf{u}_1 =$	0	1	0
$\mathbf{u}_2 =$	0	0	1
$\mathbf{u}_3 =$	1	0	0
$\mathbf{n} =$	3	0	3
$\mathbf{p} =$	8	5	3
$\mathbf{q} =$	8	7	3

12.11. For the RSSR mechanism shown in Fig. 12.28, calculate the output displacement

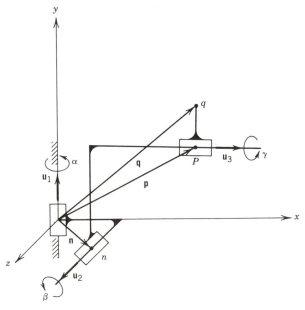

FIGURE 12.27

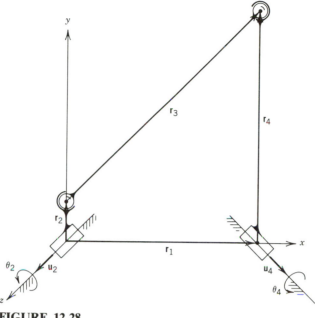

FIGURE 12.28

θ_4 for an input displacement $\theta_2 = 10°$. The following values define the mechanism in its initial position:

	i	j	k
$\mathbf{r}_1 =$	10	0	0
$\mathbf{r}_2 =$	0	2	0
$\mathbf{r}_3 =$	10	10	0
$\mathbf{r}_4 =$	0	12	0
$\mathbf{u}_2 =$	0	0	1
$\mathbf{u}_4 =$	0.707	0	0.707

Appendix One

Position Analysis of the Four-Bar Linkage Using Vectors in Complex Polar Form

A planar four-bar linkage is shown in Fig. A1.1. The requirement that the links of the mechanism must form a closed loop is expressed by the following vector equation:

$$r_2 e^{i\theta_2} + r_3 e^{i\theta_3} = r_1 + r_4 e^{i\theta_4} \tag{A1.1}$$

Solving for $r_3 e^{i\theta_3}$ gives

$$r_3 e^{i\theta_3} = r_1 + r_4 e^{i\theta_4} - r_2 e^{i\theta_2} \tag{A1.2}$$

Taking the complex conjugate of each term of Eq. A1.2 gives the following valid loop-closure equation:

$$r_3 e^{-i\theta_3} = r_1 + r_4 e^{-i\theta_4} - r_2 e^{-i\theta_2} \tag{A1.3}$$

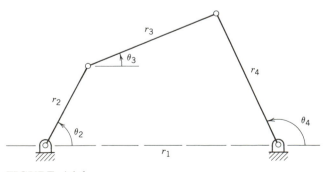

FIGURE A1.1

Multiplying Eq. A1.2 and Eq. A1.3 together gives

$$r_3^2 = r_1^2 + r_2^2 + r_4^2 + r_1 r_4 (e^{i\theta_4} + e^{-i\theta_4})$$
$$- r_1 r_2 (e^{i\theta_2} + e^{-i\theta_2}) - r_2 r_4 (e^{i\theta_2} e^{-i\theta_4} + e^{-i\theta_2} e^{i\theta_4}) \tag{A1.4}$$

Or, upon expansion using $e^{i\theta} = \cos\theta + i\sin\theta$

$$r_3^2 = r_1^2 + r_2^2 + r_4^2 + 2r_1 r_4 \cos\theta_4$$
$$- 2r_1 r_2 \cos\theta_2 - 2r_2 r_4 \cos\theta_4 \cos\theta_2 - 2r_2 r_4 \sin\theta_4 \sin\theta_2 \tag{A1.5}$$

Given a value of θ_2 and the four link lengths r_1, r_2, r_3, and r_4, it is possible to solve this equation for the unknown value of θ_4. However, the solution is not straightforward because this is a transcendental equation containing both $\sin\theta_4$ and $\cos\theta_4$.[1] One possible method of solution is to make the substitution $\sin\theta_4 = (1 - \cos^2\theta_4)^{1/2}$. An easier method of solution, however, involves the use of the following trigonometric identities:

$$\cos\theta_4 = \frac{(1 - t^2)}{(1 + t^2)}$$
$$\sin\theta_4 = \frac{2t}{(1 + t^2)} \tag{A1.6}$$

where

$$t = \tan\left(\frac{\theta_4}{2}\right)$$

Substituting these identities into Eq. A1.5, multiplying through by $(1 + t^2)$, and grouping terms gives

$$At^2 + Bt + C = 0 \tag{A1.7}$$

where

$$A = r_3^2 - r_1^2 - r_2^2 - r_4^2 + 2r_1 r_2 \cos\theta_2$$
$$+ 2r_1 r_4 - 2r_2 r_4 \cos\theta_2$$
$$B = 4r_2 r_4 \sin\theta_2$$
$$C = r_3^2 - r_1^2 - r_2^2 - r_4^2 + 2r_1 r_2 \cos\theta_2$$
$$- 2r_1 r_4 + 2r_2 r_4 \cos\theta_2$$

---

[1] An expression is transcendental if it cannot be represented by an equivalent expression containing a finite number of terms involving only the basic operations of addition, subtraction, multiplication, and division. For example, $\cos\theta$ cannot be exactly represented as a finite series containing only these four basic operations.

This quadratic equation is easily solved for t as follows:

$$t = \frac{-B \pm (B^2 - 4AC)^{1/2}}{2A}$$
(A1.8)

The two values of t found from this equation will yield two distinct values of θ_4, one corresponding to each branch of the linkage.

TABLE OF INVOLUTE FUNCTIONS

Degrees	0.0	0.1	0.2	0.3	0.4	0.5	0.6	0.7	0.8	0.9
0	0.000000	0.000000	0.000000	0.000000	0.000000	0.000000	0.000000	0.000000	0.000000	0.000001
1	0.000002	0.000002	0.000003	0.000004	0.000005	0.000006	0.000007	0.000009	0.000010	0.000012
2	0.000014	0.000016	0.000019	0.000022	0.000025	0.000028	0.000031	0.000035	0.000039	0.000043
3	0.000048	0.000053	0.000058	0.000064	0.000070	0.000076	0.000083	0.000090	0.000097	0.000105
4	0.000114	0.000122	0.000132	0.000141	0.000151	0.000162	0.000173	0.000184	0.000197	0.000209
5	0.000222	0.000236	0.000250	0.000265	0.000280	0.000296	0.000312	0.000329	0.000347	0.000366
6	0.000384	0.000404	0.000424	0.000445	0.000467	0.000489	0.000512	0.000536	0.000560	0.000586
7	0.000612	0.000638	0.000666	0.000694	0.000723	0.000753	0.000783	0.000815	0.000847	0.000880
8	0.000914	0.000949	0.000985	0.001022	0.001059	0.001098	0.001137	0.001178	0.001219	0.001262
9	0.001305	0.001349	0.001394	0.001440	0.001488	0.001536	0.001586	0.001636	0.001688	0.001740
10	0.001794	0.001849	0.001905	0.001962	0.002020	0.002079	0.002140	0.002202	0.002265	0.002329
11	0.002394	0.002461	0.002528	0.002598	0.002668	0.002739	0.002812	0.002894	0.002962	0.003039
12	0.003117	0.003197	0.003277	0.003360	0.003443	0.003529	0.003615	0.003712	0.003792	0.003883
13	0.003975	0.004069	0.004164	0.004261	0.004359	0.004459	0.004561	0.004664	0.004768	0.004874
14	0.004982	0.005091	0.005202	0.005315	0.005429	0.005545	0.005662	0.005782	0.005903	0.006025
15	0.006150	0.006276	0.006404	0.006534	0.006665	0.006799	0.006934	0.007071	0.007209	0.007350
16	0.007493	0.007637	0.007784	0.007932	0.008082	0.008234	0.008388	0.008544	0.008702	0.008863
17	0.009025	0.009189	0.009355	0.009523	0.009694	0.009866	0.010041	0.010217	0.010396	0.010577
18	0.010760	0.010946	0.011133	0.011323	0.011515	0.011709	0.011906	0.012105	0.012306	0.012509
19	0.012715	0.012923	0.013134	0.013346	0.013562	0.013779	0.013999	0.014222	0.014447	0.014674
20	0.014904	0.015137	0.015372	0.015609	0.015850	0.016092	0.016337	0.016585	0.016836	0.017089
21	0.017345	0.017603	0.017865	0.018129	0.018395	0.018665	0.018937	0.019212	0.019490	0.019770
22	0.020054	0.020340	0.020630	0.020921	0.021216	0.021514	0.021815	0.022119	0.022426	0.022736
23	0.023049	0.023365	0.023684	0.024006	0.024332	0.024660	0.024992	0.025326	0.025664	0.026005
24	0.026350	0.026697	0.027048	0.027402	0.027760	0.028121	0.028485	0.028852	0.029223	0.029598
25	0.029975	0.030357	0.030741	0.031130	0.031521	0.031917	0.032315	0.032718	0.033124	0.033534
26	0.033947	0.034364	0.034785	0.035209	0.035637	0.036069	0.036505	0.036945	0.037388	0.037835
27	0.038287	0.038696	0.039201	0.039664	0.040131	0.040602	0.041076	0.041556	0.042039	0.042526
28	0.043017	0.043513	0.044012	0.044516	0.045024	0.045537	0.046054	0.046575	0.047100	0.047630
29	0.048164	0.048702	0.049245	0.049792	0.050344	0.050901	0.051462	0.052027	0.052597	0.053172
30	0.053751	0.054336	0.054924	0.055519	0.056116	0.056720	0.057267	0.057940	0.058558	0.059181

TABLE OF INVOLUTE FUNCTIONS (*Continued*)

Degrees	0.0	0.1	0.2	0.3	0.4	0.5	0.6	0.7	0.8	0.9
31	0.059809	0.060441	0.061079	0.061721	0.062369	0.063022	0.063680	0.064343	0.065012	0.065685
32	0.066364	0.067048	0.067738	0.068432	0.069133	0.069838	0.070549	0.071266	0.071988	0.072716
33	0.073449	0.074188	0.074932	0.075683	0.076439	0.077200	0.077968	0.078741	0.079520	0.080305
34	0.081097	0.081974	0.082697	0.083506	0.084321	0.085142	0.085970	0.086804	0.087644	0.088490
35	0.089342	0.090201	0.091066	0.091938	0.092816	0.093701	0.094592	0.095490	0.096395	0.097306
36	0.098224	0.099149	0.100080	0.101019	0.101964	0.102916	0.103875	0.104841	0.105814	0.106795
37	0.107782	0.108777	0.109779	0.110788	0.111805	0.112828	0.113860	0.114899	0.115945	0.116999
38	0.118060	0.119130	0.120207	0.121291	0.122384	0.123484	0.124592	0.125709	0.126833	0.127965
39	0.129106	0.130254	0.131411	0.132576	0.133749	0.134931	0.136122	0.137320	0.138528	0.139743
40	0.140968	0.142201	0.143443	0.144694	0.145954	0.147222	0.148500	0.149787	0.151082	0.152387
41	0.153702	0.155025	0.156358	0.157700	0.159052	0.160414	0.161785	0.163165	0.164556	0.165956
42	0.167366	0.168786	0.170216	0.171656	0.173106	0.174566	0.176037	0.177518	0.179009	0.180511
43	0.182023	0.183546	0.185080	0.186625	0.188180	0.189746	0.191324	0.192912	0.194511	0.196122
44	0.197744	0.199377	0.201022	0.202678	0.204346	0.206026	0.207717	0.209420	0.211135	0.212863
45	0.214602	0.216353	0.218117	0.219893	0.221682	0.223483	0.225296	0.227123	0.228962	0.230814
46	0.232678	0.234557	0.236448	0.238352	0.240270	0.242202	0.244147	0.246105	0.248077	0.250064
47	0.252064	0.254078	0.256106	0.258149	0.260206	0.262277	0.264363	0.266463	0.268578	0.270709
48	0.272855	0.275015	0.277191	0.279381	0.281588	0.283810	0.286047	0.288300	0.290570	0.292855
49	0.295157	0.297474	0.299809	0.302160	0.304527	0.306912	0.309313	0.311731	0.314166	0.316619
50	0.319088	0.321577	0.324082	0.326605	0.329146	0.331706	0.334284	0.336879	0.339493	0.342127
51	0.344779	0.347451	0.350141	0.352850	0.355579	0.358328	0.361096	0.363885	0.366693	0.369522
52	0.372371	0.375241	0.378130	0.381041	0.383974	0.386927	0.389903	0.392899	0.395917	0.398958
53	0.402021	0.405105	0.408213	0.411343	0.414495	0.417671	0.420870	0.424094	0.427340	0.430610
54	0.433905	0.437222	0.440566	0.443933	0.447326	0.450744	0.454187	0.457655	0.461150	0.464670
55	0.468217	0.471790	0.475390	0.479017	0.482670	0.486351	0.490060	0.493797	0.497562	0.501355
56	0.505177	0.509027	0.512908	0.516817	0.520755	0.524724	0.528724	0.532753	0.536814	0.540905
57	0.545027	0.549182	0.553368	0.557586	0.561836	0.566120	0.570436	0.574789	0.579173	0.583591
58	0.588043	0.592530	0.597053	0.601609	0.606203	0.610832	0.615498	0.620200	0.624940	0.629717
59	0.634535	0.639387	0.644279	0.649210	0.654181	0.659190	0.664240	0.669331	0.674462	0.679635
60	0.684853	0.690109	0.695409	0.700751	0.706137	0.711567	0.717041	0.722561	0.728126	0.733736

Appendix Three

This Appendix contains four computer programs written directly from material in the text. The first three programs are written in FORTRAN and are functionally equivalent to BASIC programs contained in the text. The fourth program, written in BASIC, is derived from the material on force analysis using the matrix method. It contains subroutines for matrix inversion and matrix multiplication. With slight modifications, this program will handle the force analysis of any planar four-bar linkage. A brief description and a listing of each program follows.

```
*********************************************************************
* Mechanism Design - Displacement Analysis (Fortran 77 version)
* -   Uses Newton-Raphson root finding method to determine unknown
*     angles of links 3 & 4 of a four bar linkage.
* -   Mabie and Reinholtz, 4th Ed.
* -   Written by:  Steve Wampler (11/25/85)
*********************************************************************
C     Declare all variable types
      REAL    DG2RD,THETA2,THETA3,THETA4,FUNC1,FUNC2,R1,R2,R3,R4
      REAL    DF1DT3,DF1DT4,DF2DT3,DF2DT4,DEL,DELT3,DELT4
C     Set up deg. to rad. conversion factor
      DG2RD = 3.14159/180.0
C     Get mechanism information from the user
      WRITE(*,*)'ENTER ANGULAR DISPLACEMENT OF LINK 2 (DEGREES).'
      READ(*,*,ERR=400,END=400)THETA2
      THETA2 = THETA2 * DG2RD
      WRITE(*,*)'GUESS ANGULAR DISPLACEMENT OF LINK 3 (DEGREES).'
      READ(*,*,ERR=400,END=400)THETA3
      THETA3 = THETA3 * DG2RD
      WRITE(*,*)'GUESS ANGULAR DISPLACEMENT OF LINK 4 (DEGREES).'
      READ(*,*,ERR=400,END=400)THETA4
      THETA4 = THETA4 * DG2RD
      WRITE(*,*)'ENTER LENGTH OF LINKS 1,2,3 AND 4 SEPARATED BY ",".'
      READ(*,*,ERR=400,END=400)R1,R2,R3,R4
C     Print Headings
      WRITE(*,*)
      WRITE(*,*)'  THETA3    THETA4    FUNC1    FUNC2    DELTA3    DELTA4'
C     Loop until both equations (FUNC1 & FUNC2) are approx. = 0
      FUNC1 = 1.0
 100  IF (ABS(FUNC1) .LT. 0.001 .AND. ABS(FUNC2) .LT. 0.001) THEN
         GO TO 500
      END IF
C        Evaluate loop closure equations
         FUNC1 = R1+(R4*COS(THETA4))-(R2*COS(THETA2))-(R3*COS(THETA3))
         FUNC2 = (R4*SIN(THETA4))-(R2*SIN(THETA2))-(R3*SIN(THETA3))
C        Evaluate partial derivatives
         DF1DT3 = R3*SIN(THETA3)
         DF1DT4 = -R4*SIN(THETA4)
         DF2DT3 = -R3*COS(THETA3)
         DF2DT4 = R4*COS(THETA4)
C        Now solve 2 equations and 2 unknowns with cramer's rule
         DEL   = DF1DT3*DF2DT4-DF1DT4*DF2DT3
         DELT4 = (DF2DT3*FUNC1-DF1DT3*FUNC2)/DEL
         DELT3 = -(DF2DT4*FUNC1-DF1DT4*FUNC2)/DEL
C        Output the results
         WRITE(*,1000)THETA3/DG2RD,THETA4/DG2RD,FUNC1,FUNC2,
     >               DELT3/DG2RD,DELT4/DG2RD
C        Make new guesses for both THETA3 and THETA4
         THETA3 = THETA3 + DELT3
         THETA4 = THETA4 + DELT4
C     Loop back for another try
      GO TO 100
 400  WRITE(*,*)'INPUT ERROR ...'
1000  FORMAT((' ',6(F8.3,' ')))
 500  STOP
      END
```

FIGURE A3.1 FORTRAN program for the displacement analysis of a four-bar linkage using iterative methods. This program is functionally equivalent to the BASIC program given in Fig. 2.8, section 2.3 of the text.

```
*************************************************************************
* Cam Design Program (Fortran 77 version)
* - Disk cam with radial roller follower.
* - Cycloidal rise - Dwell - Cycloidal return.
* - Mabie and Reinholtz, 4th Ed.
* - Program revised by: Steve Wampler (11/25/85)
*************************************************************************
C     Declare all variable types
      REAL      BASE,L,S,V,A,T
      REAL      PI,TWOPI
      INTEGER   THETA,DGRISE,DGDWEL,DGINC
C     Set up deg. to rad. conversion factor
      TWOPI = 2.0*3.14159
C     Get cam information from the user
      WRITE(*,*)'ENTER BASE CIRCLE RADIUS'
      READ(*,*,ERR=400,END=400)BASE
      WRITE(*,*)'ENTER TOTAL FOLLOWER DISPLACEMENT'
      READ(*,*,ERR=400,END=400)L
      WRITE(*,*)'ENTER RISE ANGLE (IN DEGREES)'
      READ(*,*,ERR=400,END=400)DGRISE
      WRITE(*,*)'ENTER CAM ANGLE AT END OF DWELL (IN DEGREES)'
      READ(*,*,ERR=400,END=400)DGDWEL
      WRITE(*,*)'ENTER PRINTOUT ANGLE INC. (IN DEGREES)'
      READ(*,*,ERR=400,END=400)DGINC
C     Print Headings
      WRITE(*,*)
      WRITE(*,*)' INPUT ANG    DISPL     VELOCITY    ACCEL',
     >          '    RADIUS   CURVATURE   PRESS ANG'
      WRITE(*,*)' (THETA)       (S)         (V)        (A) ',
     >          '     (R)       (RHO)      (ALPHA) '
C     **** C-5 rise ****
      DO 50 THETA = 0,DGRISE,DGINC
         T = (TWOPI * FLOAT(THETA) / 180.0)
         S = L*((THETA/180.0)-(1.0/TWOPI)*SIN(T))
         V = (L/180.0)*(1.0-COS(T))
         A = ((TWOPI*L)/(180.0**2))*SIN(T)
C     Calculate radius (R) & curvature (RHO) & print everything
         CALL FINISH (BASE,S,V,A,THETA)
 50   CONTINUE
C     **** Dwell ****
      DO 60 THETA = DGRISE,DGDWEL,DGINC
         S = L
         V = 0.0
         A = 0.0
C     Calculate radius (R) & curvature (RHO) & print everything
         CALL FINISH (BASE,S,V,A,THETA)
 60   CONTINUE
C     **** C-6 return ****
      DO 70 THETA = DGDWEL,360,DGINC
         T = FLOAT(THETA) - 270.0
         S = L*((1.0-(T/90.0))+(1.0/TWOPI)*SIN(TWOPI*T/90.0))
         V = -(L/90.0)*(1.0-COS(TWOPI*T/90.0))
         A = -(TWOPI*L/(90.0**2))*(SIN(TWOPI*T/90.0))
C     Calculate radius (R) & curvature (RHO) & print everything
         CALL FINISH (BASE,S,V,A,THETA)
 70   CONTINUE
```

(a)

FIGURE A3.2(a) and A3.2(b) FORTRAN program for designing a disk cam with a radial roller follower. This program is functionally equivalent to the BASIC program given in Fig. 3.32, section 3.10 of the text. (*continued next page*)

```
      STOP
 400  WRITE(*,*)'INPUT ERROR ...'
      STOP
      END
*****************************************************************************
* Routine to Calculate radius (R) & Curvature (RHO) & print everything
*****************************************************************************
      SUBROUTINE FINISH (BASE,S,V,A,THETA)
      REAL     BASE,S,V,A,R,VR,AR,RHO,ALPHA,PI
      INTEGER  THETA
      PI = 3.14159
      R = BASE+S
      VR = V*180.0/PI
      AR = A*(180.0/PI)**2
      RHO = (((R**2)+(VR**2))**(3.0/2.0))/((R**2)+(2.0
     >     *(VR**2))-R*AR)
      ALPHA = (180.0/PI)*ATAN(VR/R)
      WRITE(*,1000)THETA,S,V,A,R,RHO,ALPHA
 1000 FORMAT((' ',' ',I9,6(' ',E9.3)))
      RETURN
      END
```

(b)

FIGURE A3.2*b*

```
*******************************************************************
* Mechanism Design - Dis. Vel. & Acc. Analysis (Fortran 77 version)
* -  Uses Newton-Raphson root finding method to determine unknown
*    angles of links 3 & 4 of a four bar linkage.
* -  Mabie and Reinholtz, 4th Ed.
* -  Written by:  Steve Wampler (11/25/85)
*******************************************************************
C     Declare all variable types
      REAL    DG2RD,THETA2,THETA3,THETA4,FUNC1,FUNC2,R1,R2,R3,R4
      REAL    DF1DT3,DF1DT4,DF2DT3,DF2DT4,DEL,DELT3,DELT4
      REAL    TWOPI,OMEGA2,ALPHA2
      INTEGER PASS,MAXPAS,I1,ANGINC
C     Deg. to rad. conversion factor & max. no. of iteration passes
      DG2RD = 3.14159/180.0
      MAXPAS = 10
      TWOPI = 2.0 * 3.14159
C     Get mechanism information from the user
      WRITE(*,*)'ENTER ANGULAR DISPLACEMENT OF LINKS 2,3,4 (DEGREES).'
      READ(*,*,ERR=400,END=400)THETA2,THETA3,THETA4
      THETA2 = THETA2 * DG2RD
      THETA3 = THETA3 * DG2RD
      THETA4 = THETA4 * DG2RD
      WRITE(*,*)'ENTER ANGULAR VELOCITY OF LINK 2 (RAD/SEC)'
      READ(*,*,ERR=400,END=400)OMEGA2
      WRITE(*,*)'ENTER ANGULAR ACCELERATION OF LINK 2 (RAD^2/SEC)'
      READ(*,*,ERR=400,END=400)ALPHA2
      WRITE(*,*)'ENTER LENGTH OF LINKS 1,2,3 AND 4 SEPARATED BY ",".'
      READ(*,*,ERR=400,END=400)R1,R2,R3,R4
      WRITE(*,*)'ENTER ANGULAR STEP SIZE FOR INPUT LINK',
     >          ' ROTATION (DEGREES)'
      READ(*,*,ERR=400,END=400)ANGINC
C     Print Headings
      WRITE(*,*)
      WRITE(*,*)' THETA2   THETA3   THETA4   OMEGA3   OMEGA4   ALPHA3 ',
     >          ' ALPHA4'
      WRITE(*,*)' (DEG.)   (DEG.)   (DEG.)   (RAD/S)  (RAD/S) (RAD/S^2)',
     >          ' (RAD/S^2)'
C     Let THETA2 loop through 360 degrees
      DO 350 I1 = 0,360+ANGINC,ANGINC
C     Loop until both equations (FUNC1 & FUNC2) are approx. = 0
      FUNC1 = 1.0
 100  IF ((ABS(FUNC1) .LT. 0.001 .AND. ABS(FUNC2) .LT. 0.001)
     >    .OR. PASS .GE. MAXPAS) THEN
        GO TO 300
      END IF
C     Evaluate loop closure equations
      FUNC1 = R1+(R4*COS(THETA4))-(R2*COS(THETA2))-(R3*COS(THETA3))
      FUNC2 = (R4*SIN(THETA4))-(R2*SIN(THETA2))-(R3*SIN(THETA3))
C     Evaluate partial derivatives
      DF1DT3 = R3*SIN(THETA3)
      DF1DT4 = -R4*SIN(THETA4)
      DF2DT3 = -R3*COS(THETA3)
      DF2DT4 = R4*COS(THETA4)
C     Now solve 2 equations and 2 unknowns with cramer's rule
      DEL   = DF1DT3*DF2DT4-DF1DT4*DF2DT3
      DELT4 = (DF2DT3*FUNC1-DF1DT3*FUNC2)/DEL
```

(a)

FIGURE A3.3(a), A3.3(b), and A3.3(c) FORTRAN program for the displacement, velocity, and acceleration analysis of a four-bar linkage. This program is functionally equivalent to the **BASIC** program given in Fig. 8.48, section 8.28 of the text. (*continued next page*)

```
         DELT3  = -(DF2DT4*FUNC1-DF1DT4*FUNC2)/DEL
C        Make new guesses for both THETA3 and THETA4
         THETA3 = THETA3+DELT3
         THETA4 = THETA4+DELT4
C        Count the number of iteration attempts
         PASS = PASS + 1
C        Loop back for another try
         GO TO 100
 300     IF (PASS .LT. MAXPAS) THEN
C           Go calculate the velocities and accelerations and print answers
            CALL VELACC (R2,R3,R4,THETA2,THETA3,THETA4,OMEGA2,ALPHA2)
         ELSE
C           Otherwise the mechanism must not assemble so print error
            CALL MECERR (THETA2)
         END IF
         PASS = 0
         THETA2 = THETA2 + ANGINC * DG2RD
 350     CONTINUE
         STOP
 400     WRITE(*,*)'INPUT ERROR ...'
         STOP
         END
********************************************************************************
* Subroutine to calculate velocity & acceleration and print answers
********************************************************************************
         SUBROUTINE VELACC(R2,R3,R4,THETA2,THETA3,THETA4,OMEGA2,ALPHA2)
         REAL   DG2RD,A,B,C,D,E,F,R2,R3,R4,THETA2,THETA3,THETA4
         REAL   OMEGA2,OMEGA3,OMEGA4,ALPHA2,ALPHA3,ALPHA4,CPRIME,FPRIME
         DG2RD = 3.14159/180.0
         A = -R3*SIN(THETA3)
         B = R4*SIN(THETA4)
         C = R2*SIN(THETA2)*OMEGA2
         D = R3*COS(THETA3)
         E = -R4*(COS(THETA4))
         F = -R2*COS(THETA2)*OMEGA2
C        Calculate angular velocities
         OMEGA3 = (F*B-E*C)/(D*B-E*A)
         OMEGA4 = (D*C-F*A)/(D*B-E*A)
C        Calculate angular accelerations
         CPRIME = R2*COS(THETA2)*OMEGA2**2+R2*SIN(THETA2)*ALPHA2
     >            +R3*COS(THETA3)*OMEGA3**2-R4*COS(THETA4)*OMEGA4**2
         FPRIME = R2*SIN(THETA2)*OMEGA2**2-R2*COS(THETA2)*ALPHA2
     >            +R3*SIN(THETA3)*OMEGA3**2-R4*SIN(THETA4)*OMEGA4**2
         ALPHA3 = (FPRIME*B-E*CPRIME)/(D*B-E*A)
         ALPHA4 = (D*CPRIME-FPRIME*A)/(D*B-E*A)
C        Print the results
         WRITE(*,1000)THETA2/DG2RD,THETA3/DG2RD,THETA4/DG2RD,OMEGA3,
     >        OMEGA4,ALPHA3,ALPHA4
 1000    FORMAT(('  ',7(F7.2,'  ')))
         RETURN
         END
********************************************************************************
* Subroutine to print "Mechanism does not assemble ...".
********************************************************************************
         SUBROUTINE MECERR (THETA2)
         REAL    THETA2,DG2RD
```

(b)

FIGURE A3.3*b*

```
      DG2RD = 3.14159/180.0
      WRITE(*,1000)THETA2/DG2RD
 1000 FORMAT(' ','MECHANISM DOES NOT ASSEMBLE AT THETA2 =',F6.1,' DEG.')
      RETURN
      END
```

(c)

FIGURE A3.3*c*

```
10 '**********************************************************************
20 '* Basic program for FORCE ANALYSIS
30 '* - Uses the matrix method to determine the forces on
40 '*     each link of a 4-bar linkage in addition to the driving
50 '*     torque applied to link 2.  The mass center, mass center
60 '*     acceleration, mass, and mass moment of inertia must be
70 '* - given  for each link.  Mabie and Reinholtz, 4th Ed.
80 '* - Program written by - Steve Wampler (6/20/85)
90 '**********************************************************************
100 '------------------------------------------------------------------
110 ' Force calculation main program.
120 '------------------------------------------------------------------
130 CLS:PRINT TAB(17)"4 Bar Linkage Force Analysis - Matrix Method"
140 SIZE=9:DIM MATRIX(9,9),INV.MATRIX(9,18) ' save some memory
150 GOSUB 220 ' set up CG position matrix
160 GOSUB 400 ' define mechanism
170 GOSUB 1000 ' invert CG position matrix
180 IF DET.FLAG=0 THEN GOTO 210           ' error in matrix inversion
190 GOSUB 1900 ' multiply column matrix by inverted matrix
200 GOSUB 710   ' print results
210 PRINT:END
220 '------------------------------------------------------------------
230 ' Set up CG position matrix.
240 '------------------------------------------------------------------
250 FOR ROW=1 TO 9                 ' row counter
260   FOR COLUMN=1 TO 9            ' column counter
270     READ MATRIX(ROW,COLUMN) ' get an element from DATA below
280   NEXT COLUMN
290 NEXT ROW
300 RETURN
310 DATA -1, 0, 1, 0, 0, 0, 0, 0, 0:' <=== This  is the  matrix without
320 DATA  0,-1, 0, 1, 0, 0, 0, 0, 0:'      the CG  position.  This
330 DATA  0, 0, 0, 0, 0, 0, 0, 0, 1:'      matrix is transformed to the
340 DATA  0, 0,-1, 0, 1, 0, 0, 0, 0:'      array named MATRIX using the
350 DATA  0, 0, 0,-1, 0, 1, 0, 0, 0:'      FOR/NEXT loops above. The CG
360 DATA  0, 0, 0, 0, 0, 0, 0, 0, 0:'      positions are  inserted
370 DATA  0, 0, 0, 0,-1, 0, 1, 0, 0:'      into the matrix later in the
380 DATA  0, 0, 0, 0, 0,-1, 0, 1, 0:'      program.
390 DATA  0, 0, 0, 0, 0, 0, 0, 0, 0
400 '------------------------------------------------------------------
410 ' Define mechanism - Edit this section to change mechanism.
420 '------------------------------------------------------------------
430 R21X=0     :MATRIX(3,2)=-R21X' pos. of link 2 CG to joint 1 in x-dir
440 R21Y=0     :MATRIX(3,1)=R21Y ' pos. of link 2 CG to joint 1 in y-dir
450 R22X=2.4   :MATRIX(3,4)=R22X ' pos. of link 2 CG to joint 2 in x-dir
460 R22Y=1.81  :MATRIX(3,3)=-R22Y' pos. of link 2 CG to joint 2 in y-dir
470 R32X=-3.68 :MATRIX(6,4)=-R32X' pos. of link 3 CG to joint 2 in x-dir
480 R32Y=1.56  :MATRIX(6,3)=R32Y ' pos. of link 3 CG to joint 2 in y-dir
490 R33X=3.68  :MATRIX(6,6)=R33X ' pos. of link 3 CG to joint 3 in x-dir
500 R33Y=-1.56 :MATRIX(6,5)=-R33Y' pos. of link 3 CG to joint 3 in y-dir
510 R43X=-2.28 :MATRIX(9,6)=-R43X' pos. of link 4 CG to joint 3 in x-dir
520 R43Y=-2.27 :MATRIX(9,5)=R43Y ' pos. of link 4 CG to joint 3 in y-dir
530 R44X=.46   :MATRIX(9,8)=R44X ' pos. of link 4 CG to joint 4 in x-dir
540 R44Y=5.25  :MATRIX(9,7)=-R44Y' pos. of link 4 CG to joint 4 in y-dir
550 M2=10/32.2 ' mass of link 2
560 M3=4/32.2  ' mass of link 3
```

(a)

FIGURE A3.4(a), A3.4(b), A3.4(c) and A3.4(d)   BASIC program for force analysis of the four-bar linkage using the matrix method, as described in section 9.7. The numerical values within the program are from Example Problem 9.3. This program contains general-purpose subroutines for matrix inversion and matrix multiplication. (*continued next page*)

```
570 M4=8/32.2   ' mass of link 4
580 I2=.017*12  ' mass moment of inertia of link 2
590 I3=.006*12  ' mass moment of inertia of link 3
600 I4=.026*12  ' mass moment of inertia of link 4
610 AG2X=0      :INERTIA.MATRIX(1)=M2*AG2X ' acc. of link 2 in x-dir.
620 AG2Y=0      :INERTIA.MATRIX(2)=M2*AG2Y ' acc. of link 2 in y-dir.
630 AG3X=-91.08:INERTIA.MATRIX(4)=M3*AG3X ' acc. of link 3 in x-dir.
640 AG3Y=-9.72 :INERTIA.MATRIX(5)=M3*AG3Y ' acc. of link 3 in y-dir.
650 AG4X=-54.08:INERTIA.MATRIX(7)=M4*AG4X ' acc. of link 4 in x-dir.
660 AG4Y=31.73 :INERTIA.MATRIX(8)=M4*AG4Y ' acc. of link 4 in y-dir.
670 ALPHA2=0          :INERTIA.MATRIX(3)=I2*ALPHA2 ' ang. acc. of link 2
680 ALPHA3=241        :INERTIA.MATRIX(6)=I3*ALPHA3 ' ang. acc. of link 3
690 ALPHA4=-129       :INERTIA.MATRIX(9)=I4*ALPHA4 ' ang. acc. of link 4
700 RETURN
710 '-------------------------------------------------------------------
720 ' Print results.
730 '-------------------------------------------------------------------
740 I=0:PRINT:PRINT TAB(22)"Results":PRINT
750 FOR LINK.J=1 TO 4                              ' link counter
760   IF LINK.J=4 THEN LINK.I=1 ELSE LINK.I=LINK.J+1
770   FOR DIRECTION=1 TO 2
780     IF DIRECTION=1 THEN DIR$="x" ELSE DIR$="y"
790     PRINT "Force of link"LINK.I"on link"LINK.J"in the "DIR$;
800     I=I+1 ' matrix counter
810     PRINT " direction =";USING " ###.####";REACT.MATRIX(I)
820   NEXT DIRECTION
830   PRINT "Resultant force of link"LINK.I"on link"LINK.J" =";
840   PRINT USING" ###.####";SQR(REACT.MATRIX(I-1)^2+REACT.MATRIX(I)^2)
850   PRINT
860 NEXT LINK.J
870 PRINT "The required input torque applied to link 2  =";
880 PRINT USING " ###.####";REACT.MATRIX(I+1)
890 RETURN
1000 '*****************************************************************
1010 '* MATRIX INVERSION AND MATRIX MULTIPLICATION SUBROUTINES
1020 '* -  The subroutine will invert the matrix stored in the
1030 '*    array MATRIX and return the inversion in the array INV.MATRIX.
1040 '*    The variable SIZE must equal the number of rows contained in
1050 '*    the array MATRIX.  The following BASIC statement must be
1060 '*    executed within the calling program before this inversion
1070 '*    subroutine is called:
1080 '*       DIM MATRIX(SIZE,SIZE),INV.MATRIX(SIZE,2*SIZE) ' save memory
1090 '* -  The multiplication subroutine will multiply the matrices
1100 '*    INV.MATRIX and INERTIA.MATRIX and return the results in
1110 '*    REACT.MATRIX.  The variable SIZE must equal the number of rows
1120 '*    contained in the array MATRIX.
1130 '* -  Adapted from:  P. M. Wolfe and C. P. Koelling, BASIC
1140 '*                    Engineering and Scientific Programs for the
1150 '*                    IBM-PC, Robert J. Brady Company, 1983
1160 '* -  Mabie and Reinholtz, 4th Ed.
1170 '* -  Program revised by - Steve Wampler (6/20/85)
1180 '*****************************************************************
1190 '-------------------------------------------------------------------
1200 ' Matrix inversion main program
1210 '-------------------------------------------------------------------
1220 ICOL=2*SIZE:IROW=SIZE                    ' set parameter for inverse
```

(b)

FIGURE A3.4*b*

```
1230 GOSUB 1340          ' make matrix upper triangular
1240 IF DET.FLAG=0 THEN PRINT "Error - MATRIX IS SINGULAR":GOTO 1330
1250 GOSUB 1710          ' make matrix lower triangular
1260 PRINT:PRINT TAB(31)"Inverted Matrix":PRINT
1270 FOR I=1 TO IROW
1280   FOR J=IROW+1 TO ICOL
1290     PRINT USING " ###.###";INV.MATRIX(I,J);
1300   NEXT J
1310   PRINT
1320 NEXT I
1330 RETURN
1340 '-----------------------------------------------------------------
1350 ' Convert INV.MATRIX to upper triangle.
1360 '-----------------------------------------------------------------
1370 PRINT:PRINT TAB(18)"Link Center of Gravity Displacement Matrix":PRINT
1380 FOR I=1 TO IROW
1390   FOR J=1 TO IROW
1400     INV.MATRIX(I,J)=MATRIX(I,J)
1410     IF I=J THEN INV.MATRIX(I,J+IROW)=1
1420     PRINT USING " ###.###";INV.MATRIX(I,J);
1430   NEXT J
1440   PRINT
1450 NEXT I
1460 PRINT:PRINT "Calculating Inverse of Matrix ..."
1470 DET.FLAG=1
1480 FOR I=1 TO IROW-1
1490   IF INV.MATRIX(I,I)=0 THEN GOSUB 1590      ' check for 0 det.
1500   IF DET.FLAG=0 THEN RETURN               ' error so exit
1510   FOR J=I+1 TO IROW
1520     XM=INV.MATRIX(J,I)/INV.MATRIX(I,I)
1530     FOR K=1 TO ICOL
1540       INV.MATRIX(J,K)=INV.MATRIX(J,K)-XM*INV.MATRIX(I,K)
1550     NEXT K
1560   NEXT J
1570 NEXT I
1580 RETURN
1590 '-----------------------------------------------------------------
1600 ' Check for zero determinant.
1610 '-----------------------------------------------------------------
1620 FOR J=I+1 TO IROW                      ' check ith column
1630   IF INV.MATRIX(J,I)=0 THEN GOTO 1680
1640   FOR K=1 TO ICOL
1650     INV.MATRIX(I,K)=INV.MATRIX(I,K)+INV.MATRIX(J,K)
1660   NEXT K
1670   RETURN
1680 NEXT J
1690 DET.FLAG=0
1700 RETURN
1710 '-----------------------------------------------------------------
1720 ' Convert INV.MATRIX to lower triangle.
1730 '-----------------------------------------------------------------
1740 FOR IJ=1 TO IROW-1
1750   IK=IROW-IJ+1
1760   FOR I=1 TO IK-1
1770     XM=INV.MATRIX(I,IK)/INV.MATRIX(IK,IK)
1780     FOR J=I+1 TO ICOL
```

(c)

FIGURE A3.4c

```
1790      INV.MATRIX(I,J)=INV.MATRIX(I,J)-XM*INV.MATRIX(IK,J)
1800    NEXT J
1810   NEXT I
1820 NEXT IJ
1830 FOR I=1 TO IROW
1840   DIV=INV.MATRIX(I,I)
1850   FOR J=1 TO ICOL
1860     INV.MATRIX(I,J)=INV.MATRIX(I,J)/DIV
1870   NEXT J
1880 NEXT I
1890 RETURN
1900 '------------------------------------------------------------------------
1910 ' Matrix multiplication subroutine
1920 '------------------------------------------------------------------------
1930 PRINT:PRINT "Calculating forces and link 2 input torque ..."
1940 IROW=SIZE:ICOL=SIZE              'set up counters
1950 FOR I=1 TO IROW                  ' count rows
1960   REACT.MATRIX(I)=0              ' zero matrix element
1970   FOR K=1 TO ICOL               ' count col.s
1980     ADD.TO=INV.MATRIX(I,ICOL+K)*INERTIA.MATRIX(K)
1990     REACT.MATRIX(I)=REACT.MATRIX(I)+ADD.TO
2000   NEXT K
2010 NEXT I
2020 RETURN
```

(d)

FIGURE A3.4*d*

Answers
to Selected Problems

Chapter Four

4.3 $R_B = 3.739$ in., $t_B = 0.153$ in.

4.5 $t_b = 0.240$ in.

4.7 $\alpha = 18°$, $\beta = 14\frac{1}{2}°$

4.11 $a = 0.094$ in.

4.13 $R_1 = 1.125$ in., $R_{b1} = 1.019$ in., $a = 0.125$ in., $t = 0.196$ in., $R_2 = 2.813$ in., $R_{b2} = 2.549$ in., $b = 0.156$ in.

4.15 $14\frac{1}{2}°$, $m_p = 2.626$; $20°$, $m_p = 1.981$; $25°$, $m_p = 1.662$

4.17 $p_b = 0.738$ in.

4.19 $N_1 = 20$, $N_2 = 60$, $P_d = 8$; or $N_1 = 25$, $N_2 = 75$, $P_d = 10$; or $N_1 = 30$, $N_2 = 90$, $P_d = 12$

4.25 $k_1 = 5.728$, $k_2 = 1.656$

4.29 $B = 0.0362$ in.

4.31   (*a*) $R'_1 = 1.809$ in., $R'_2 = 3.016$ in.; (*b*) $\phi' = 15.61°$; (*c*) $B = 0.0135$ in.

4.33 $\phi' = 24.02°$, $C' = 4.115$ in.

4.35 $14\frac{1}{2}°$, $B = 0.0028$ in.; $20°$, $B = 0.0038$ in.; $25°$, $B = 0.0048$ in.

Metric

4.3m $R_B = 94.97$ mm, $t_b = 3.885$ mm

4.5m $t_b = 6.102$ mm

4.11m $a = 4.445$ mm

4.13m $m_p = 1.63$

4.29m $B = 0.9244$ mm

4.31m   (*a*) $R'_1 = 45.244$ mm, $R'_2 = 75.406$ mm; (*b*) $\phi' = 20.83°$; (*c*) 0.4845 mm

4.33m $\phi' = 24.82°$, $C' = 104.05$ mm

4.35m $14\frac{1}{2}°$, $B = 0.0705$ mm; $20°$, $B = 0.0963$ mm; $25°$, $B = 0.1225$ mm

Chapter Five

5.3 $R = 1.857$ in., $t = 0.1703$ in.

5.5 $e = 0.1497$ in.

5.7 $t = 0.2645$ in.

5.11 $R_{o_1} = 1.2107$ in., $R_{o_2} = 1.4314$ in., $h_t = 0.3694$ in., $m_p = 1.303$

5.13 $e_1 = 0.0500$ in., $e_2 = -0.0500$ in., $\phi' = 20°$

5.15 $e_1 = 0.0396$ in., $e_2 = 0.0923$ in.

5.19 $m_p = 1.558$, $m_p = 1.584$

5.23   RECESS/APPROACH $= 2.328$

5.25 $m_p = 1.546$ (SEMI RECESS), $m_p = 1.369$ (FULL RECESS)

5.27 $e^* = 0.0084$ in., $R_g = 1.2036$ in., $t_g = 0.1599$ in.

5.29 $e^* = 0.0247$ in. (no undercut), $t_g = 0.4285$ in., $t_b = 0.4646$ in.

5.31 $e = -0.0151$ in.

5.33 $\phi_r = 25.42°$, $C' = 1.6257$ in., $(e_1 + e_2) = 1.06 \, \Delta C$

5.35 $C' = 4.5607$ in., $R_{o_1} = 2.3940$ in., $R_{o_2} = 2.8340$ in., $h_t = 0.7506$ in., $m_p = 1.38$

5.37 $e_1 = e_1^* = 0.0433$ in., $e_2 = 0.3020$ in.

5.39 Let $N_1 = 35$, $N_2 = 44$, let $e_1 = -0.0100$ in., $e_2 = -0.0370$ in., $D_{o_1} = 3.674$ in., $D_{o_2} = 4.520$ in., $h_t = 0.2127$ in., $m_p = 2.30$

Metric

5.3m $R = 45.500$ mm, $t = 4.1711$ mm

5.5m $e = 1.0649$ mm

5.7m $t = 6.3453$ mm

5.11m $\phi' = 28.85°$, $e_1 = 3.645$ mm, $e_2 = 1.974$ mm, $R_{o_1} = 47.926$ mm, $R_{o_2} = 79.255$ mm, $h_t = 12.781$ mm, $m_p = 1.22$

5.13m $e_1 = 1.547$ mm, $e_2 = -1.547$ mm, $\phi' = 20°$

5.15m $e_1 = 1.0699$ mm, $e_2 = 2.4911$ mm

5.19m $m_p = 1.56$, $m_p = 1.58$ (STANDARD GEARS)

5.21m $B = 0.13980$ mm

5.23m RECESS/APPROACH $= 2.326$

5.25m $m_p = 1.55$ (SEMI RECESS), $m_p = 1.37$ (FULL RECESS)

5.27m $e^* = -0.928$ mm, $t_g = 3.525$ mm, $R_g = 29.589$ mm

5.29m $e^* = -1.380$ mm (no undercut), $t_g = 10.601$ mm, $t_b = 12.412$ mm

5.31m $e = 0.193$ mm

5.33m $\phi_r = 25.39°$, $C' = 39.006$ mm, $(e_1 + e_2) = 1.07 \, \Delta C$

5.35m $C' = 55.585$ mm, $R_{o_1} = 29.129$ mm, $R_{o_2} = 34.372$ mm, $h_t = 8.916$ mm, $m_p = 1.31$

5.39m Let $N_1 = 35$, $N_2 = 44$, let $e_1 = -1.000$ mm, $\therefore e_2 = 1.204$ mm; $D_{o_1} = 90.712$ mm, $D_{o_2} = 117.62$ mm, $h_t = 5.73$ mm, $m_p = 1.70$

Chapter Six

6.3 $\Gamma_1 = 30°$, $\Sigma = 120°$

6.5 $\Gamma_1 = 37.87°$, $\Gamma_2 = 52.13°$; $a_G = 0.1364$ in., $a_P = 0.1969$ in.; $b_G = 0.2303$ in., $b_P = 0.1697$ in.; $(F < 0.95$ in.$) \therefore$ let $F = 0.875$ in.

6.7 $\Gamma_1 = 17.77°$, $\Gamma_2 = 27.23°$; $a_G = 0.1653$ in., $a_P = 0.2347$ in.; $b_G = 0.2743$ in., $b_P = 0.2049$ in; $(F < 1.74$ in.$) \therefore$ let $F = 150$ in.

6.11 $e = 0.02091$ in.

6.13 $\psi = 21.05°$; $N_1 = 28$, $N_2 = 98$; $(F > 1.006$ in.$) \therefore$ let $F = 1.125$ in.

6.15 (a) Hob: $\Delta C = 0.340$ in. (b) Fellows cutter: $\Delta C = 0$

6.17 $N_1 = 15$, $N_2 = 27$, $N_3 = 30$; $\psi = 20.36°$; $(F > 1.298$ in.$) \therefore$ let $F = 1.3125$ in. $D_{o_1} = 2.200$ in., $D_{o_2} = 3.800$ in., $D_{o_3} = 4.200$ in.; $\Delta C_{13} = 0.400$ in.

6.19 $N_1 = 32$, $N_2 = 40$, $N_3 = 80$; $\psi_1 = \psi_2 = 27.26°$, $\psi_3 = 62.74°$; $D_1 = 4.000$ in., $D_2 = 5.000$ in., $D_3 = 19.41$ in.; $C_{23} = 12.200$ in.

6.23 Use hob B; $N_1 = 32$, $N_2 = 64$; $\psi = 25.84°$; $F = 0.414$ in.; $D_{o_1} = 1.878$ in., $D_{o_2} = 3.300$ in.

6.25   (*a*) Hob: $P_{nd} = 7.07$, ∴ would require a special hob;   (*b*) Fellows cutter: $P_d = 5$ ∴ OK standard cutter

6.29 $D_1 = 6.928$ in., $D_2 = 10.0722$ in.; $N_2 = 45$; $\psi_2 = 26.67°$; $\Sigma = 56.67°$

6.37 $N_1 = 4$, $N_2 = 72$; $D_1 = 2.541$ in., $D_2 = 11.459$ in.

6.39 $N_1 = 3$, $N_2 = 60$; $D_1 = 2.451$ in., $D_2 = 9.549$ in.

6.41 $L = 2.000$ in.; $D_1 = 1.749$ in.

6.43 $D_1 = 1.621$ in., $D_2 = 3.879$ in.; $L = 1.854$ in.; $p_x = 0.371$ in.

Metric

6.3m $\Gamma_1 = 30°$, $\Sigma = 120°$

6.5m $\Gamma_1 = 37.87°$, $\Gamma_2 = 52.13°$; $a_G = 3.4613$ mm, $a_P = 4.9987$ mm; $b_G = 5.8439$ mm, $b_P = 4.3065$ mm; ($F < 21.71$ mm) ∴ let $F = 21.5$ mm

6.7m $\Gamma_1 = 17.76°$, $\Gamma_2 = 27.24°$; $a_G = 4.1977$ mm, $a_P = 5.9623$ mm; $b_G = 6.9673$ mm, $b_P = 5.2027$ mm; ($F < 39.969$ mm) ∴ let $F = 39.5$ mm

6.11m $e = 0.5018$ mm

6.13m $\psi = 21.04°$; $N_1 = 28$, $N_2 = 98$; ($F > 25.158$ mm) ∴ let $F = 25.2$ mm

6.15m   (*a*) Hob: $\Delta C = 8.01$ mm;   (*b*) Fellows cutter: $\Delta C = 0$

6.17m $N_1 = 15$, $N_2 = 27$, $N_3 = 30$; $\psi = 25.83°$ ($F > 24.88$ mm) ∴ let $F = 25$ mm; $D_{o_1} = 55.995$ mm, $D_{o_2} = 95.991$ mm, $D_{o_3} = 105.99$ mm; $\Delta C_{13} = 10.01$ mm

6.19m $N_1 = 32$, $N_2 = 40$, $N_3 = 80$; $\psi_1 = \psi_2 = 29.99°$, $\psi_3 = 60.01°$; $D_1 = 101.6$ mm, $D_2 = 127.0$ mm, $D_3 = 440.12$ mm; $C_{23} = 283.56$ mm

6.23m   Use hob A; $N_1 = 27$, $N_2 = 54$; $\psi = 32.47°$; $F = 8.169$ mm; $D_{o_1} = 51.01$ mm, $D_{o_2} = 99.01$ mm

6.29m $D_1 = 173.2$ mm, $D_2 = 258.6$ mm; $N_2 = 45$; $\psi_2 = 29.53°$; $\Sigma = 59.53°$

6.37m $N_1 = 2$, $N_2 = 35$; $D_1 = 64.2$ mm, $D_2 = 291.3$ mm

6.39m $N_1 = 2$, $N_2 = 40$; $D_1 = 81.66$ mm, $D_2 = 222.34$ mm

6.41m $L = 55.956$ mm; $D_1 = 46.207$ mm

6.43m $D_1 = 39.70$ mm, $D_2 = 139.10$ mm; $L = 63.494$ mm; $p_x = 10.582$ mm

Chapter Seven

7.1 $\omega_9 = 7.5$ rpm, $V_{10} = 25.53$ ft/min downward

7.3   (*a*) $\omega_2/\omega_3 = 0.803$;   (*b*) Gear 6 is left hand, Gear 1 turns cw when viewed toward motor

7.5   (*a*) Hob left hand;   (*b*) $\omega_7/\omega_5 = 0.700$

7.7 $x = 0.0855$ in. to the right, $y = 0.1111$ in. to the right

7.9   (*a*) Spindle speeds of 183.6 rpm and 30.51 rpm;   (*b*) $N_1 = 16$, $N_5 = 40$

7.11   (*a*) $\omega_{41} = 128.6$ rpm ccw;   (*b*) $\omega_{51} = 90$ rpm cw

7.13 $\omega_{71} = 140$ rpm

7.15 $\omega_{41} = 905.5$ rpm

7.17 $\omega_2/\omega_4 = 1$; $\omega_2/\omega_4 = 0.5$

7.19 $\omega_A/\omega_B = \infty$

7.21 $\omega_{71} = 474$ rpm

7.23 $\omega_A/\omega_B = 55.57$

7.25 $\omega_{51} = 1498$ rpm

7.27 $\omega_{51} = 156.0$ rpm

7.29 $\omega_{51} = 1667$ rpm

7.31 $\omega_B = 269.4$ rpm

7.33 $\omega_C = 292.9$ rpm

7.35 $n_{max} = 6.97$ planets, $n = 2, 3,$ or 6 equally spaced planets

7.37 $n_{max} = 7.01$ planets for gear 1, 2, and 3; $n_{max} = 7.8$ planets for gears 5, 6, and 7; $\therefore$ 2 equally spaced compound planets can be used

7.39 (a) $N_1 = 114$, $N_2 = 19$, $N_3 = 76$; $D_1 = 285$ mm; (b) three equally spaced planets cannot be used

7.41 (a) $N_1 = 102$, $N_2 = 17$, $N_3 = 68$; $D_1 = 12.75$ in.; (b) three equally spaced planets cannot be used

7.47 $P_{cir} = +10$ hp

7.49 $P_{cir} = -60$ hp

Chapter Eight

8.1 $V = 1.57 \times 10^4$ in/s, $A = 2.47 \times 10^7$ in/s^2

8.3 $n = 3357.5$ rpm, $V = 75.39$ ft/s $= 4523$ ft/min

8.5 $\omega_R = 2.11$ rad/s cw

8.7 $V = 5796$ mm/s, $A = 116,489$ mm/s^2

8.9 $V_B = 274.5$ m/s, $V_P = 268$ m/s

8.11 (a) $\omega_4 = 0$, $\alpha_4 = 1.12$ rad/s^2 ccw

8.13 (a) $V_C = 8.55 \times 10^4$ mm/s, $A_C = 1.96 \times 10^7$ mm/s^2
(b) $\omega_3 = 152.8$ rad/s ccw, $\omega_4 = 111.6$ rad/s cw

8.15 $V_B = 73.19$ mm/s, $\omega_4 = 0.244$ rad/s cw
$A_B = 69.07$ mm/s, $\alpha_4 = 0.222$ rad/s^2 cw

8.17 $\alpha_2 = 38,970$ rad/s^2 ccw

8.19 (a) $A = 51.24$ mm/s^2; (c) $A = 115$ mm/s^2

8.21 $V = 79.63$ in/s, $A = 1131.80$ in/s^2

8.23 (a) $V = 846.2$ mm/s, $\omega = 4.17$ rad/s ccw, $A = 13,185$ mm/s^2, $\alpha = 62.6$ rad/s^2 cw; (b) $V = 846.2$ mm/s, $\omega = 4.17$ rad/s cw, $A = 16,200$ mm/s^2, $\alpha = 77.8$ rad/s^2 cw

8.25 $V = 215$ mm/s, $A = 560$ mm/s^2, $\omega = 0.727$ rad/s ccw, $\alpha = 0.240$ rad/s^2 ccw

8.27 $\omega = 4.68$ rad/s cw, $\alpha = 27.36$ rad/s ccw

8.29 $V = 131.73$ in/s, $A = 844,370$ in/s^2

8.31 $V_Q = 47.3$ m/s, $A_Q = 771.3$ m/s, $\omega_3 = 4.46$ rad/s cw, $\alpha_3 = 264.9$ rad/s^2 cw

8.33 $V = 179$ in/s, $A = 5386$ in/s^2

8.37 $V_{DB} = 21$ mm/s, $A_{DB} = 565$ mm/s^2

8.39 $V_B = 3.1$ in/s, $V_C = 1.9$ in/s, $V_D = 5.0$ in/s, $A_B = 5.1$ in/s^2, $A_C = 3.7$ in/s^2, $A_D = 1.3$ in/s^2

8.41 $V = 665$ mm/s, $\omega = 2.66$ rad/s cw, $A = 3580$ mm/s^2, $\alpha = 19.5$ rad/s^2 ccw

8.43 $V_B = 5.0$ in/s

8.49 $\omega_4 = 15.9$ rad/s ccw

8.51 $V_D = 24.5$ in/s

8.53 $V_F = 0.45$ in., $\omega_4 = 0.061$ rad/s cw

8.57 $A_{g_3} = 4.20 \times 10^6$ mm/s², $\beta = 309.3°$

Chapter Nine

9.1 $\omega = 713.4$ rad/s

9.3 $S_b = (\omega^2\rho/6)(R_o - R_i)[R_o(2k + 1) + R_i(k + 2)]$

9.5 $F = 117{,}800$ lb, $M = 3360$ lb · in

9.7 $F_{o_2} = 167.5$ N, $F_{o_3} = 179.9$ N, $F_{o_4} = 83.88$ N

9.9 $F_A = 135$ lb, $F_{12} = 656$ lb

9.11 $F_{o_3} = 7.38$ N

9.13 $F_A = 204$ lb, $F_B = 103$ lb, $F_C = 103$ lb, $F_{o_2} = 204$ lb, $F_{o_4} = 108$ lb, $T_S = 612$ lb · in

9.15 $F_A = 820$ N, $F_B = 820$ N, $F_C = 295$ N, $F_D = 1000$ N, $F_{o_2} = 820$ N, $F_{o_4} = 295$ N

9.17 $n = 338$ rpm

9.19 $Q = 394$ lb

9.21 $F = 18280$ N

9.23 $A_{g_2} = 0$, $A_{g_3} = 5660$ m/s², $A_{g_4} = 7130$ m/s², $\alpha_3 = 46{,}067$ rad/s² ccw, $\alpha_4 = 39{,}719$ rad/s² ccw, $F_{o_2} = 0$, $F_{o_3} = 20546$ N, $F_{o_4} = 64{,}669$ N

9.25 $F_A = 588$ lb, $F_B = 405$ lb, $F_{12} = 588$ lb, $T_S = 1147$ lb · in

9.27 $T_S = 3334$ N · m cw

9.29 $T_S = 32600$ lb · in cw

9.31 $T_S = 284$ N · m ccw

9.33 (*b*) $I = 0.0007$ kg · m²

9.35 Not kinetically equivalent

9.37 (*a*) $A_A = 8954$ m/s², $A_B = 6614$ m/s²; (*b*) $M_B = 0.2267$ kg, $l_B = 0.152$ m, $M_P = 0.6805$ kg, $l_P = 0.0507$ m; (*c*) $M_A = 0.6802$ kg, $M_B = 0.2270$ kg; (*d*) $F_{o_4} = 8225$ N, $F_{B_3} = 1501$ N, $F_{A_3} = 6091$ N; (*e*) $F_{14} = 1250$ N, $F_{12} = 9800$ N, $T_S = 193.8$ N · m cw

9.41 $T = 1700$ lb · in ccw

9.43 at $\theta = 0°$, $T = -281$ N · m; at $\theta = 240°$, $T = -75$ N · m; at $\theta = 480°$, $T = -147$ N · m; at $\theta = 720°$, $T = -281$ N · m

9.45 ω_{max} at $450°$, ω_{min} at $90°$

9.47 $A = 266$ N · m

9.49 (*a*) $T_{av} = 509$ ft · lb; (*b*) $H_P = 29.1$; (*c*) $I = 20.3$ slug · ft²

9.51 $F_{25} = 305$ N

9.53 $T_S = 422$ lb · in ccw

9.55 $n_{max} = 236$ rpm

9.57 $\omega = 1.37$ rad/s

9.59 $F = 1055$ lb
9.61   (*a*) 24%; (*b*) 6.6%
9.63 $I = \tau^2 Wr^2/4\pi^2 l$

Chapter Ten

10.1 $(Wr)_4 = 82$ lb · in, $r_4 = 10$ in., $W_4 = 8.2$ lb
10.5 $(Wr)_4 = 93.3$ lb · in, $W_4 = 23.3$ lb @ 210°, $W_3 = 49.3$ lb @ 52.4°
10.7 $F = [(Wr)\omega^2]/g = 1.78$ lb at 1000 rpm, $F = 178$ lb at 10,000 rpm
10.9 $r = 8.934 \times 10^{-6}$ m, $(Wr) = 0.0402$ N · m
10.13 $F_{\text{total}} = 548.7$ N
10.15 $T_{\text{total}} = 1692$ lb · in
10.19 $S = 2225$ N, $a_{R_2} = -296.4$ mm to left of cylinder 1
10.21 $S = 0$, $C = -4MR^2a/L(2\omega)^2$
10.23 $S = -2$, $a_R = +1$; when $\theta_1 = 45°$ or 225°, the resultant primary force equals zero
10.27 Add 6.856 lb at point A' where $\overline{O_2A'} = \frac{1}{2}$ in and add 11.144 lb at point B' where $\overline{O_4B'} = 1$ in. Lengths $\overline{O_2A'}$ and $\overline{O_4B'}$ are on extensions of links 2 and 4, respectively, in the negative direction.

Chapter Eleven

11.1 $a = 1.123$ in, $b = 1.975$ in, $c = 1.600$ in
11.6 $a = 38$ mm, $b = 55$ mm, $c = 55$ mm, $d = 51$ mm

Chapter Twelve

12.1   RSSR (2), RCCC (1), RSSR-SC (1), RSCR (1)
12.3 $M = 1$
12.5   must consider clearance in joints
12.8 $\mathbf{q'} = 7\mathbf{i} + 5\mathbf{j} + 5\mathbf{k}$
12.11 $\theta_4 = 2.520°$ or 147.347° (two branches)

Index

635